Colloidal Organization

Tsuneo Okubo
*Institute for Colloidal Organization,
Hatoyama, Uji, Kyoto, Japan*

ELSEVIER

AMSTERDAM • BOSTON • HEIDELBERG • LONDON • NEW YORK • OXFORD
PARIS • SAN DIEGO • SAN FRANCISCO • SINGAPORE • SYDNEY • TOKYO

Elsevier
Radarweg 29, PO Box 211, 1000 AE Amsterdam, Netherlands
The Boulevard, Langford Lane, Kidlington, Oxford OX5 1GB, UK
225 Wyman Street, Waltham, MA 02451, USA

ISBN: 978-0-12-802163-7

British Library Cataloguing-in-Publication Data
A catalogue record for this book is available from the British Library

Library of Congress Cataloging-in-Publication Data
A catalog record for this book is available from the Library of Congress

For Information on all Elsevier publications
visit our website at http://store.elsevier.com/

Contents

Preface to the Edition

As is well known, colloidal and surface science was born in our usual life in nature and is one of the oldest sciences. A large number of colloidal systems exist in daily life such as clouds, fog, muddy water, milk, paint, and stained glass, as is described in this book. General sizes of colloidal particles range from several 10 nm to several millimeters. Most colloidal particles are not recognized with the naked eye, but are often observable through an optical microscope. Colloidal particles are small, but the ratios of the surface areas against their volumes are significant. Thus colloidal science developed rapidly in the field of surface science.

Professor Emer. Dr. Heinz Hoffmann (Bayreuth) kindly informed me of the following story. Dr. Wo. Ostwald, who made major contributions to colloid and surface science, wrote in his book *Licht und Farbe in Kolloiden*[1] that colloidal dispersion is not beautiful. During his age, backscattered and transmitted light and Mie scattering light were not enough to impress him with a sense of beauty.[2] I, on the other hand, was deeply impressed with the beauty of colloidal crystals. Colloidal science is now one of the most beautiful worlds of science. The term *beautiful* is not used so often in science. However, I like to use *beautiful* sometimes. Furthermore, I believe that the experience of a deep emotion during observation of new findings in nature in our life is also beautiful.

Colloidal organization has broad meaning, but *colloidal crystallization* and *ordered drying patterns* of colloidal dispersions are mainly described in this book. These two fields are typical systems representing colloidal organization, I believe. In a narrow sense, colloidal crystals are "ordered structures of colloidal particles in a suspension state." However, we should note that there exist eight kinds of colloidal dispersion systems: gas-in-liquid, gas-in-solid, liquid-in-gas, liquid-in-liquid, liquid-in-solid, solid-in-gas, solid-in-liquid, and solid-in-solid. Thus eight kinds of colloidal crystal systems are considered in a wide sense. Plasma crystals, opals, emulsion crystals, and gel crystals are typical examples of colloidal crystals in a wide sense. Ordered drying patterns are considered to be one of the colloidal crystals in the frozen (or fixed) states in a wide sense. When liquid-state colloidal crystals are dried on a substrate, drying processes from typical colloidal crystals to the solid-state colloidal crystals can be studied.

The main cause of colloidal organization is originated from the cooperative (or synchronized) and weak interactions among particles, substrate (or cell wall) and medium molecules. Particle–particle interactions are influenced greatly by the electrical double layers extended around the particles, especially in colloidal crystallization in the aqueous media. Particle–medium interactions are important for colloidal crystallization and pattern formation through the electrical double layers and convectional and sedimentary motion of particles. Particle–substrate interactions are also important for colloidal organization. Without a vessel or interface, crystallization of suspensions does not take place spontaneously. Most dense crystal planes orient along the cell wall. Furthermore, colloidal crystallization is enhanced in a plane cell, but disturbed in a narrow capillary tube, for example. Among many contributions to colloidal organization, the most important factor is the cooperative thermal motion of colloidal particles, including all the relating components, colloidal particles, substrate, ionic species, and medium molecules. Atmospheric conditions such as temperature, humidity, pressure, external fields, and other factors are also important for the thermal motion. As is well known, thermal energies k_BT (k_B is the Boltzmann constant and T is the absolute temperature) given for all the components are essential factors for determining the properties of substances. Considering the previously described features of colloidal organization phenomena, careful experiments are necessary to obtain reliable results. For example, we must deionize the colloidal samples with coexistence of ion-exchange resins for more than 1year to obtain the exhaustively deionized state of colloidal samples!

One of the main purposes of this book is to show the beautiful world of the colloidal organization to the readers. I believe careful experiments and observation are most important for understanding phenomena of colloidal organization. Explanation should follow the reliable observation. In this book, I have tried not to use complex equations, although I worry whether this was accomplished.

I deeply appreciate the guest researchers of the Institute for Colloidal Organization: Professor Akira Tsuchida and Hiroshi Kimura (Gifu University), Hiromi Kitano (University of Toyama), Kohji Yoshinaga (Kyushu Institute of Technology) and Etsuo Kokufuta (University of Tsukuba). The late Professor Emeritus Sei Hachisu (University of Tsukuba) is appreciated for his continuing encouragements and advice through our work. The late Professor Emeritus Ichiro Sakurada and other professors of Kyoto University are acknowledged for their kind comments. The late Professor Emeritus Ronald H. Ottewill (Bristol University) and Robert L. Lowell (University of Massachusetts) are acknowledged for their encouragement of our work. Professor Zoltan A. Schelly (University of Texas, Arlington), Thomas Palberg (University of Mainz), Maria Stoimenova (Institute of Physical Chemistry, Sofia, Bulgaria) and Peter A. Lovell (Manchester

University) are thanked sincerely for their discussions and comments. I acknowledge Dr. Anita Koch, Sarah J. Watson and Anitha Sivaraj of Elsevier for their kind support from arrangement to printing of this book. I thank my wife, Keiko, for her understanding and cooperation during the course of my research and writing.

Institute for Colloidal Organization
Tsuneo Okubo, in Uji, Kyoto
April 22, 2015

References

1. Ostwald W. *Licht und Farbe in Kolloiden.* Dresden, Germany; 1924. Verlag Th. Steinkopff.
2. Hoffmann H. Hundred years of colloid science fascinating phenomena in surfactant solutions. *Ber Bunsenges Phys Chem* 1994;**98**:1433–55.

1 Introduction

1.1 Common Colloidal Dispersions

Colloidal and surface chemistry have always existed. Rain, fog, and mist are typical examples of liquid-in-gas–type colloidal dispersion. These types of colloidal dispersions are further distinguished depending on their sizes. In Japan, for example, they are named *moya* ("dry fog," smaller than 10 μm in diameter), *kiri* (≈10-100 μm), *kirisame* (≈100-300 μm), *sitosito-ame* (≈0.3-1 mm), *ame* ("rain"; ≈1 mm), and *suko-ru* ("squall"; larger than 1 mm). *Ice-fog* is the dispersion of ice particles in the air (solid-in-gas type). *Cloud* refers to the concentration of water or ice particles (≈2-40 μm) in the air. Gravity causes clouds to descend at the velocity of approximately 20 mm/min without wind; however, even a faint ascending air current can keep a cloud aloft for a long time. *Snow* is crystallized ice from saturated water vapor in a cloud and is a typical example of gas-phase crystallization. The maximum crystal growth rate of snow is approximately 30 μm/s and is surprisingly similar to that of colloidal crystallization.

Observation of snow crystals has a long history. Johannes Kepler (1571-1630) discussed why snow crystals are hexagonal (1611). Rene Descartes (1596-1650) observed and sketched many shapes of snow crystals (1637). Wilson Alwyn Bentley (1865-1931) and William Jackson Humphreys (1862-1949) published a famous book titled *Snow Crystals* (1931),[1] which included many beautiful pictures of snow crystals.

Ukichiro Nakaya (1900-1962), a famous scientist who studied snow, said, "Snow is a letter from the sky,"[2,3] meaning that the state of the sky becomes clear by studying the morphologic characteristics of snow crystals. He made the Nakaya diagram showing the relationship between various crystal structures of snow as a function of the degree of supersaturation and temperature of water vapor in the sky (Fig. 1.1). For example, hexagonal polyhedric snow crystals grow in a low supersaturation state of vapor, whereas various crystal structures, including needlelike, folding fan–like, sheathlike, and dendritic crystals, form in a state of high supersaturation. Fig. 1.1 is newly redrawn by the author based on several types of Nakaya diagrams reported previously.

Colloidal Organization. http://dx.doi.org/10.1016/B978-0-12-802163-7.00001-5

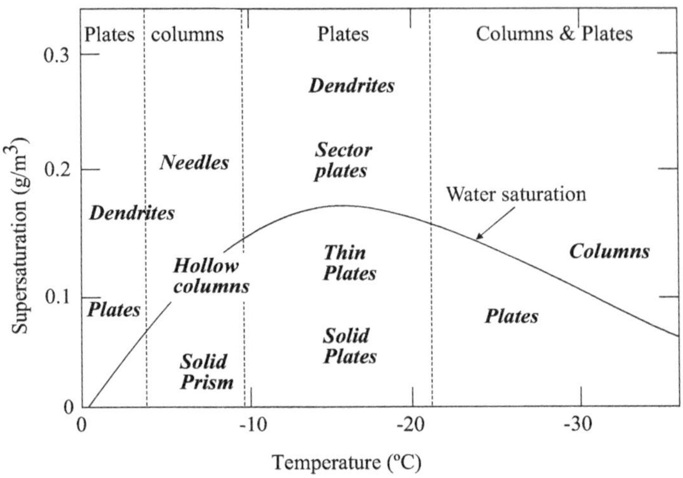

Figure 1.1 Schematic representation of a Nakaya diagram of snow.

Smoke, the dispersion of the liquid-in-gas or solid-in-gas colloid type, is also familiar. *Smog* is a new word combing *smoke* and *fog,* and relates to the pollution problem. Mist and fume are also typical examples of colloidal dispersion systems in solid-in-gas and liquid-in-gas systems. Furthermore, *aerosol* is a general name for colloidal dispersions of solid-in-gas and liquid-in-gas systems. A large, red sunset may be beautiful, but this also means air pollution is a serious problem in the presence of colloidal particles.

Colloidal dispersion systems in the liquid phase are most familiar. For example, turbid water in a pond contains colloidal particles of soil. Mayonnaise and milk are typical examples of liquid-in-liquid–type colloidal systems. Most paints are solid-in-liquid–type dispersions. Coffee,[4,5] black tea,[6] green tea,[7] and Miso soup[8] are also solid-in-liquid–type colloidal dispersions. The monodispersed colloidal spheres are now available commercially thanks to the recent advancement in the field of synthetic colloidal chemistry. Fig. 1.2 shows typical colloidal crystals of polystyrene spheres (D1K88, 137 $\pm$ 16 nm in diameter) in the stock (*A*) and diluted (*B*) suspensions in a bottle and a test tube, respectively. Iridescent colors are beautiful. To remove most of the ionic impurities in the suspension as completely as possible, the stock suspension was deionized with the mixture of ion-exchange resins 10 years previously. The sample suspensions were prepared by diluting the stock suspension with deionized pure water and the ion-exchange resins were introduced again. The author found the viscosity of the sample suspension containing the resins increased with time and reached saturation after several years. Complete deionization of the sample spheres took such a long time because deionization is one of the equilibrium reactions proceeding between colloidal spheres and solid resins dispersed in the aqueous phase. Colorful single crystals are observed in Fig. 1.2, B.

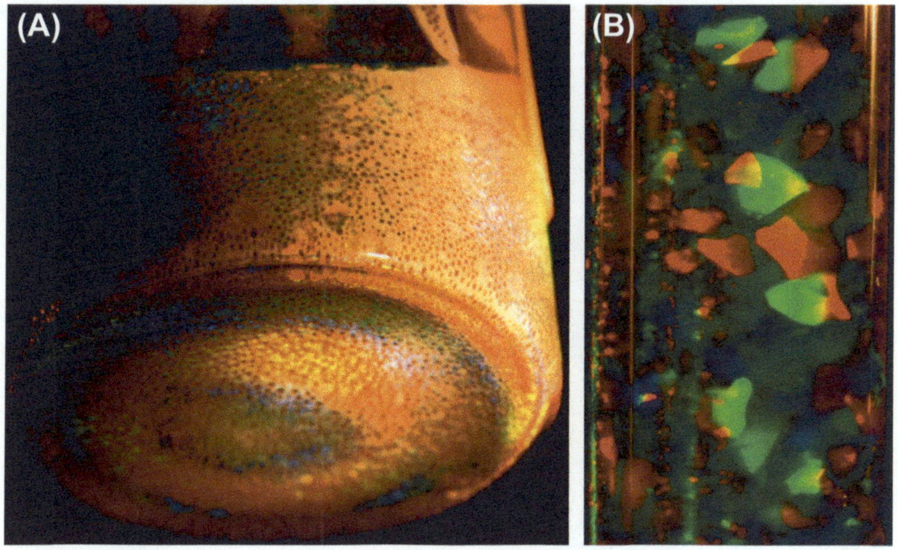

Figure 1.2 Colloidal crystals of polystyrene spheres (D1K88, 137 $\pm$ 16 nm in diameter) in the stock (A, ϕ = 0.062) and diluted (B, 0.00053) suspensions.

Fig. 1.3 shows one of the typical microscopic pictures of colloidal green tea leaves. In most Japanese hotels, a guest will find in her or his room a tea set to prepare green tea; the tea set is composed of green tea leaves in a paper filtering bag, a tea cup, and a water heater, for example. One day, after preparing tea and leaving it on the desk for several minutes, I was excited to find a broad ring forming in the cup. Fig. 1.4 shows typical examples of the broad ring patterns of green tea.[7] The broad rings of the hills created by accumulated green tea particles stayed on the inclined wall of the cup, and did not drop into the deepest central area. This was my first finding of

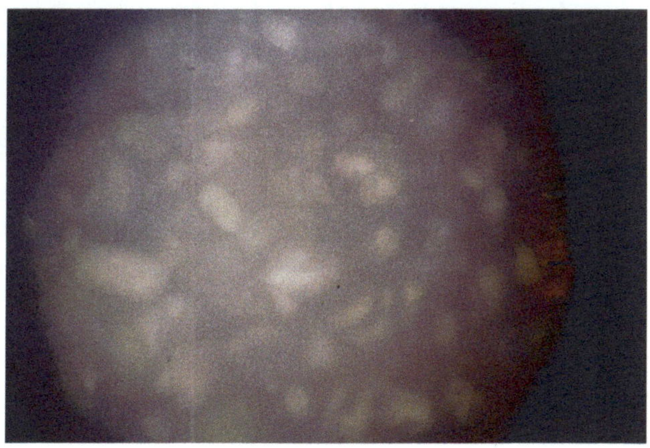

Figure 1.3 Microscopic observation of green tea leaves. Objective lens $\times$40, eye lens $\times$10; full scale is 25 μm. *(From Okubo T: Sedimentation and drying dissipative structures of green tea. Colloid Polym Sci 285:331-337, 2006.)*

Figure 1.4 Sedimentary patterns of green tea in the several cities in Japan. **A,** Sapporo, **B,** Yonezawa, **C,** Kumamoto, and **D,** Ishigaki in Japan. *(**C** and **D** from Okubo T:* Sedimentation and drying dissipative structures of green tea. Colloid Polym Sci *285:331-337, 2006.)*

sedimentation patterns. The sedimentary patterns were always observed irrespective of the kind of particles, cup, and observed places. The smaller the colloidal particles, the larger the broad rings that form up the inclined wall. Therefore approximate information on the size distribution of the particles is available from the sedimentary broad ring patterns. The details are described in Chapter 5.

Biologic polymeric substances are grouped into three categories: proteins (polypeptides), polysaccharides, and polynucleotides. Many are assemblies of anionic polyelectrolytes. Spherical proteins of hemoglobin and fibrous collagen or keratin exist in the state of solid-in-liquid–type colloidal systems.

Sponge, hydrated silica gel, and stained glass are typical examples of gas-in-solid, liquid-in-solid, and solid-in-solid systems, respectively. Stained glass is the beautiful colloidal system composed of metal particles 100 to 300 nm in diameter dispersed in a glass solid.[9] An example of stained glass is shown in Fig. 1.5. Stained glass can be seen almost anywhere. Only light in a certain range of wavelengths is transmitted through the stained glass by the quantum effect, and the transmitted light is colorful (e.g., reddish, greenish, or yellowish, depending on the kind of the metal particles). Usually, a stained glass window in a church is beautiful when we see it in the daytime. Many years ago, I was impressed to find beautiful stained glass when I looked at a lighted church in the evening. Table 1.1 describes examples of colloidal

Figure 1.5
A typical example of stained glass in Notre-Dame Cathedral, Strasbourg, France.

Table 1.1 Types of Colloidal Dispersions

		Dispersion Solutes		
		Gas	Liquid	Solid
Dispersion medium	Gas	———	**Aerosol** Fog, mist, cloud	**Aerosol** Smoke, smog
	Liquid	**Foam** Foam	**Emulsion** Mayonnaise, milk	**Solid dispersion** Paint, coffee
	Solid	**Solid foam** Sponge	**Solid emulsion** Hydrated silica-gel	**Solid dispersion** Stained glass

systems grouped by the combination of the interfaces formed between the solute and the medium. Gas-in-gas systems do not exist in equilibrium, because gas and gas are mixed homogeneously and no interfaces between gas and gas exist.

We should note here that there exist many ordering phenomena relating to the colloidal organization in nature. Fig. 1.6 shows dewdrops observed on a window glass within a house in the winter season. Here, the distribution looks somewhat ordered. Interfacial tensions between air and water, water and glass, and air and glass control the shape, size, and distribution of the droplets. Watching the synchronized self-organization processes is interesting. Self-organization phenomena are classified into two groups: (1)

Figure 1.6
Patterns of dewdrops formed on a window glass in winter.

dissipative structures in a nonequilibrium state and (2) consecutive structures in equilibrium. Most self-organization phenomena in nature are in the first group, and are exemplified by the fractal networks of rivers and mountains, and the chevron patterns that appear in a bacterium colony of *Escherichia coli.* In the laboratory, examples include the hexagonal convection cell (Benard cell), Belousov-Zhabotinsky reaction patterns, and the film formed by drying liquids, among others. The ordering of dewdrops in Fig. 1.6 are grouped into the dissipative structures. However, there have been few laboratory studies of the dissipative structure. Most of the self-organized structures studied have been the second group of conservative structures in equilibrium. Fast micellization of surfactant molecules and crystallization of proteins and metals are the typical examples for the second group. To understand the complex ordering phenomena in nature, the systematic study of the dissipative structures in laboratory scale is very important.

It is well known that the ratio of surface area of a substance against its volume increases substantially with decreasing size. Surface area and volume of a sphere in radius r are $4\pi r^2$ and $(4/3)\pi r^3$, respectively, and the area per volume of a sphere is $4\pi r^2/[(4/3)\pi r^3] = 3/r$. The ratio will be very large when the value of r comes close to zero. Thus more significant surface phenomena occur as the size of a solute decreases. This is a main reason why colloidal particles are studied so often in surface chemistry. However, we should note that the surface forces per unit of surface area, of course, do not depend on the size of solutes. Even for a very large solute, such as a man coming in contact with water in a bathtub, for example, several hundred nanometers of "electrical double layers" form around the surface of his body. Of course, the thickness of the double layers are very small, within approximately 1 μm, and do not affect him at all. However, when any kind of colloidal particles of approximately 100 nm in size are dispersed in water, they do not make direct contact with the inner wall of the container. This is because the repulsion forces between the particles and the wall are brought about by the electrical double layers formed from the interfaces of the particle and the wall in the water phase. We should note here that interfaces between solutes and medium molecules will disappear when the size of the solute further decreases and becomes similar in size to the medium (solvent) molecules, water, for example. In this condition, both solute and medium molecules move vigorously by thermal energy and mix homogeneously to each other without interface formation. The size of the solute must be larger by several fold at least compared with the size of the medium molecules for the interface between solutes and the medium molecules to occur; thermal movement of the solute becomes slow compared with that of medium molecules.

The study of colloid and surface chemistry developed in nature. The study of physics, physical chemistry, biology, and polymer chemistry probably originated from colloidal and surface chemistry. Colloid and surface chemistry is certainly the mother of many other sciences. Important contributions of colloid and surface science to the development of other sciences can also be understood from changes in the names of international journals relating to colloid and surface science. For example, between 1947 and 1950, the *Journal of Physical Chemistry,* published by the American Chemical Society, was known as the *Journal of Physical and Colloid Chemistry. Zeitshrift fur Chemie und Industrie der Kolloide* was founded in 1906 in Germany, and changed its name to *Kolloid Zeitshrift* in 1913, *Kolloid Zeitshrift und Zeitschrift fur Polymere* in 1962, and further to *Colloid and Polymer Science* in 1974.

1.2 Pioneers in the Field of Colloid and Surface Science

Here we briefly consider great scientists who developed colloid and surface science. Thomas Young (1773-1829), known for his study on surface tension and contact angle (1805), and Robert Brown (1773-1858), known for Brownian motion (1827), are famous also as founders of colloid and surface science. The studies of Michael Faraday (1791-1867) on gold colloids (1857) and John Tyndall (1820-1893) on Tyndall phenomena are well known. Lord Rayleigh (1842-1919) published theories of light-scattering in 1871. Franz Hofmeister (1850-1922) presented the Hofmeister series (or Lyotropic series) (1888).[10] Electro-kinetic phenomena were studied by Gustar Heinrich Wiedemann (1826-1899; 1852), Georg Herman Quincke (1834-1924; 1861), and Herman Ludwig Ferdinant Helmholtz (1821-1894; 1879). Study of the monolayer begun by Lord Rayleigh (1842-1919; 1891) and the absorption equation by Josiah Willard Gibbs (1839-1903; 1878) are also famous. In the period around 1900, Richard Adolf Zsigmondy (1865-1929; 1903) and Friedrich Wilhelm Ostwald (1853-1932; 1907) began ultramicroscopic observation of colloidal dispersion. *Zeitshrift fur Chemie und Industrie der Kolloids* was founded during this period (1906), as described previously. A mechanism of monolayered adsorption was proposed by Herbert Max Finlay Freundlich (1880-1941; 1907) and Irving Langmuir (1881-1957; 1916). Many famous works were published in this era. Several that should be noted include viscosity theory by Albert Einstein (1879-1955; 1906), the sedimentation equilibrium method by Marian von Smoluchowski (1872-1917; 1906), the motion of colloidal particles ($\approx$1908-1926) by Jean Baptiste Perrin (1870-1942), and micelle formation by James William McBain (1882-1953; 1911). The concept of electrical double layers was one of the most important contributions in colloid and surface science. Louis Georges Gouy (1854-1926; 1909) and David Leonard Chapman (1869-1958; 1913) contributed greatly in this field, and calculation of ion-atmosphere was made in 1923 by Peter Joseph Wilhelm Debye (1884-1966) and Erich Armand Joseph Huckel (1896-1980). The contributions of Hermann Staudinger (1881-1965; 1920) and Theodor Svedberg (1884-1971; 1926) are important especially relating to polymer chemistry. The study of the multilayered membrane by Katharine B. Blodget (1898-1979; 1934) is also important. Furthermore, Arne Wilhelm Kaurin Tiselius (1902-1971) studied electrophoresis (1938). Stephen Brunauer, Paul Hugh Emmett (1900-1985), and Edward Teller (1908-2003) are famous for studying multiple adsorption, known as *BET theory* for their names (1938). X-ray—scattering and light-scattering studies of colloidal dispersions by Andre Guinier (1911-2000; 1939), Paul Mead Doty (1920-2011), and Bruno H. Zimm (1920-2005; 1947) are also important. The DLVO theory on the stability of colloidal dispersion, which is named for Boris Vladimirovich Derjaguin (1902-1994), Lev Davidovich Landau (1908-1968; 1941), Evert Johannes Willem Verwey (1905-1981) and Jan Theodoor Gerard Overbeek (1911-2007; 1948), is epoch-making in the history of colloid science.

I further wish to acknowledge two Japanese colloid scientists relating to colloidal organization, Torahiko Terada (1878-1935), late professor of physics at the University of Tokyo, and Sei Hachisu (1919-1999), late professor emeritus at the University of Tsukuba. Terada was a famous physicist in the field of acoustics and geophysics and also an essayist.[11] Furthermore, he also contributed to colloid and interface science. He studied convectional patterns using Chinese black ink before Benard published his Benard pattern of convection. He also studied toy fireworks ("Senko-Hanabi" in Japanese) from a standpoint of colloid chemistry and published his findings in *Chemische*

Berichte. It should be mentioned here that his doctoral thesis was a physical analysis of sounds of the bamboo flute known as *shakuhachi.*

Hachisu is one of the pioneers in the field of colloidal crystal.[12] He observed a colloidal crystal array and colloidal alloy structure in a suspension state for the first time by using a metallurgical microscope. Real-time movement of colloidal particles in the crystal structures was observed directly. He also clarified that colloidal crystal forms by the Alder transition and further that colloidal crystal is one of the excellent models of metals, proteins, and ice in nature, although the interparticle interaction of colloidal crystal is repulsion, contrary to the usual crystal interaction of attraction.

References

1. Bentley WA, Humphreys WJ. *Snow crystals.* New York: McGraw Hill; 1931.
2. Nakaya U. *Flowers in winter* (in Japanese). Tokyo: Iwanami; 1938.
3. Nakaya U. *Snow crystal, natural and artificial.* Cambridge, Mass: Harvard University Press; 1954.
4. Deegan RD, Bakajin O, Dupont TF, Huber G, Nagel SR, Witten TA. *Capillary flow as the cause of ring strains from dried liquid drops. Nature* 1997;**389**:827–9.
5. Okubo T, Okamoto J, Tsuchida A. *Convectional, sedimentation and drying dissipative structures of coffee in the presence of cream and in its absence. Colloid Polym Sci* 2009; **287**:351–65.
6. Okubo T, Okamoto J, Tsuchida A. *Convectional, sedimentation and drying dissipative structures of black tea in the presence of cream and in its absence. Colloid Polym Sci* 2009;**287**:645–57.
7. Okubo T. *Sedimentation and drying dissipative structures of green tea. Colloid Polym Sci* 2006;**285**:331–7.
8. Okubo T. *Convectional and sedimentation dissipative patterns of miso-soup. Colloid Polym Sci* 2009;**287**:167–78.
9. Okubo T. *Beautiful world of colloids and interfaces* (in Japanese). Gifu, Japan: Matsuo Press; 2001.
10. Hofmeister F. *Zur Lehre von der Wirkung der Salze. Arch Exp Pathol Pharmacol* 1888;**24**: 247–60.
11. Nakaya U. *Memoirs of Torahiko Terada* (in Japanese). Tokyo: Kobunsya; 1947.
12. Hachisu S, Yoshimura S. *Optical demonstration of crystalline superstructures in binary mixtures of latex globules. Nature* 1980;**283**:188–9.

2

Fundamentals of Colloid and Surface Chemistry

2.1 Substances are the Sum of their Elementary Particles

All kinds of substances (or materials) are composed of a huge number (in the order of Avogadro's number, ideally) of elementary particles (atoms, molecules, colloidal particles, etc.). Furthermore, all the elementary particles, including substances, move with the translational and rotational modes by the thermal energy k_BT given to each particle. The thermal energy can never be zero anywhere the substances and their elementary particles exist. Here, k_B is the Boltzmann constant and T is the absolute temperature. For the colloidal particles larger than 1 μm in diameter, the translational thermal movement is often observable with the naked eyes through an optical microscope. Translational-diffusion (D_t) and rotational-diffusion (D_r) coefficients of a spherical particle are given by Equation 1 (the Einstein-Stokes equation) and 2, respectively.

$$D_t = k_BT/3\pi\eta d \qquad 1$$

$$D_r = k_BT/\pi\eta d^3 \qquad 2$$

Here, d is the diameter and η is the viscosity of the medium. It should be noted that the numerator in the equations indicates the driving forces for the thermal movement and the denominator the retarding forces against the thermal motion, respectively.

Substances exist in one of three states—gas, liquid (normal and ordered liquids), and solid (crystal and amorphous solid)—depending on the distribution of the elementary particles as a function of time. In the gas state, all the particles move freely by the thermal energy k_BT given to each particle in the modes of translation and rotation. When an elementary particle hits another particle or the wall of the container, it is simply reflected. In the liquid state, all the elementary particles also move freely, but keep the nearest-neighbored interparticle distances are close to each other. The elementary particles in

Colloidal Organization. http://dx.doi.org/10.1016/B978-0-12-802163-7.00002-7

the solid state also move in the translational (or vibrational) and rotational modes with thermal energy, but keep their equilibrium points. The distinction between crystal and an amorphous solid is simple: The former distributes in order and the latter is in disorder. Irrespective of the kind of elementary particles, all the substances have three states depending on these conditions. Thus *colloidal gases, colloidal liquids, colloidal crystals,* and *colloidal amorphous-solids* are sound scientific terms describing substances in which the elementary particles are colloidal particles. Fig. 2.1 shows the gas, liquid, and crystal phases in aqueous dispersion of colloidal spheres (open circles), where the electrical double layers composed of simple ions shown by dots are also shown. The details on the double layers are described in Section 2.7 later in this chapter. Recently, several researchers have identified the immobilized or fixed crystal arrays of colloidal particles as *colloidal crystals.* However, thermal motion of the fixed particles is extremely limited. The fixed arrays are therefore not colloidal crystals in a strict sense, but may be called *frozen colloidal crystals.*

Three states of substances exist also for one- and two-dimensional distributions of elementary particles. Furthermore, three states are often discussed conveniently using the radial distribution function, $f(r)$, which is a spectrum of the frequencies of all the elementary particles existing in the unit range of the distance (r) from a certain particle in a substance as a function of the r values. Typical examples of the function $f(r)$ of colloidal dispersions are shown schematically in Fig. 2.2 corresponding to the real distribution in Fig. 2.1, although the distribution of amorphous solid is not shown in Fig. 2.1. Fig. 2.2 shows gas (*A*), liquid (*B*), crystal (*C*), and amorphous solid (*D*). These radial distribution functions correspond to the Fourier transforms of the real distributions of elementary particles. However, $f(r)$ does not contain the time parameter, then the spectrum gives information on the distribution only at a certain moment. The radial distribution functions are obtained directly from the x-ray-, neutron-, or light-scattering measurements.

Fig. 2.3 shows a schematic picture of colloidal crystal, which is one of the typical phenomena of colloidal organization. Here, the colloidal spheres are shown in the solid sphere and surrounded with the electrical double layers shown by the large circles. Many single crystals are packed densely and separated by the grain boundaries. The effective size of the colloidal particles is more or less enlarged by the contribution of the double layers, and the "excluded volume" (size) of colloidal spheres increases. Colloidal crystallization takes place by the thermal movement of the enlarged colloidal spheres, including the electrical double layers. In other words, colloidal crystals form by the balancing of the thermal movement of spheres and the intersphere *repulsive* forces in a closed vessel. Surprisingly, interparticle interactions between colloidal crystals and the general

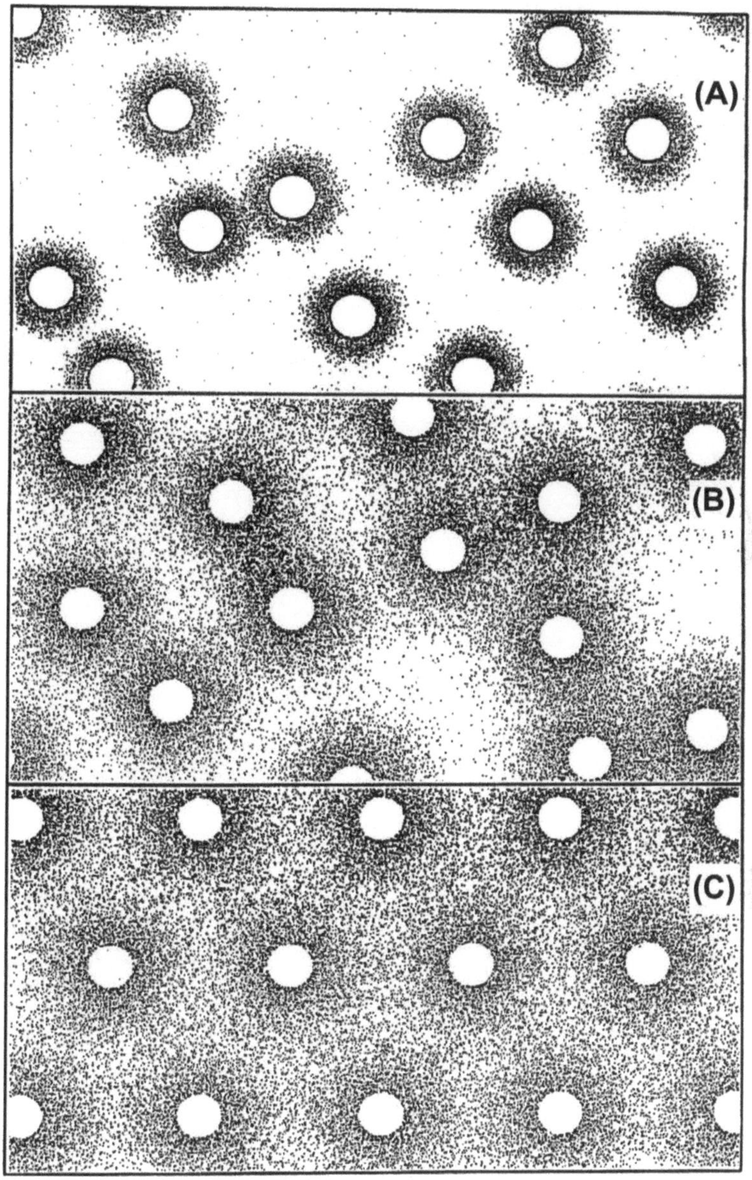

Figure 2.1 Schematic pictures of colloidal gas (*A*), liquid (*B*), and crystal (*C*). Dots show simple ions forming the electrical double layers. *Reprinted with permission from Accounts Chemical Research, copyright 1988, American Chemical Society.*

crystals (e.g., metals, proteins, and ice) are diagonally opposite, repulsive and attractive, respectively. However, their morphologic characteristics are similar; in other words, single crystals are packed densely and separated with the grain boundaries. Crystallization kinetics are also similar to each other.

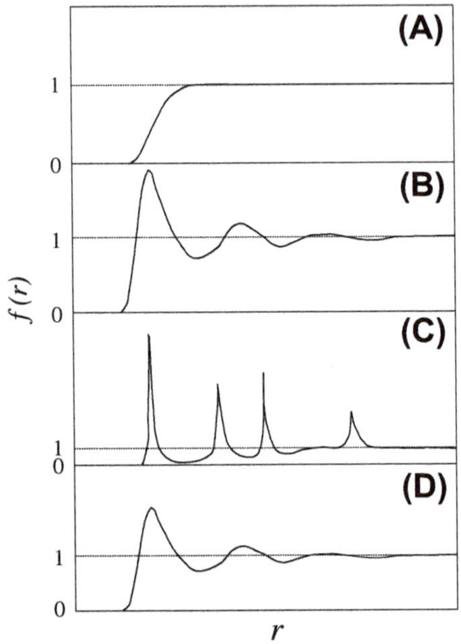

Figure 2.2
Radial distribution function, $f(r)$ of colloidal gas (*A*), liquid (*B*), crystal (*C*), and amorphous solid (*D*).

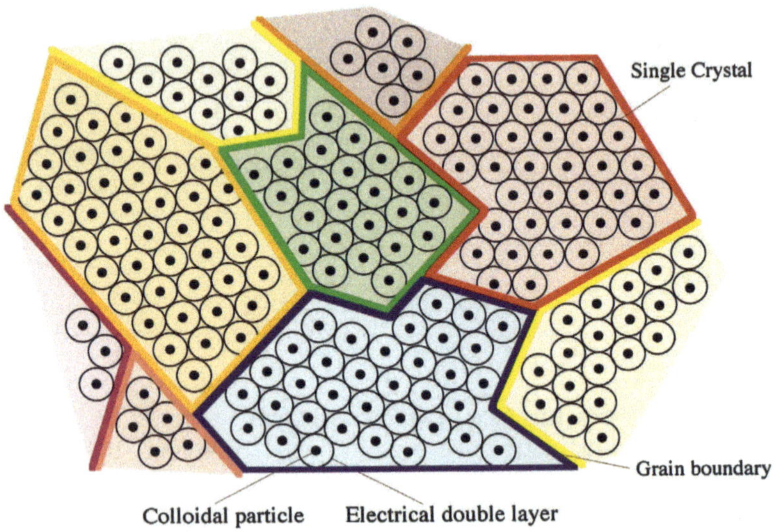

Figure 2.3
Schematic representation of colloidal crystals.

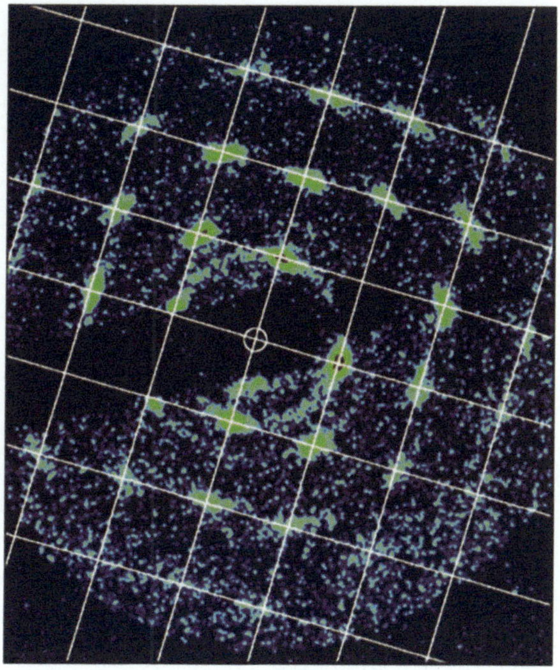

Figure 2.4
Plasma crystal of Be^{+}-ions. $N = 8 \times 10^4$. *Reprinted with permission from* Science, *copyright 1998, AAAS.*

Gas, liquid, and solid structures form in the appropriate conditions irrespective of size and shape of the elementary particles. Fig. 2.4 shows a typical example of a plasma crystal reported by Itano et al.[1] Beryllium (Be^{+}) cations are the elementary particles, which are oriented like a body-centered cubic (*bcc*) lattice. The number density is quite low: $8 \times 10^4/cm^3$. The *bcc* lattices align along (110) the axis as a spot at each intersection of the grid lines overlaid on the image. Plasma crystal is very soft because of its low density in the elementary particles, and the structure is distorted easily even by the gravity. Microgravity experiments are therefore exciting to study in plasma crystal.

Let's consider crystals further. To demonstrate in a classroom, ask the students to sit in their chairs in a lecture room and assign each student to play the role of an elementary particle. The students have their own equilibrium positions on their chairs and are probably moving within 10% of the distance with the adjoining students. Then the two-dimensional "student crystal" forms! However, we should note that number of elementary particles of the student crystal is quite few compared with that of general crystals in the order of Avogadro's number. When I visited Bryce Canyon National Park in Utah and looked at the view as shown in Fig. 2.5, I said "Amazing! It's crystal!" Depending on the choice of elementary particles as a high "statue" or a "block" piled up to

Figure 2.5
Bryce canyon national park, Utah, United States.

compose a statue, two- or three-dimensional crystals are imaged, respectively. These statues or blocks fluctuate according to the external forces such as wind and at least by the internal thermal movement of silica atoms, for example, although the fluctuation is extremely small. Thus this scenery itself is considered to be a giant crystal. We should note here that the kind of the medium of the elementary particles is not considered for the criterion of the three states of gas, liquid, and solid (crystal or amorphous-solid).

2.2 Properties of Substances are Originated from the Thermal Movement of the Elementary Particles

The elastic modulus is one of the most fundamental properties of substances, including colloidal dispersions, and has been measured by various techniques. Let's consider the colloidal dispersions. The static elastic moduli of colloidal crystals of monodisperse polystyrene spheres were evaluated by Crandall and Williams[2] from reflection spectrum measurements at the sedimentation equilibrium. On the other hand, dynamic (shear) elastic modulus has been measured by Goodwin and Smith,[3] for example, by shear wave propagation experiments. The order of magnitude of the elastic modulus is given in terms of the magnitude of the thermal fluctuation δ of a particle as the following[2,4]:

$$G \sim f/l \sim (k_B T/<\delta^2>)/l \qquad 3$$

where f is the force constant, l is the intersphere distance, and δ is the thermal fluctuation of a particle in the effective potential valley. k_B and T are the Boltzmann constant and the absolute temperature. Introducing a nondimensional parameter g for $<\delta^2>^{1/2}/l$, the modulus is obtained as a linear function of N,

$$G \sim Nk_BT/g^2 \quad (4)$$

When g is unity, Equation 4 gives the rigidity of an ideal gas having the same sphere concentration. Lindemann's rule of crystal melting tells us that $g < 0.1$ (typically ≈ 0.03) holds for a stable solid, including crystal. G values of the substances are estimated roughly using Equation 4. We should note here that the rigidity of G is simply given by the thermal energy k_BT, in addition to the number density of the elementary particles (N) and the fluctuation parameter (g). It is interesting to note here that Equation 4 holds true for all kinds of substances in gas, liquid, and solid states. Rigidity of the actual substance is further influenced by the additional interparticle interactions.

Fig. 2.6 shows the log G versus log N plots of several groups of substances. The solid, dotted, and broken lines show the calculations when g-factors are 1, 0.1, and 0.01, respectively. The figure demonstrates clearly that the G values are determined mainly by the number density of elementary particles of a substance, and further by the fluctuation parameter. The absolute temperature will not change G-values so effectively compared

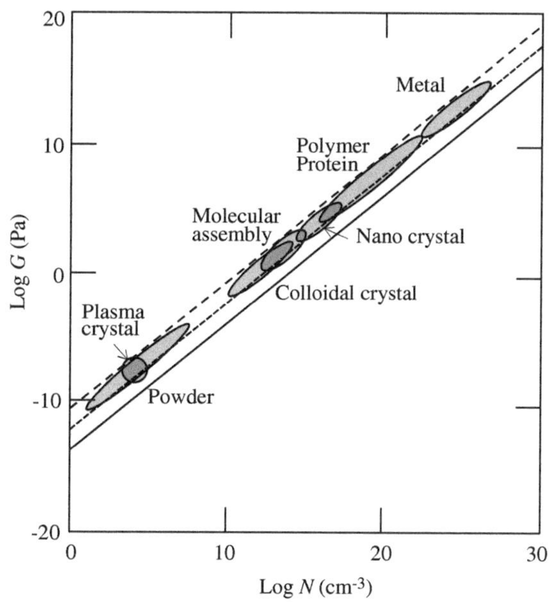

Figure 2.6 Relationship of log G with log N for various substances.

with N values. Thus the rigidity of the substances composed of large elementary particles is quite low in general.

2.3 Instability at Interfaces Plays an Important Role for the Interfacial Phenomena

The *interface* means the boundary region where two different substances are in contact. When two different phases contact each other, the elementary particles of each substance in the interface region is apt to go back to its own bulk phases as soon as possible, because the Gibbs free energy level is raised and unstable in the interface region. Equation 5 holds in the process of the interface formation.

$$\Delta G = \Delta H - T\Delta S > 0, \quad (5)$$

Here, ΔG, ΔH, and ΔS indicate the differences in the Gibbs free energy, enthalpy, and entropy values between final and initial states in the process of interface formation. ΔG always becomes positive at the formation of interfaces, as is shown in Equation 6 schematically. When a block is cut off, two interfaces are born.

6

In general, the interfaces are unstable compared with the bulk phase and $\Delta G > 0$ holds true. However, we should pay attention to the fact that a decrease in ΔG is more or less apt to take place spontaneously in nature, resulting in the partial decrease of the raised ΔG. Adsorption, interfacial tension, orientation of the amphiphilic (amphipathic) molecules at the interface and formation of the electrical double layers are the typical examples of the spontaneous decrease in ΔG.

Interfacial tension is induced when water contacts with air phase, for example. This tension occurs because water molecules at the interface region are apt to go back to their own water bulk phase. Thus many kinds of adsorption phenomena take place to reduce the Gibbs free energy level raised by the interface formation. Adsorption is, of course, the inverse reaction against the interface formation reaction shown in Equation 7. Langmuir's monolayer adsorption is one of the most fundamental interfacial phenomena. When the amount of bound-state adsorbates and the concentration of the free-state adsorbates are given by Q and C, respectively, Equation 8 holds true, where K is the association constant between the adsorbent and the adsorbates.

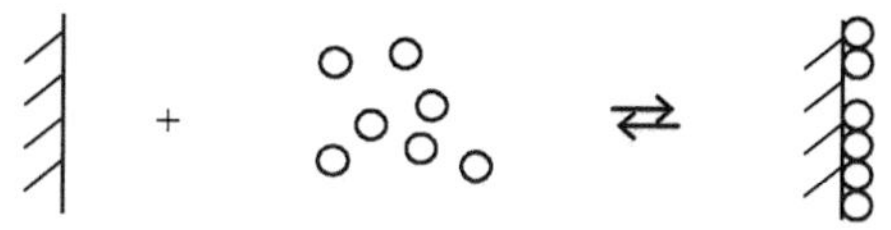 7

$$Q = (Q_m KC)/(1 + KC) \quad 8$$

Q_m is the maximum amount of adsorbates when C exists in an excess amount compared with the adsorption sites of the adsorbents. Equation 9 is easily derived by inverting both the left and right terms of Equation 8. The relations of Q versus C and $1/Q$ versus $1/C$ are shown schematically in Fig. 2.7, *A*.

$$1/Q = 1/Q_m + 1/(Q_m KC) \quad 9$$

Here, Q_m is the maximum adsorbed amount of adsorbates on the adsorbents. Equation 9 shows that Q_m and K are evaluated from the intercept and the slope in the $1/Q$ versus $1/C$ plots, respectively, as is shown in Fig. 2.7, *A*. It should be noted here that the Langmuir's monolayer adsorption in Equation 9 does not always hold true for many actual adsorption systems. In many cases, multiple-layered adsorptions discussed by Brunauer, Emett, and Teller (BET theory) take place instead.

Interestingly, Langmuir's adsorption Equation 8 is quite similar to the Michaelis-Menthen equation for enzyme reaction analysis. Substrate (*S*) is bound to enzyme (*E*) and the Michelis-Menthen complex (*E*S*) forms in the first step in general. When *E*S* complexes form, the product (*P*) forms rapidly and the enzyme is regenerated.

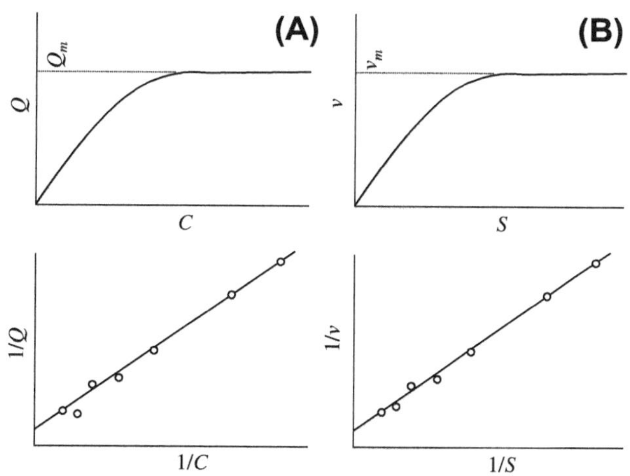

Figure 2.7
Comparison of Langmuir adsorption with enzyme reaction.

$$E + S \underset{k_{-1}}{\overset{k_1}{\rightleftarrows}} E * S \overset{k2}{\rightleftarrows} E + P \quad 10$$

$$v = v_m(1/K_M)[S]/\{1 + (1/K_M)[S]\} \quad 11$$

Equation 12 also holds.

$$1/v = 1/v_m + (K_M/v_m)/[S] \quad 12$$

Here, K_M is the dissociation constant contrary to the association constant K in Equation 8. The velocities observed in the enzyme reaction and the maximum velocity in the presence of excess amount of substrates are represented as v and v_m, respectively. The similarity between Equations 8 and 11 (or Equations 9 and 12) is understandable when the formation of the Michaelis-Menthen complex is the rate determiner and corresponds to the adsorption reaction of enzyme with substrates. Most enzymes have both the binding and catalytic sites, and product formation takes place rapidly with rate k_2 when Michelis-Menthen complexes, $E*S$, form. Thus most biological enzyme reactions and antigen–antibody complexation reactions are considered to be interfacial adsorption phenomena. Fig. 2.8 shows the inclusional complexation of a substrate with an enzyme taking place efficiently with the additional help of the attractive interaction such as electrostatic, hydrogen-bonding, and hydrophobic forces, for example. It should be noted here that most enzyme reactions do not always follow Equations 10 and 11 strictly; in other words, enzyme reaction velocity does not always saturate in the presence of an excess amount of substrate, but still increases. This enzyme reaction is also similar to the BET adsorption mechanisms. We should note here that adsorbents, enzymes, and antibodies may correspond to colloidal particles.

Adsorption of the substances at the air-solution interface is evaluated using Gibbs's adsorption equation, Equation 13:

$$\Gamma = (d\gamma/d\mu_2)_T = -(1/RT)(d\gamma/d \ln a_2) \quad 13$$

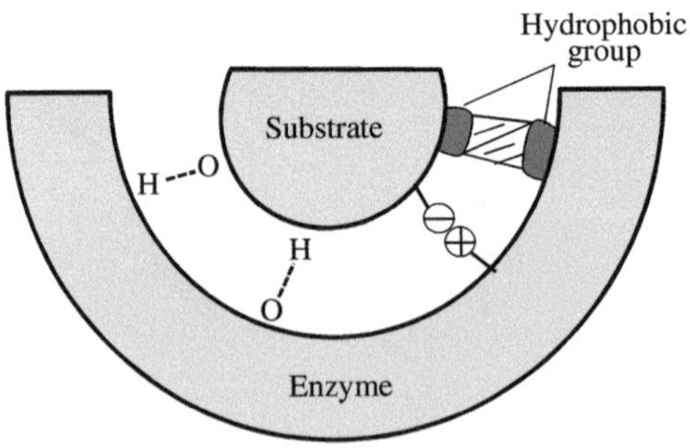

Figure 2.8 Schematic pictures of enzyme substrate complexation.

Here, Γ is the excess concentration of solutes at the interface, γ is the interfacial tension, μ_2 is the chemical potential of solutes, and a_2 is the activity of solutes. When solutes are surface-active agents, their surface tensions decrease linearly with increasing logarithms of solute concentration below the critical micelle concentration (*cmc*). Activity of solutes is approximated by the concentration for the single molecules in the diluted solution. Then, Γ becomes positive. This means that the surface-active agents exist in excess compared with their bulk phase, because the term ($d\gamma/d \ln a_2$) is negative. For the simple electrolytes such as sodium chloride (NaCl) and calcium chloride ($CaCl_2$), however, Γ becomes negative, which means that these salt concentrations at the air–solution interface are lower than those of the bulk solution. Thus positive and negative adsorptions take place for surface-active agents and simple electrolytes, respectively. There are several methods to measure surface tension, including the liquid upward method in a capillary, the ring tension method, the liquid drop methods (weight or shape of a drop), and others. Surface-tension measurements of a series of synthetic and biologic polyelectrolytes and further colloidal dispersions have been reported.[5-18]

We should note here that the adsorption takes place between the colloidal particles and the low-molecular-weight substances or linear-type polymers, although the latter two substances have no interfaces with solvent. Linear-type polymers are also known to be adsorbed on the solid interfaces in the modes of (A) tail, (B) train, and (C) loop, as is shown in Fig. 2.9. There exist many driving forces besides the spontaneous adsorption of polymers onto colloidal particles, such as electrostatic, hydrophobic, charge-transfer-type, and hydrogen-bonding-type attractive forces.

Adsorption of polyelectrolytes on the colloidal surfaces in polar liquid media takes place synchronously.[19-21] Figs. 2.10 and 2.11 show the transitional changes in the zeta-potential and the effective diameter, including adsorption layer, when cationic and anionic polyelectrolytes were added into aqueous suspensions of polystyrene spheres (Fig. 2.10) and colloidal silica spheres (Fig. 2.11), respectively.[20] Charge reversal of the colloidal surface

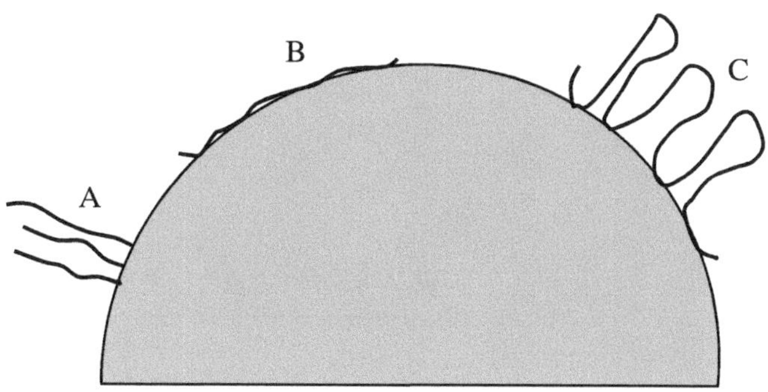

Figure 2.9 Three types of adsorption of linear-type polymer to solid surface: **A**, tail, **B**, train, and **C**, loop.

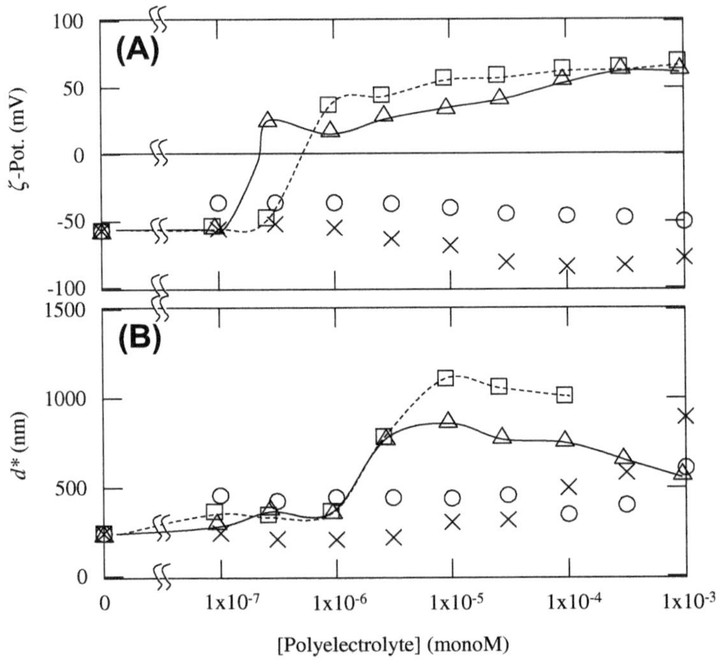

Figure 2.10 ζ-Potential (*A*) and effective diameter *d** (*B*) of D1A19 polystyrene spheres in the presence of NaPAA (circles), NaPSS (crosses), C2PVP (triangles), and C16BzPVP (squares) at 25° C. $\phi = 1 \times 10^{-5}$ in volume fraction, ELS measurements. *Redrawn based on Okubo T, Suda M:* J Colloid Interf Sci *213:565-571, 1999, with permission from the* Journal of Colloid Interface Science, *copyright 1999, Elsevier.*

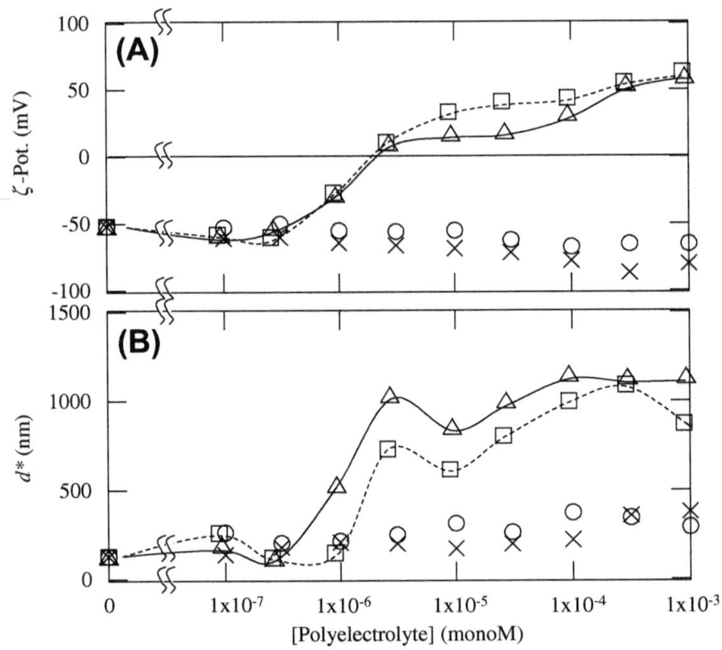

Figure 2.11 ζ-Potential (*A*) and effective diameter *d** (*B*) of CS91 colloidal silica spheres in the presence of NaPAA (circles), NaPSS (crosses), C2PVP (triangles), and C16BzPVP (squares) at 25° C. $\phi = 7 \times 10^{-5}$ in volume fraction, ELS measurements. *Redrawn based on Okubo T, Suda M:* J Colloid Interf Sci *213:565-571, 1999, with permission from the* Journal of Colloid Interface Science, *copyright 1999, Academic Press.*

from negative to positive takes place abruptly above the critical concentration of the polycations added (i.e., avalanche-type adsorption[20]). The effective diameter of colloidal spheres including the adsorbed layers also increased by four- to tenfold compared with the bare colloidal size. In the presence of a large amount of polycations, aggregation of colloidal spheres mediated by the layers of adsorbed polycations may occur. Weak adsorption also takes place on the colloidal spheres in the presence of an excess amount of polyanions caused by the hydrophobic and dipole–dipole attraction interactions between polyanions and also anionic colloidal spheres.

The synchronous alternative multilayered adsorption of polycations and polyanions on the colloidal spheres will be deduced rather easily when the charge reversal phenomena described previously are taken into account.[21-43] Fig. 2.12 shows a typical example of the alternative changes in zeta potential and thickness of the adsorbed layers, when polycations poly-4-*N*-butyl pyridinium bromide (C4PVP) and polyanions sodium polystyrene sulfonate (NaPSS) were added alternatively onto the colloidal silica spheres. The adsorbed layers are packed more densely as the alternative additional processes proceed, as is shown in Fig. 2.12. The most plausible scheme of the multilayered adsorption is shown in Fig. 2.13.

Flocculation of colloidal particles takes place by the addition of high-molecular-weight and linear-type polymers, polyethyleneimine hydrobromide (PEI*HBr) and polyacrylamide (PAA_m), for example. PEI*HBr and PAA_m have been used to treat wastewater for many years. My group prepared a series of carbamoylated polyethyleneimine (CPEI) by the carbamoylation reaction of PEI with cyanate ions shown in Equation 14.[44-47]

$$\sim(CH_2CH_2NH_2^+)\sim + OCN^- = \sim(CH_2CH_2N(CONH_2))\sim \qquad 14$$

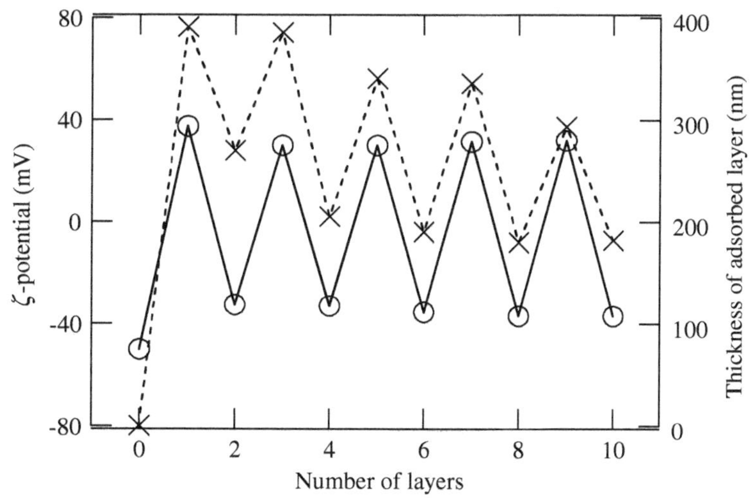

Figure 2.12 Alternative adsorption of C4PVP (8×10^{-6} monoM) and NaPSS (8×10^{-6} monoM) on CS91 colloidal silica spheres (110 nm in diameter, $\phi = 7 \times 10^{-5}$) at 25° C. Circles: ζ-potential, crosses: thickness of adsorbed layer.

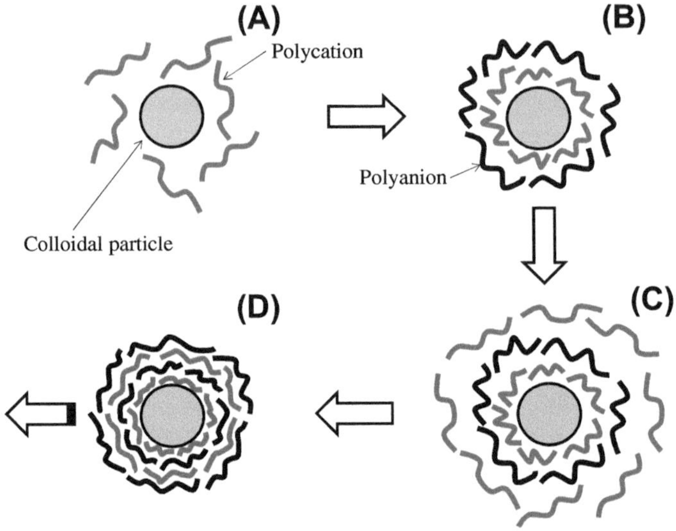

Figure 2.13 Scheme of the alternative adsorptions of C4PVP and NaPSS on CS91 spheres. *From Figure 2 of Okubo T, Suda M:* Colloid Polym Sci *277:813-817, 1999, with permission from* Colloid Polymer Science, *copyright 1999 Springer.*

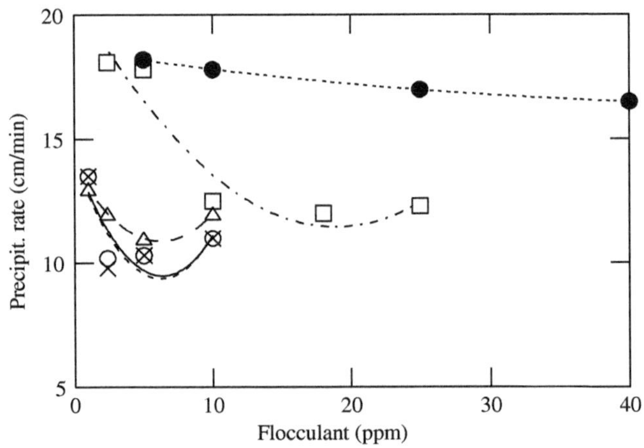

Figure 2.14 Precipitation rates of aqueous kaolin dispersion (1%, pH = 8) by the addition of carbamoylated polyethylene imine. Circles: degree of carbamoylation = 0%, crosses: 19.9%, triangles: 58%, squares: 94%, solid circles: 98%.

As is shown in Fig. 2.14, the flocculation rate of kaolin dispersion (1%) increased with increasing degree of carbamoylation. Furthermore, the supernatant fluid became more transparent by the addition of more CPEI than PEI. Flocculation by CPEI is interesting, because CPEI has the dual characteristics of PEI and PAAm.

Most catalytic reactions take place by the synchronous combination of physical- and chemical-adsorptions. Let's consider here the dissociation reaction of A_2 molecules to 2A atoms, as shown in Equation 15:

$$A_2 \rightleftarrows 2A \qquad 15$$

The dissociation energy of A_2 is usually extremely high compared with the thermal energy k_BT, and the chemical adsorption reaction, which introduces the chemical reaction in Equation 15, never occurs, as is shown in Fig. 2.15. In the presence of the metallic catalysts having appropriate interfaces for the reaction in Equation 15, however, the physical adsorption of the A_2 molecules on the catalyst's interface will take place favorably and further continue to the chemical adsorption with the help of thermal energy k_BT only. Jumping of the A_2 from the physically adsorbed position to the intercept position between the curves of physical- and chemical-adsorptions will be low enough to proceed toward the chemical adsorption with the help of thermal energy. In other words, the dissociation reaction of A_2 takes place easily by the synchronous combination of the physical and chemical adsorption at the interface of the catalyst.

Another example demonstrating the importance of the interface phenomena relating to the colloidal organization is the formation of the "electrical double layers" around the colloidal interfaces with polar solvents, typically water molecules. First, the interfaces of the substance contacting with water are spontaneously charged negatively. Second, the ionic species in the water phase gather toward the negatively charged colloidal interfaces. Third, the electrical double layers form. These charging and double-layer formation processes at the interfaces are essentially important for the colloidal organization. The details of these events are described in Sections 2.5 through 2.7 later in this chapter.

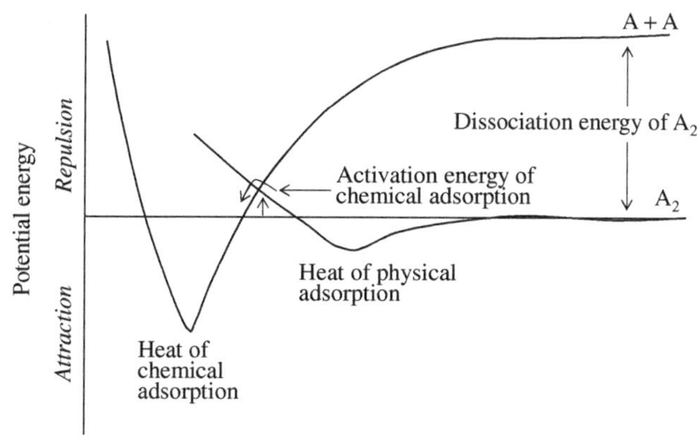

Figure 2.15 Synchronization of physical and chemical adsorption in the dissociation reaction of A_2 molecules.

2.4 Various Types of Solutes in Aqueous Media

A huge number of solutes in aqueous media can be categorized into four groups (Fig. 2.16): (1) simple substances (low molecular-weight substances), (2) linear-type polymers, (3) gels and molecular assemblies, and (4) colloidal particles. In aqueous media, polar and ionic substances and small colloidal particles are solubilized or dispersed stably. Boundary regions among four kinds of substances are vague. Interfaces with medium molecules are recognized for colloidal particles larger than approximately 20 nm in diameter and also for the gels having the similar sizes and high degree of cross-linking. In general, the electrical double layers form for the substances of groups 3 and 4 larger than approximately 20 nm in diameter.

Let's discuss further the relationship among the substances of groups 2, 3, and 4. In general, polymer solution (group 2), which is composed of linear-type polymers and their solvents, is transparent. Transparency occurs primarily because the unit's size of the dynamic thermal motion of the polymers is close to that of their monomer units and very small compared with the light wavelengths and rather close to size of solvents. Thus there exist no interfaces between the polymers and solvents—in other words, the monomer units and solvent molecules are able to exchange their positions freely by their thermal motion. On the other hand, colloidal dispersions

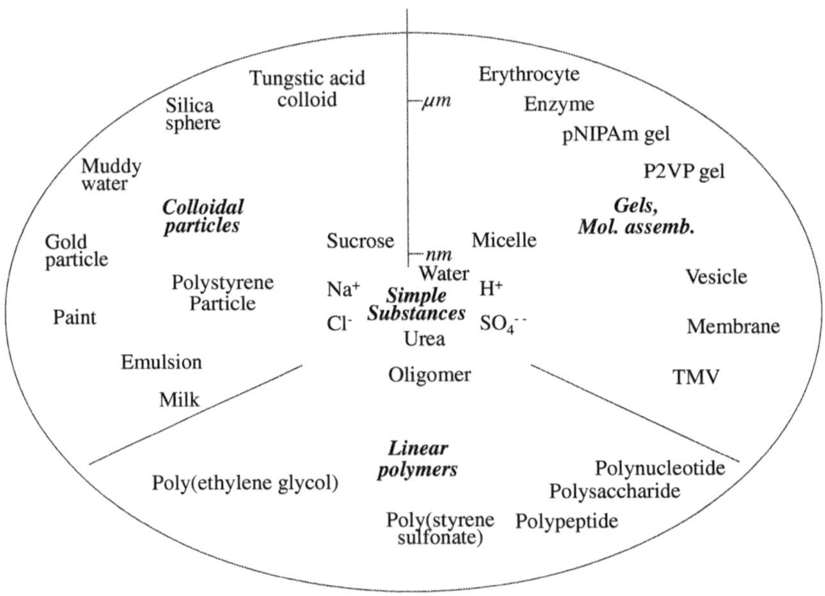

Figure 2.16 Various types of substances in water.

of hard spheres (group 4) are composed of the large solid particles and very small medium molecules in most cases. The size of the typical colloidal particles is close to or even larger than the light wavelengths, 100- to 1000-fold larger than that of the medium molecules. Thus colloidal dispersion is turbid except for those in the refractive index-matched conditions between particles and medium. Turbid dispersion supports the existence of the interfaces in a sense of colloid chemistry. Most dispersions of gel particles (group 3) are also turbid, and their nature looks to be just in between the linear-type polymer solutions and colloidal dispersions. When the degree of cross-linking is quite low, the gels must behave like a polymer solution, whereas the gel will behave like colloidal particles as their degree of cross-linking increases.

Micelles are the assemblies of surface-active agents containing both hydrophilic and hydrophobic groups in a molecule (Fig. 2.17, *A*). Hydrophilic groups are ionizable groups (such as sulfate, sulfonate, and amine) or water-soluble neutral ones (such as oligomers of ethylene glycol), whereas hydrocarbon short chains are the hydrophobic groups. Depending on the relative distribution of the hydrophilic and hydrophobic groups, various modes of assembling and orientation take place to reduce Gibbs free energy in the total system as much as possible. At the air–water interface, the surfactant molecules orient and two-dimensional gas, liquid, and crystal (rarely amorphous solid) structures of surfactants form (see Fig. 2.17, *B*). Surfactants existing in water form spherical micelles (see Fig. 2.17, *D*). Layered or rodlike micelles form depending on the kind of surfactant molecules and their concentrations (see Fig. 2.17, *E* and *F*). It should be noted that the micellization always takes place at a concentration higher than

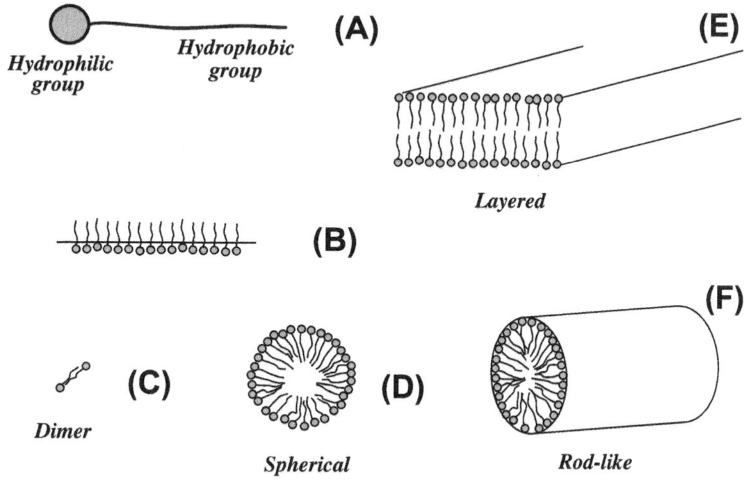

Figure 2.17 Schematic representation of various types of micelles (*1*).

the critical micelle concentration (*cmc*). Furthermore, these micelles maintain equilibrium with corresponding monomer molecules. Very fast in–out reactions have been found with the use of various techniques.[48] Micellization reaction corresponds to the interface formation, and so many studies on micellization are helpful for understanding the mechanism of colloidal organization. Further complex micelles such as lamellae (Fig. 2.18, *A*), hexagonal (Fig. 2.18, *B*), and bicontinuous cubic structures formed for aqueous solutions of potassium myristate at high concentration and at high temperature. Fig. 2.18, *C*, demonstrates excess amount of hydrophobic moieties compared with the hydrophilic moieties, and various ordered phases appeared as a function of concentration (20 to 100 wt%) and temperature (≈50° C–380° C).

Membranes also have the interfaces and one of the important groups for studying colloidal organization. Lipid bilayer membranes and liposomes (or vesicles) are the typical examples. Liposomes are further grouped into small unilamellar vesicles (SUVs; ≈0.025-0.1 μm), large unilamellar vesicles (LUVs; ≈0.2-2 μm), and multilamellar vesicles (MLVs; ≈0.3-1 μm).[49] These structured substances are unique in shape. These various membrane structure formations are accompanied with the lowering of the interface energy.

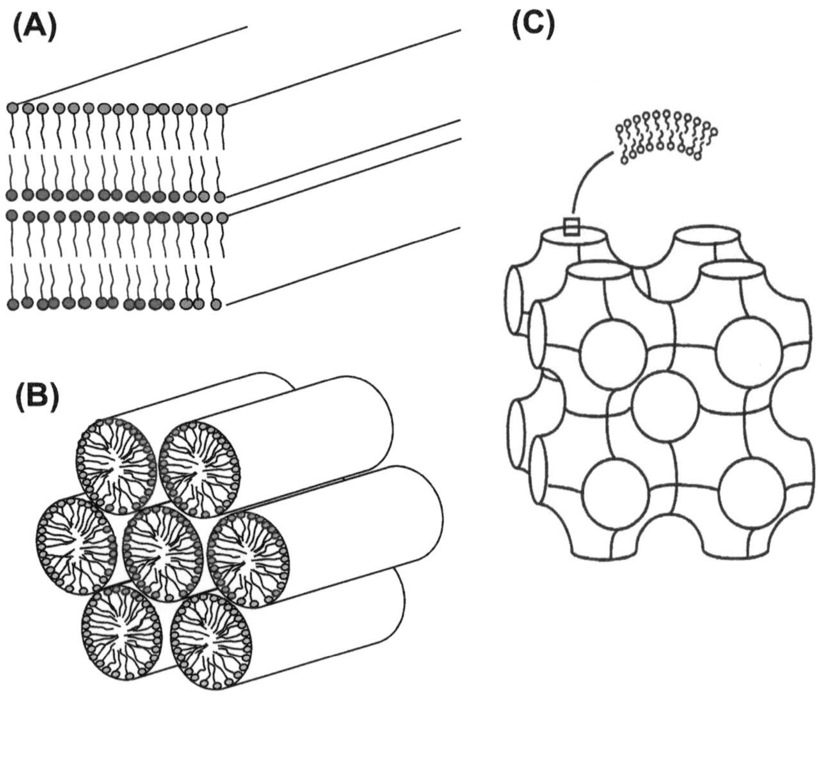

Figure 2.18 Schematic representation of various types of micelles (*2*). **A,** Lamella structure (neat phase); **B,** hexagonal structured rods (middle phase); **C,** bicontinuous cubic lattice structure.

Recently, syntheses of lactosomes and protosomes have been made as the biological colloids for the carrier.[50]

Preparation of Langmuir-Blodget (LB) membrane has been made by AA-type and AB-type sequential depositing of one or more monolayers of surface-active substance onto a porous or nonporous support.[51] Emulsion is the liquid–liquid type colloidal system. Emulsions prepared by agitation are considered to be in high state of Gibbs free energy compared with emulsions before agitation, and are apt to separate to each macroscopic phase after elapsed time to reduce the interfacial area as much as possible. Oil in water (O/W) and water in oil (W/O) are the typical examples of emulsions; examples are milk and salad dressing, respectively. *Phase inversion* refers to a phenomenon occurring when O/W emulsion reverts to W/O and vice-versa by agitation. Emulsification by phase inversion is widely used in the fields of cosmetic, pharmaceutical, and food production.[52-56]

2.5 Most Colloidal Particles are Charged Negatively in Water

We should note that most colloidal particles in water get "negative charges" at their interfaces to lower the enlarged interfaces of Gibbs free energy by two mechanisms: by the dissociation of ionizable groups and by the preferential adsorption of ions from suspension. Table 2.1 shows the sign of the charged surfaces of typical colloidal substances in water medium. Several kinds of cationic charged particles are shown in the table. However, most cationic particles are artificial substances. Interestingly, part of the cationic particles is transferred to the anionic charged particles after time, from several weeks to probably several million years, has elapsed. It is highly plausible that most of the anionic charged colloidal particles are dispersed stably in water. Several researchers know that some of the cationic charged polymer particles synthesized by using the cationic initiator molecules are converted to the anionic ones with the elapsed times. Furthermore, most of the metallic particles such as gold and silver particles, for example, which should be neutral in air, are charged with a certain amount of negative charges in the aqueous suspension state.

Table 2.1 Sign of Charged Particles of Typical Substances in Aqueous Medium

Positively Charged Particles	Negatively Charged Particles
$Fe(OH)_3$, $Cr(OH)_3$, $Cd(OH)_3$, $Al(OH)_3$, TiO_2, ZrO_2	Pt, Au, Ag, S, AsS_3, Sb_2S_3, SiO_2 (quartz, glass, ceramics), SnO_2, V_2O_5, air bubbles, filter paper, wool, cotton

$Al(OH)_3$, aluminum hydroxide; *Au*, gold; *Ag*, silver; *AsS_3*, arsenic trisulfide; *$Cd(OH)_3$*, cadmium hydroxide; *$Cr(OH)_3$*, chromium hydroxide; *$Fe(OH)_3$*, iron oxide-hydroxide; *Pt*, platinum; *S*, sulfur; *Sb_2S_3*; antimony trisulfide; *SiO_2*, silicon dioxide; *SnO_2*, tin dioxide; *TiO_2*, titanium dioxide; *V_2O_5*, vanadium oxide; *ZrO_2*, zirconium dioxide.

Why does the negative charging of solid surfaces take place in aqueous media? There are several interpretations for the phenomena. First, Cohen's (1909) rule, which says that when both the solutes and medium molecules are insulators, substances having higher permittivity become positively charged at the interface, holds true in aqueous suspensions as well. For example, permittivity of glass and water is known to be approximately 5 or 6 and 78, respectively; colloidal silica particles in water are charged negatively. Permittivity of benzene is 2.3, and colloidal silica is charged positively in benzene, and so on. Second, most colloidal particles themselves have negative charges. Third, most colloidal particles adsorb the negative charges on their interfaces in the condition of dispersing in water. These interpretations are still not definite. Mechanisms of the charging of colloidal particles are very important for colloidal organization, but are not still fixed. We should note here that these charging events take place not only for colloidal solid particles, but also gas and nonaqueous liquid particles, when they contact with water or other polar medium molecules.

2.6 Ion-Binding of Colloidal Particles

Characterization of the surface charges is very important for study of colloidal organization, especially in aqueous media. The stoichiometric charge density of colloidal particles can be determined conveniently by the conductometric titration.[57] The titration curves have been shown in several papers.[58-61] Potentiometric (pH) determination of the charge density has been reported,[62] but may not be so convenient and accurate compared with the conductance method, in general.

For colloidal particles, irrespective of their shapes, the counter-ion association (i.e., the strong associative interaction between the charges on the particles and their counter-ions, which deprives some of the counter-ions of freedom of thermal motion) has been a subject of theoretical and experimental study. The effective charge number (β) as a function of the stoichiometric charges (Z), which is determined from conductance,[59,63,64] electrophoretic mobility,[65] dielectric,[66] and osmotic pressure measurements,[67] decreases as the analytical (stoichiometric) charge number of a particle (Z) increases. β values of the typical colloidal spheres are compiled in Table 2.2. The relationship between β and Z is shown in Equation 16 and Fig. 2.19.[59]

$$\text{Log}\,\beta = 2.61 - 0.87\log Z, \text{for } Z > 3000 \qquad 16$$

Note that the β values for colloidal spheres having very large Z values become extremely small. For example, 99.9% of total counter-ions are associated with the charges on the sphere surface when Z is 10^6. Charge

Table 2.2 Characteristics Including β Values of Typical Anionic Colloidal Spheres

Sphere	Diameter	Charge Density* (nm)	$\beta^{\dagger}$ ($\mu C/cm^2$)	$\beta^{\ddagger}$	Z
Snowtex-S	8 ± 1	0.32	0.88	———	4
Ludox-AM	13.5 ± 1	0.54	0.85	———	19
D1B89	38 ± 1	21.5	0.031	———	6100
Snowtex-20L	45 ± 5	0.68	0.47	———	271
#8691	70 ± 1	2.0	0.32	———	1880
D1C25	85 ± 6	2.1	0.26	———	3010
D1C27	91 ± 6	2.8	0.16	0.32	4550
D1B22	109 ± 3	3.6	0.092	0.15	8490
SS29	126 ± 5	8.9	0.029	———	27600
D1B72	173 ± 7	1.0	0.16	0.24	6040
D1B41	215 ± 5	1.3	0.15	———	12200
N200	220 ± 5	4.2	0.04	0.035	39800
D1B33	269 ± 3	2.9	0.05	0.029	41000
SS39	350 ± 5	10.6	0.0092	———	2.60×10^5
C351	351 ± 5	51.7	0.0032	———	1.25×10^6
#103	363 ± 30	3010	———	3.4×10^{-5}	7.8×10^7
G5301	369 ± 10	7.2	———	8.9×10^{-3}	1.9×10^5
G0509	410 ± 5	51.7	0.015	———	1.7×10^5
D1A92	497 ± 6	2.8	———	0.014	1.3×10^5
N601	600 ± 12	19.5	0.024	0.0014	1.37×10^6
N1000	1020 ± 20	2.6	———	0.0054	5.3×10^5
H1001	1022 ± 5	171	0.0029	6.9×10^{-5}	3.5×10^7
D1A12	2020 ± 14	6.2	———	8.5×10^{-4}	5.0×10^6
D1A35	2950 ± 130	14.1	———	2.3×10^{-4}	2.4×10^7

**Strongly acidic groups.*
†Conductance measurements.
‡Electrophoretic measurements.
Snowtex-S, Snowtex-20L: colloidal silica spheres from Nissan Chemical Co. Tokyo.
Ludox-AM: colloidal silica spheres from Du Pont, Japan.
#series of codes: polystyrene spheres from Polyscience Co.
D1 series of codes: polystyrene spheres from Dow Chemical Co.
SS series of codes: polystyrene spheres synthesized in my laboratory.
N and C series of codes: polystyrene spheres from Sekisui Chemical Co.
G series of codes: polystyrene spheres from Japan Synthetic Rubber Co.

renormalization effect has been discussed by Alexander et al.[68] and Palberg et al.[69,70] Counter-ion binding of colloidal spheres was also discussed by Gilanyi.[71] We should note here that 1-β term corresponds to the degree of the counter-ion binding at the surface of the Helmholtz layer of the electrical double layers.

It is well known that a cation and an anion are not able to dissociate when the distance between these two ions becomes shorter than the Bjerrum length, *B*, given in Equation 17.

$$B = (e^2/\varepsilon k_B T)^{-1/2} \qquad 17$$

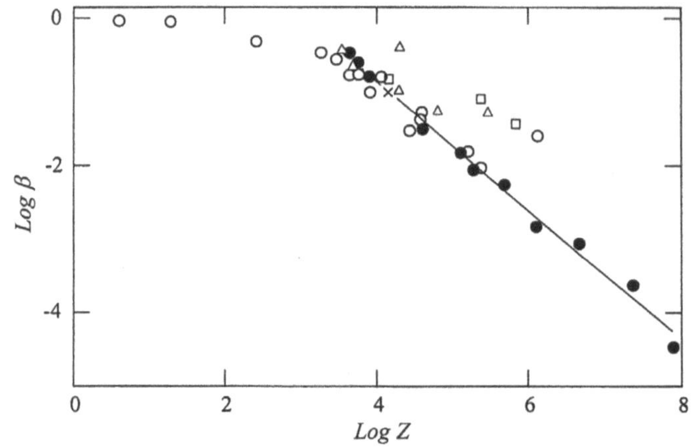

Figure 2.19
Effective charged fraction (β) of various ions plotted as a function of stoichiometric charge number (Z). Circles, triangles: conductance method; squares: transference experiments; cross: theory; solid straight line: fitting line of the experimental data. *From Fig. 6 of Okubo T:* J Colloid Interf Sci *125:380-385, 1988, with permission from the* Journal of Colloid Interface Science, *copyright 1988 Academic Press.*

Here, e is the elementary charge, and ε is the permittivity of medium. When a cation and an anion come too close, the electrostatic attraction becomes strong enough to prohibit the dissociation, and the ion-binding takes place. Manning[72,73] and Oosawa[74] proposed independently that the degree of the ion-binding of linear-type ionic polymers (polyelectrolytes) are predicted simply by the ratio (ξ) of the Bjerum length (B) and the inter-ionic distance (b).

$$\beta = 1/\xi \quad 18$$

$$\xi = e^2/\varepsilon k_B T = B/b \quad 19$$

It should be noted here that most physicochemical properties of polyelectrolyte solutions such as mean and single-ion activity coefficient, solubility, partial molal volume, heat of dilution, diffusion coefficient, and second-viral coefficient are basically interpreted properly with use of Manning's theory.[75-77]

2.7 Electrical Double Layers

Most colloidal particles get negative charges on their surfaces as described previously, and these excess charges accumulate near the colloidal surface, forming electrical double layers. The counter-ions in the diffuse region are distributed according to a balance between the thermal diffusive forces and the forces of electrical attraction with colloidal particles. The importance of the electrical double layers for the colloidal crystallization has been clarified by many researchers.[63,78-109] The giant size of single crystals has been formed in the exhaustively deionized and diluted suspension of colloidal spheres, in which the number of the nuclei for crystallization become quite few[99-109] and also the effective size of colloidal spheres including the double layers becomes large enough for the enlarged spheres to contact each other. Hachisu et al.[79,110] clarified that the colloidal crystallization takes place by the Kirkwood-Alder transition. When extra repulsive interaction (like electrostatic repulsion) is in effect among colloidal particles, in addition to the repulsion forces from their Brownian movement of colloidal particles (the excluded-volume effect), colloidal crystallization takes place easily even at the very low particle concentrations. The thickness of the electrical double layers is estimated approximately with the Debye-screening length, l_D, given by Equation 20. Debye length corresponds to the distance from the colloidal surface, where the electrostatic potential decreases to 1/e compared with that at the slipping zone (Fig. 2.20), where e is the base of the natural logarithm, 2.718.

$$l_D = (4\pi e^2 n/\varepsilon k_B T)^{-1/2} \tag{20}$$

Here, e is the electronic charge, and ε is the dielectric constant of the solvent, n is the concentration of "diffusible" or "free-state" cations and anions in suspension. Thus, n is the sum of the concentrations of diffusible counter-ions, foreign salt, and both H^+ and OH^- from the dissociation of water. The maximum value of l_D in water is approximately 1 μm, which is estimated from Equation 20 by taking $n = 2 \times 10^{-7}$ (M) $\times N_A \times 10^{-3}$ cm^{-3} at 25° C, where

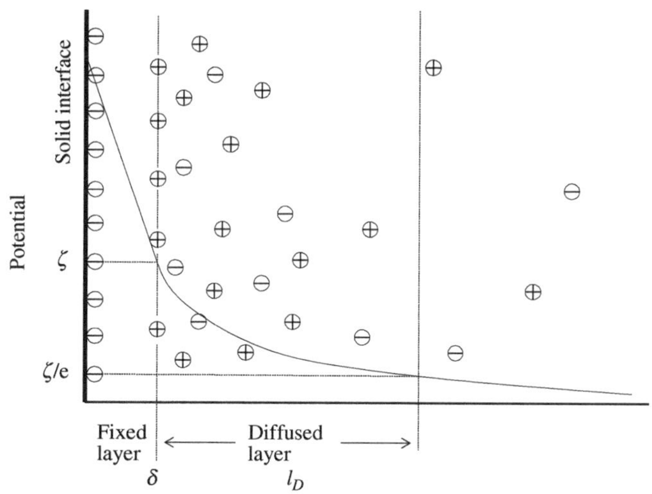

Figure 2.20 Schematic picture showing the electrical double layers.

N_A is the Avogadro number. It should be noted that the electrical double layers form always without exception in the colloidal dispersions in a polar liquid.

In the deionized state, the electrical double layer is very thick, and the interparticle repulsive forces prevail over a long distance, as long as "micrometers," although the forces become very weak. Formation of the crystal-like ordering of monodisperse spheres is explained nicely with the "effective soft-sphere model." The effective diameter, d_{eff}, of spheres includes the Debye-screening length, l_D, and is given by the diameter plus twice the Debye length (i.e., $d_{eff} = d + 2 \times l_D$). When d_{eff} is shorter than the observed intersphere distance, l, a gas distribution is observed. When d_{eff} is comparable to or a bit shorter than the intersphere distance, the distribution of spheres is usually liquid. When the effective diameter is close to or larger than the observed intersphere distance, crystal ordering forms. The effective soft-sphere model has been supported by many researchers from the systematic comparison of d_{eff} with l values.[87,96,101,104-107]

Translational diffusion coefficients of colloidal spheres (D_t) decrease substantially in the deionized suspension, because diameter (d) in Equation 1 is replaced by d_{eff} ($= d + 2 \times l_D$). Figs. 2.21 and 2.22 demonstrate the important role of the electrical double layers on the diffusional properties.[111] Here, the translational diffusion constants of monodispersed polystyrene spheres in a gas state were observed directly by using a metallurgic microscope assisted with a video image data analyzer. The D_t values increased substantially as the NaCl concentration increased, and were insensitive to the size of the

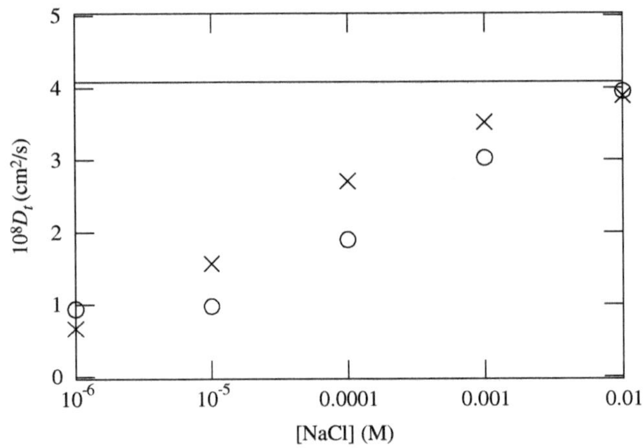

Figure 2.21
Comparison of observed and calculated values of D_t. SSH3 spheres, 120 nm in diameter, $\phi = 0.0001$ in volume fraction; circles: observed; crosses: calculated using d_{eff} ($= d + 2 \times l_D$); solid line: calculated using d.
Redrawn from Fig. 1 of Okubo T: J Phys Chem B *93:4352-4354, with permission from the* Journal of Physical Chemistry, *copyright 1989, American Chemical Society.*

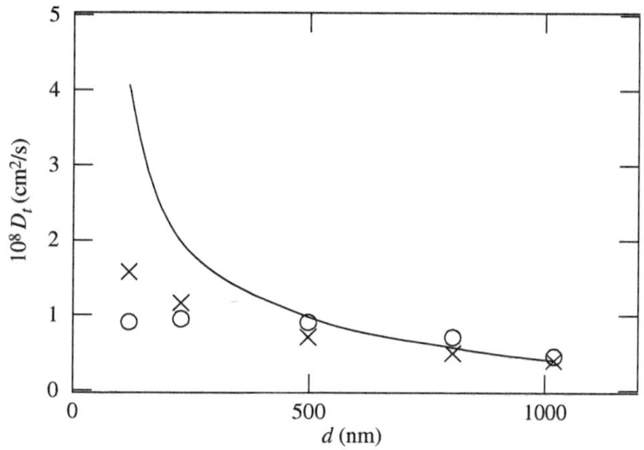

Figure 2.22
Comparison of observed and calculated values of D_t. Polystyrene spheres, $\phi = 0.0001$ in volume fraction, [NaCl] $= 1 \times 10^{-5}$ M; circles: observed; crosses: calculated using d_{eff} ($= d + 2 \times l_D$); solid line: calculated using *d*. *Redrawn using Figure 2 of Okubo T:* J Phys Chem B *93:4352-4354, with permission from the* Journal of Physical Chemistry, *copyright 1989, American Chemical Society.*

spheres between 120 and 1020 nm in diameter. In Figs. 2.21 and 2.22, agreement between the observation and the prediction from d_{eff} is excellent in the presence of high salt concentration and for the spheres larger than 500 nm in diameter. However, observed data differ from those predictions from *d* values, especially at low concentrations of NaCl and in small spheres, respectively. These differences are explained by the important role of the electrical double layers; in other words, observation comes to be close to the prediction when effective size of spheres ($d_{eff} = d + 2 \times l_D$) is taken instead of real *d* value.

For the anisotropic-shaped particles, effective shape and size, including the electrical double layers, must change greatly by the deionization of the dispersion. Two examples are shown in Figs. 2.23 and 2.24. Bowl-shaped particles were found to orient in the sedimentation equilibrium state, as shown in Fig. 2.23, *A*. These array particle formations occur to maximize the packing efficiency.[112,113] Fig. 2.23, *B*, shows clearly that the bowl-shaped particles included one or two small spherical particles at the low concentration of foreign salt in suspension state.[112,113] The fact that the effective shape and size of the particles change depending on the ionic concentration of dispersion is exciting. Furthermore, change of the effective size and shape with ionic concentration was observed for the spindle-shaped particles.[114] Interestingly, anisotropic arrays shifted to the isotropic hexagonal arrays as ionic concentration decreased. At the exhaustively deionized state, free rotation of the spindle particles was observed through a microscope, as is shown in Fig. 2.24.

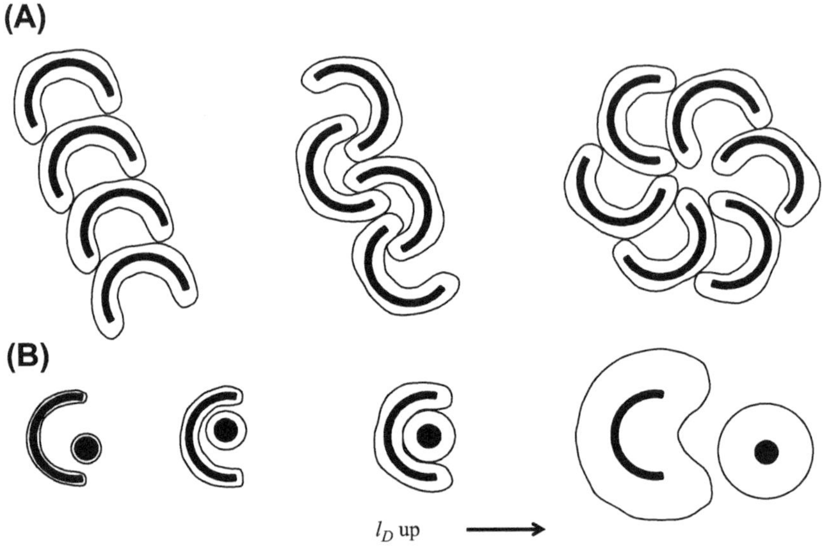

Figure 2.23
Important role of the electrical double layers on orientation and inclusion as a function of ionic concentration.
A, Bowl-shape particles;
B, bowl-shape particles and spheres.

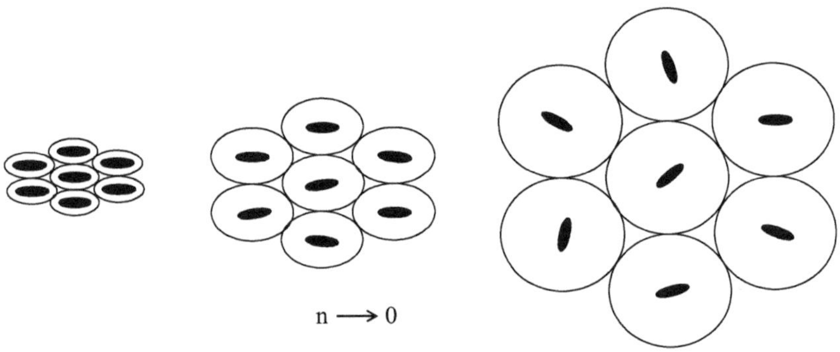

Figure 2.24
Importance of the electrical double layers upon the orientation of spindle-shaped particles. *Redrawn with permission from* Colloid Polymer Science, *copyright 1988, Springer.*

2.8 Stability of Colloidal Dispersion and DLVO Theory

Stability of colloidal dispersions is discussed by the Derjaguin-Landau and Verwey-Overbeek (DLVO) groups, resulting in DLVO theory.[115] In the dispersion state, van der Waals attraction (V_A) and electrostatic repulsion between the colloidal spheres originated from the overlapping of the electrical double layers (V_R) are given by Equations 21 and 22, respectively.

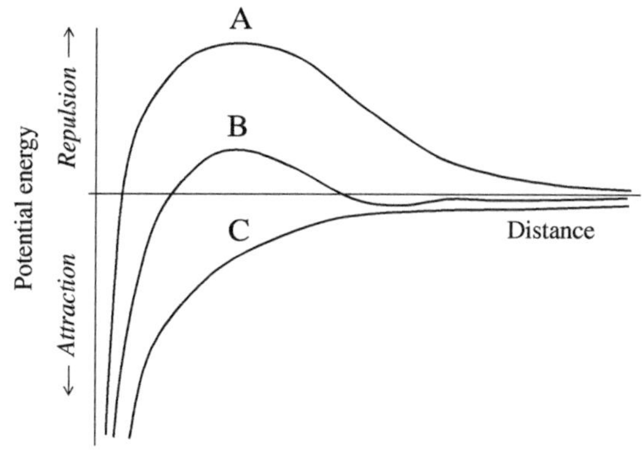

Figure 2.25 Typical interparticle potential energy curves. **A,** Stable dispersion; **B,** slow aggregation; **C,** fast aggregation.

$$V_A = -(Ad/24l) \qquad 21$$

$$V_R = (1/4)\varepsilon d\zeta^2 \exp(-\kappa l) \qquad 22$$

Here, d is the diameter of colloidal spheres, l is the inter-particle distance, A is the Hamaker constant, which changes depending on the kind of colloidal particle. The dielectric constant of the solvent is represented by ε, ζ is the zeta-potential, and κ is the reciprocal ionic concentration in the medium phase. V_R-values were evaluated theoretically by the Derjaguin-Landau and Verwey-Overbeek (DLVO) groups of Russia and the Netherlands independently. Fig. 2.25 shows the typical potential energy curves obtained from DLVO theory. The maximum height of the curve, *A,* is higher than the thermal energy, k_BT, and the interparticle repulsion is strong enough to keep dispersion stable for a long time. In the deionized condition, Debye-length (l_D) increases, repulsive potential height becomes higher than the thermal energy, and the dispersion becomes stable. In a small amount of salt (Fig. 2.25, *B*), slow aggregation takes place, because the height becomes close to the thermal energy. In the presence of a large amount of salt, the potential curve occurs as shown in Fig. 2.25, *C,* where the repulsive peak disappears and rapid aggregation takes place in general.

2.9 Diffusion, Rotational Relaxation, Light Scattering, Viscosity, and Viscoelasticity

The diffusion coefficients, *D,* were evaluated from the cumulative analysis of the dynamic light-scattering measurements. The *D* values of colloidal particles decreased as the scattering vector (q) increased, whereas the background intensity of scattered light for the crystals increased as q increased. Furthermore, the diffusion coefficients from the suspensions showing peaks in the light-scattering curve were substantially lower. According to Pusey and Tough,[116] the effective diffusion coefficient (D_{eff}) is given by

$$D_{eff} = D_0/S(q) \quad 23$$

where D_o is the diffusion coefficient of spheres in gas suspension and $S(q)$ is the structural factor. The constancy of $D_{eff} \times S(q)$ has been satisfactory for colloidal "crystals" and liquids, within experimental error. The interparticle structural factor, $S(q)$, which is important to know the dynamic particle distribution and the interparticle interaction, has been observed from light-, X-ray, and neutron-scattering measurements for the deionized and salt-containing colloidal suspensions.[117-127] The effect of polydispersity of colloidal spheres on the structural factor was discussed in detail.[128]

Fluctuation of colloidal spheres in the crystal structure is discussed based on the structural relaxation time (τ). Two kinds of τ values, fast (τ_f) and slow steps (τ_s), are observed by the stopped-flow technique. The former is given by Equation 24.

$$\tau_f \sim l^2/6D_t \quad 24$$

where l is the mean intersphere distance and D_t is the cooperative translational diffusion constant in strongly interacting structure. The D_t value of spheres is given by the Stokes-Einstein equation for self-diffusion.

$$D_t = k_B T/(3\eta\pi d_{eff}) \quad 25$$

Here, η is the viscosity of the solvent. It should be noted that Equation 25 is similar to Equation 1, but d is replaced by d_{eff}. The essential point in the fluctuation of spheres is that the D_t and τ_f values observed in the colloidal crystals agree with the calculated values from Equations 24 and 25 when d_{eff} is used instead of d (true diameter of colloidal sphere in Equation 25).[129,130] Agreement between the calculation and observation in the deionized suspensions of monodisperse polystyrene spheres was excellent. It should be noted here that d_{eff} value, including the Debye screening length, corresponds to the hydrodynamic length. Ottewill et al.[131] observed particle motion directly with microscopy and concluded that the observed translational diffusion constants agreed with the value calculated from Equation 25 using the real value of sphere diameter, d. However, the researchers do not deny the tendency of D_{obs} to be less than D_{calc} (i.e., $D_{obs} = 3.6 \times 10^{-9}$ cm^{2}s^{-1} and $D_{calc} = 4.0 \times 10^{-9}$ cm^{2}s^{-1} for $d = 0.458$ μm). The author clarified that the diffusion coefficients are influenced greatly by the thickness of the Debye screening length as described previously using Equations 24 and 25.[111]

The τ_f-value is further given by the simple relation[132,133]

$$\tau_f \sim \eta/G \quad 26$$

where *G* denotes the rigidity or dynamic (local shear) elastic modulus of the suspension. *G* ranges from 0.01 Pa to 10 Pa for colloidal crystals, and these values are not far from those of previous reports by the dynamic method.[82,134-137] Effect of the electrical double layers on the translational diffusion constant of colloidal spheres has been discussed using photon correlation spectroscopy[138,139] and theoretically.[140] The diffusion constant showed a minimum when the double layer thickness was comparable to the radius of the particle.[86]

The rotational diffusion coefficient (D_r) is one of the fundamental properties of spherical and anisotropic colloidal particles in dispersion and also of flexible biopolyelectrolytes. However, the experimental method for its determination is not always easy. Flow birefringence, non-Newtonian flow, electric birefringence, fluorescence depolarization, and dielectric dispersion are typical methods that have been studied so far.[141-144] The stopped-flow (SF) technique as a unique and convenient method for determining the rotational relaxation times of anisotropic colloidal particles such as ellipsoidal tungstic acid particles was proposed[114,145-148] (Fig. 2.26). Optical, conductometric, and other properties become different during the flow between the parallel and perpendicular directions to the flow. The birefringence effects are observed even for isotropic spherical particles, because the electrical double layers forming around the spheres are distorted from spherical to flamelike shape by the flow of the suspension.

The rotational diffusion coefficient (D_r) is obtained directly from the observed rotational relaxation time (τ) by using Equation 27.

$$D_r = 1/6\tau \tag{27}$$

For ellipsoidal particles, the theoretical values of D_r are calculated from Perrin's equations, Equations 28 to 31, without any assumptions when three axes (*a*, *b*, and *c*) are given.

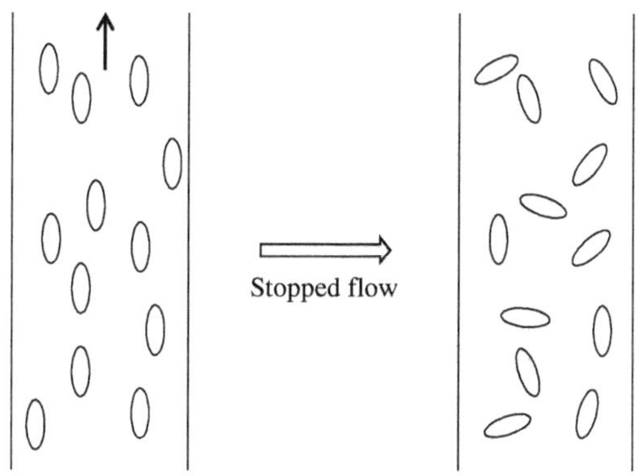

Figure 2.26 Flow-birefringence of anisotropic particles (ellipsoids in this figure).

$$D_r = \left\{3k_BT(a^2P + b^2Q)\right\} / 16\pi\eta(a^2 + b^2) \quad 28$$

$$P = \int ds/(a^2 + s)\left\{(a^2 + s)(b^2 + s)(c^2 + s)\right\}^{1/2} \quad 29$$

$$a^2P + b^2Q + c^2R = \int ds / \left\{(a^2 + s)(b^2 + s)(c^2 + s)\right\}^{1/2} \quad 30$$

$$P + Q + R = 2/abc\,, \quad a < b < c \quad 31$$

Here, a, b, and c denote the lengths of the minor, middle, and major axes of an ellipsoidal particles. The constants, P, Q, and R, are derived from Equations 29 through 31. The Boltzmann constant, absolute temperature, and the viscosity of solvent, are represented by k_B, T, and η, respectively. Rotational diffusion coefficient (D_r) observed in aqueous dispersions is slightly smaller than the calculation, especially at the diluted concentrations.[145] The differences were explained well by the contribution of the electrical double layers formed around the particles. Rotational diffusion of anisotropic-shaped colloidal particles in microgravity was studied by free-fall experiments.[149] Influence of the electrical double layers was also clarified.

Light-scattering intensity, I, in the presence of colloidal particles is given by Rayleigh's equation, Equation 32.[150]

$$I = kNV^2/\lambda^4 \quad 32$$

I is proportional to the particle number per unit volume (number density, N), square of the particle volume and inverse of λ^4 (λ: wavelength of light). Constant k is the function of the scattered angle against incident light. The I value of the forward direction is called *turbidity*. Light-scattering intensity measurements as a function of scattering angles give information on the particle size and its dispersion particle factor and further structural information of the particle's structural factor. The intensity of scattered light $I(q)$ is given by Equation 33.

$$I(q)|\infty|P(q)S(q) \quad 33$$

Here, q is the scattering vector whose magnitude is equal to $(4\pi/\lambda)\sin(\theta/2)$, where λ is the wavelength of the incident beam in medium. $P(q)$ is the single-particle form factor and $S(q)$ is the structure factor. $P(q)$ can be estimated experimentally from the $I(q)$ measurements on the same sample dispersion, but in the presence of salt, for example. For spherical particles, $P(q)$ was given as Equation 34 theoretically.[150]

$$P(q) = \left\{3[\sin(qd_0/2)\cos(qd_0/2)]/(qd_0/2)^3\right\}^2 \quad 34$$

$S(q)$ is the structural factor, which is important to obtain the structural information of the colloidal dispersion (i.e., gas, liquid, crystal, or amorphous-solid structures). It should be noted that $S(q)$ versus q plots correspond to the radial distribution function shown in Fig. 2.2. The Bragg distance (l_B) is observable from the peak scattering vector (q_m) using the Bragg equation:

$$l_B = 2\pi/q_m \tag{35}$$

For colloidal crystal, lattice assignment of the respective peaks, q_m, to the crystal planes is made from comparison of the observed peak vectors with the calculation from Equation 36:

$$q_{hkl} = 2\pi/a\left(h^2 + k^2 + l^2\right)^{1/2} \tag{36}$$

Here, h, k, and l are Miller's indices and a is the unit cell size.

The autocorrelation function of the scattered field, $g^{(1)}(\tau)$ obtained from the dynamic light-scattering measurements is represented by a single exponential type of decay function, Equation 37, from which the translational diffusion coefficient, D_o is obtained.

$$g^{(1)}(t) = \exp(-q^2 D_0 \tau) \tag{37}$$

From D_o we may calculate the hydrodynamic diameter, d_H, of colloidal sphere using the Stokes-Einstein equation:

$$d_H = k_B T/3\pi\eta D_o \tag{38}$$

Here, η is the medium viscosity. According to Pusey and Tough,[116] Equation 23 was also approved experimentally.[114,152-154] As is shown in Fig. 2.27, the q value's corresponding peaks in I values and valleys in D_{eff} observed for colloidal crystals of colloidal silica spheres coincided with each other.[154] The diffusion coefficients of the colloidal liquids and crystals were characterized by the two and three different coefficients: short time (D_s) and long time (D_l) diffusion coefficients for liquid and D_s, D_l and fastest coefficient (D_f) for solid (or crystal). The relations, $D_s = D_o$ and $D_l \approx D_s / 3$ for liquid and $D_f > D_s > D_l$ for crystal. Furthermore, knowledge of the fluctuation of colloidal particles around their equilibrium positions in solid phase is also available from the dynamic light-scattering measurements. Dynamic light-scattering measurements have been made in detail for colloidal crystals by many researchers.[155-162] Particle motion was studied also by tracer diffusion method for concentrated colloidal suspensions.[163]

I have measured viscosities of colloidal liquids and crystals in exhaustively deionized aqueous suspensions, for the first time, with the use of an Ostwald rotational viscometer.[164-169] The viscosity of the deionized suspension was very sensitive to the degree of deionization of the suspension and decreased substantially when the suspension was contaminated with ionic species. Typical plots of the reduced viscosity η_{sp}/c of colloidal crystal suspensions versus sphere concentration are shown in Fig. 2.28. The peaks correspond to

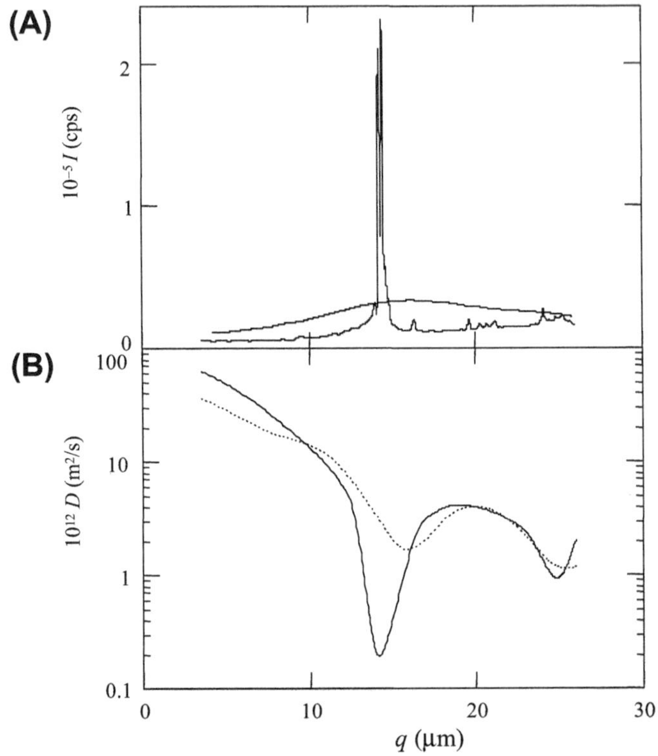

Figure 2.27 Scattering intensity (I) and diffusion coefficient of CS91 dispersion as a function of scattering vector at 24° C. Sharp peaks (A), solid curve (B): with resins; broad curve (A), dotted curve (B): without resins. *Redrawn using the data of Okubo T, Kiriyama K:* J Mol Liquids *72:347-364, 1997, with permission from the* Journal of Molecular Liquids, *copyright 1997, Elsevier.*

the crystal-liquid transition in the suspension. The reduced viscosity, η_{sp}/c (specific viscosity, η_{sp}, divided by particle concentration, c) of colloidal liquids was also much higher than that predicted by Einstein's equation and decreased sharply with increasing ion concentration. The specific viscosity of colloidal spheres can be given by Equation 39.

$$\eta_{sp} = \eta/\eta_0 - 1 = k_1\phi + k_2\phi_2 + ... \qquad 39$$

Here η and η_o are the viscosities of the colloidal suspension and the solvent. The concentration of the colloidal particles in volume fraction is represented by ϕ. The Einstein coefficient 2.5 is k_1, and k_2 denotes the electrostatic interaction parameter, which is very large for the deionized suspensions. The deviation of the observed η_{sp} values from Einstein's equation was significant in the deionized liquids. However, the η_{sp} versus ϕ relationship agreed very well with the Einstein equation in the presence of an excess amount of sodium chloride. Qualitatively speaking, the large k_2 observed implies that the electrostatic repulsion interaction is significant and the shells of the electrical double layers around the spheres are very thick, because the effective hydrodynamic diameter (d_{eff}) of a colloidal

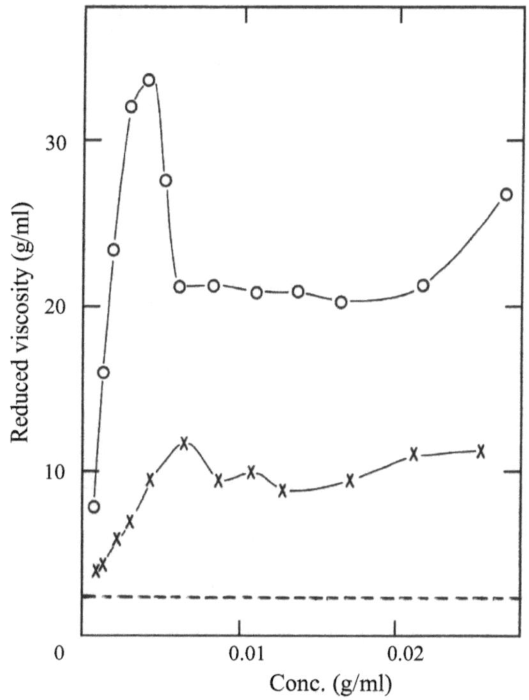

Figure 2.28 Viscosity of deionized aqueous colloidal suspensions. Circles D1B22; crosses D1B41; broken line Einstein coefficient. *Modified from Fig. 9 of Okubo T:* J Chem Phys *87:6733-6739, 1987, with permission from the* Journal of Chemical Physics, *copyright 1987, American Institute of Physics.*

sphere is given by Equation 40. Furthermore, η_{sp}/c increased as the charge density decreased or the polydispersity index increased. Details on the viscosity of the structured suspensions are described in Chapter 3.

$$d_{eff} = d + 2 \times l_D \quad (40)$$

Here, d is the real diameter of a sphere and l_D is Debye length given by Equation 20. It should be noted here that the reduced viscosity of aqueous colloidal sphere suspensions were explained successfully by the contribution from the electrostatic interaction between charged spheres theoretically.[170]

Viscoelastic properties of the structured colloidal dispersions are important. A coaxial cylindrical-type rheometer is appropriate for rheologic experiments, because the viscosity of the colloidal dispersions is usually very low except at high-particle concentrations. For the colloidal crystal samples, simultaneous measurements of rheologic plus small-angle neutron scattering measurements[171-174] and rheologic plus reflection spectroscopy[175-177] were reported.

2.10 Electrokinetic Phenomena

At the interfaces of colloidal particles contacting with polar medium, the electrical double layers form. When the electrical field is applied to the

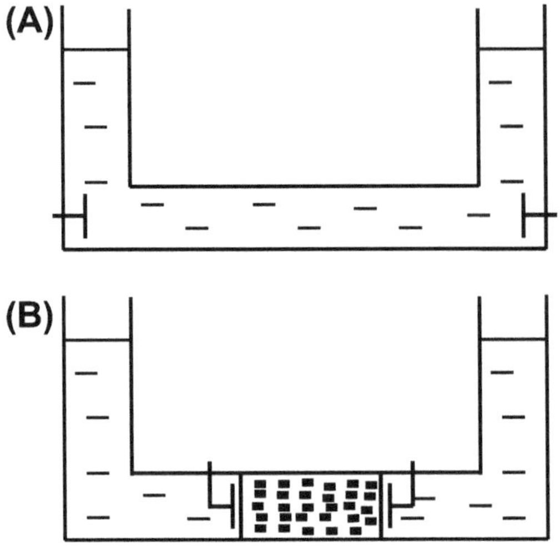

Figure 2.29 Two types of the cells used for electrokinetic measurements. **A,** Electrophoresis, **B,** Electro-osmosis.

colloidal dispersion, electrophoresis (i.e., movement of the particles) takes place. When charges of colloidal particles are negative, the particles move toward the anode (Fig. 2.29, *A*). Mobility (μ) is given by Equation 41, the Smoluchowski equation.[178]

$$\mu = v/E = \zeta\varepsilon/(4\pi\eta) \qquad 41$$

Here, ε is the dielectric constant of the medium, η is the viscosity of the medium, E is the electric field strength applied to the cell, and ζ is the zeta-potential of the particles. Using Equation 41, zeta-potential of the particles is evaluated from the measurements of mobility. Many kinds of zeta-meters are available commercially. Effective charge numbers (α) and their ratios (β) against the stoichiometric charge numbers (Z) ($\beta = \alpha/Z$) are evaluated from the electrophoretic mobility measurements.[65] The β-values of colloidal particles were also estimated from the transference measurements.[178] Recently, the albumin/globulin ratio has been measured from the electrophoretic mobility measurements using a tiny amount of blood.

When the electric field is applied to the cell of Fig. 2.29, *B,* colloidal particles are not able to move through the semipermeable layer and only the medium molecules and simple ions move instead. This phenomenon is called *electro-osmosis.* The flow of medium molecules is the opposite direction of that of the colloidal particles in electrophoresis shown in the cell depicted in Fig. 2.29, *A.* The flow rate μ_{EO} is given by Equation 42.

$$\mu_{EO} = -\varepsilon\zeta/\eta \qquad 42$$

In the cell depicted in Fig. 2.29, *B,* when the medium molecules are forced to flow through the polar porous solid by application of pressure difference, Δp between the two solid surfaces, streaming current (I_{stream}) and streaming potential ($\Delta\phi$) appear, following Equations 43 and 44.

$$I_{stream} = (\varepsilon\zeta/\eta)(\Delta p/L)A \qquad 43$$

$$\Delta\phi = (\varepsilon\zeta/\eta\kappa)\Delta p \qquad 44$$

Here, L and A are the length and cross-sectional area of the channel and $1/\kappa$ is the Debye length. The electric potential is also obtained from the sedimentation of the colloidal particles and supersonic vibration potential of the dispersion.

2.11 Kinetic Analyses in Colloidal Organization

2.11.1 Reaction Kinetics

Kinetic analyses of association or dissociation of colloidal spheres are important for understanding of colloidal organization. Let's consider the association reaction between unlikely charged colloidal particles.

$$A + B \underset{k_b}{\overset{k_f}{\rightleftarrows}} C \qquad 45$$

This association reaction appears to proceed irreversibly to the formation of the aggregates. However, in general, all the chemical and physical reactions can be safely assumed to be reversible equilibrium reactions. The principle that "all the actual reactions are composed of the equilibrium elementary reactions" has been assumed hitherto. The reaction in Equation 45 is one of the interionic reactions and is analyzed kinetically with the use of conductance stopped-flow technique.[180] The relaxation traces of the conductance for the cationic and anionic colloidal spheres were clearly observed. In most cases, one or two additional relaxations having different time scales from the fastest relaxation (τ) were observed. The fastest step corresponds to the binary association of colloidal spheres of opposite charge (i.e., a dimerization process), which has been clarified by the independent microscopic observations assisted by an image-processing system.[181] After the start of the association (which is several times longer than the relaxation time of the dimerization process), further association processes of the dimeric spheres with other spheres to form trimers, tetramers, and so on, were confirmed by microscopic observations. The slowest relaxation step might be a reorientation process to form a stable complex, although direct evidence for this has not been observed experimentally.

Using a pseudo–first-order analysis of the τ-values, the first-order rate constant, τ^{-1}, at various concentrations of the excess amount of anionic spheres is observed. Equation 46 gives τ^{-1} from a relaxation analysis method.

$$\tau^{-1} = k_f[\mathrm{B}] + k_b \quad (46)$$

From the plots of τ^{-1} versus [B], k_f is evaluated from the slope. Table 2.3 shows the temperature dependency of k_f for the association reaction between cationic colloidal spheres (300 $\pm$ 15 nm in diameter, 4 μC/cm^2 in charge density) and anionic colloidal spheres (300 $\pm$ 10 nm, $-$11 μC/cm^2). The theoretical rate constants for diffusion-controlled association reaction between neutral spheres are calculated by Equations 47 and 48.[182,183]

$$k_f = 4\pi N_A(D_{tA} + D_{tB})l_{AB}/1000 \quad (47)$$

$$k_b = 3(D_{tA} + D_{tB})/l_{AB}^2 \quad (48)$$

where D_t denotes the translational diffusion coefficient of a sphere given by Equation 1. N_A and l_{AB} are Avogadro's number and the distance of closest approach A and B ($r_A + r_B$) in the case of rigid spheres. The radii of spheres A and B are r_A and r_B. The experimental and theoretical k_f values are of the same order of magnitude, and the observed k_f values confirm that the association of the spheres is close to that of a diffusion-controlled reaction.

The influence of the electrostatic interaction between the spheres is taken into account by the following equations[184-186]:

$$k_f = f_{el}k_{fo} \quad (49)$$

$$f_{el} = Z/[\exp(Z) - 1] \quad (50)$$

$$Z = Z_AZ_Be^2/\varepsilon l_{AB}k_BT \quad (51)$$

Here, Z_A and Z_B are the stoichiometric charge number of spheres A and B. The theoretical k_f values are much larger than the experimental ones; 2.7×10^{17} M^{-1}s^{-1} was obtained for the theoretical k_f for the association of cationic spheres ($Z_A = 72000$) with anionic ones ($Z_B = 2 \times 10^5$) at 25° C. In Equations 49 through 51, f_{el} is the electrostatic factor; k_{fo} is the reaction rate constant without electrostatic contributions; $Z = U/k_BT$ (where U is the potential

Table 2.3 Kinetic Data of the Association Reaction between Cationic and Anionic Colloidal Spheres

Temp. (° C)	k_f (1/Ms) obs.	Reference	Theory	k_b (1/s) obs.	Reference	Theory
15	———	1.9×10^9	5.6×10^9	≈ 0	≈ 0	82
20	2.9×10^9	2.4×10^9	6.5×10^9	≈ 0	≈ 0	95
25	3.8×10^9	3.0×10^9	7.4×10^9	≈ 0	≈ 0	10^9

energy); Z_A and Z_B are the total stoichiometric charges of spheres A and B, respectively; ε is the dielectric constant of the medium; and e the elementary charge. As was discussed in Section 2.6, counter-ion binding by the colloidal spheres is significant, especially for large spheres.[59,65] Only a few percent of the total stoichiometric charge number of counter-ions (H^+) are free (or diffusible). Z_A and Z_B are assumed to be 1730 and 2000 for cationic and anionic spheres, by assuming that the fractions of free counter-ions (β) are 0.024 and 0.01, respectively.[57,63] The theoretical value of k_f is now estimated to be $6.4 \times 10^{13}\ M^{-1}s^{-1}$, which is still much larger than the experimental result. This is probably because the point charge model assumed in the theory is far from reality for very large colloidal spheres. The number of effective charges responsible for the attractive interaction between oppositely charged spheres may be even less than the number of net charges of each sphere. One of the simplest interpretations is that, in the association processes, only the charges on the relatively narrow surfaces near the collision center play a decisive role; the charge on the surfaces diametrically opposite from the collision center may not be influenced because of the large size of the surfaces. However, note that the experimental findings support the importance of the electrostatic interaction in the association of oppositely charged spheres. The kinetic analyses of the association reaction between the anionic colloidal spheres and the bovine serum albumin (BSA) molecules were also made.[180] The k_f values were much smaller than those corresponding to the diffusion-controlled rates. Both k_f and k_b decreased and the association constant (k_f/k_b) increased as size of the anionic spheres increased. Furthermore, k_f increased sharply, whereas k_b remained constant as pH values of the reaction mixtures decreased. We should note here that the kinetic parameters, such as k_f and k_b on the colloidal associations including adsorption and aggregation, often afford more useful information on the mechanisms than the static thermodynamic parameters. Furthermore, kinetic parameters on the intersphere reactions must be strongly influenced by the relative size ratios of reactant A and B spheres. However, these studies have not been made often. Kinetic studies on the heterocoagulation reactions between large and small spheres are also interesting.

Several important kinetic parameters for the colloidal particles relating to reactions such as adsorption and aggregation must be described. Let's consider the second-order physical and chemical reactions, Equation 45, for example. The reaction rate v_f and v_b are given by Equations 52 and 53.

$$v_f = k_f[A][B] \tag{52}$$

$$v_b = k_b[C] \tag{53}$$

[A], [B] and [C], denote the concentrations of the reactants A and B, and of product C, respectively. In equilibrium state, $v_f = v_b$ holds, and then the equilibrium association constant K_{ass} is given by Equation 54.

$$K_{ass} = [C]/\{[A][B]\} = k_f/k_b \quad 54$$

K_{ass} and both k_f and k_b are called *static* and *dynamic parameters of reaction* (Equation 45). It should be noted here that Equation 55 always holds true for the equilibrium elementary reactions.

$$\Delta G = -RT \ln K_{ass} \quad 55$$

The difference in the Gibbs free energies of reactants and products is ΔG $\{= G_C - (G_A + G_B)\}$; ΔG is evaluated directly from the association constant K_{ass}, as is shown in Fig. 2.30.

Following the activated-complex theory studied theoretically by H. Eyring, physical and chemical reactions take place when the reactants attain the free energy, $\Delta G^{\neq}$, enough to overcome the barrier wall at the activated-complex ($X^{\neq}$) in the forward reaction, as is shown in Fig. 2.30.

$$A + B \underset{k_{-1}}{\overset{k_1}{\rightleftarrows}} X^{\neq} \overset{k_2}{\rightleftarrows} C \quad 56$$

$$\Delta G^{\neq} = -RT \ln K^{\neq} \quad 57$$

According to the activated-complex theory, rate constant k is given by Equation 58.

$$k = (RT/N_A h)K^{\neq} \quad 58$$

Here, h is the Planck constant. Arrhenius Equation 59 showing the temperature dependency of k is also important for the kinetic analyses.

$$k = A \exp(-E/RT) \quad 59$$

A is the frequency factor and E is the energy of activation. These parameters are obtained from the slope and interception with the y-axis in the ln k versus

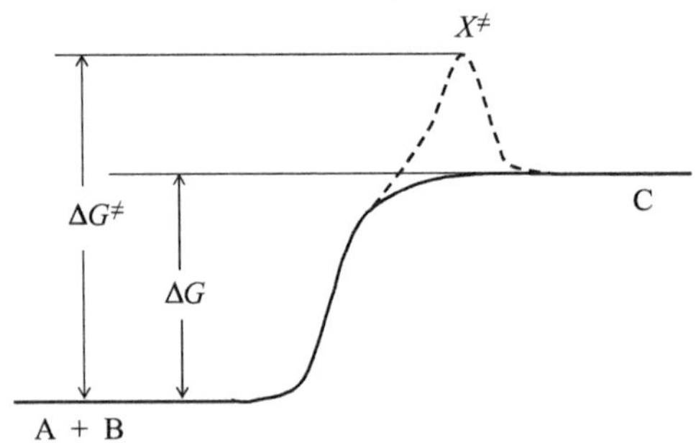

Figure 2.30 Schematic representation of the differences in the Gibb's free energy ΔG and $\Delta G^{\neq}$.

$1/T$ plots. In the liquid and gas phases, E is given by Equations 60 and 61, respectively.

$$\text{Liquid}: E = \Delta H^{\neq} + RT \tag{60}$$

$$\text{Gas}: E = \Delta H^{\neq} + 2RT \tag{61}$$

From Equations 5, 57, 58, and 60, $\Delta G^{\neq}$, $\Delta H^{\neq}$ and $\Delta S^{\neq}$ are given as Equations 62 through 64, respectively, for a liquid system.

$$\Delta G^{\neq} = RT \ln(RT/N_A h) - RT \ln k \tag{62}$$

$$\Delta H^{\neq} = E - RT \tag{63}$$

$$\Delta S^{\neq} = (\Delta H^{\neq} - \Delta G^{\neq})/T \tag{64}$$

The Gibbs free energy of activation, enthalpy of activation, and the entropy of activation are represented by $\Delta G^{\neq}$, $\Delta H^{\neq}$ and $\Delta S^{\neq}$, respectively, and they are typical and important parameters indicating the kinetic characters of the reaction.

2.11.2 Crystallization Kinetics

It is important to know the ordering processes for kinetic study on colloidal organization. Main studies have been made on the colloidal crystallization. In general, crystallization takes place when the barrier of free energy difference from liquid and crystal states is overcome by supersaturation. An example of my analyses using reflection spectroscopy for colloidal crystallization is summarized. Most kinetic measurements on colloidal crystallization showed an induction period, after which the crystal growth starts, especially in diluted suspensions. This observation supports that the kinetics of colloidal crystallization is explainable by the classical diffusive crystallization theory, including nucleation and crystal growth processes.[187-190]

The number of nuclei that germinate per unit of time, the nucleation rate, v_n, is given by:

$$v_n = N_n/t_i, \tag{65}$$

where N_n is the total number of nuclei formed during the nucleation process and t_i is the induction period, which corresponds to the time of the nucleation process. Here, we assume that the number of nuclei equals the number of a single crystal, N_c, which is given for a cubic lattice by:

$$N_c = \sqrt{2}L^3/l^3, \tag{66}$$

where L and l are the mean size of single crystals formed and the nearest-neighbor intersphere distance, respectively. The total number of colloidal spheres (N_T) in a unit volume is $\phi/[(4/3)\pi(d_o/2)^3]$. Here, d_o is the diameter of sphere particles. Then v_n is given by:

$$v_n = N_T/N_C t_i = \phi l^3 \Big/ \left[\left(4\sqrt{2}/3\right)\pi(d_0/2)^3 L^3 t_i\right], \quad 67$$

where ϕ is the volume fraction of sphere particles.

The size of colloidal single crystals formed from the homogeneous nucleation, L, is estimated from the peak intensity, I, of the reflection spectra[191]

$$I \propto N_{cryst} L^3 \propto L^3 \quad 68$$

where N_{cryst} is the number of single crystals in the reflection volume, which is directly proportional to the number concentration of crystals in the final stages of the crystallization process, being equal to the total number of nuclei formed in the whole course of crystallization.

Colloidal crystallization processes are composed of the fast growth of a metastable crystal (step 1) and the slow reorientation toward the stable crystal (step 2). The first crystal growth rate of a metastable crystal, v_1, is given by[192,193]:

$$v_1 = v_{1\infty}[1 - \exp(-\sigma)]. \quad 69$$

Here, σ is the relative supersaturation given by $(\phi - \phi_c)/\phi_c$. The critical concentration of melting is represented by ϕ_c and $v_{1\infty}$ is the maximum crystal growth rate. Equation 69 is further simplified as shown in Equation 70:

$$v_1 = v_{1\infty} - v_{1\infty}\phi_c/\phi. \quad 70$$

Equation 70 means that the growth rate should decrease linearly as the reciprocal sphere concentration increases, which was supported experimentally, but only in a very low sphere concentration range.[190] Equation 70 holds true in many cases in our experiments for typical colloidal hard spheres hitherto.[187-190,194-198] Equation 71 gives v_∞.

$$v_{1\infty} = 4D_0/\xi \quad 71$$

Here, D_o is the diffusion coefficient of spheres and ξ is the mean diffusion length (path) in the crystallization suspension.

The peak in the reflection spectroscopy became sharp with time during the course of crystallization, which corresponds to an increase in crystal size. The background intensity decreased during the course of crystallization, which is ascribed to the diminution in the multiple scattering of the suspension. The peak shift reflects the fact that metastable and loose crystals are formed in the first crystallization step and then the crystals become stable and compact. The nucleation rate, v_n, was estimated using Equations 68, 69, and observed t_i. As sphere concentration increased, v_n increased substantially. When ϕ increased by 10^2-fold, v_n increased by 10^{10}-fold.

The fast and slow crystal growth rates, v_1 and v_2, were evaluated for colloidal crystallization. The v_1 value increases first from 5 μm/s to 20 μm/s and then decreases back to 5 μm/s, passing a maximum. The v_1 values above $\phi = 0.01$ remain 5 μm/s and are insensitive to sphere concentration. The v_1 values in the diluted suspensions decrease linearly as the reciprocal concentration of spheres increases.[190] A clear explanation for the crystallization rate first going up and then going down as sphere concentration increases has not yet been obtained. However, the main cause is the dynamic nature of phase transition—in other words, spheres in the colloidal lattices interact strongly with each other, and further, interact in a cooperative manner through the electrical double layers. This dynamic phase transition is different from the classical explanation, which is the transitional diffusion of monomers. It should be noted here that the importance of the dynamic phase transition has been reported for colloidal crystals and for other crystals such as metals and polymers.

The slow step of crystallization decreases very sharply as sphere concentration increases. The v_2 value decreases from 3 μm/s to 0.7 nm/s when ϕ increases from 0.004 to 0.04.[190] The second slow crystallization step is ascribed to the reorientation of the metastable crystals formed in the first step toward the stable ones matched with the other crystals, mediated by the grain boundaries and the cell wall. When sphere concentration is high, the number of the metastable crystals formed is huge and their size becomes quite small. Then a long time is required for the crystals to be matched with other crystals and the cell wall. The occurrence of Ostwald ripening is also highly plausible in the slow process, although direct observation was not successful until Wong et al.[199] observed the ripening for colloidal crystallization processes.

2.12 Close-Up Color Pictures, Optical Microscopy, and Reflection Spectroscopy

Photographs of colloidal crystals and colloidal liquids in a test tube, an optical plane cell having various optical lengths, such as a vial, for example, are made with a camera, a macro lens, and life-size converter, which are available commercially. Velvia film (Fujichrome, RVP135, ISO = 50) was often used in my laboratory to obtain color pictures. Close-up video films using a charge-coupled device (CCD) camera are also convenient to observe crystallization processes visually. A pocket-type flashlight is best for the light source. A flashlight containing a plenty of reddish wavelengths makes a beautiful color picture.[200-204] In general, large colloidal single crystals are broken easily even by the heat accompanied with the light source. Giant single crystals are easily broken by the very weak and nonrecognized vibration of an observation desk, because their rigidities are extremely low.

Figure 2.31 shows a typical example of close-up color pictures during the course of crystal growth of colloidal silica spheres (CS81, 103 $\pm$ 13.2 nm) in a test tube (outer diameter: 12 mm$^{\phi}$, inner diameter: 11 mm$^{\phi}$) at 25° C. Within 170 seconds, very small single crystals form but move so vigorously that taking still pictures of the single crystals is difficult. When the positions of each single crystal is fixed, it is possible to take pictures. At the initial stage, the crystal growth rate is unexpectedly fast, and kinetic analysis from the close-up pictures is impossible in most cases. CCD camera observation and the reflection spectroscopy is appropriate instead. The extremely diluted dispersions exhibiting single crystals are transparent. In these pictures, a background is dark, but the single crystals in the dispersion are actually packed densely, separated with grain boundaries. With a slight rotation of the test tube, the bright crystal pattern changes rapidly. A change in colors from bluish to reddish within a single crystal is due to the change in the incident light angles through the curved cell wall. For the concentrated dispersions of colloidal crystals, incident light must hit the front of the cell obliquely, because the light does not go through the dispersion except through several Bragg layers.

For preparation of the exhaustively deionized dispersions, stock suspensions around 10 vol% must be deionized for a long time, more than 1 year and ideally more than 10 years. Completeness of the deionization is known from the change in the critical concentration of melting (or crystallization). It is noted here that the conductance measurements of the deionized suspension gives very small specific conductivity, even smaller than that of the deionized and purified water. Furthermore, the deionized suspension shielded in a test tube must be open to insert an electrode into the suspension. The conductance measurements, therefore, do not afford any evidence for the "exhaustively deionized state." The critical concentration decreases steadily as the deionization proceeds. In our laboratory, the ion-exchange resins were always Bio-Rad, AG501-X8(D), 20-50 mesh. This resin was recommended to me by the late professor emeritus Sei Hachisu of the University of Tsukuba.

Hachisu also made microscopic observations of colloidal crystal structures in aqueous dispersion for the first time using a reversed-type metallurgical microscope (Axiomat IAC, Carl-Zeiss, Germany). The suspension cell was a glass tube patched with a cover glass at one end. A mixed bed of ion-exchange resins (Bio-Rad) was packed in a small nylon mesh bag and suspended from the top into the sample suspension in the cell.[79] This treatment allowed continuous removal of the ionic impurities coming from the glass (alkali), bag, air, and other factors. In many cases, the resins must coexist for several weeks to achieve the sedimentation equilibrium state. The pictures shown in Fig. 2.32 are typical examples of the microscopic observation of the single crystals of CS81 spheres

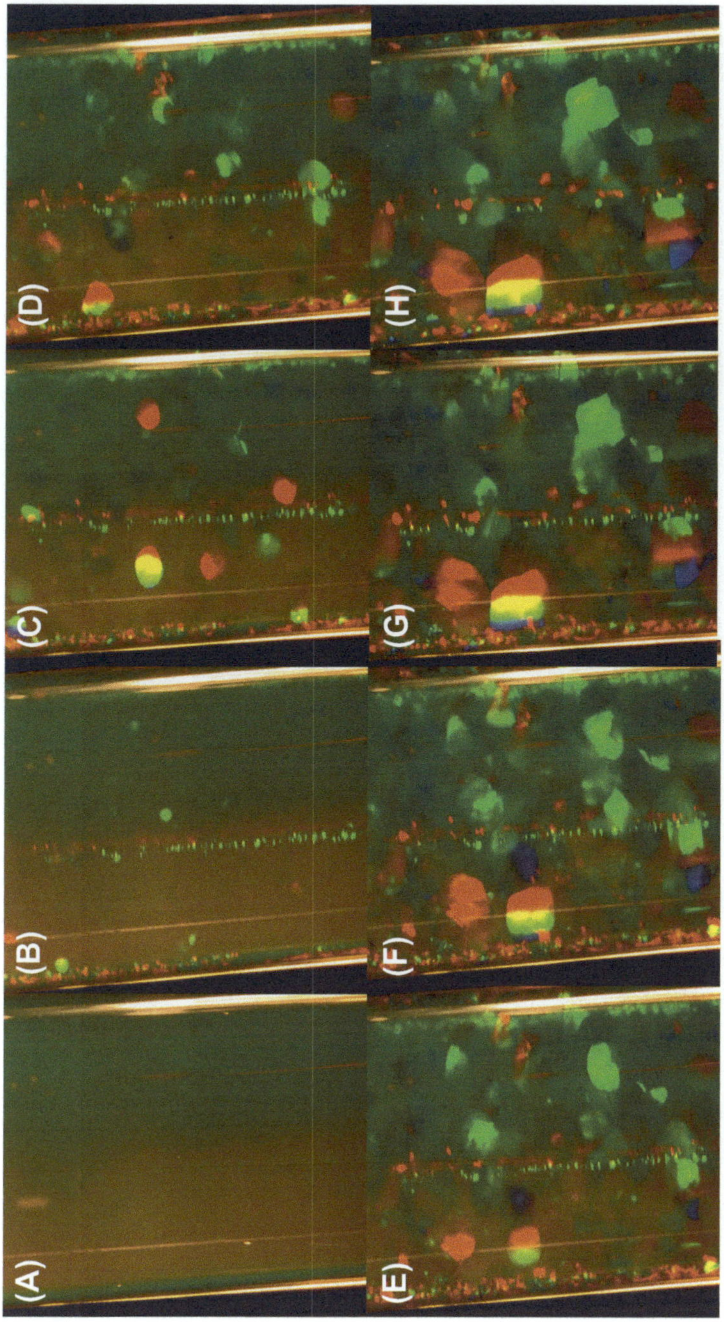

Figure 2.31 Close-up color pictures during crystal growth of CS81 spheres at 25° C. $\phi = 0.001$, **A,** 5 seconds after inverted mixing; **B,** 60 seconds; **C,** 110 seconds; **D,** 140 seconds; **E,** 170 seconds; **F,** 4 minutes; **G,** 18 minutes; **H,** 130 minutes.

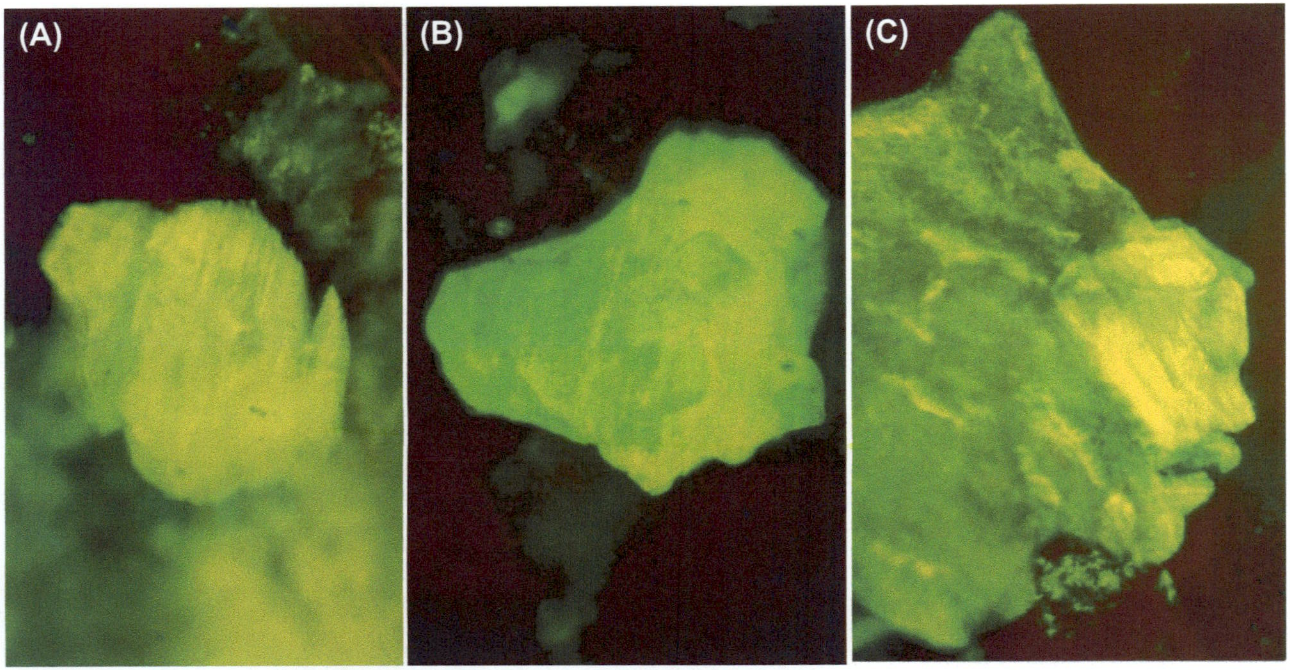

Figure 2.32 Microscopic observation of colloidal single crystals of CS81 spheres in the presence of NaPES. $\phi = 0.0360$; [NaPES] = 3.06×10^{-5} monoM; frame size = 0.82 mm × 1.2 mm. *From experiments described in Okubo T, Fujita H, Kiriyama K, Yamaoka H:* Colloid Polym Sci *274:73-80, 1996.*

in the presence of a small amount of sodium polyethylene sulfonate (NaPES).[205] In Fig. 2.32, pictures were taken without ion-exchange resins.

Optical reflection spectroscopy is one of the most convenient techniques for analyses of colloidal organized structures, especially colloidal crystals and liquids, because the interparticle distances in the structures are just in the range of usual light wavelengths. There are two set-ups for the reflection spectrum measurements, as shown in Fig. 2.33. According to Hiltner and Krieger,[206] the nearest-neighbored interparticle distance in the face-centered cubic (*fcc*) or *bcc* lattices, $l_{obs,f}$ or $l_{obs,b}$ is given by Equation 72 for types Aa, Ab, and B, and Equation 73 for type Ac:

$$l_{obs,f}\left(l_{obs,b}\right) = (2/8)^{1/2} \times \left[\lambda_P / \left(n_s^2 - n_a^2 \cos\theta\right)^{1/2}\right] \tag{72}$$

$$l_{obs,f}\left(l_{obs,b}\right) = (2/8)^{1/2} \times \left[\lambda_P / \left(n_s^2 - n_g^2 \cos\theta\right)^{1/2}\right] \tag{73}$$

Here, λ_p denotes the wavelength of the primary reflection peak. The refractive indices of sample dispersion, air, and quartz glass are designated by n_s, n_a, and n_g, respectively. When sample suspensions are rather diluted, n_s is taken as 1.333 at 25° C (water). The incident angles θ are 45 degrees and 90 degrees for systems A and B, respectively.[207] A typical example of the reflection spectra of colloidal crystals of gel spheres are shown in Fig. 2.34.[208] It should be mentioned here that the transmitted-light spectrum is taken using a spectrophotometer that is commercially available, and the extinction spectrum peak corresponds to that in the optical system of Fig. 2.33, *B*.[209,210]

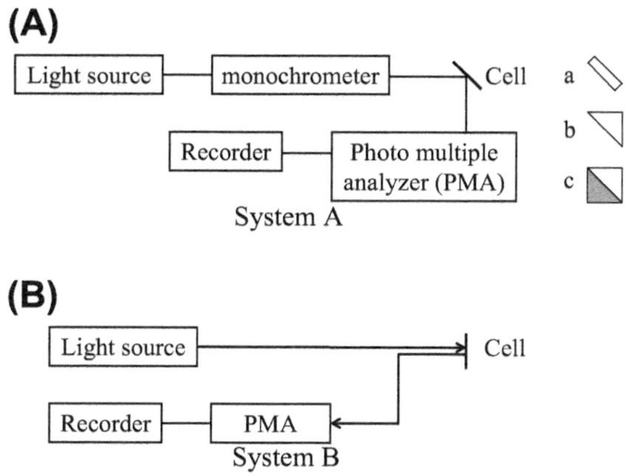

Figure 2.33
Two systems for the reflection spectrum measurements. **A,** Fluorescence spectrophotometer is used and the cells are types *a, b,* and *c.* Incident light angle is 45 degrees. **B,** An example of time-resolved reflection spectrophotometer at 90 degrees of the incident angle.

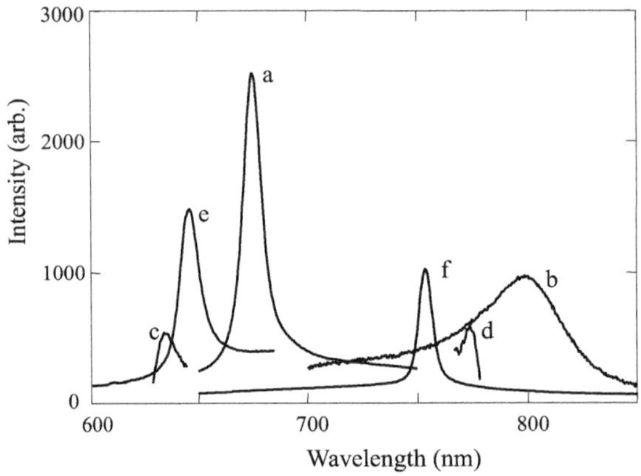

Figure 2.34
Typical examples of the reflection spectra of poly-N-isopropyl acrylamide (pNIPAm[200-5]) gel crystals at 25° C. *Modified from Okubo T, Suzuki D, Yamagata T, Katsuno A, Sakurai M, Kimura H, Tsuchida A:* Colloid Polym Sci *289:291-299, 2011, with permission from* Colloid Polymer Science, *copyright 2011, Springer.*

2.13 Microgravity Experiments

Microgravity effects on the physicochemical properties of most colloidal dispersion systems have not been discussed often hitherto. We should note that the gravitational effect becomes significant for the translational diffusion of colloidal particles larger than 0.1 μm and heavier than 0.1 in the relative density of the particles with reference to medium ($\Delta\rho$). The Peclet number, P_e, the ratio of the time of sedimentation of a sphere in gravity against that of translational Brownian diffusion, is one of the convenient parameters to use in understanding the gravitational effect compared with Brownian diffusion.[211-214]

$$p_e = rS/D_t = (4\pi r^4 g\Delta\rho)/(3k_BT) \qquad 74$$

Here, r is the sphere radius, and S and D_t are the sedimentation and translational diffusion coefficients. When $\Delta\rho$ values are 0.1 and 1, for example, P_e values are evaluated to be 10^{-8} ($r = 10$ nm), 10^{-4} (100 nm), 1 (1 μm) and 10^4 (10 μm), and 10^{-7} (10 nm), 10^{-3} (100 nm), 10 (1 μm) and 10^5 (10 μm), respectively. For example, sedimentation of colloidal single crystals is quite fast in normal gravity compared with that calculated from the Stokes-Einstein equation for each particle. When the lattice structure of single crystals is assumed to be simple-cubic and the nearest neighbored intersphere distance is 500 nm, for example, the number of spheres composing a single crystal, 0.1 mm × 0.1 mm × 0.1 mm in volume, is estimated to be 8×10^6. Thus it will be easily understood that each single crystal, which is composed of a huge number of colloidal particles, sediment quite quickly compared with the respective spheres.

To compensate the sedimentation effect in normal gravity, some researchers have studied colloidal dispersions where the densities of the particles and medium molecules are matched, using sucrose-water mixtures as medium, or a direct current (*dc*) electric field applied perpendicularly for anionic charged colloidal particles. However, these ideas do not really mimic the microgravity environment, because the gravitational convection flow of dispersion is not eliminated and extra interactions of the particles with the medium are added. Thus studies of colloidal dispersions in microgravity are considered to be important for the determination of precise physicochemical parameters, and also for syntheses of highly monodisperse colloidal spheres.

Microgravity has been achieved by several methods: (1) free-fall experiments of a capsule through an evacuated tube (time duration, 3 to 10 seconds, quality of microgravity, approximately 10^{-5} to 10^{-4} g), (2) parabolic flight in an aircraft (20 to 30 seconds, approximately 0.01-0.05 g), (3) a rocket flight (10 to 20 minutes, approximately 10^{-4} g); (4) space shuttle flight (several days, approximately 10^{-3} g), and (5) space station experiments (several years, approximately 10^{-3} g).

Microgravity effects on the colloidal dispersions especially relating to the colloidal organization have been studied on (1) colloidal crystallization, (2) syntheses of colloidal particles, and (3) physicochemical properties of colloidal dispersions. Microgravity experiments on colloidal crystallization are described in Chapter 3. Studies on colloidal particle synthesis and the physicochemical properties of colloidal dispersions are reviewed here briefly. Vanderhoff et al.[215] succeeded in preparing monodispersed polystyrene spheres, 5 to 18 μm in diameter, with narrower particle size distributions than those prepared in normal gravity. A series of polyacrylamide gels have been synthesized in microgravity.[216] These microgravity experiments are, however, restricted to the studies on morphologic characteristics (size, shape) and their reaction yields. Kinetic analyses of polymerization reactions have not been reported to date.

Polymerization rates of colloidal silica spheres from tetraethylortho silicate (TEOS), ammonia and water in ethanol were studied in microgravity using parabolic flight of an air craft.[217] The specific gravity of colloidal silica spheres is approximately 2.2, high enough compared with that of the solvent, ethanol (0.82). Thus the spheres formed are sedimented in normal gravity and the reaction mixtures must be agitated throughout the polymerization process. In microgravity, on the other hand, the products do not sediment and the reaction proceeds homogeneously without stirring. Another advantage of microgravity experiments is the substantial decrease in convection of the reaction mixtures. Polymerization rates were lowered greatly when the reaction mixtures experienced microgravity in the beginning stages in the reaction processes. It is highly plausible that a decrease in the formation rate of the primary particles and also a decrease in their number in microgravity influenced the polymerization (coalescence) rates. Increase in the induction times, when the primary particles form, were also observed in microgravity. One of the main reasons for these observations is the fact that the translational Brownian movement of product spheres is free from downward translational movement in microgravity. Lack of convection in the reaction processes in microgravity is another important factor. Interfacial polycondensation of nylon-6,6 was studied in microgravity by the free-fall experiments.[218] Size and the aspect ratio of the product vesicles decreased in microgravity.

Experimental clarification of the microgravity effects on the translational diffusion coefficient of colloidal particles is important. Krutzer et al.[219] reported that the colloidal particles larger than 2 μm coagulate faster in pseudomicrogravity, which is created by the density-matching method. Static and dynamic light-scattering measurements in microgravity on colloidal gelation have been made on the US space shuttle and Russian orbiter Mir.[220] However, the results are not published yet, except for the details of the instrument used.[221]

Free-fall experiments of anisotropic-shaped tungstic acid particles were made to determine the accurate rotational diffusion coefficients in microgravity.[222] The Peclet number of tungstic acid particles is roughly estimated to be 230, which shows that the contribution of sedimentation in normal gravity is much larger than the contribution of the Brownian translational diffusion. Clearly, sedimentation of nonspherical particles is complicated by the coupling of the translational and rotational motions. The direction of sedimentation of anisotropic-shaped particles depends on their orientation. Brownian motion changes the orientation of the particles and, hence, their sedimentation speeds and directions. However, on average, the particles move in the direction of gravity only. The rotational relaxation times (τ) were evaluated from the relaxation traces of the optical transmittance of the dispersion using the stopped-flow technique. Four flow directions of the sample dispersion in the observation cell are shown in Fig. 2.35 (abbreviated *up, down, back,* and *side*). τ values were observed in microgravity (mcg) and in normal gravity (1 g) (Fig. 2.36). The probing light always passed through the

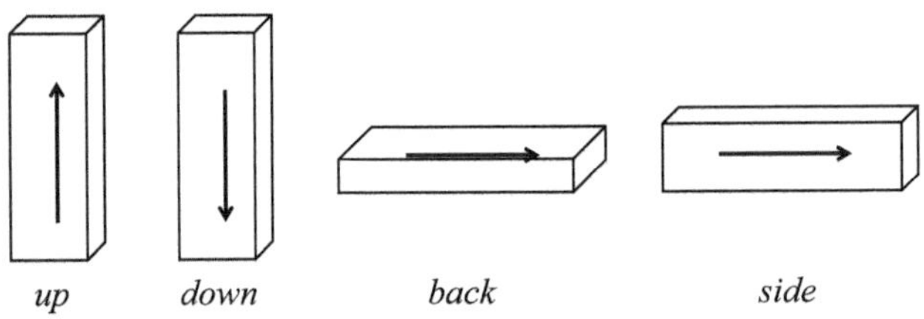

Figure 2.35
Four flow directions of the samples through the cell. *Redrawn from Okubo T, Tsuchida A, Yoshimi H, Maeda H:* Colloid Polym Sci *277:601-606, 1999, with permission from* Colloid Polymer Science, *copyright 1999, Springer.*

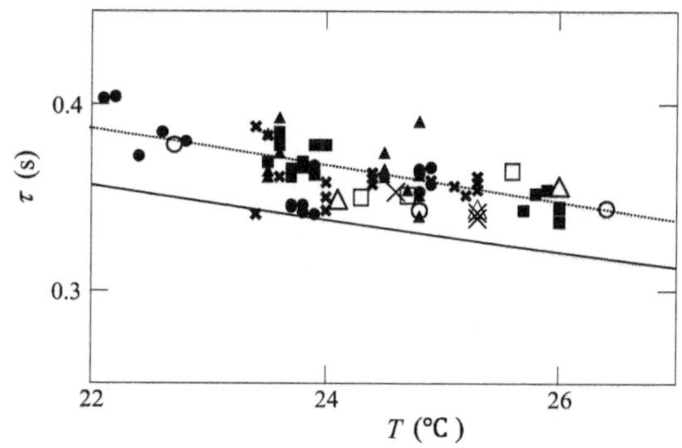

Figure 2.36
Rotational relaxation times (τ) of tungstic acid particles (W29), 0.20 wt%; large circles: up, 0 g; large crosses: down, 0 g; large triangles: back, 0 g; large squares: side, 0 g; solid circles: up, 1 g; small crosses: down, 1 g; solid triangles: back, 1 g; solid squares: side, 1 g; solid line calculated from Perrin's equation, $l_D = 0$ nm; broken line: calculated, $l_D = 21$ nm. *Redrawn from Okubo T, Tsuchida A, Yoshimi H, Maeda H:* Colloid Polym Sci *277:601-606, 1999, with permission from* Colloid Polymer Science, *copyright 1999, Springer.*

cell perpendicular to the largest planes of the cell. Fig. 2.37 shows the τ values thus obtained at mcg and 1 g. All the solid symbols show the τ values observed at 1 g. A decreasing tendency of τ was observed as the suspension temperature increased. Clearly, the observed τ values at 1 g scattered significantly, which is explained by the convection of the dispersion. τ values at 1 g were approximately 6% larger than those evaluated theoretically using Perrin's equations. This difference demonstrates the importance of the electrical double layers formed around the colloidal particles. The broken curve shows the τ values when the Debye length was taken to be 21 nm. Conductivity measurements of the dispersion gave 36.7 μS/cm (23.3° C) at $\phi = 0.20$ wt%, which corresponds to 2.5×10^{-4} mol/l KCl and a Debye length of 27 nm. Agreement between the observed and calculated values of the double layers is satisfactory. Clearly, the τ values in microgravity were not scattered as much as those in normal gravity, and there was no difference between the τ values observed at 1 g and mcg.

Studies on the microgravity effects on other thermodynamic parameters of colloidal systems, such as activity, surface tension, and adsorption of colloidal dispersions, are exciting, because the fundamental relationship shown in Equation 75 holds true, where acceleration of gravity, *g*, is included.[223]

$$dU = T\Delta S - PdV + \sigma dA + g(h - h_0)dM - (r^2\omega^2/2)dM + \mu_i dn_i \qquad 75$$

Here, U is the internal energy, T, S, P, V, σ, A, g, h, h_o, M, r, ω, μ_i, and n_i are the temperature, entropy, pressure, volume, surface tension, area, acceleration of gravity, height, reference height, molecular weight, radius, angular velocity, chemical potential of component i, and number of moles of component i, respectively.

2.14 Why Does Colloidal Organization Take Place?

One of the essentially important factors for colloidal organization is the translational and rotational thermal movement of colloidal particles themselves. As is shown in Fig. 2.37, other important factors for the organization are (1) packing effects, (2) wall effects, (3) interparticle interactions, and (4) external field effects. In general, these four factors play their role simultaneously and synchronously with the help of the thermal motion of the elementary particles. Furthermore, we should note here that these effects are weak in their forces and that synchronous organization occurs by the weak interaction forces in most cases.

First, *packing effects* means that the most effective packing with the smallest nonpacked space makes ordered structure by the particles' thermal motion. Depending on the shape, size, and distribution, the characteristic structures of colloidal particles will form. The packing effects are able to play by the excluded volume effects originated from the vigorous moving colloids and the interparticle repulsion forces. Furthermore, we should note these packing effects always take place in a closed vessel. At first glance, the air-suspension interfaces look for free space to diffuse. However, forces of the interfacial tension at the air–suspension interface are strong enough to protect the outward diffusion of colloidal particles from the interface. Various colloidal alloy crystals form mainly by the packing effects. Many researchers have clarified that colloidal crystallization, for example, forms by Brownian movement of colloidal particles and the interparticle repulsion according to the principle of most dense packing with minimum dead space.[87,90,92,97] In other words, particles form crystal-like distribution spontaneously with the help of Brownian thermal movement of particles by maximizing packing density.[224] Outer shapes of the synthetic and biological colloidal particles play an important role for the structural distribution of the particles. Three-dimensional (3D) calculation of the volume density of each particle is not always so easy. The monodispersity in shape and size is

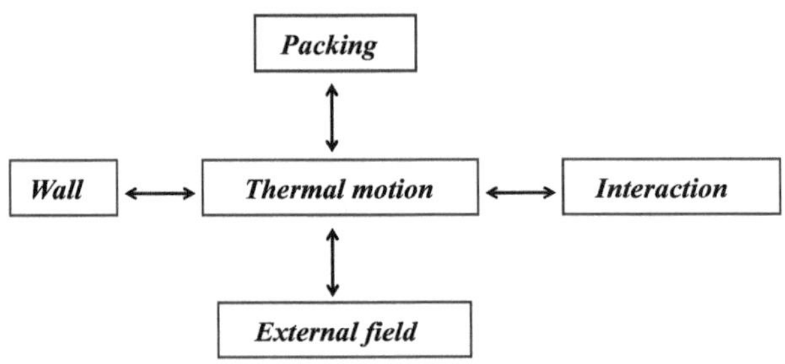

Figure 2.37 Essential factors for colloidal organization.

also very important for ordering. Five kinds of regular polyhedrons exist according to Euler's theorem on the polyhedral. Hales proved Kepler's proposal that the volume density of the most dense 3D packing of spheres, *fcc*, is $\pi/18^{1/2}$ (= 0.74048 . . .).[225,226] The volume density of the most densely packed two-dimensional (2D) hexagonal structure was solved to be $\pi/12^{1/2}$ (= 0.906899 . . .).[227,228] For 2D ellipses, it becomes difficult to calculate the volume density of the densest packing. Several different kinds of packing densities have been calculated.[229] For a 3D ellipsoid, density is rarely discussed theoretically. Some protein molecules form 2D close packing ordering. For example, a spherical molecular chaperone and rectangular catalase molecules form hexagonal and rectangular 2D sheet structures in a solid state.[230] Hachisu et al.[79,110] clarified that colloidal crystallization takes place by the Kirkwood-Alder transition. When extra repulsive interaction, such as electrostatic interparticle repulsion, is in effect among colloidal particles in addition to the repulsion forces from their Brownian movement (excluded volume effect), colloidal crystallization takes place more easily even at the very low particle concentrations. An increase in particle concentrations resulting from sedimentation, filtering, evaporation of solvent, and other factors, also contribute to the ordering. In these cases, cooperative function of the packing, interaction, and external field effects become important for enhancing the ordered structures. Fig. 2.38 shows my idea (not an observation) for switching the inclusion reactions resulting in regulating the colloidal distribution by thickening or thinning the double layers. By the deionization of the suspension with ion-exchange resins or the salt addition, suspension properties such as turbidity and viscosity must change sharply and reversibly.

Second, the wall effects come from the interactions between the cell wall (or air–suspension interface) and the colloidal particles. Interestingly, it is well known that the most dense lattice plane orients along the cell wall in

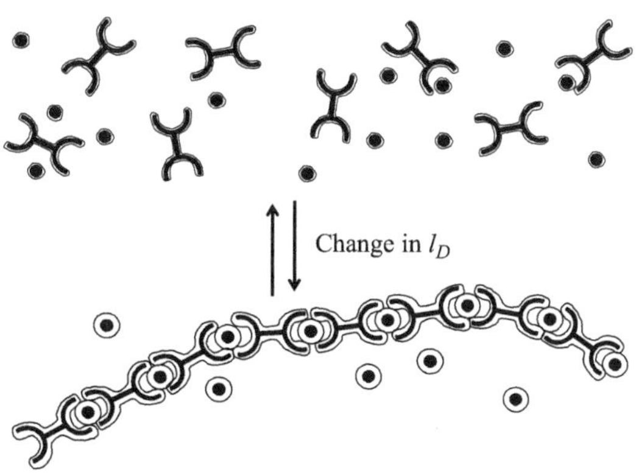

Figure 2.38
An idea of change of the colloidal distribution by thickening or thinning of the electrical double layers.

the colloidal crystallization. Furthermore, transferring the information of the colloidal solutes such as size, shape, and molecular structures toward the drying structures has been found to take place mainly by the wall effects from the solute–substrate interactions (see Chapter 5).

Third, the intercolloidal particles' interaction effect is also one of the main factors for colloidal organization. The repulsive interaction by the excluded volume effects and by the electrostatic repulsion caused by the electrical double layers around colloidal particles are the typical examples. We should note here that the repulsive interaction through formation of the electrical double layers is also related to the packing effects, because the effective size and shape of the colloidal particles change systematically by thickening or thinning of the double layers. Specific adsorption and inclusion-type interparticle interactions will also be grouped in this category. Recently, many researchers have become interested in the organization of various types of Janus particles having two different interactive groups, for example, polar and nonpolar, hydrophobic and hydrophilic sites, and cationic and anionic groups.

Fourth, the external field effect is one of the most important. Electric field effects introduced by application of alternating current (*ac*) and direct current (*dc*) and square wave electric fields on the colloidal dispersions are quite interesting to study. Microgravity experiments have been made to study the effects of gravity on colloidal organization. Sedimentation, centrifugal forces, and pressure effects also afford valuable information on colloidal organization. Mixing of the dispersion may be grouped in this category. Segregation effects of colloidal size and shape are very important for the organization phenomena. Flow birefringence effects of anisotropic-shaped colloidal particles have been studied by detecting optical and conductive proofs. We should note here that significant flow birefringence takes place even for spherical particles, especially in the deionized aqueous media, because the electrical double layers form and are deformed from a spherical shape by the weak shearing forces induced by the flow process. Shearing forces induced by sliding the two glass plates in between the colloidal suspensions being set are well known to enhance the ordering of colloidal particles. With use of polymerization techniques, snow doll-like, raspberry-like (or comfit-like), and other anisotropic-shaped particles are now available commercially or in the laboratory scale.

We should pay attention here to why the repulsive interparticle interaction is important in colloidal organization instead of attractive forces. The latter is quite common for most crystals in nature, for example, metals, proteins, and ice. For usual crystals in nature, the elementary particles are very small and the contribution of both the translational and rotational diffusion constants given in Equations 1 and 2 are so large that the ordering will be quite difficult without strong interparticle attractive forces. On the other hand, the diffusive forces of large colloidal particles are too weak to form ordering without some other additional interparticle repulsion forces in a closed vessel. Without additive repulsion interactions, the colloidal particles will associate easily to each other. Keeping the balance between the attractive and repulsive interelementary particle interactions is important for the ordering phenomena.

2.15 From Atoms to Galaxies

Let's discuss the intimate relationship among the structures of atoms, colloidal dispersions, and galaxies. A schematic image of an atom is given in Fig. 2.39 from the standpoint of quantum chemistry, where the probabilities of the existence of electrons are distributed around the atomic nucleus at the center. The number of nuclei is single but very heavy compared with the cloud of electrons. Many researchers believe that the atomic structure is similar to that of colloidal particle in aqueous media; in other words, a colloidal particle is composed of a heavy single colloidal particle and a large number of lightweight single ions, as is shown in Figs. 2.1 and 2.39. Fig. 2.40 shows an image of the Andromeda galaxy, which is the galaxy nearest to the earth's galaxy, the Milky Way. A galaxy is also composed of a heavy black hole and a huge number of stars. When a galaxy is called a *macro-cosmos,* then atoms or colloidal particles are *micro-cosmos.* Fig. 2.41 shows the distribution of galaxies in the deep cosmos as recorded by the Hubble space telescope. This picture showing the distribution of galaxies looks like colloidal dispersion. Fig. 2.42 is a microscopic picture of colloidal alloy crystals composed of aqueous colloidal mixtures of polystyrene spheres of D1C25 (85 nm in diameter) and D1B72 (173 nm) in deionized aqueous suspension. The picture looks like that of the cosmos.

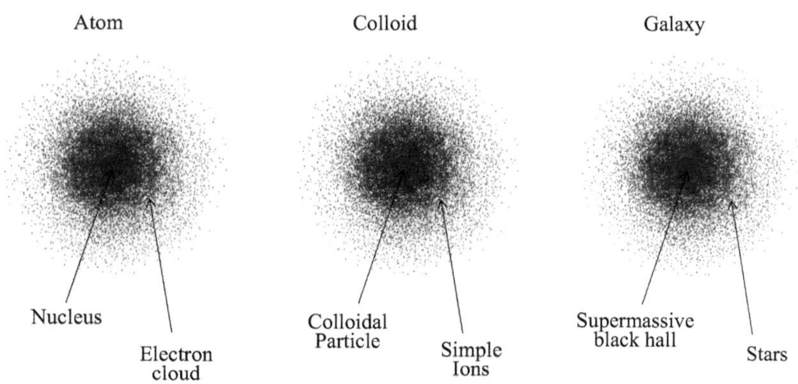

Figure 2.39 Simplified sketches of an atom, a colloidal sphere, and a galaxy.

Figure 2.40
Andromeda galaxy, photographed by the Hubble space telescope.

Figure 2.41
Distribution of galaxies in deep cosmos, photographed by the Hubble space telescope.

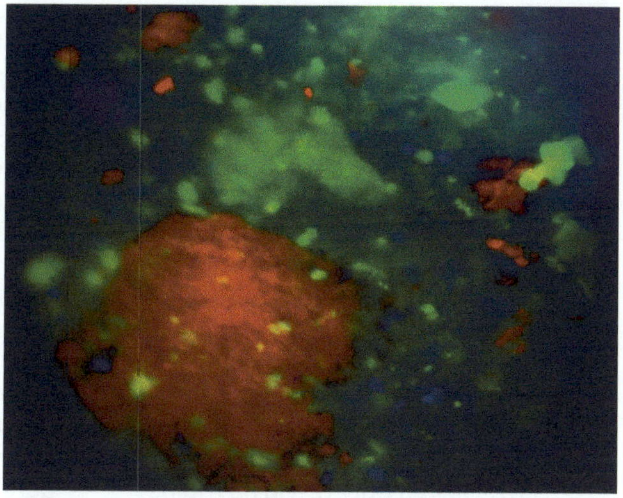

Figure 2.42 Microscopic observation of colloidal single crystals of D1C25 + D1B72 (code 537) mixtures.

2.16 Synthetic and Biological Colloidal Particles

2.16.1 Syntheses of Colloidal Particles

A variety of synthetic methods have been developed using gas-, liquid-, and solid-phase processes. First, the gas-phase method includes flame pyrolysis (fumed silica, titanium dioxide [TiO_2]), furnace flow reactors (silver, lead sulfide, lead, and silicon), laser-induced vaporization and pyrolysis (silica and iron), thermal and microwave plasmas (metals and metal oxides), sputtering (metals), and laser ablation. The vapors deposition is followed by reduction, oxidation, and deposition of the vapor on the surfaces. Second, the liquid-phase (colloidal) methods are based on precipitation processes in solutions. Particles of various sizes and morphologic characteristics are synthesized by controlling the nucleation and growth kinetics in bulk solutions and also in the special systems such as micelles. Many kinds of organic particles, metals, and metal oxides have been produced by the liquid-phase method. Third, the molecular self-assembly method takes place spontaneously from molecules. This is typical for the production of polymer particles starting from amphiphilic block copolymer molecules, for example. Controlling the size and shape is easy in this method. Fourth, the solid–liquid interface syntheses of colloidal particles using heterogeneous nucleation sites are also available. For example, by using patterned surfaces, nucleation and growth processes can be controlled. Fifth, the phase equilibrium method is one of the new techniques developed recently. Sixth, the mechanical size reduction by grinding and milling is a traditional method and usually is made with the coexistence of a colloidal stabilizer to avoid particle aggregation.

To study the colloidal organization phenomena, monodispersities in shape and size of the colloidal particles are most important. Several kinds of monodispersed colloidal spheres have been prepared in an aqueous media from much smaller preparticles by the cooperative and weak attraction forces between preparticles. Monodispersed colloidal silica spheres were prepared by Stoeber et al. using the cooperative association (agglomeration) of the monodispersed prespheres of several nanometers in their size.[231,232] Similar sized colloidal spheres formed from the polymer complex of poly (acrylic acid) with poly (ethylene glycol) formed colloidal crystals emitting iridescent bluish colors.[233] The iridescent colors support that the particles are monodispersed at approximately within 10% in their size. The main driving force of formation of the similar-sized aggregates is the cooperative interpolymer *hydrogen bonding* interaction and the thermal repulsion forces. Similar-sized aggregates formed from the thermosensitive gel spheres of poly (*N*-isopropyl acrylamide)[234-237] and from cationic gel spheres of lightly cross-linked poly (2-vinylpyridine)[238-240] during the course of dryness on the glass substrates. The important role of the weak but cooperative hydrophobic interactions among the gel spheres and the gel spheres—substrate interactions are highly plausible for the formation of the similar-sized aggregates for the gel systems. It should be recalled that the aqueous suspensions of very small particles (several nanometers), in general, are not so stable because of the strong interparticle van de Walls attraction forces, and polydispersed aggregates form easily with time. Gold particles of 0.5 to 1.2 μm in mean diameter were prepared by the reduction of a chloroauric acid solution with salicylic acid; they formed colloidal crystals and distorted colloidal crystal structures in the sedimentation equilibrium.[241] The similar-sized particles formed were clearly composed of the very small preparticles in the microscopic picture of the exhaustively deionized aqueous suspension (see Fig. 2, *A*, in Okubo [1987][241]). Similar-sized diamond particles also formed through formation of the monodisperse agglomerates.[242]

There are several kinds of balancing of the attraction and repulsion. First, existence of the secondary minimum in the potential curve was proposed as attraction for the reversible agglomeration. Jeffrey and Ottewill[243] reported that the reversible aggregation of large polystyrene spheres in the presence of NaCl was broadly in agreement with expectations from the secondary minimum of the simplified DLVO theory. Second, the cooperative balance between the repulsion that originates from the thermal motion of the preparticles and the weak intersphere hydrophobic attraction is proposed. A typical example is the formation of similar-sized agglomerates from anionic (or cationic) gel spheres during the course of drying on the substrates.[234-240] For very small preparticles, the interparticle van der Waals attraction becomes strong, and the agglomeration does not take place in general; the aggregates form instead of agglomerates. Third, the balancing of the hydrogen-bonding attraction with the thermal motion repulsion is also important for similar-sized and reversible agglomerates and then aggregates. For example, the double-helix complexes of DNA form by the hydrogen bonding between the complementary nucleic acid bases and further bigger complexes form.[244] Study of the criteria determining whether the pre-particles proceed to form highly polydispersed aggregates or similar sized agglomerates is important and exciting.

2.16.2 Inorganic Colloidal Particles

Monodispersed colloidal particles of metals,[245] metal oxides,[246] and metal sulfides[247] have been synthesized intensively. Colloidal silica spheres have often been used for the study of colloidal organization phenomena such as colloidal crystallization and drying patterns for more than 20 years.[248,249] Recently, highly monodispersed colloidal silica spheres ranging from 5 nm to 10 mm in diameter have become available commercially (Fig. 2.43, *A*). Colloidal silica spheres coated with other functional groups such as dyes, fluorescent probes, and so on, using the reactive silanol groups have been studied.[250-253] Colloidal silica spheres modified with linear-type polymers and their colloidal crystallization in organic solvent have been studied intensively by Yoshinaga et al.[254-259] The dispersion property of colloidal silica spheres coated with a layer of end-grafted polydimethylsiloxane was studied in heptanes by Nommensen.[260] Titania colloidal spheres are synthesized in a diffusion flame reactor and by other methods.[261-265] Spherical

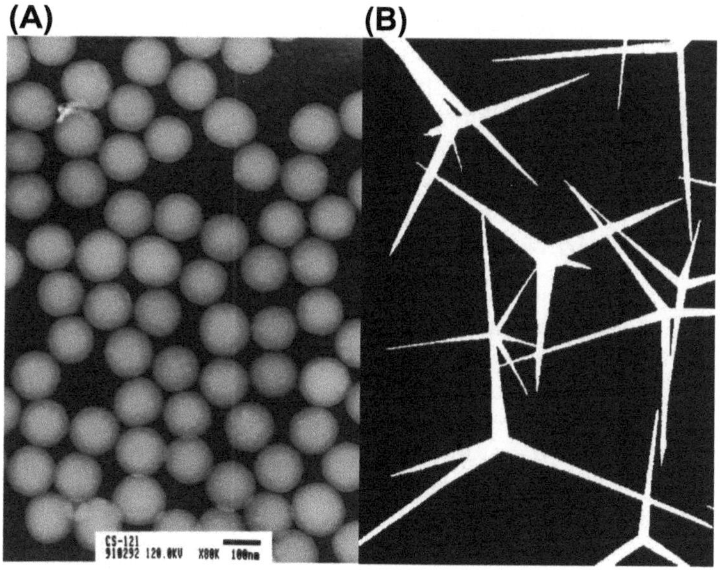

Figure 2.43
A, Colloidal silica spheres, 3 μm in diameter. **B,** Particles of ZnO.

(polyhedral in the strictest sense) gold particles ranging from 2 nm to several μm in diameter are famous as the hydrophobic colloidal spheres and they aggregate by the addition of salts following the Shultz-Hardy rule.[266,267] Similar-sized aggregates (approximately 1 μm) from gold preparticles were obtained.[268] Gold spheres have been also synthesized.[269,270] Studies using silver spheres were reported.[271,272] Small silver spheres capped with dodecanethiol were prepared and they made ordered arrays after drying.[273] Syntheses and their organization of the similar-sized agglomerates (1.5 μm in diameter) from the diamond preparticles ($\approx$ 4-6 nm in diameter) have been reported.[274,275] Spherical particles of unique morphologic characteristics of boehmite, α-AlOOH, were obtained by aging acidic solution of aluminium perchlorate.[276-278] Colloidal spheres of CdS and ZnS have been prepared.[279-281] Charge reversal of CdS colloidal spheres under the Langmuir-Blodget membrane were reported.[282] Molybdenum sulfide particles were synthesized in microemulsion.[283] Colloidal copper and copper oxide dispersions were studied.[284,285] Colloidal spheres of PbI_2 were prepared.[286] Core-shell type (CdSe)ZnS particles ($\approx$ 2.3-5.5 nm) were studied by Dabbousi et al.,[287] for example. Core-shell microspheres consisting of single-walled carbon nanotubes and $CaCO_3$ were synthesized.[288] Recently, Fujii et al. synthesized new types of microcapsules of silica spheres by the fabrication of liquid marbles stabilized with polymer particles.[289] Future study of colloidal organization will be interesting in relation to colloidal organization.

Symmetrical polyhedron-shaped particles are prepared mainly from metals. Cubic colloidal particles ($\approx$ 20-50 nm) composed of MgO, CdO, PbSe, PbTe, and αFe_2O_3 have been synthesized,[290,291] Cubic-shaped platinum particles were prepared by El-Sayed et al.[292] Hexahedron-shaped particles of WO_3, InS_3, and SnO_2 were also obtained.[293] Octahedron of In_2O_3 was prepared.[294] Octahedron-shaped particles of $Fe_3(SO_4)_2(OH)_5.2H_2O$ was obtained by heating acidic solutions of ferric sulfate.[277,278] Flowerlike particles formed from $FeCl_2$ and bipyridine are beautiful. Whisker-shaped particles of ZnO (5 μm in wing length) were synthesized by Matsushita Sangyo Co. (Osaka) (see Fig. 2.43, *B*). Hexahedral copper oxide particles, CuO, were obtained.[246] Furthermore, Au, Ag, Ni, and Al are known to produce octahedron-shaped particles.[273]

Anisotropic-shaped colloidal particles are important for studying colloidal organization. Spindle-shaped colloidal silica particles have been prepared.[293] Rod-shaped[294-297] and platelike particles of gold[298,299] are available in laboratory scale. Stacking aggregates of platelike gold particles were reported.[299] Platelike particles of nickel (II) hydroxide coated with polyacrylate were stable in aqueous suspension.[300] Platinum colloidal particles of different shapes were synthesized.[301] Rod-shaped (or ellipsoidal) particles of β-FeOOH are obtained from $FeCl_3$ after approximately 6 to 10 months at room temperature as a red suspension at the bottom of container.[302] The late professor S. Hachisu informed me that the flow-induced crystallization took place for the aqueous suspensions of β-FeOOH. However, reproducibility was not high. Rodlike boehmite (γ-AlOOH) particles have been used for study of phase transition.[303] α-Alumina, α-Al_2O_3 particles (average size ≈200 nm) with an irregular shape are available commercially. Silica-coated boehmite rods (e.g., 230 nm in length, 9 nm in diameter)[304] were synthesized according to the method of Buining.[305] Sterically stabilized gibbsite, $Al(OH)_3$ platelets are also used for the organization phenomena.[306] Ellipsoidal TiO_2[307] and rodlike rutite-type particles[308] were prepared. Bentonite (or montmorillonite) particles are platelike but their monodispersity in size is not so high.[58] However, monodispersed montmorillonite (bentonite) particles were obtained by the fractional repeated centrifugation technique.[309-312] An ellipsoidal, plate-shaped particle of tungstic acid, $H_2WO_4 \cdot H_2O$, is highly monodisperse[313-316] (Fig. 2.44, *A*). But these particles disappeared after several weeks in the deionized state with coexistence of ion-exchange resins, because this particle is one of the sparingly water-soluble electrolytes in water (like silver chloride), and tungstic acid anions were ion-exchanged during a long period. Rod-shaped or spindle-shaped V_2O_5 particles have been prepared[317] (see Fig. 2.44, *B*). Disklike manganese oxide particles were synthesized and their ordered arrays were studied.[318] Monodisperse cylindrical (or rod-shaped) particles of nickel dimethylglyoxime were obtained and an ordered array of the particles was observed in the sedimentary equilibrium state[319] (see Fig. 2.44, *C*). Disklike pseudo cubic,[320] ellipsoidal,[321]

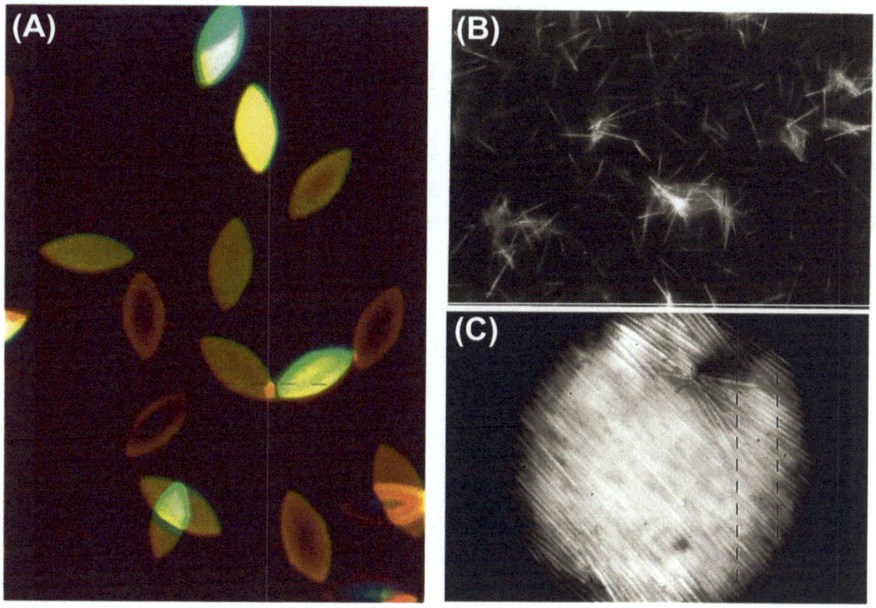

Figure 2.44
A, Spindle-shaped tungstic acid particles. **B,** Dispersion of V_2O_5 rods in aqueous dispersion state. **C,** Deionized dispersion of nickel dimethylglyoxime particles in aqueous dispersion state and in the sedimentation equilibrium

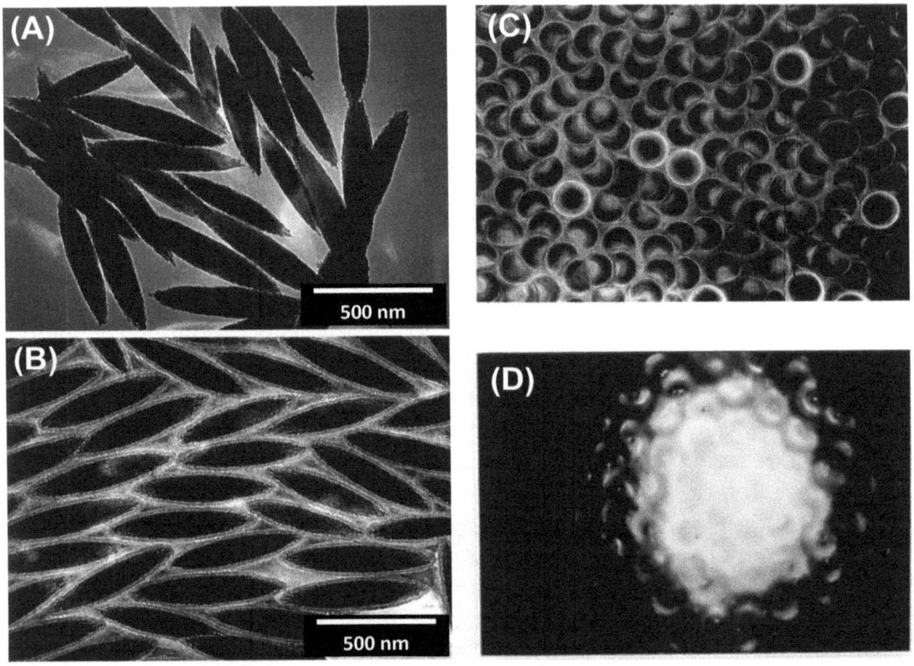

Figure 2.45
Electron microscopy of spindle-shaped hematite particles (*A*) and particles coated with a polymer brush (*B*); ordered distribution of bowl-shaped styrene-methylmethacrylate-methacrylate copolymer (outer and inner diameters are 9 and 6 μm) in sedimentation equilibrium (*C*), $\phi^* = 0.0199$ vol%, 1 day after the cell was set, and inclusional association of bowl-shaped particles ($\phi^* = 0.0199$ vol%) with polystyrene spheres (1.99 vol%, 2 μm in diameter) in sedimentation equilibrium (*D*). (**A** and **B** from Fig. 1 of Ohno K, Tsuchida A, Okubo T: *Colloid Polym* Sci 292:723-731, 2014; **C** and **D** from pictures in Okubo T, Aotani S: *Naturwiss* 75:145-146, 1988; Okubo T, Aotani S: *Colloid Polym Sci* 266:1049-1057, 1988, copyright 1988 and 2014, Springer.

peanut-shaped,[322] and hexagonal plate–shaped hematite (αFe_2O_3) particles[312] were prepared by the sol-gel method. Spindle-shaped hematite particles (Fig. 2.45, *A*) and the particles coated with polymer brush (see Fig. 2.45, *B*) were synthesized and their crystallization was studied.[323] A disklike particle (3 nm in diameter and 0.1 nm thick) of synthetic trioctahedric hectorite has been studied.[324-326] Laponite, synthetic hectorite clay, is a disklike particle.[327,328]

2.16.3 Polymeric Colloidal Particles

Monodispersity in size of spherical particles plays an important and basic role for the colloidal organization. Highly monodisperse polystyrene and polymethyl methacrylate spheres (PMMA; size ranging from ≈20 nm to several mm in diameter) are now available commercially. Polystyrene spheres such as Dow polystyrene and Duke Scientific sulfonated polystyrene spheres are often used by researchers as the standard samples.[118,131,329-334] Anionic and cationic polystyrene spheres are synthesized by using anionic and cationic initiators, respectively. However, we should note that cationic initiators are not so stable in general. Various kinds of core-shell spheres, for which hard core is coated with soft shell and vice-versa, have been prepared by the copolymerization method, for example.[63,332,333] PMMA stabilized by thin layers ($\approx$10 nm) of poly-12-hydroxystearic acid spheres were used in a mixture of decalin and tetralin (or carbon disulfide) to match the refractive index of the spheres in the field of the hard colloidal crystallization.[117,335-339] Monodisperse poly(1H,1H-heptafluorobutyl methacrylate) spheres of 138 nm in diameter were synthesized by Asher et al.[340] Several kinds of fluorine-containing polymer spheres—polytetrafluoro ethylene (PTFE), copolymer of tetrafluoroethylene and perfluorovinylether, and copolymer of perfluoroethylene and perfluoropropylene—were also used for study of colloidal organization.[341,342] Colloidal spheres having cores rich in polystyrene and shells rich in hydrated poly(2-hydroxyethyl methacrylate) were synthesized.[343-345] Heterocoagulation of large cationic spheres with small anionic polyacrylate beads followed by heat annealing produced core-shell particles.[346] Hollow polyelectrolyte capsules were prepared by means of stepwise adsorption of polyelectrolytes followed by dissolution of the templating core.[347] Furthermore, vesicle-templated polymer hollow spheres were synthesized.[348] Polyelectrolyte multilayer coatings on the colloidal particles have been studied extensively.[349-352]

Anisotropic-shaped colloidal particles are most interesting for study of the organization phenomena, including crystallization. Anisotropic-shaped polymer particles, such as symmetric doublets,[353] asymmetric doublets,[353] egg-shaped,[354] ellipsoidal,[353] beadlike,[355] snow doll-like, dumbbell-like, raspberry-like,[356,357] and comfitlike (*framboidal*),[358] golf ball–like,[359] red-blood corpuscle–like,[360,361] disklike, and polyhedral[362,363] particles are now available in laboratory scale. Bowl-shaped particles were used for study of the orientation and inclusion phenomena in colloidal systems[112,113,364] (see Figs. 2.45, *C* and *D*). Preparation of monodisperse ellipsoidal polystyrene particles has been prepared by the unique method of stretching the polystyrene spheres dispersed in the film of polyvinyl alcohol[365,366] (see Fig. 2.46).

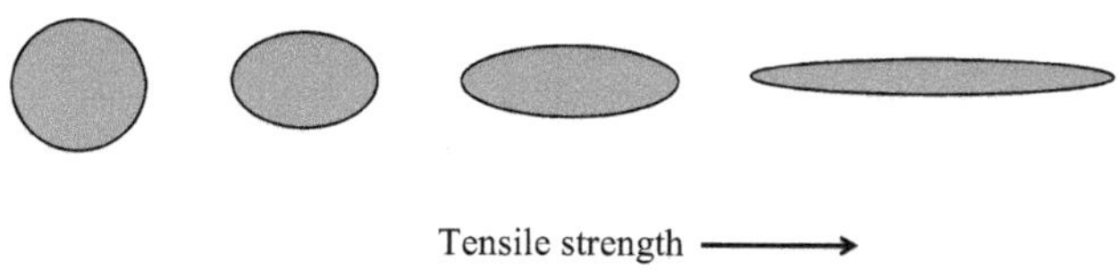

Figure 2.46
Preparation of monodispersed ellipsoidal polystyrene particles by stretching polymer spheres dispersed in the film of polyvinyl alcohol.

Monodispersed colloidal gel spheres have been synthesized. Gel spheres of poly-*N*-isopropylacrylamide are famous as typical samples of thermosensitive hydrogels.[367]

2.16.4 Biological Colloidal Particles

Several biological colloidal particles exist. Tobacco mosaic virus (TMV) (cylinder-like, 18 nm in diameter and 300 nm in length) is typical as rodlike colloidal particles.[368-371] Further large assemblies of TMV was also studied.[372] Cowpea mosaic virus (CPMV) is also rod-shaped.[373] Purified samples of bacterial cellulose (*nata de coco*) are rod-shaped, approximately 1 to 2 μm in length, and 10 × 50 nm in cross-section, and their suspension shows many iridescent birefringence phenomena.[374] Various kinds of 2D assemblies of H-shaped streptavidin mutants form depending on the pH-values.[375] Filamentous bacteriophage *fd* virus is also interesting as one of the rodlike biological polymers.[376] Microcrystalline cellulose is rodlike but rather polydisperse ellipsoidal.[58,377] Microcrystalline cellulose was stabilized in aqueous media by grafting the cellulose with polyethylene glycol or carboxylate.[378] Typical collagen is a rodlike polymer with a 280-nm length and a diameter of 1.2 nm, for example, and is composed of three amino acid chains wound into coaxial helices.[320,379,380] Temperature-induced sol-gel transition of the small chains of collagens,[381] arrow-root starch,[382] and others is very interesting phenomena. The shape of the *lycopodium* spore is the combination of hemisphere and tetrahedron, but is close to a sphere (30 μm in size).[383] Pollen particles from flowers (≈10-50 μm in size) are also a typical example of biological colloidal samples. However, many of them are not so stable and are not dispersed homogeneously in water. Diatoms are typical examples of inorganic and natural-occurring colloidal particles.[384] Their shape and surface structure are beautiful, and may be spherical, ellipsoidal, linear, triangle, crescent-like, or violin-like, among other shapes. Fossil colloidal spheres of radiolarian are rather monodispersed in size.[385] Diatoms and radiolarians are originated from algae and plankton, respectively. However, their main chemical components are the same: silica. Marine colloidal particles such as Marine clay particles will be also grouped into one of the natural colloidal particles.

References

1. Itano WM, Bolinger JJ, Tan JN, Jelenkovic B, Huang XP, Wineland DJ. *Science* 1998;**279**:686–9.
2. Crandall RS, Williams R. *Science* 1977;**198**:293–5.
3. Goodwin JW, Smith RW. *Faraday Discuss Chem Soc* 1974;**57**:126–9.
4. Mitaku S, Ohtsuki T, Kishimoto A, Okano K. *Biophys Chem* 1980;**11**:411–6.
5. Neumann AW, Moscarello MA, Epand RM. *Biopolymers* 1973;**12**:1945–57.

6. Badley RA, Carruthers L, Phillips MC. *Biochim Biophys Acta* 1977;**495**:110–8.
7. Buckinham JH, Lucassen J, Hollway F. *J Colloid Interf Sci* 1978;**67**:423–31.
8. Birdi KS, Jeppesen J. *Volloid Polym Sci* 1978;**256**:261–5.
9. Keshavarz E, Nakai S. *Biochim Biophys Acta* 1979;**576**:269–79.
10. Sato H, Ueberreiter K. *Makromol Chem* 1979;**180**:829–35.
11. Ishimuro Y, Ueberreiter K. *Colloid Polym Sci* 1980;**258**:928–31.
12. Ishimuro Y, Ueberreiter K. *Colloid Polym Sci* 1980;**258**:1052–4.
13. Arakawa T, Timasheff SN. *Biochemistry* 1982;**21**:6545–52.
14. Ricci E, Sangiorgi R, Passerone A. *J Colloid Interf Sci* 1984;**102**:295–7.
15. Van Oss CJ, Moore LL, Good RJ, Chaudhury MK. *J Protein Chem* 1985;**4**:245–63.
16. Okubo T. *J Colloid Interf Sci* 1988;**125**:386–98.
17. Okubo T, Kobatashi K. *J Colloid Interf Sci* 1998;**205**:433–42.
18. Okubo T. *J Colloid Interf Sci* 1995;**171**:55–62.
19. Okubo T. *Polym Bull* 1990;**23**:211–8.
20. Okubo T, Suda M. *J Colloid Interf Sci* 1999;**213**:565–71.
21. Decher G, Hong JD. *Makromol Chem Symp* 1991;**46**:321–7.
22. Decher G, Hong JD. *Ber Bunseng Phys Chem* 1991;**95**:1430–4.
23. Decher G, Hong JD, Schmitt J. *Thin Solid Films* 1992;**210/211**:831–5.
24. Schmitt J, Grunewald T, Losche M. *Macromolecules* 1993;**26**:7058–63.
25. Decher G, Lvov Y, Schmitt J. *Thin Solid Films* 1994;**244**:772–7.
26. Decher G, Lehr B, Lowack K, Lvov Y, Schmitt J. *Biosens Bioelectron* 1994;**9**:677–84.
27. Ferreiva M, Cheung JH, Rubner MF. *Thin Solid Films* 1994;**244**:806–9.
28. Gao M, Zhang X, Yang Y, Yan B, Shen J. *J Chem Soc Chem Commun* 1994:2777–8.
29. Kleinfeld ER, Ferguson GS. *Science* 1994;**265**:370–2.
30. Keller SW, Kim HN, Mallouk TE. *J Am Chem Soc* 1994;**116**:8817–8.
31. Kong W, Zhang X, Gao ML, Zhou H, Li W, Shen JC. *Macromol Rapid Commun* 1994;**15**:405–9.
32. Korneev D, Lvov Y, Decher G, Schmitt J, Yaradaikin S. *Thin Solid Films* 1995;**B213/214**: 954–6.
33. Onoda M, Yoshino K. *Jpn J Appl Phys* 1995;**34**:L260–3.
34. Savemi F, Maasen E, Tieke B, Jordan G, Rammensee W. *Langmuir* 1995;**11**:1068–71.
35. Lvov Y, Ariga K, Kunitake T. *J Am Chem Soc* 1995;**117**:6117–23.
36. Sukhorukov GB, Schmitt J, Decher G. *Ber Bunsenges Phys Chem* 1996;**100**:948–53.
37. Caruso F, Donath E, Mohwald H. *J Phys Chem* 1998;**102**:2011–6.
38. Okubo T, Suda M. *Colloid Polym Sci* 1999;**277**:813–7.
39. Okubo T, Suda M. *Colloid Polym Sci* 2000;**278**:380–4.
40. Okubo T, Suda M. *Colloid Polym Sci* 2002;**280**:533–8.
41. Okubo T, Suda M. *Colloid Polym Sci* 2003;**281**:782–7.
42. Okubo T, Suda M. *Colloid Polym Sci* 2004;**282**:518–23.
43. Okubo T, Suda M, Tsuchida A. *Colloid Polym Sci* 2005;**284**:284–92.
44. Ise N, Moritani H, Okubo T. *Polymer* 1972;**13**:187–8.
45. Ise N, Okubo T. *Polymer* 1972;**13**:552.
46. Okubo T, Ise N. *Bull Chem Soc Japan* 1973;**46**:2493–7.
47. Ise N, Okubo T. In: Bikalis NM, editor. *Water-soluble polymers.* Plenum; 1973. p. 243–53.
48. Okubo T, Kitano H, Ishiwatari T, Ise N. *Proc Roy Soc London* 1979;**A366**:81–90.
49. Cullis PR, Fenske DB, Hope MJ. In: Vance DE, Vance JE, editors. *Biochemistry of lipids, lipoproteins and membranes.* Amsterdam: Elsevier; 1996. p. 1–34. chapter 1.
50. Makino A, Kizaka-Kondoh S, Yamahara R, Hara I, Kanzaki T, Ozeki E, et al. *Biomaterials* 2009;**30**:2126–5160.
51. Girard-Egrot AP, Blum LJ. In: Martin D, editor. *Nanobiotechnology of biomimetic membranes.* New York: Springer; 2007. p. 23–74. chap 2.
52. Shulman JH, Cookbain EG. *Trans Faraday Soc* 1940;**35**:651–61.

53. Shulman JH, Cookbain EG. *Trans Faraday Soc* 1940;**35**:661–8.
54. Clayton W. *In The theory of emulsions and emulsification.* London: J & A Churchill; 1923.
55. Becher P. *In Emulsions: theory and practice.* New York: Reinhold; 1957.
56. Walstra P. In: Becher P, editor. *Encyclopedia of emulsion technology*, vol. 4. New York: Dekker; 1996. p. 1–62.
57. Vanderhoff JW, van den Hul HJ. *J Macromol Sci Chem A* 1973;**7**:677–707.
58. Okubo T. *Colloid Polym Sci* 1990;**268**:1159–66.
59. Okubo T. *J Colloid Interf Sci* 1988;**125**:380–5.
60. Okubo T, Aotani S. Microscopic observation of ordered colloids in sedimentation equilibrium and the importance of the Debye-screening length. 9. Compressed crystal of giant colloidal spheres (diam. = 2~16 micrometer). *Colloid Polym Sci* 1988;**266**:1042–8.
61. Okubo T. J Chem Soc Faraday Trans *1* 1989;**85**:455–66.
62. Bolt GH. *J Am Chem Soc* 1957;**61**:1166–9.
63. Vanderhoff W, van de Hul HJ, Tausk RJM. Overbeek JThG. In: Goldfinger G, editor. *Clean surfaces: Their preparation and characterization for interfacial studies.* New York: Marcel Dekker; 1970. p. 15–44.
64. Schaffer DW. *J Chem Phys* 1977;**66**:3980–4.
65. Okubo T. *Ber Bunsenges Phys Chem* 1987;**91**:1064–9.
66. Midmore BR, Diggins D, Hunter RJ. *J Colloid Interf Sci* 1989;**129**:153–61.
67. Rymden R. *J Colloid Interf Sci* 1988;**124**:396–402.
68. Alexander S, Chaikin PM, Grant P, Morales GJ, Pincus P, Hone D. *J Chem Phys* 1984;**80**:5776–81.
69. Palberg T, Kottal J, Bitzer F, Simon R, Wurthard M, Leiderer P. *J Colloid Interf Sci* 1995;**169**:85–9.
70. Garbow N, Evers M, Palberg T, Okubo T. *J Phys Condens Matter* 2004;**16**:3835–42.
71. Gilanyi T. *J Colloid Interf Sci* 1988;**125**:641–8.
72. Manning GS. *J Chem Phys* 1969;**51**:924–33.
73. Manning GS. *J Chem Phys* 1969;**51**:3249–52.
74. Oosawa F. *Polyelectrolytes.* New York: Marcel Dekker; 1971.
75. Manning GS, Zimm BH. *J Chem Phys* 1965;**43**:4250–9.
76. Manning GS. *J Chem Phys* 1965;**43**:4260–7.
77. Ise N, Okubo T. Thermodynamic and kinetic properties of polyelectrolyte solutions. A unified interpretation in terms of Manning's theory. *Macromolecules* 1978;**11**:439–47.
78. Hiltner PA, Papir YS, Krieger IM. *J Phys Chem* 1971;**75**:1881–6.
79. Kose A, Ozaki M, Takano K, Kobayashi Y, Hachisu S. *J Colloid Interf Sci* 1973;**44**:330–8.
80. Williams R, Crandall RS, Wojtowicz P. *J Phys Rev Lett* 1976;**37**:348–51.
81. Mitaku S, Ohtsuki T, Kishimoto A, Okano K. *Biophys Chem* 1980;**11**:411–6.
82. Lindsay HM, Chaikin PM. *J Chem Phys* 1982;**76**:3774–81.
83. Pieranski P. *Contemp Phys* 1983;**24**:25–73.
84. Ottewill RH. *Ber Bunsenges Phys Chem* 1985;**89**:517–25.
85. Aastuen DJW, Clark NA, Cotter LK, Ackerson BJ. *Phys Rev Lett* 1986;**57**:1733–6.
86. Pusey PN, van Megen W. *Nature* (London) 1986;**320**:340–2.
87. Okubo T. *Acc Chem Res* 1988;**21**:281–6.
88. Russel WB, Saville DA, Schowalter WR. *Colloidal dispersions.* Cambridge, UK: Cambridge University Press; 1989. p. 329–365.
89. Sood AK. *Solid State Phys* 1991;**45**:1–73.
90. Ackerson BJ, Clark NA. *Phys Rev Lett* 1981;**46**:123–6.
91. Okubo T. *J Chem Soc Faraday Trans* 1 1988;**84**:1163–9.
92. Monovoukas Y, Gast AP. *J Colloid Interf Sci* 1989;**128**:533–48.
93. Aastuen DJW, Clark NA, Swindall JC, Muzny CD. *Phase Transitions* 1990;**21**:139–55.
94. Okubo T. *J Chem Phys* 1991;**95**:3690–7.

95. Okubo T. *Naturwiss* 1992;**79**:317–20.
96. Okubo T. *Prog Polym Sci* 1993;**18**:481–517.
97. Simon R, Palberg T, Leiderer P. *J Chem Phys* 1993;**99**:3030–6.
98. Okubo T. *Colloid Polym Sci* 1993;**271**:190–6.
99. Okubo T. *Langmuir* 1994;**10**:1695–702.
100. Okubo T. *Langmuir* 1994;**10**:3529–35.
101. Fujita H, Ametani K. *Jpn J Appl Phys* 1977;**16**:1091–6.
102. Tobochnik J, Chapin PM. *J Chem Phys* 1988;**88**:5824–30.
103. Thirumalai D. *J Phys Chem* 1989;**93**:5637–44.
104. Okubo T. In: Schmitz KS, editor. *Macro-ion characterization. From dilute solutions to complex fluids.* Washington, DC: American Chemical Society; 1994. p. 364–80.
105. Okubo T. Importance of the electrical double layers in structural and diffusional properties of deionized colloidal suspension. *Colloids Surf A* 1996;**109**:77–88.
106. Okubo T, Tsuchida A. *Forma* 2002;**17**:141–53.
107. Okubo T. In: Kinoshita S, Yoshioka S, editors. Colloidal crystal. *Structural colors in biological systems.* Osaka, Japan: Osaka University Press; 2005. p. 267–86.
108. Okubo T, Yoshimi H, Shimizu T, Ottewill RH. Giant colloidal crystals of fluorine-containing polymer spheres in the exhaustively deionized aqueous suspensions and in the presence of sodium chloride. *Colloid Polym Sci* 2000;**278**:469–74.
109. Okubo T. *Polym J* 2008;**40**:882–90.
110. Kose A, Hachisu S. *J Colloid Interf Sci* 1974;**46**:460–9.
111. Okubo T. *J Phys Chem* B 1989;**93**:4352–4.
112. Okubo T, Aotani S. *Naturwissd* 1988;**75**:145–6.
113. Okubo T, Aotani S. *Colloid Polym Sci* 1988;**266**:1049–57.
114. Okubo T, Shimizu T. *J Colloid Interf Sci* 1990;**135**:300–3.
115. Verwey EJW. *Overbeek JThG: Theory of the stability of lyophobic colloids.* Amsterdam: Elsevier; 1948.
116. Pusey PN, Tough RJA. In: Pecora R, editor. *Particle interactions: dynamic light scattering.* London: Plenum; 1985. p. 85–179.
117. Goodwin JW, Ottewill RH. *J Chem Soc Faraday Trans* 1991;**87**:357–69.
118. Ottewill RH. *Ber Bunsenges Phys Chem* 1985;**89**:517–25.
119. Okubo T. *J Chem Phys* 1989;**90**:2408–15.
120. Bartlet P, Ottewill RH, Pusey PN. *J Chem Phys* 1990;**93**:1299–312.
121. Ackerson BJ, Clark NA. *Faraday Discuss Chem Soc* 1983;**76**:219–28.
122. Brown JC, Pusey PN, Goodwin JW, Ottewill RH. *J Phys* (Paris) 1975;**A8**:664–82.
123. Cebula DJ, Goodwin JW, Jeferey GC, Ottewill RH, Parentich A, Richardson RA. *Faraday Discuss Chem Soc* 1983;**76**:37–52.
124. Philipse AP, Vrij A. *J Chem Phys* 1988;**88**:6459–70.
125. Smith GS, Sirota EB, Satinya CR, Plano RJ, Clark NA. *J Chem Phys* 1990;**92**:4519–29.
126. Ashdown S, Markovic I, Ottewill RH, Lindner P, Obertur RC, Rennie AR. *Langmuir* 1990;**6**:303–7.
127. Reus V, Belloni L, Zemb T, Lutterbach N, Versmold H. *J Phys II* (France) 1997;**7**:603–26.
128. D'Aguanno B, Klein R. *J Chem Soc Faraday Trans* 1991;**87**:379–90.
129. Okubo T. *J Colloid Interf Sci* 1987;**117**:165–71.
130. Okubo T. *J Chem Phys* 1987;**87**:3022–8.
131. Cornel RM, Goodwin JW, Ottewill RH. *J Colloid Interf Sci* 1979;**71**:254–66.
132. Clark NA, Ackerson BJ. *Phys Rev Lett* 1980;**44**:1005–8.
133. Lindsay HM, Dozier PWD, Chaikin M, Klein P. *J Phys* 1986;**A19**:2583–9.
134. Mitaku S, Ohtsuki T, Enari K, Kishimoto A, Okano K. *Jap J Appl Phys* 1978;**17**:305–13.
135. Ohtsuki T, Mitaku S, Okano K. *Jap J Appl Phys* 1978;**17**:627–35.
136. Ohtsuki T, Kishimoto A, Mitaku S, Okano K. *Jap J Appl Phys* 1981;**20**:509–18.

137. Joanicot HM, Jorand M, Pieranski P, Rothen F. *J Phys* (Paris) 1984;**45**:1413–21.
138. Schmacher GA, van de Ven TGM. *Faraday Discuss Chem Soc* 1987;**83**:75–81.
139. Schmacher GA, van de Ven TGM. *J Chem Soc Faraday Trans* 1991;**87**:971–6.
140. Ohshima H, Healy TW, White LR, O'Brien RW. *J Chem Soc Faraday Tras* 2 1984;**80**: 1299–317.
141. Cohn EJ, Edsall JT. *Proteins, amino acids and peptides*. New York: Reinhold; 1943.
142. Weber G. *Adv Protein Chem* 1953;**8**:415–59.
143. Jerrard HG. *Chem Rev* 1959;**59**:345–428.
144. Frey M, Wahl P, Benoit HJ. *J Chim Phys* 1964;**61**:1005.
145. Okubo T. *J Am Chem Soc* 1987;**109**:1913–6.
146. Okubo T. *J Phys Chem* 1987;**91**:1977–80.
147. Okubo T. *J Chem Soc Faraday Trans* 1 1988;**84**:703–12.
148. Okubo T. *Macromolecules* 1989;**22**:1818–21.
149. Okubo T, Tsuchida A, Yoshimi H, Maeda H. *Colloid Polym Sci* 1999;**277**:601–6.
150. Rayleigh L. *Proc Roy Soc* London 1911;**A84**:24–38.
151. Okubo T, Kiriyama K, Nemoto N, Hashimoto H. *Colloid Polym Sci* 1996;**274**:93–104.
152. Okubo T, Kiriyama K. *Ber Bunsenges Phys Chem* 1996;**100**:849–56.
153. Okubo T, Kiriyama K. *J Mol Liquids* 1997;**72**:347–64.
154. Okubo T, Tsuchida A. *Colloid Polym Sci* 2002;**280**:438–45.
155. van Veluwen A, Lekerkerker HNW, de Kruif LG, Vrij A. *J Chem Phys* 1987;**87**:4873–80.
156. Brown JC, Pusey PN, Goodwin JW, Ottewill RH. *J Phys* (Paris) 1975;**A8**:664–82.
157. Gruner F, Lehmann W. *J Phys* (Paris) 1979;**A12**:L303–7.
158. Hansen JP, Hayter JB. *Molec Phys* 1982;**46**:651–6.
159. Hess W, Klein R. *Adv Phys* 1983;**32**:173–283.
160. Klein R, Hess W. *Faraday Discuss Chem Soc* 1983;**76**:137–50.
161. van Megen W, Snook I. *Faraday Discuss Chem Soc* 1983;**76**:151–63.
162. Hartl W, Vertsmold HZ. *Phys Chem* (Neue Folge) 1987;**153**:1–4.
163. Ottewill RH, Williams NStJ. *Nature* (London) 1987;**325**:232–4.
164. Okubo T. *Ber Bunsenges Phys Chem* 1988;**92**:504–11.
165. Okubo T. *J Chem Phys* 1987;**87**:6733–9.
166. Okubo T. *Naturwiss* 1988;**75**:91–3.
167. Okubo T. The viscosity of deionized colloidal spheres. Importance of surface charge density and monodispersity of spheres. *Polym Bull* 1988;**20**:269–76.
168. Okubo T. *J Phys Chem* 1990;**94**:1962–6.
169. Okubo T. *Colloid Polym Sci* 1993;**271**:873–83.
170. Nishida K, Kiriyama K, Kanaya T, Kaji K, Okubo T. *J Polym Sci B Phys Ed* 2004;**42**: 1068–74.
171. Ashdown S, Markovic I, Ottewill RH, Lindner P, Oberthur RC, Rennie AR. *Langmuir* 1990;**6**:303–7.
172. Chen LB, Chow MK, Ackerson BJ, Zukoski CF. *Langmuir* 1994;**10**:2817–29.
173. Chen LB, Zukoski CF. *Phys Rev Lett* 1990;**65**:44–7.
174. Chen LB, Zukoski CF, Ackerson BJ, Hanley HJM, Straty GC, Barker J, et al. *Phys Rev Lett* 1992;**69**:688–91.
175. Okubo T, Kimura H, Hatta T, Kawai T. *Phys Chem Chem Phys* 2002;**4**:2260–3.
176. Okubo T, Kimura H, Hatta T. *Studies Surf Catalysis* 2001;**132**:387–90.
177. Okubo T, Kimura H, Kawai T, Niimi H. *Langmuir* 2003;**19**:6014–8.
178. von Smoluchowski M. *Bull Int Acad Sci Cracovie* 1903;**184**.
179. Ito K, Ise N, Okubo T. *J Chem Phys* 1985;**82**:5732–6.
180. Okubo T, Kitano H, Iwai S. *J Chem Soc Faraday Trans* 1 1988;**84**:4317–25.
181. Kitano H, Iwai S, Ise N, Okubo T. *J Am Chem Soc* 2004;**109**:6641–4.
182. Bird RB, Stewart WB, Lightfoot EN. *Transport phenomena*. New York: Wiley; 1960.
183. Einstein A. *Ann Phys* 1905;**17**:549–60.

184. Eigen M. *Quantum statistical mechanisms in natural science*. New York: Plenum; 1974.
185. Debye P. *Trans Electrochem Soc* 1942;**82**:265–72.
186. Alberty RA, Hammes GG. *J Phys Chem* 1958;**62**:154–9.
187. Okubo T, Okada S, Tsuchida A. *J Colloid Interf Sci* 1997;**189**:337–47.
188. Okubo T, Okada S. *J Colloid Interf Sci* 1997;**192**:490–6.
189. Okubo T, Tsuchida A, Kato T. *Colloid Polym Sci* 1999;**277**:191–6.
190. Okubo T, Ishiki H. *J Colloid Interf Sci* 2000;**228**:151–6.
191. Dhont JKG, Smits C, Lekkerkerker HNW. *J Colloid Interf Sci* 1992;**152**:386–401.
192. Wilson HW. *Phil Mag* 1900;**50**:238–50.
193. Frenkel J. *Phys Z Sowjetunion* 1932;**1**:498–500.
194. Okubo T. *Curr Topics Colloid Interf Sci* 1997;**1**:169–93.
195. Okubo T, Okada S. *J Colloid Interf Sci* 1998;**204**:198–204.
196. Okubo T, Ishiki H. *J Colloid Interf Sci* 1999;**211**:151–9.
197. Okubo T, Tsuchida A, Okada S, Fujitsuna K, Ishikawa M, Morita T, et al. *Colloid Surf A* 1999;**160**:311–20.
198. Okubo T, Ishiki H, Chiyoda M, Yoshinaga K. *Colloid Polym Sci* 2002;**280**:290–5.
199. Wong EM, Bonevich JE, Pearson PC. *J Phys Chem* 1998;**B102**:7770–5.
200. Okubo T. *Naturwiss* 1992;**79**:317–20.
201. Okubo T. *J Colloid Interf Sci* 1992;**153**:587–91.
202. Okubo T. *Colloid Polym Sci* 1993;**271**:190–6.
203. Okubo T. *Langmuir* 1994;**10**:1695–702.
204. Okubo T. *Langmuir* 1994;**10**:3529–35.
205. Okubo T, Fujita H, Kiriyama K, Yamaoka H. *Colloid Polym Sci* 1996;**274**:73–80.
206. Hiltner PA, Krieger IM. *J Phys Chem* 1969;**73**:2386–9.
207. Okubo T. *J Chem Soc Faraday Trans* 1 1986;**82**:3163–73.
208. Okubo T, Suzuki D, Yamagata T, Katsuno A, Sakurai M, Kimura H, et al. *Colloid Polym Sci* 2011;**289**:291–9.
209. Okubo T. *J Chem Soc Faraday Trans* 1 1986;**82**:3175–83.
210. Okubo T. *J Chem Soc Faraday Trans* 1 1986;**82**:3185–96.
211. Hunter RJ. *Zeta potential in colloid science: principle and applications*. London: Academic Press; 1981. p. 117–118.
212. Russel WB, Saville DA, Schowalter WR. *Colloidal dispersions*. Cambridge, UK: Cambridge University Press; 1989. p. 394–428.
213. van de Ven TGM. *Colloidal hydrodynamics*. London: Academic Press; 1989. p. 78–96.
214. Hiemenz PC, Rajagopalan R. *Principles of colloid and surface chemistry*. 3rd ed. New York: Marcel Dekker; 1997. p. 176–181.
215. Vanderhoff JW, El-Aasser MS, Micale FJ. PMSE Proc Am Chem Soc, Div Polym Mat Sci Eng. *Polym Kinet Process Develop* 1986;**54**:584.
216. Briskman V, Kostarev K, Leontyev V, et al. *Proc 48th Intern Astron Cong*. Turin, Italy: Intern Astron Fed; 1997.
217. Okubo T, Tsuchida A, Kobayashi K, Kuno A, Morita T, Fujishima M, et al. *Colloid Polym Sci* 1999;**277**:474–8.
218. Tsuchida A, Yoshimi H, Kiriyama S, Ohiwa K, Okubo T. *Colloid Polym Sci* 2003;**281**: 760–5.
219. Krutzer LLM, Folkersma R, van Diemen AJG, Stein HN. *Adv Colloid Interf Sci* 1993;**46**: 59–71.
220. Ansari RR, Suh KI. *J Jpn Soc Microgravity Appl* 1998;**15**(Suppl II):186–93.
221. Lant CT, Smart AE, Cannell DS, et al. *Appl Opt* 1997;**36**:7501–7.
222. Okubo T, Tsuchida A, Yoshimi H, Maeda H. *Colloid Polym Sci* 1999;**277**:601–6.
223. Lewis GN, Randall M. *Thermodynamics*. New York: McGraw-Hill; 1961. p. 513–514.
224. Conway JH, Sloane NJA. *Sphere packing, lattices and groups*. 3rd ed. New York: Springer-Verlag; 1998.

225. Hales TC. *Discu Comput Geom* 1997;**17**:1–51.
226. Hales TC. *Discu Comput Geom* 1997;**18**:135–49.
227. Fejes Toth L. *Lagerungen in der Ebene, auf der Kugel und im Raum.* 2nd ed. Heidelberg: Springer; 1972.
228. Thue A. *Christiania Vid Selsk Skr* 1910;**1**:3–9.
229. Tanemura M, Matsumoto T. *Kristallographie* 1997;**212**:637–47.
230. Aoyama K, Ogawa K, Kimura Y, Fujiyoshi Y. *Ultramicroscopy* 1995;**57**:345–54.
231. Stoeber W, Fink A, Bohn E. *J Colloid Interf Sci* 1968;**26**:62–9.
232. Iler RK, editor. *The chemistry of silica.* New York: Wiley-Interscience; June 6, 1979 [chapter 4].
233. Okubo T, Okamoto J, Tsuchida A. *Colloid Polymer Sci* 2010;**288**:189–97.
234. Okubo T, Suzuki D, Yamagata T, Katsuno A, Mizutani M, Kimura H, et al. *Colloid Polymer Sci* 2011;**289**:807–16.
235. Okubo T, Suzuki D, Tsuchida A. *Colloid Polymer Sci* 2012;**290**:411–21.
236. Okubo T, Suzuki D, Tsuchida A. *Colloid Polymer Sci* 2012;**290**:867–77.
237. Okubo T, Suzuki D, Tsuchida A. *Colloid Polymer Sci* 2012;**290**:1901–11.
238. Okubo T, Fujii S, Aono K, Nakamura Y. *Colloid Polymer Sci* 2013;**291**:1019–30.
239. Okubo T, Fujii S, Nakamura Y. *Colloid Polym Sci* 2013;**291**:2805–13.
240. Fujii S, Yamashita Y, Nakamura Y, Okubo T. *Colloid Polym Sci* 2014;**292**:2621–31.
241. Okubo T. *Colloid Polymer Sci* 1987;**265**:522–7.
242. Okubo T. *Colloid Polym Sci* 2013;**291**:1623–9.
243. Jeffrey GC, Ottewill RH. *Colloid Polym Sci* 1988;**266**:173–9.
244. Crick FHC, Watson JD. *Proc Roy Soc London* 1954;**223A**:80–96.
245. Zsigmondy R. *Z Physik Chem* 1906;**56**:65–9.
246. Matijevic E. *Acc Chem Res* 1981;**14**:22–9.
247. Matijevic E, Wilhelmy DM. *J Colloid Interf Sci* 1982;**86**:476–84.
248. Okubo T. *J Chem Phys* 1991;**96**:2261–8.
249. Guez HM, Meseguer F, Lopez C, Mifsud A, Moya JS. *Langmuir* 1997;**13**:6009–11.
250. van Helden AK, Jansen JW, Vrij A. *J Colloid Interf Sci* 1981;**81**:354–68.
251. Emmett S, Lubetkin SD, Vincent B. *Colloids Surf* 1989;**42**:139–53.
252. Veihaegh NAM, van Blaaderen A. *Langmuir* 1994;**10**:1427–38.
253. Imhof A, van Blaaderen A, Dhont JKG. *Langmuir* 1994;**10**:3477–84.
254. Yoshinaga K, Nakanishi K. *Composite Interf* 1994;**2**:95–103.
255. Yoshinaga K, Chiyoda M, Yoneda A, Nishida H, Komatsu M. *Colloid Polym Sci* 1999;**277**:479–82.
256. Yoshinaga K, Fujiwara K, Mouri E, Ishii M, Nakamura H. *Langmuir* 2005;**21**:4471–7.
257. Yoshinaga K, Shigeta M, Komune S, Mouri E. *Colloids Surf B Biointerf* 2007;**54**:108–13.
258. Ma Z, Watanabe M, Mouri E, Yoshinaga K. *Colloid Polym Sci* 2010;**288**:519–25.
259. Ma Z, Watanabe M, Mouri E, Yoshinaga K. *Colloid Polym Sci* 2011;**289**:85–91.
260. Nommensen PA, Duits MHG, Lopulissa JS, van den Ende D, Mellema J. *Prog Colloid Polym Sci* 1998;**110**:144–9.
261. Pratsinis SE, Zhu W, Vemury S. *Powder Technol* 1996;**86**:87–93.
262. Zhu W, Pratosinis SE. In: Chow GM, Gonzalves KE, editors. *Nanotechnology.* Washington, DC: ACS Symp 622; 1996. p. 64–78.
263. Lee JH, Beaucage G, Pratsinis SE, Vemury S. *Langmuir* 1998;**14**:5751–6.
264. Zang L, Qu P, Zhao J, Shen T, Hidaka H. *Chem Lett* 1997;**26**:791–2.
265. Qu P, Zhao J, Zang L, Shen T, Hidaka H. *Colloids Surf A* 1998;**138**:39–50.
266. Uyeda N, Nishino M, Suito E. *J Colloid Interf Sci* 1973;**43**:264–76.
267. Liz-Marzan LM, Glersig M, Mulvaney P. *Langmuir* 1996;**12**:4329–35.
268. Okubo T. *Colloid Polym Sci* 1987;**265**:522–7.
269. Mayya KS, Patil V, Sastry M. *Langmuir* 1997;**13**:2575–7.
270. Santhanam V, Liu J, Agarwal R, Andres RP. *Langmuir* 2003;**19**:7881–7.

271. Patil V, Mayya KS, Pradhan SD, Sastry M. *J Am Chem Soc* 1997;**119**:9281–2.
272. Kim SH, Medeiros-Ribeiro G, Ohlberg DAA, Williams RS, Heath JR. *J Phys Chem B* 1999;**103**:10341–7.
273. Korgel BA, Fullam S, Connolly S, Fitzmaurice D. *J Phys Chem B* 1998;**102**:8379–88.
274. Okubo T. *Colloid Polym Sci* 2013;**291**:1623–9.
275. Okubo T. *Colloid Polym Sci* 2013;**291**:1887–93.
276. Scott WB, Matijevile E. *J Colloid Interf Sci* 1978;**66**:447–581.
277. Matijevic E, Sapieszko RS, Melvilb JB. *J Colloid Interf Sci* 1975;**50**:567–81.
278. Sapieszko RS, Patel RC, Matijevic E. *J Phys Chem* 1977;**81**:1061–8.
279. Sugimoto T, Wang Y, Itoh H, Muramatsu A. *Colloids Surf A* 1998;**134**:265–79.
280. Sugimoto T, Dirige GE, Muramatsu A. *J Colloid Interf Sci* 1996;**180**:305–8.
281. Sondi I, Siiman O, Koester S, Matijevic E. *Langmuir* 2000;**16**:3107–18.
282. Mayya KS, Patil V, Kumar PM, Sastry M. *Thin Solid Films* 1998;**312**:300–5.
283. Boakye E, Radovic LR, Osseo-Asare K. *J Colloid Interf Sci* 1994;**163**:120–9.
284. Pileni MP. *Langmuir* 1997;**13**:3266–76.
285. Muramatsu A, Sugimoto T. *J Colloid Interf Sci* 1997;**189**:167–73.
286. Sandroff CJ, Hwang DM, Chung WM. *Phys Rev B* 1986;**33**:5953–5.
287. Dabbousi BO, Viejo JR, Mikulec FV, Heine JR, Mattoussi H. *J Phys Chem B* 1997;**101**: 9463–75.
288. Tajima T, Tsutsui A, Fujii T, Takada J, Takaguchi Y. *Polym J* 2012;**44**:620–4.
289. Ueno K, Hamasaki S, Wanless EJ, Nakamura Y, Fujii S. *Langmuir* 2014;**30**:3051–9.
290. Sugimoto T, Sakata K. *J Colloid Interf Sci* 1992;**152**:587–90.
291. Uyeda R. Morphology of crystals, Part B. In: Sunagawa I, editor. Tokyo: Terra; 1987. p. 367–508.
292. Petroski JM, Green TC, El-Sayed MA. *J Phys Chem* 2001;**105**:5542–7.
293. Ogino K, Kuronuma I. *J Colloid Interf Sci* 1976;**56**:629–30.
294. Hachisu S, Kose A, Kobayashi Y, Takano K. *J Colloid Interf Sci* 1976;**55**:499–509.
295. Yu YY, Chang SS, Lee CL, Wang CRC. *J Phys Chem B* 1997;**101**:6661–4.
296. Link S, Burda C, Mohamed MB, Nikoobakht B, El-Sayed MA. *J Phys Chem A* 1999;**103**: 1165–70.
297. Chang SS, Shih CW, Chen CD, Lai WC, Wang CRC. *Langmuir* 1999;**15**:701–9.
298. Okamoto S, Hachisu S. *J Colloid Interf Sci* 1973;**43**:30–5.
299. van de Zande BMI, Dhont JKG, Bohmer MR, Fokkink LGJ, Schonenberger C. *Langmuir* 2000;**16**:451–8.
300. Brown ABD, Clarke SM, Rennie AR. *Langmuir* 1998;**14**:3129–32.
301. Ahmadi TS, Wang ZL, Green TC, Henglein A, El-Sayed MA. *Science* 1996;**272**:1924–7.
302. Watson Jr JHL, Cardel RR, Heller W. *J Phys Chem* 1962;**66**:1757–63.
303. Buining PA, Philipse AP, Lekkerkerker HNW. *Langmuir* 1994;**10**:2106–14.
304. Koenderink GH, Vliegenthart GA, Kluijtmans SGJM, van Blaaderen A, Philipse AP, Lekkerkerker HNW. *Langmuir* 1999;**15**:4693–6.
305. Buining PA, Patmamanoharan C, Jansen JHB, Lekkerkerker HNW. *J Am Ceram Soc* 1991;**74**:1303–7.
306. van der Kooij FM, Lekkerkerker HNW. *J Phys Chem B* 1998;**102**:7829–32.
307. Sugimoto T, Okada K, Itoh H. *J Colloid Interf Sci* 1997;**193**:140–3.
308. Sedlacik M, Mrlik M, Kozakova Z, Pavlinek V, Kuritka I. *Colloid Polym Sci* 2013;**291**: 1105–11.
309. Matsumoto M. *Biophys Chem* 1996;**58**:173–83.
310. Matsumoto M. *Colloids Surf A* 1999;**148**:75–81.
311. Gong T, Wu DT, Marr WMM. *Langmuir* 2002;**18**:10064–7.
312. Rowan DG, Hansen JP. *Langmuir* 2002;**18**:2063–8.
313. Zocher VH, Jacobsohn K. *Kolloid Beih* 1929;**28**:167–206.
314. Okubo T. *J Am Chem Soc* 1987;**109**:1913–6.

315. Okubo T. *J Phys Chem* 1987;**91**:1977–80.
316. Okubo T. *J Chem Soc Faraday Trans* 1 1988;**84**:703–12.
317. Hachisu S, Furusawa K. *Sci Light* 1963;**12**:1–8.
318. Brock SL, Sanabria M, Suib SL, Urban V, Thiyagarajan P, Potter DI. *J Phys Chem B* 1999;**103**:7416–28.
319. Okubo T. *Angew Chem Intern Ed Engl* 1987;**26**:765–7.
320. Ramachandran GN, Kartha G. *Nature* 1954;**174**:269–70.
321. Sugimoto T, Khan MM, Muramatsu A. *Colloids Surf A* 1993;**70**:167–9.
322. Sugimoto T, Waki S, Itoh H, Muramatsu A. *Colloids Surf A* 1996;**109**:155–65.
323. Ohno K, Tsuchida A, Okubo T. *Colloid Polym Sci* 2014;**292**:723–31.
324. Mourchid A, Delville A, Lambard J, Lecolier E, Levitz P. *Langmuir* 1995;**11**:1942–50.
325. Grillo I, Levitz P, Zemb Th. *Langmuir* 2000;**16**:4830–9.
326. Kimura H, Sakurai M, Sugiyama T, Tsuchida A, Okubo T, Masuko T. *Rheol Acta* 2011;**50**:159–68.
327. Morvan M, Espinat D, Lambard J, Zemb T. *Colloids Surf A* 1994;**82**:193–203.
328. Kroon M, Vos WL, Wegdam GH. *Phys Rev E* 1998;**57**:1962–70.
329. Ottewill RH, Shaw JN. *Kolloid Z Z Polym* 1966;**218**:34–40.
330. Hiltner PA, Krieger IM. *J Phys Chem* 1969;**73**:2386–9.
331. Fitch RM, editor. *Polymer colloids*. New York: Plenum; 1971.
332. Murray CA, Wenk RA. *Phys Rev Lett* 1989;**62**:1643–6.
333. Fitch RM, editor. *Polymer colloids. A comprehensive introduction*. San Diego: Academic; 1997.
334. Okubo T. *Colloid Polym Sci* 2008;**286**:1307–15.
335. Ottewill RH. *Langmuir* 1989;**5**:4–9.
336. van Megen W, Underwood SM, Pusey PN. *Phys Rev Lett* 1991;**67**:1586–9.
337. He Y, Ackerson BJ, van Megen W, Underwood SM, Schatzel K. *Phys Rev E* 1996;**54**: 5286–97.
338. Harland JL, van Megen W. *Phys Rev E* 1997;**55**:3054–67.
339. Moussaid A, Pusey PN. *Phys Rev E* 1999;**60**:5670–6.
340. Guisheng P, Kesavamoorthy R, Asher SA. *J Am Chem Soc* 1998;**120**:6518–24.
341. Lutterbach N, Versmold H, Reus V, Belloni L, Zemb Th. *Langmuir* 1999;**15**:337–44.
342. Okubo T, Yoshimi H, Shimizu T, Ottewill RH. *Colloid Polym Sci* 2000;**278**:469–74.
343. Okubo M, Yamamoto Y, Kamei S. *Colloid Polym Sci* 1989;**267**:861–5.
344. Cardoso AH, Leite CAP, Galembeck F. *Langmuir* 1999;**15**:4447–53.
345. Chen Y, Ford WT, Materer NF, Teeters D. *J Am Chem Soc* 2000;**122**:10472–3.
346. Li H, Kumacheva E. *Colloid Polym Sci* 2003;**281**:191–205.
347. Sukhorukov GB, Brumen M, Donath E, Mohwald H. *J Phys Chem B* 1999;**103**:6434–40.
348. Hotz J, Meier W. *Langmuir* 1998;**14**:1031–6.
349. Caruso F, Donath E, Mohwald H. *J Phys Chem B* 1998;**102**:2011–6.
350. Decher G, Schmitt J. *Prog Colloid Polym Sci* 1992;**89**:160–4.
351. Ferreira M, Rubner MF. *Macromolecules* 1995;**28**:7107–14.
352. Dai Z, Mohwald H, Tiersch B, Dahne L. *Langmuir* 2002;**18**:9533–8.
353. Sheu HR, Tseng CM, El-Aasser MS, Micale FJ, Vanderhoff JW. *Grad Res Progr Reps* 1985;**24**:6–7. ibid. 25:7–8, 1985.
354. Vanderhoff JW, El-Aasser MS, Fitch RM, editors. *Future directions in polymer colloids*. Dordrecht: Martinus Nijhoff; 1987. p. 23–46.
355. Saadat Y, Hosseinzadeh S, Abdolbaghi S, Afshar-Taromi F. *Colloid Polym Sci* 2013;**291**: 1113–20.
356. Huang H, Liu H. *J Polym Sci A Polym Chem* 2010;**48**:5198–205.
357. Okubo M, Katsuta Y, Matsumoto T. *J Polym Sci Polym Let Ed* 1980;**18**:481–6.
358. Okubo M, Fujibayashi T, Yamada M, Minami H. *Colloid Polym Sci* 2005;**283**:1041–5.
359. Okubo M, Takekoh R, Suzuki A. *Colloid Polym Sci* 2002;**280**:1057–61.

360. Okubo M, Minami H. *Colloid Polym Sci* 1997;**275**:992–7.
361. Okubo M, Minami H. *Macromol Symp* 2000;**150**:201–10.
362. Okubo M, Fujibayashi T, Terada A. *Colloid Polym Sci* 2005;**283**:793–8.
363. Fujibayashi T, Okubo M. *Langmuir* 2007;**23**:7958–62.
364. Riegel IC, Eisenberg A, Petzhold CL, Samios D. *Langmuir* 2002;**18**:3358–63.
365. Nagy M, Keller A. *Polym Comm* 1989;**30**:130–2.
366. Ho CC, Keller A, Odell JA, Ottewill RH. *Colloid Polym Sci* 1993;**271**:469–79.
367. Pelton R. *Adv Colloid Interf Sci* 2000;**85**:1–33.
368. Parsegian VA, Brenner SL. *Nature* 1976;**259**:632–5.
369. Butler PJG, Lomonossoff GP. *Biophys J* 1980;**32**:295–312.
370. Jeng TW, Growther RA, Stubbs G, Chiu W. *J Mol Biol* 1989;**205**:251–7.
371. Maier EE, Krause R, Deggelmann M, Hagenbuchle M, Weber R, Fraden S. *Macromolecules* 1992;**25**:1125–33.
372. Niu Z, Li S, Bruckman MA, Lee LA, Lee B, Pingali SV, et al. In: Nagarajan R, Hatton TA, editors. *Nanoparticles: synthesis, stabilization, passivation, and functionalization*. Washington, DC: ACS Symposium Series 996; 2008. p. 369–85.
373. Ratna BR, Soto CM, Danner L, Blum AS, Fang J, Lin T, et al. *Langmuir* 2003;**19**:489–90.
374. Araki J, Kuga S. *Langmuir* 2001;**17**:4493–6.
375. Wang SW, Robertson CR, Gast AP. *J Phys Chem B* 1999;**103**:7751–61.
376. Adams M, Dogic Z, Keller SL, Fraden S. *Nature* 1998;**393**:349–52.
377. Araki J, Wada M, Kuga S, Okano T. *Langmuir* 2000;**16**:2413–5.
378. Araki J, Wada M, Kuga S. *Langmuir* 2001;**17**:21–7.
379. Hall CE, Doty P. *J Am Chem Soc* 1958;**80**:1269–74.
380. Hulmes DJS, Miller A. *Nature* 1979;**282**:878–80.
381. Okubo T, Tsuchida A, Kokufuta E. *Colloid Polym Sci, in press* 2005. http://dx.doi.org/10.1007/s00396-015-3558-z.
382. Okubo T, Tsuchida A. *Colloid Polym Sci,* 2014;**292**:3187–93.
383. Okubo T, Kokufuta E, Nakamuro M, Yoshinaga K, Mizutani M, Tsuchida A. *Colloids Surf B* 2010;**80**:193–9.
384. Round FE, Crawford RM, Mann DG. *The diatoms*. Cambridge, UK: Cambridge University Press; 1990.
385. Haeckel E. *Art forms from the ocean*. Munich: Prestel; 2005.

3 Colloidal Crystallization

3.1 Colloidal Crystals

Keen attention has been paid to colloidal crystals (i.e., crystal-like distribution of colloidal particles in dispersions of aqueous and organic solvents.)[1-13] Many researchers have studied structural colors, interparticle interaction, crystal structure, the morphologic characteristics of single crystals, phase transition, crystallization kinetics of nucleation and crystal growth, physicochemical properties (rigidity, viscosity, etc.), structural change induced by the external fields (such as gravity, centrifugal and electric fields, and shearing forces), and technical application (photonic crystals, colloidal lithography, etc.).

As described in Chapter 2, all kinds of substances or materials are composed of a huge number (in the order of Avogadro's number, ideally) of elementary particles (atoms, molecules, colloidal particles, etc.). Substances exist in three states: gas, liquid (normal and ordered liquids), and solid (crystal and amorphous solid), depending on the distribution of the elementary particles. In the gas state, all the elementary particles move freely by the thermal energy, k_BT, given to each elementary particle in the modes of translation and rotation. Here, k_B is the Boltzmann constant and T is the absolute temperature. When an elementary particle hits other particles or the wall of the container, it is simply reflected. In the liquid state, all the elementary particles also move freely but keeping the condition that the nearest-neighbored interelementary particle distances are close to each other. The elementary particles in the solid state also move in the vibrational and rotational modes with thermal energy, but keep their own equilibrium points. Distinction between crystal and amorphous solids is simple: The former distributes in order and the latter is in disorder. Irrespective of the kind of elementary particles, most of the substances have three states, depending on the conditions previously described. Thus *colloidal gases, colloidal liquids, colloidal crystals,* and *colloidal amorphous solids* are sound scientific terms describing substances in which the elementary particles are themselves colloidal particles.

Two groups of colloidal crystals have been studied intensively hitherto: those in (1) diluted and deionized aqueous suspensions (soft system)[14-33] and

Colloidal Organization. http://dx.doi.org/10.1016/B978-0-12-802163-7.00003-9

(2) concentrated suspensions mainly in the refractive index-matched organic solvents (hard system).[34-39] The soft crystals are convenient models to study both the soft- and hard-sphere systems, depending on the ionic concentrations of suspension (i.e., soft crystals form in the exhaustively deionized state and hard crystals form in the presence of substances like salt, sodium chloride [NaCl], for example). The hard crystal is best for study of crystallization by only the interparticle repulsive interaction because of the thermal movement of colloidal particles (i.e., the excluded volume effects or steric stability effects). On the other hand, soft crystals form by electrostatic interparticle repulsion in addition to excluded volume effects. We should note here that in the usual crystals, metals, proteins, and ice, for example, the interparticle attractive interactions, such as chemical and hydrogen bonds, play an important role in crystallization.

Most colloidal particles in water get negative charges on their interfaces with water to lower the raised-interface Gibbs free energy by two mechanisms: the dissociation of ionizable groups and the preferential adsorption of ions from suspension. These ionic groups leave their counter-ions, and these excess charges accumulate near the interface, forming an electrical double layer. The counter-ions in the diffuse region are distributed according to a balance between the thermal diffusive forces and the forces of electrical attraction with colloidal particles. The importance of the electrical double layers for the colloidal crystallization of a soft system has been clarified by many researchers.[1-33] Large single crystals have formed in the exhaustively deionized and diluted suspension of colloidal spheres, where the nuclei for crystallization are few.[23-33] Figs. 3.1 through 3.4 show the typical examples of colloidal single crystals with coexistence of the mixtures of cationic and anionic ion-exchange resins. The interparticle distances of colloidal crystals are in the order of light wavelengths and each single crystal reflects light in different colors by the Bragg diffraction, which makes colloidal crystals beautiful and fantastic. Table 3.1 compiles size and charge density values of the typical colloidal spheres appearing in this chapter.

When colloidal particle size is large and concentration is high, iridescent colors of the colloidal crystal samples come only from the several Bragg reflection layers along the cell surface; they look metallic, as is shown in Fig. 3.1. Color changes are due to the change in the incident light angles from a pocket flashlight source through the curved test tube. The iridescent colors of the colloidal crystals of poly(methyl methacrylate) (PMMA) spheres, with a diameter of approximately 100 to 150 nm, in the medium concentrations (of the order of several volume percents) are shown in Fig. 3.2. Colors changed systematically depending on their sphere sizes and concentrations. Here, bluish and brownish granules included in the vials are the mixtures of the cation- and anion-exchange resins. When sizes and concentrations of colloidal spheres are approximately 100 nm and

Table 3.1 Characteristics of Typical Colloidal Spheres of Polystyrene and Silica

Spheres	d(nm) ± s(nm)	s/d	Charge Density (strongly acidic) (μC/cm^2)
Polystyrene Spheres			
D1B89	38 ± 7.5	0.20	3.8
D1C25	85 ± 6	0.071	1.5
D1W52	88 ± 6.2	0.071	1.14
D1C27	91 ± 6	0.066	2.0
SS18	95 ± 10	0.11	1.8
D1B22	109 ± 3	0.028	3.6
D1B76	109 ± 3	0.028	2.1
D1P30	109 ± 27.8	0.255	0.62
LSN	109 ± 5	0.046	3.3
D1K88	137 ± 16	0.12	0.58
D1B72	173 ± 7	0.040	1.0
VN1001	192 ± 6	0.031	0.50
D1B28	212 ± 3	0.014	1.29
D1B41	215 ± 5	0.023	1.3
D1A19	220 ± 6.5	0.030	1.32
N200	220 ± 5	0.023	4.2
D1B62	238 ± 3.4	0.014	1.82
D1B33	269 ± 3	0.011	2.9
C301	300 ± 10	0.033	34.3 (carboxylic acid-type)
D1B52	305 ± 8.4	0.028	1.75
C351	351 ± 5	0.014	51.7 (carboxylic acid-type)
G5301	369 ± 10	0.027	7.2
D4N7W	460 ± 5	0.011	2.0
D1A92	497 ± 6	0.012	2.8
N550	563 ± 8	0.014	6.6
N601	600 ± 12	0.020	19.5
N800	780 ± 9	0.012	1.2
D1C11	804 ± 5	0.0062	2.4
N1000	1020 ± 20	0.020	2.6
H1001	1022 ± 5	0.0049	171 (carboxylic acid-type)
D1A12	2020 ± 14	0.0069	6.2
D1A35	2950 ± 130	0.044	14.2
Colloidal Silica Spheres			
CS06	6 ± 2	0.33	0.23
Snowtex-S	8 ± 1	0.13	0.32
Ludox-SM	12 ± 1	0.083	0.59
Ludox-AM	13.5 ± 1	0.074	0.54
CS15	15 ± 3	0.20	0.24
Ludox-TM	25 ± 2	0.08	0.33
CS22P	29 ± 6	0.21	0.48
Cataloid45	45 ± 5	0.11	0.35
Snowtex20L	45 ± 5	0.11	0.95
CS45	56.3 ± 7.6	0.13	0.30
CS61	81.2 ± 11.5	0.142	0.60
CS81, CS82	103 ± 13	0.13	0.38
CS91	110 ± 4.5	0.041	0.48
CS20	133 ± 28	0.21	0.38

(Continued)

Table 3.1 Characteristics of Typical Colloidal Spheres of Polystyrene and Silica—cont'd

Spheres	d(nm) ± s(nm)	s/d	Charge Density (strongly acidic) (μC/cm²)
CS121	136 ± 10.9	0.080	0.40
CS22	178 ± 9	0.051	0.52
CS161	184 ± 18.6	0.101	0.47
CS301	311 ± 22	0.071	0.40
CS501	507 ± 51	0.10	0.59
CS1001	1090 ± 45	0.041	—
Poly(methyl methacrylate) Spheres			
PMA31~34	94 ± 13.2	0.14	—
GW1	100 ± 7	0.07	—
PMA41,42	149 ± 17.7	0.12	—
GW6	200 ± 10	0.05	—
GW8	300 ± 12	0.04	—
PM1000	1000 ± 20	0.02	—
Fluorine-containing spheres			
PFA		120	—
PTP		130	—
PTFEA		130	—
PTFEB		210	—

D1 series (D1B89, D1C25, for example): From Dow Chemical Co.
SS series: Author's laboratory.
LSN: Author's laboratory. Radical copolymerization of styrene, styrene sulfonate and a small amount of vinylnaphthalene without emulsifier.
VN1001: Nippon Zeon Co., Tokyo.
N series: Sekisui Chemical Co., Tokyo.
C series: Sekisui Chemical Co., Tokyo.
H1001, G5301 (a copolymer of styrene and styrene sulfonate), N1000: JSR Co., Tokyo.
CS series: Catalysts & Chemicals Ind. Co.
Snowtex series: Nissan Chemical Co.
Ludox series: Du Pont, Japan.
Cataloid series: Catalysts & Chemicals Ind. Co.
PM and GW series: Soken Chemical & Eng. Co., Tokyo.
PFA: Copolymer of tetrafluoroethylene and perfluorovinylether
PTP: Copolymer of tetrafluoroethylene and perfluoropropylene
PTFEA, PTFEB: poly(tetrafluoro ethylene)

extremely low, and the suspensions are deionized exhaustively with the presence of the ion-exchange resins, very large colloidal single crystals appear as is demonstrated in Figs. 3.3 and 3.4. The dispersions were prepared in a plane cell and a test tube, respectively. A plane cell was an optical cell for the fluorescence measurements of 10 mm in the optical path. Inner and outer sizes of the test tube were 11 mm$^{\phi}$ and 12 mm$^{\phi}$ (or sometimes 14 mm$^{\phi}$ and 15 mm$^{\phi}$), respectively. Large colloidal single crystals appeared in these experimental conditions, because very few nuclei form. In a plane cell (Fig. 3.3), colors within a single crystal did not change, because the incident

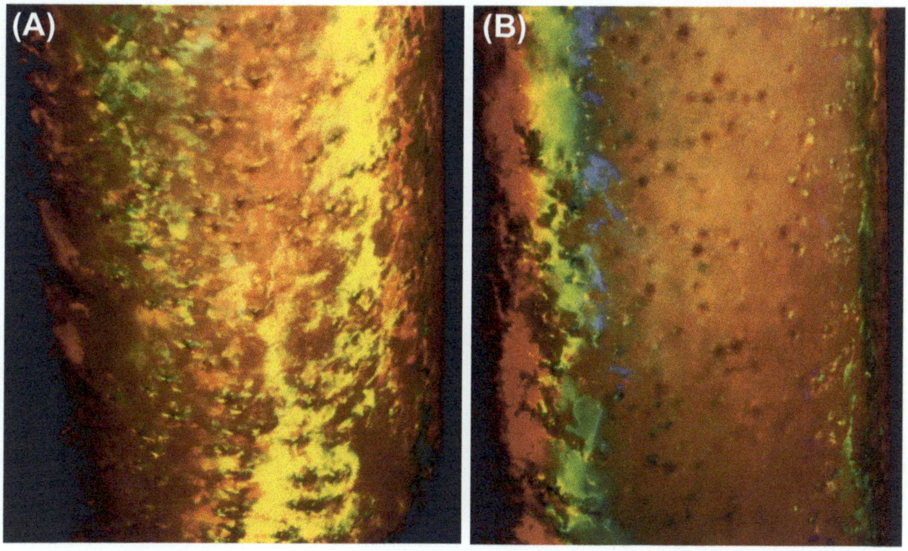

Figure 3.1 Close-up color pictures of colloidal crystals in a test tube at 25° C. Polystyrene spheres (D1B28), ϕ (volume fraction) = 0.088.

Figure 3.2 Close-up color pictures of colloidal crystals in a vial at 25° C. From the left of the back row: poly(methyl methacrylate) spheres (PMA31, ϕ = 0.13), PMA32 (ϕ = 0.043), PMA33 (ϕ = 0.013), and PMA34 (ϕ = 0.0013). From the left of the front row: PMA41 (ϕ = 0.12) and PMA42 (ϕ = 0.040).

light angle kept constant the whole crystal having the same crystal planes. In a test tube (Fig. 3.4), on the other hand, colors changed even within a single crystal according to the change in the incident light angles coming through the curved wall of the test tube. It should be noted that the iridescent colors originated from the Bragg diffraction condition of light

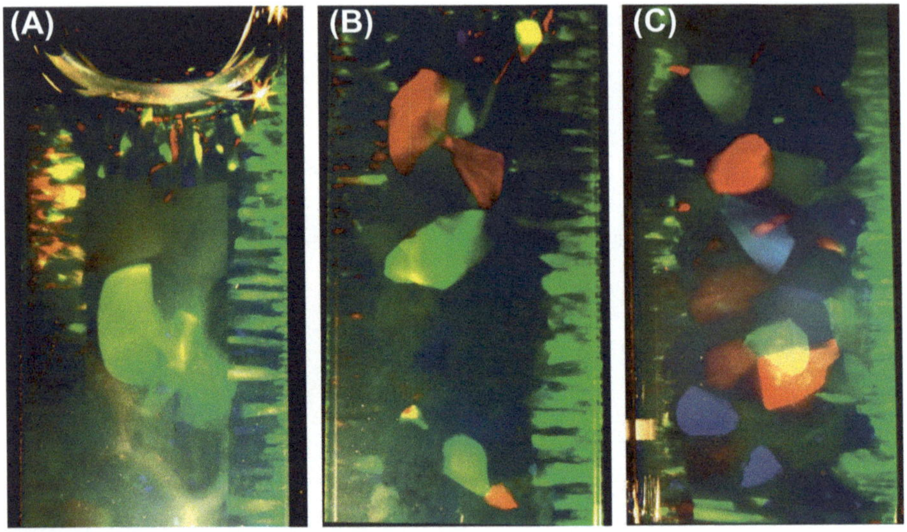

Figure 3.3 Close-up color pictures of colloidal crystals in a plane cell at 25° C. Colloidal silica spheres (CS81), $\phi = 0.00095$ (*A, B*), 0.00122 (*C*). With resins 7 days after suspension preparation and 11 minutes (*A*) and 20 minutes (*B, C*) after inverted mixing.

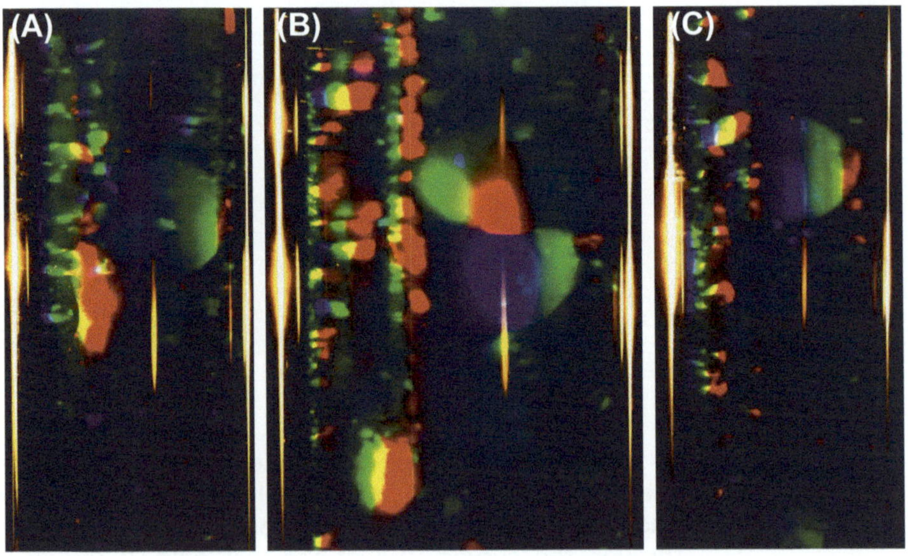

Figure 3.4 Close-up color pictures of colloidal crystals in a test tube at 25° C. Colloidal silica spheres (CS91). **A,** $\phi = 0.00060$; **B,** 0.00068; **C,** 0.00083.

and lattice constant (interparticle distance of crystal). Furthermore, single crystal size was highly polydispersed, especially for blocklike crystals from the homogeneous nucleation mechanisms, as is shown in Figs. 3.3 and 3.4. There are several possible causes for the high polydispersity in size. One is the fact that the nucleation processes prior to the crystal growth process do not proceed simultaneously. In the exhaustively deionized state, the number of the nuclei formed are few and they are localized; they would be influenced greatly by the interaction with the neighboring crystals previously formed from the heterogeneous and homogeneous nucleation mechanisms.

In this chapter, general features of colloidal particles and their crystallization in terms of their structural, kinetic, and dynamic properties such as morphologic characteristics, crystallization kinetics, physicochemical properties, and electrooptics phenomena especially in exhaustively deionized aqueous media are reviewed and discussed based on the careful experiments.

3.2 Why Colloidal Crystals Form

Many researchers have clarified that the colloidal crystals form by Brownian movement of colloidal size of particles and by the interparticle repulsion according to the principle of maximizing free volume (or maximizing packing density) accompanied with entropy gain.[11,20,25-31] In other words, particles form crystal-like distribution spontaneously with the help of Brownian thermal movement of particles by maximizing packing density.[40] Hales proved Kepler's proposal that the volume density of the most dense three-dimensional (3D) packing of sphere, face-centered cubic (*fcc*) is $\pi/18^{1/2}$ ($= 0.74048\ldots$).[41,42] The volume density of the most densely two-dimensional (2D) packing (hexagonal structure) was found to be $\pi/12^{1/2}$ ($= 0.906899\ldots$).[43,44] For 2D ellipses, it comes to be very difficult to calculate the volume density of most densely packed sphere particles. Densities of the seven kinds of packing have been calculated.[45] The density of 3D ellipsoids is rarely discussed theoretically. It should be noted that some protein molecules are apt to form 2D close packing ordering. For example, spherical molecular chaperone and rectangular catalase molecules form hexagonal and rectangular 2D sheet structures in the solid state.[46]

Hachisu et al.[3,47] clarified that the colloidal crystallization takes place by the Kirkwood-Alder transition. When extra repulsive interaction (like electrostatic repulsion) is in effect among colloidal particles, in addition to the repulsion forces from their Brownian movement (excluded volume effect), colloidal crystallization takes place easily even at low particle concentrations. The thickness of the electrical double layers is estimated approximately with the Debye screening length, l_D, given by Equation 1. Debye length corresponds to the distance from the colloidal surface, where the electrostatic potential decreases to 1/2.718 compared with that at the slipping zone.

$$l_D = \left(4\pi e^2 n/\varepsilon k_B T\right)^{-1/2} \qquad 1$$

Here, e is the electronic charge, ε is the dielectric constant of the solvent, and n is the concentration of diffusible or free-state cations and anions in suspension. Thus, n is the sum of the concentrations of diffusible counter-ions, foreign salt, and both H^+ and OH^- from the dissociation of water. The maximum value of l_D in water is approximately 1 μm, which is estimated from Equation 1 by taking $n = 2 \times 10^{-7}$ (M) $\times N_A \times 10^{-3}$ cm^{-3}, where N_A is the Avogadro number. It should be noted that the electrical double layers form always in the colloidal dispersions of solid particles in the polar liquid.

In the deionized state, the electrical double layer is extended, and the interparticle repulsive forces prevail over a long distance, as long as micrometers, though the forces become very weak. Formation of the crystal-like ordering is explained nicely by the effective soft-sphere model. The effective diameter, d_{eff}, of spheres includes the Debye screening length, l_D, and given by the diameter plus twice the Debye length. When d_{eff} is shorter than the observed intersphere distance, l, a gaslike distribution is observed. When d_{eff} is comparable to or a bit shorter than the intersphere distance, the distribution of spheres is usually liquidlike. When the effective diameter is close to or larger than the observed intersphere distance, crystal-like ordering forms, as is shown schematically in Fig. 3.5. The effective soft-sphere model has been supported by many researchers from the systematic comparison of d_{eff} with l values.[11,20,25,28-31] We should note again that the colloidal gases, colloidal liquids, colloidal crystals, and colloidal amorphous solids are analogous to and models of the usual gases, liquids, crystals, and amorphous solids.[3,48]

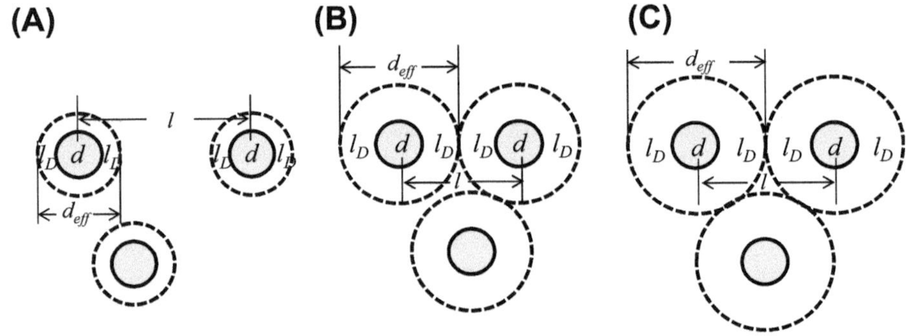

Figure 3.5
Schematic representation of colloidal gas (*A*), liquid (*B*), and crystal (*C*). The nearest-neighbored intersphere distance is indicated by l, d is the true diameter, and d_{eff} is the effective diameter including thickness of the electrical double layers (Debye screening length, l_D).

We should mention here the influence of the charge density of colloidal particles on crystallization. For the highly charged colloidal spheres, the Debye length, l_D, decreases, because the diffusive ionic concentration from the counter-ions, which contributes to the n value in Equation 1, increases. Thus colloidal crystallization takes place much more easily for the slightly charged colloidal particles than for the highly charged particles.

The lattice spacing of colloidal crystals changes exclusively according to the particle concentration, and supports definitely the importance of the electrostatic interparticle repulsive forces in addition to the excluded volume effects (i.e., repulsion induced by the Brownian movement of the particles themselves). Fig. 3.6 shows a comparison of the observed nearest-neighbored intersphere distance with the calculation from the sphere concentration. Agreement between l_{obs} and l_o is excellent in a wide range of intersphere distances and sphere concentrations. Thus Fig. 3.6 demonstrates that colloidal crystallization takes place by the intersphere repulsion exclusively, and never attraction. Itano et al.[49] have reported that the Coulomb crystals form for the trapped spherical plasma with a single sign (plus) of charges (see Fig. 2-4). This observation also supports the important role of the electrostatic repulsion forces for the crystallization in addition to the Brownian movement of the respective plasmas (excluding volume effect). Many researchers have proposed intersphere attraction forces instead of repulsion for colloidal crystallization. However, this idea has proven erroneous.[11,50] Colloidal crystallization takes place by the interparticle repulsion

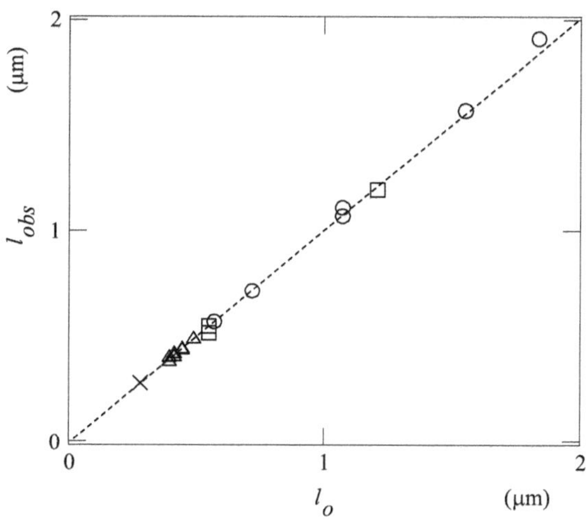

Figure 3.6
Comparison of the nearest-neighbored intersphere distances observed (l_{obs}) with the calculation (l_o) from the sphere size and concentration. *Circles, squares:* light-scattering measurements; *crosses, triangles:* reflection spectroscopy. *From Okubo T:* Polym J *40:882-890, 2008.*

exclusively, irrespective of particle concentration, which ranges from diluted to highly concentrated suspensions. We should note here again that typical crystals in nature such as ice, metals, and proteins form by attraction, not repulsion like colloidal crystals.

The repulsive interparticle interaction as opposed to attractive forces is important in colloidal crystallization. The latter is common for most crystals in nature in metals, proteins, and ice, for example. Qualitatively speaking, the elementary particles of common crystals in nature as described in Chapter 2 are small and the contribution of both the transitional and rotational diffusion constants given in Equations 2 and 3 are so large that the ordering of the strongly diffusible particles will be quite difficult without strong interparticle attractive forces.

$$D_t = k_B T/(3\eta\pi d) \qquad 2$$

$$D_r = k_B T/\pi\eta d^3 \qquad 3$$

Here, η is the viscosity of the solvent. On the other hand, the diffusive forces of very large colloidal particles are now too weak to form ordering without some other interparticle repulsion forces. Without additive repulsion interactions, the colloidal particles will associate easily to each other. Keeping the balance between attractive and repulsive interelementary particle interactions is important for the ordering.

3.3 Morphologic Characteristics and Phase Diagram of Colloidal Single Crystals

Figs. 3.1 through 3.4 show in close-up color photographs typical examples of single crystals of monodispersed polystyrene, PMMA, and silica spheres appearing in aqueous dispersions. The stock suspensions were deionized exhaustively by the introduction of cation-exchange and anion-exchange resins for more than 10 years. Cells were kept in a test tube and a quartz glass optical cell for fluorescence measurements. The resins coexisted in the bottom of the cell (the bottoms are cut out of the pictures). Sphere concentrations decreased from 0.13 in Fig. 3.1 to 0.00060 in volume fraction (in Figs. 3.3 and 3.4). In the diluted dispersions, very few nuclei form and the single crystals become very large. The iridescent colored colloidal single crystals can be observed with the naked eye and are beautiful. It should be noted here that the effective polydispersity of colloidal spheres decreases sharply in the deionized aqueous suspension, because the effective diameter $d + 2 \times l_D$ becomes large compared with the real diameter, d. Here l_D is given by Equation 1. The influence of the polydispersity of colloidal spheres on the critical concentration of crystal-liquid phase transition has been discussed.[51-56] The morphologic characteristics of colloidal crystal are similar as that of other typical crystals. Most metal crystals are composed of single crystals surrounded by the grain boundaries. Lattice spacing of colloidal crystals, which was studied with reflection spectroscopy in detail,[28-31,33] was very long, several thousand times greater compared with that of metals, and just in the range of light wavelengths. Two kinds of single crystals are observable in the diluted and deionized suspensions (e.g., see Fig. 3.3). Red, blue, yellow, and other colored block–like ones in the figure are formed from the "homogeneous nucleation mechanism" in the bulk phase far from the cell wall. Green and pillarlike crystals grow from the heterogeneous nucleation mechanism along the cell wall.

As is demonstrated in Fig. 3.7, single crystals originating from a heterogeneous nucleation mechanism grew first along the cell wall, and then large, blocklike single crystals from the homogeneous nucleation mechanism formed far from the cell wall in the exhaustively deionized and diluted suspension. When the test tube containing large single crystals was rotated slightly and very slowly, these patterns of single crystals changed rapidly, as is shown in Figs. 3.7, *F, G,* and *H.* These patterns of single crystals appear or disappear depending on whether the Bragg diffraction fits or not. Furthermore, the angles of the incident and reflection lights change by the rotation of test tube cell, and then the color of the single crystals from the

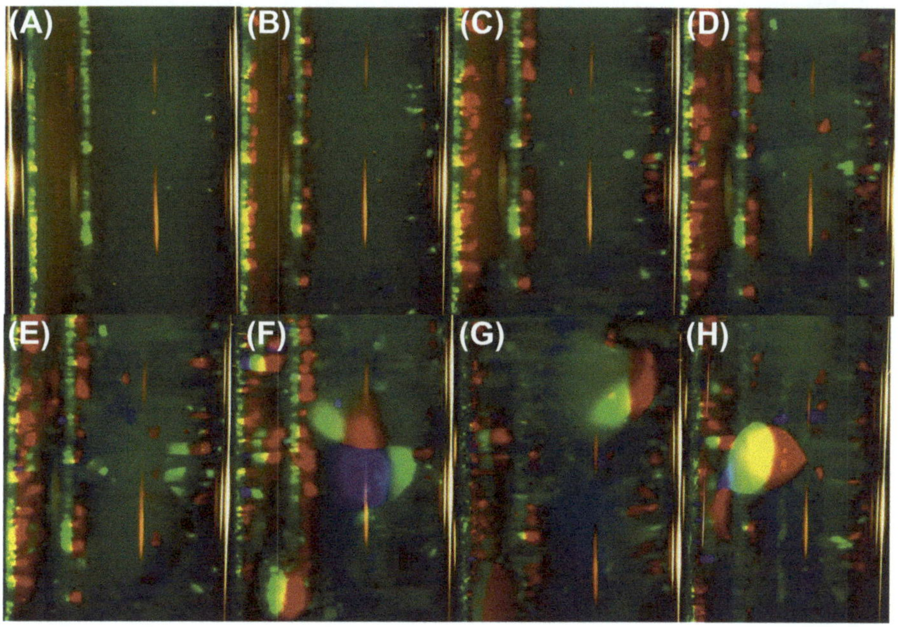

Figure 3.7 Close-up color pictures of colloidal crystallization of CS91 spheres in a test tube, $\phi = 0.000682$; **A,** 7 minutes after mixing; **B,** 15 minutes; **C,** 26 minutes; **D,** 40 minutes; **E,** 62 minutes; **F, G, H,** the cell was rotated slightly.

homogeneous nucleation changes and repeats the appearance and disappearance of the single crystals. The color changes from bluish to reddish in a single crystal are due to the change in the incident light angles through the curved wall of the test tube cell. It should be noted that each single crystal is surrounded by the grain boundaries and packed densely.[57,58] When the incident light angle changes slowly, the packed state of the single crystals is observable.[58]

The critical concentration of melting of colloidal crystals in the exhaustively deionized aqueous suspension, ϕ_c in volume fraction, is very sensitive to the "degree of deionization" of the suspensions. Table 3.2 shows the typical examples of the changes of ϕ_c as a function of time elapsed. According to my experience, the stock suspension must be deionized with the introduction of mixed beads of cation-exchange and anion-exchange resins for at least 1 month, and ideally several years. Furthermore, the sample suspensions prepared by dilution of the stock suspension with pure water must be deionized further with the introduction of the ion-exchange resins more than 3 weeks before the measurements are taken. Sphere size dependencies of the most reliable ϕ_c values for exhaustively deionized suspensions of monodisperse polystyrene spheres (*circles*),[23] colloidal silica spheres (*crosses*),[23] and fluorine-containing spheres (*triangles*)[32] are shown in Fig. 3.8. The ϕ_c values are located at approximately 0.0003 in volume fraction irrespective of sphere diameters in the range from 90 to 200 nm. The sharp increase in ϕ_c was observed, however, for spheres smaller than 80 nm. Small colloidal spheres are not definitely large enough compared with the sizes of medium molecules (water), hydronium ions, and the other simple ions for the formation of the electrical double layers.

3.3.1 Influence of Polydispersity of Colloidal Spheres

The critical concentration of melting (ϕ_c) in deionized suspension increased sharply as the polydispersity of the colloidal spheres increased. In the

Table 3.2 Change in the Critical Concentration of Melting in Volume Fraction (ϕ_c) after the Sample Suspension Preparation from the Deionized Stock Suspension and Purified Water at 25° C*

		with resins			
Sphere	**without resins**	**After 1 hour**	**5 hour**	**3 day**	**> 3 week**
D1C25	0.005	0.002	0.0023	0.0015	0.0014
D1C27	0.002	0.001	0.001	0.0004	0.0003
D1B76	0.005	0.001	——	——	0.00016
D1B72	0.025	0.0015	——	——	0.00038
CS91	——	——	0.0011	0.00096	0.0004
CS22	——	——	0.0015	0.0008	0.00028

Compiled data from Okubo T: Langmuir 10:1695-1702, 1994, Tables 2 and 3.

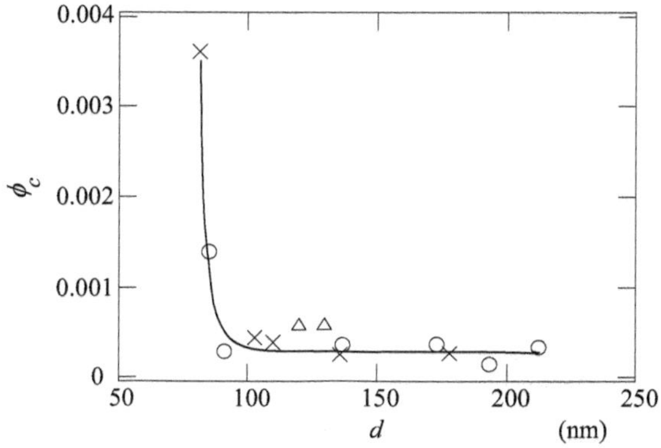

Figure 3.8
Critical concentration of melting (ϕ_c) for colloidal crystals of polystyrene spheres (*circles*), colloidal silica spheres (*crosses*), and fluorine-containing spheres (*triangles*) as a function of sphere size in diameter at 25° C.

crystallization in deionized suspensions, the effective polydispersity given by s/d_{eff} is more important than the real polydispersity of the spheres s/d. Here d_{eff} is given by $d + 2 \times l_D$. The Debye screening length is l_D. Thus s/d_{eff} values decrease with increasing degree of deionization, because the l_D increases by the progress of the deionization. For example, the s/d value of D1P30 is very high at 0.26, and crystallization does not take place at any sphere concentrations in the condition without deionization with ion-exchange resins. However, the exhaustively deionized suspension of D1P30 crystallizes at the diluted suspension state. In general, the colloidal crystallization occurs when s/d_{eff} is smaller than 0.1, which has been supported by the experiments on the critical concentration.[11,20,33,59-63] The s/d_{eff} value as a criterion for crystallization is supported by Lindemann's rule on crystal melting, which states that the fluctuation parameter must be smaller than 0.1 for crystallization.[5,64] When l_D is assumed to be 84.5 nm, s/d_{eff} of D1P30 ($s/d = 0.255$ from Table 3.1) becomes 0.1. Thus colloidal crystallization of highly polydispersed D1P30 takes place if the suspension is deionized, because the maximum value of l_D is 1000 nm in the exhaustively deionized condition and 84.5 nm of l_D is attained quite easily. In conclusion, ϕ_c values are sensitive to the polydispersity of the colloidal spheres, including the thickness of the electrical double layers. Crystallization takes place even for the highly dispersed colloidal spheres, when s/d_{eff} becomes smaller than 0.1 by the deionization of the suspension. Fig. 3.9 demonstrates the formation of large single crystals of D1P30 spheres in the exhaustively deionized suspension state, although the suspensions were highly turbid compared with D1B76 ($s/d = 0.028$). ϕ_c of D1P30 ($s/d =$ 0.255) was found to be approximately 0.0005, which was very low but approximately threefold larger than that of highly monodispersed D1B76 spheres ($s/d = 0.028$, $\phi_c = 0.00016$).

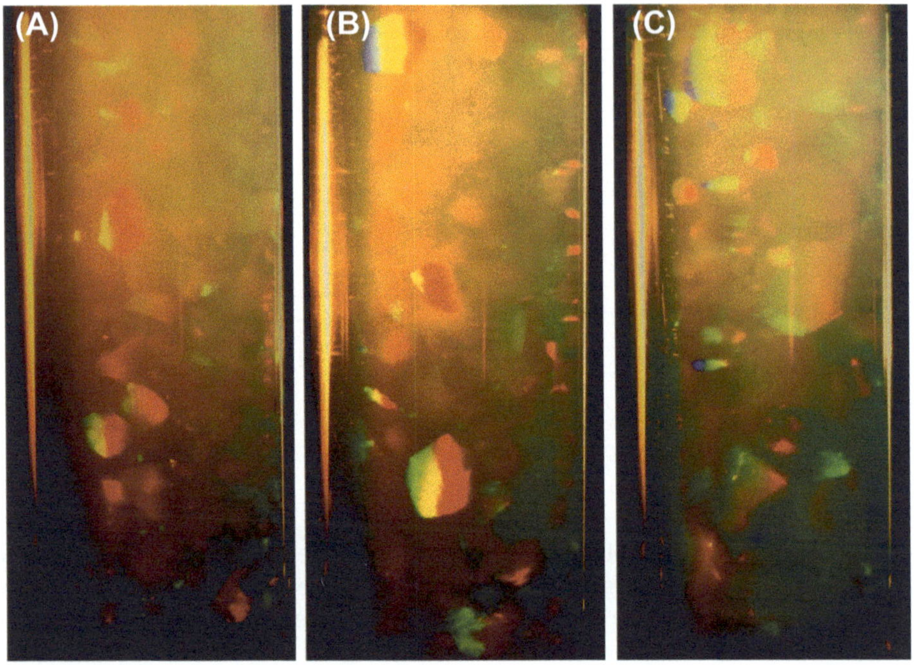

Figure 3.9 Close-up color pictures of colloidal crystals in a test tube at 25° C. Polystyrene spheres (D1P30): **A,** $\phi = 0.00088$; **B,** 0.00077; **C,** 0.00055.

3.3.2 Influence of Salt Addition

Colloidal crystallization takes place even in the presence of foreign salt when the condition for the crystallization holds true (i.e., d_{eff} is close to or larger than the observed intersphere distance l). Typical microscopic and close-up color pictures showing colloidal single crystals that appear in the presence of simple electrolytes and polyelectrolytes are given in Figs. 3.10 and 3.11, respectively. In the presence of foreign salts, the size of single crystals often becomes smaller than that of crystals in the exhaustively deionized aqueous suspension. However, when the colloidal sphere concentrations were carefully set slightly higher than the critical concentration of melting (ϕ_c), large single crystals appeared even in the presence of salts. Fig. 3.10, *A* and *B*, show two examples of the microscopic color pictures of the single crystals of CS81 spheres from the homogeneous nucleation mechanism in the presence of 2×10^{-5} M of NaCl.[65] The sphere concentration was 0.036 in volume fraction. The size of single crystals was approximately 1 mm. The difference in the iridescent colors is due to the change in the incident light angles or the direction of observation against the Bragg planes. Crystal lattice structures of each single crystal should be the same among the crystals observed in Figs. 3.10, *A* and *B*. Separation of the single crystals into flakes was often observed

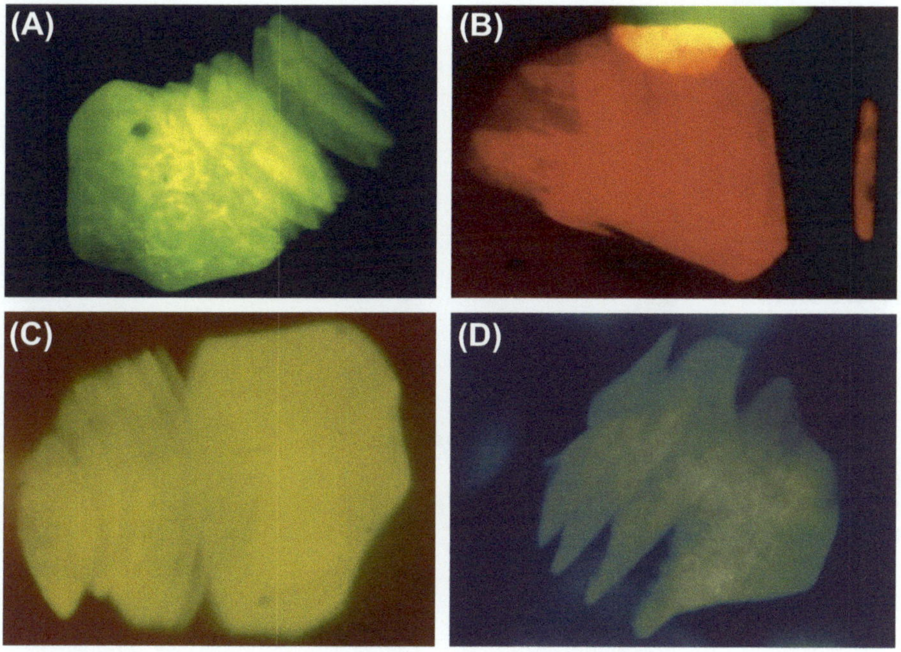

Figure 3.10 Microscopic observation of colloidal single crystals of CS81 spheres in the presence of NaCl (*A*, *B*) and NaPES (*C*, *D*). $\phi = 0.0360$; **A,** NaCl = 2×10^{-5} M, frame size = 1.2 mm × 1.76 mm; **B,** 2.00×10^{-5} M, 0.82 mm × 1.2 mm; **C,** [NaPES] = 1.70×10^{-5} monoM, 1.2mm × 1.76 mm; **D,** 2.38×10^{-5} monoM, 1.2mm × 1.76 mm. *From Okubo T, Fujita H, Kiriyama K, Yamaoka H:* Colloid Polym Sci *274:73-80, 1996.*

when the samples in a test tube were left to stand on a desk for more than 10 hours. This supports the finding that the faint shearing forces on the test tube from very weak mechanical vibration of the desk are enough to break the soft single crystals. Separation takes place often along the (1,1,1), (2,2,2) . . . planes, because these crystal planes are most dense in sphere distribution. It should be noted here that the single crystals were melted away easily and instantly into the super-cooled liquid structures, when the suspension was mixed. However, the crystallization started again when the test tube was left to stand. This kind of melting and crystallization process repeated every time. However, different morphologic characteristics in size, shape, and distributions of single crystals appeared every time. The photos shown in Figs. 3.10, *C* and *D*, show the single crystals of CS81 spheres in the presence of 1.70×10^{-5} monoM and 2.38×10^{-5} monoM of sodium polyethylene sulfonate (NaPES).[65] The sheet separation of single crystals also took place. Fig. 3.11 shows the close-up color pictures of colloidal crystals of CS81 spheres in the presence of NaCl (*A*), calcium chloride ($CaCl_2$; *B*), lanthanum (III) chloride ($LaCl_3$; *C*), polybrene (cationic ion polymer; *D*), poly-*N*-ethylpyridinium bromide (*E*), and a copolymer of *N*-*n*-cetylpyridinium bromide (5%) and *N*-benzylpyridinium chloride (95%) (*F*). It is highly interesting to observe the crystallization even combination between

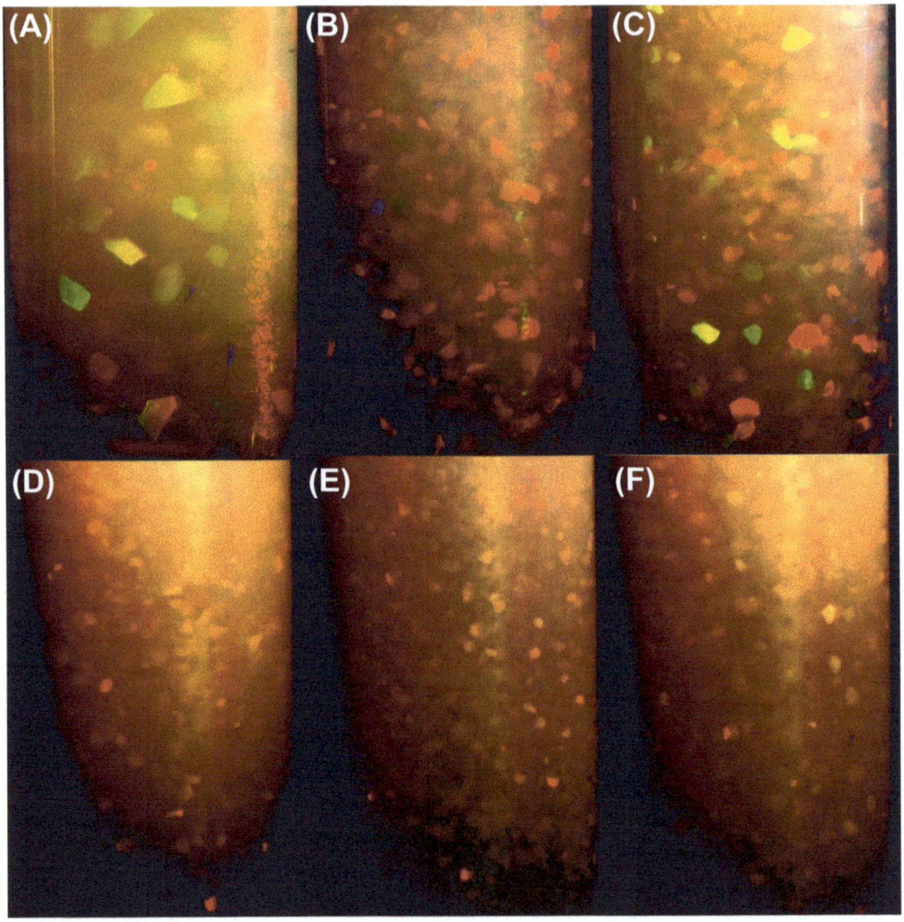

Figure 3.11 Close-up color pictures of colloidal single crystals of CS81 spheres at 25° C. **A,** CS81 (ϕ = 0.0120) + NaCl (2.86 × 10^{-6}M); **B,** CS81 (0.0240) + $CaCl_2$ (5 × 10^{-6}M); **C,** CS81 (0.0120) + $LaCl_3$ (7.14 × 10^{-7}M); **D,** CS81 (0.0360) + Polybrene (1.71 × 10^{-5} monoM); **E,** CS81 (0.0360) + C2PVP (2.4 × 10^{-5} monoM); **F,** CS81 (0.0360) + C16BzPVP (2.33 × 10^{-5} monoM) at 25° C.

anionic-charged CS81 spheres and polycations. It is highly plausible that most polycations are adsorbed tightly on the colloidal sphere surfaces in the colloidal crystal structures and charge of the colloidal silica spheres are inverted from anionic to cationic.

Fig. 3.12 shows a typical example of the influence of foreign salt on the ϕ_c values for the colloidal crystals of silica spheres CS81. Clearly, deionization of the suspension sharply decreases ϕ_c value by one tenth. On the other hand, ϕ_c values increase significantly by the addition of a slight amount of NaCl. These changes in ϕ_c values are consistent with the thickening and thinning of the electric double layers by the deionization and by salt addition, respectively.

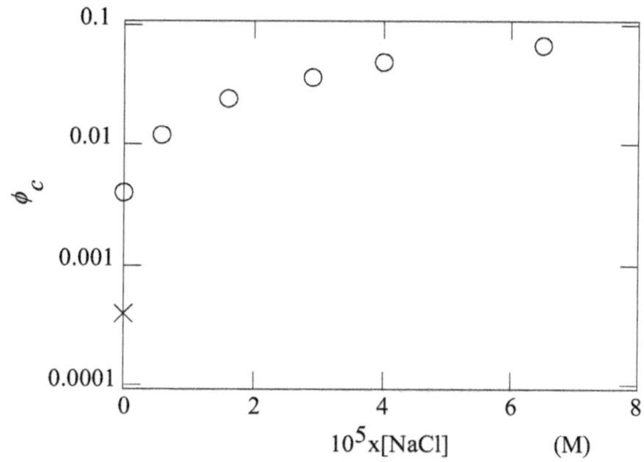

Figure 3.12 Critical concentration of melting (ϕ_c) of colloidal crystals of CS81 spheres in the presence of NaCl (*circles*) and in the exhaustively deionized suspension (*cross*) at 25° C.

The "critical salt concentrations of melting" at the same ϕ_c values of colloidal crystals were in the order[65]:

$$LaCl_3 < CaCl_2 < PB < NaCl < C2PVP \leqq C16BzPVP \leqq CTAB < Na_2SO_4 < NaPES < SDS \quad 4$$

Here, PB denotes polybrene, C2PVP is poly-*N*-ethylpyridinium bromide, C16BzPVP is a co-polymer of *N*-benzylpyridinium chloride and *N*-hexadecylpyridinium bromide, CTAB is *n*-cetyltrimethylammonium bromide, NaPES is sodium polyethylene sulfonate, and SDS is sodium dodecyl sulfate. It should be noted that the order of the critical sphere concentration of melting (ϕ_c) is in the opposite order of Equation 4 (i.e., the ϕ_c is largest in the presence of $LaCl_3$ and smallest with SDS, as is shown in Equation 5). The order of Equation 5 teaches us that the colloidal crystals are broken most effectively by the addition of $LaCl_3$, whereas SDS is weakest. Thus the order of Equation 5 shows clearly the magnitude of strength of the melting power.

$$LaCl_3 > CaCl_2 > PB > NaCl > C2PVP \geqq C16BzPVP \geqq CTAB > Na_2SO_4 > NaPES > SDS \quad 5$$

The relation of Equation 5 is simplified in the order of Equation 6.

$$\text{Simple cations} > \text{polycations} > \text{simple anions} > \text{polyanions} \quad 6$$

When the simple multiple cations such as $LaCl_3$ and $CaCl_2$ are added into the colloidal crystal suspension, the release of the hydronium ions, which originate from the colloidal anionic charges, should occur by the counter-ion binding effect and result in the melting of the crystals by the thinning of the double layers. The weak release of hydronium ions by the addition of the polycations are ascertained experimentally.[65] The effect of the addition of polyanions to the anionic colloidal crystals will be weak because the strong repulsion works between polyanions and colloidal polyanions.

3.3.3 Influence of Organic Solvents

Fig. 3.13 shows close-up color pictures of typical examples of colloidal crystals of CS91 spheres in the organic solvents.[24] Crystals in 100% methanol, ethanol, and ethylene glycol are shown in pictures Figs. 3.13, *A, B,* and *C,* respectively. Figs. 3.13, *D* and *E,* are the aqueous mixtures of 40% methanol and 40% ethanol, respectively. Figs. 3.13, *F* and *G,* are the 40% propanol and 20% *n*-butanol aqueous mixtures. All the suspensions were coexisted with the ion-exchange resins. In pure methanol and ethanol, the colloidal crystals are turbid and have small sizes (≈0.2-1 mm), but strongly brilliant single crystals appeared. In ethylene glycol, basic iridescent colors were strongly bluish. Even in aqueous mixtures of *n*-propanol and *n*-butanol, the strong iridescent-colored single crystals were recognized with the naked eyes, as shown in Fig. 3.13, *F* and *G.*

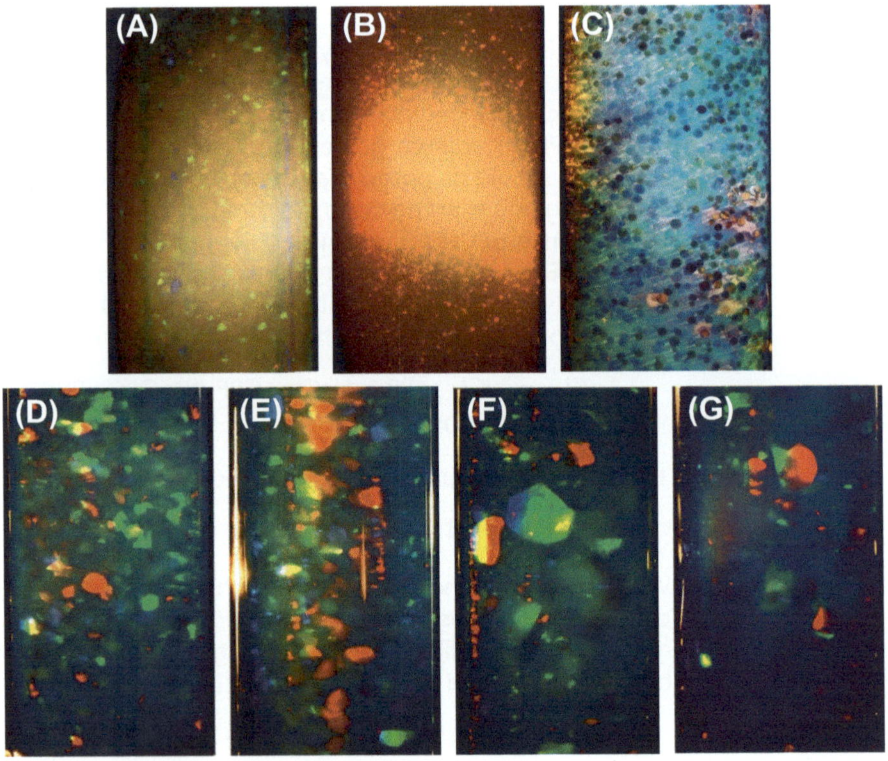

Figure 3.13 Colloidal crystals of CS91 spheres in organic solvents and in their aqueous mixtures at 25° C. **A,** 100% methyl alcohol, $\phi = 0.0177$; **B,** 100% ethyl alcohol, $\phi = 0.0382$; **C,** 100% ethylene glycol, $\phi = 0.135$; **D,** 40% methyl alcohol + 60% water, $\phi = 0.00151$; **E,** 40% ethyl alcohol + 60% water, $\phi = 0.00189$; **F,** 20% propyl alcohol + 80% water, $\phi = 0.000984$; **G,** 5% *n*-butyl alcohol + 95% water, $\phi = 0.00076$.

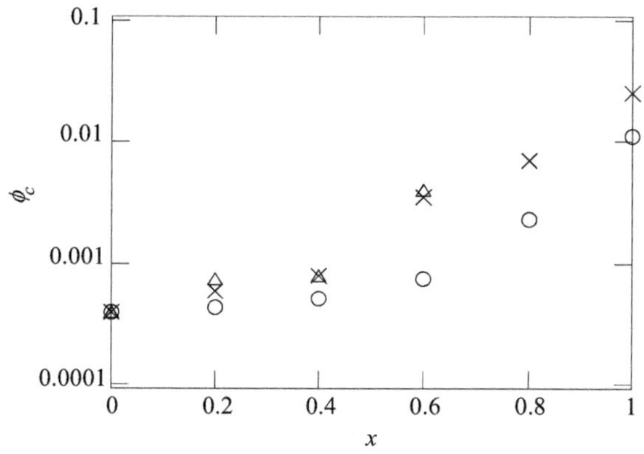

Figure 3.14
Critical concentration of melting (ϕ_c) of colloidal crystals of CS91 spheres in aqueous organic solvent mixtures at 25° C. Volume fraction of organic solvent (x), methyl alcohol (*circles*), ethyl alcohol (*crosses*), *n*-propyl alcohol (*triangles*).

Fig. 3.14 shows the typical examples of ϕ_C values of CS91 spheres as a function of the ratio of the organic solvents (methanol, ethanol and *n*-propanol) in aqueous mixtures.[24] With increasing volume fraction of the organic solvent (x), ϕ_C values increased sharply in general. Even in the purely dehydrated methyl alcohol and ethyl alcohol, colloidal crystallization took place at the low concentrations of colloidal spheres, approximately 1 to 2 vol%, when the suspensions were deionized exhaustively with the coexistence of the ion-exchange resins. Colloidal crystallization takes place even in purely organic solvents, especially when they are polar. It should be noted here that the ϕ_C values increased sharply as the fraction of organic solvent increased. Furthermore, the order of Equation 7 in the ϕ_C values seems to hold.

$$\text{Water} << \text{methanol} < \text{ethanol} \leqq \text{propanol} \approx \text{n-butanol} \quad 7$$

Note that the reverse order is obtained for the dielectric constant of the solvents (ε).

$$\text{Water } (\varepsilon = 79 \text{ at } 25°\text{C}) >> \text{methanol}(32) > \text{ethanol}(24) > \text{propanol}(20) > \text{n-butanol}(18) \quad 8$$

As is clear in Equation 1, l_D values are given by Equation 9.

$$l_D = \text{const.} \left(\varepsilon^{1/2} / n^{1/2}\right) \quad 9$$

The fraction (β) of the effectively dissociated ions in organic solvent mixtures differs from that in water. However, we neglect here the change in the n values. Then Equations 10 and 11 are obtained.

$$l_D = b\varepsilon^{1/2} \quad 10$$

$$\phi_c^{-1/3} \propto l \approx d_{eff} = a + b\varepsilon^{1/2} \quad 11$$

Here, a and b are the constants, and l is the intersphere distance. The plots of $\phi_c^{-1/3}$ against $\varepsilon^{1/2}$ display excellent linear relationship for all the data, when the other factors such as the solvation and change in β values are taken into account.[24] Thus the phase diagram of colloidal crystals in the alcoholic aqueous mixtures are explained by the difference in the dielectric constants of the dispersion media.

The ϕ_c values of CS121 without resins were between 0.02 and 0.08, as is shown in Table 3.3. When the resins coexist in methanol and ethanol, ϕ_c values were clearly lowered to 0.011 and 0.025, respectively, as is shown in Fig. 3.14. When the dielectric constants of the organic solvents are higher than approximately 20, colloidal crystallization of CS121 spheres are clearly contributed by the two factors, the extended electrical double layers and the excluded volume effects.[66] Crystallization of polystyrene spheres in *N, N*-dimethyl formamide (dielectric constant = 36.7) and *N, N*-dimethylsulfoxide (46.1) was observed at approximately 0.1 in volume fraction. Table 3.3 supports the fact that colloidal crystals in the organic solvents form with the help of the formation of the electric double layers in addition to the excluded volume effects. The exceptional organic solvents were toluene, benzene, and carbon tetrachloride in Table 3.3, where ϕ_c is larger than 0.5 and the contribution from the double layers are not supported (i.e., the crystallization should take place only by the excluded-volume effects, assisted by the Brownian movements of colloidal spheres).

3.3.4 Phase Diagram and Size of Single Crystals

The size of single crystals increased quite sharply when sphere concentration decreased in the exhaustively deionized suspensions.[11,20,23] One of the typical examples of the mean size of the single crystals against sphere concentration is shown in Fig. 3.15 for D1B76 spheres. This observation is understandable, because crystal size should increase with a decreasing number of nuclei, and also with decreasing nucleation rate. Nucleation rate, of course, decreases as sphere concentration decreases. The nucleation rate (J) for the homogeneous nucleation mechanism is given by Equation 12.

$$J = K \exp(\Delta G/k_B T) \quad 12$$

Table 3.3 ϕ_c Values of Colloidal Crystals in Organic Solvents at 25° C without Ion-Exchange Resins

Solvent	Dielectric Constant	CS121[*]	CS81[†]	PS[−‡]	PS[+§]
DMSO	46.6			<0.1	—
Acrylonitrile	≈38	0.021±0.001	—	—	—
Acetonitrile	37.5	0.021±0.001	—	—	—
DMF	36.7	0.070±0.003	—	<0.10	>0.10
Methanol	32.6	0.045±0.002	0.011	—	—
Benzonitrile	26.5	—	—	—	0.125
Ethanol	24.6	0.074±0.003	0.025	—	—
Acetone	20.7	0.035±0.001	—	—	—
Isopropyl alcohol	19.9	0.084±0.003	—	—	—
Cyclohexane	18.3	—	—	—	0.14
Benzaldehyde	17.8	—	—	—	0.16
Acetophenon	17.4	—	—	—	0.14
3-Pentanone	17.0	—	—	—	0.17
Benzyl alcohol	13.1	—	—	0.31	0.31
m-Cresol	11.8	—	—	0.30	—
Methylbenzoate	6.59	—	—	0.20	0.21
Ethylacetate	6.02	—	—	—	0.33
Chloroform	4.81	—	—	—	0.28
Anisole	4.76	—	—	—	0.185
Toluene	2.38	—	—	>0.50	—
Benzene	2.28	—	—	>0.50	>0.50
Carbon tetrachloride	2.24	—	—	>0.50	—
1,4-Dioxane	2.21	—	—	0.17	0.185

[*]*From Yoshinaga K, Chiyoda M, Ishiki H, Okubo T: Colloids Surf A 204:285-293, 2002.*
[†]*From Okubo T:* Langmuir *10:3529-3535, 1994.*
[‡]*Anionic polystyrene spheres, from Hiltner PA, Papir YS, Krieger IM:* J Phys Chem *75:1881-1886, 1971.*
[§]*Cationic polystyrene spheres, from Hiltner PA, Papir YS, Krieger IM:* J Phys Chem *75:1881-1886, 1971.*
DMF, Dimethyl formamide; *DMSO,* dimethylsufloxide.

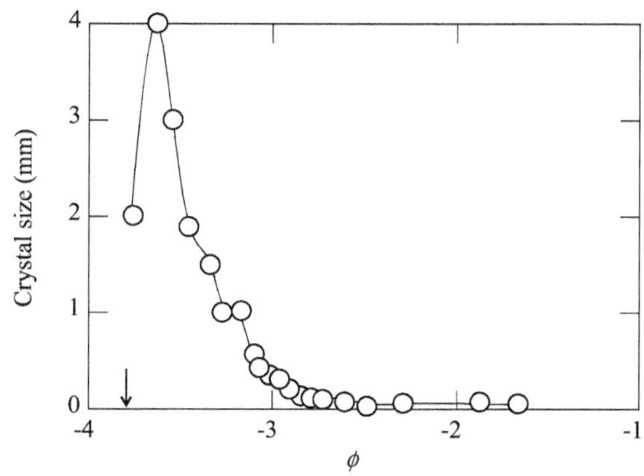

Figure 3.15 Mean single crystal sizes as a function of sphere concentration. Polystyrene spheres (D1B76). 109 nm in diameter, critical concentration of melting: 0.00016 (ϕ_c). *From Okubo T:* Prog Polym Sci *18:481-517, 1993.*

Here, K is the kinetic coefficient, ΔG is the critical activation (Gibbs) free-energy for nucleation. ΔG is given by Equation 13.[67]

$$\Delta G \propto (\ln r)^{-2} \qquad 13$$

Here, r is ϕ/ϕ_C, and ϕ_C is the critical concentration of melting. Both Equations 12 and 13 support the observation shown in Fig. 3.15: The size of single crystals decrease as $\phi - \phi_C$ increases, because the size should increase with decreasing nucleation rates. When the sphere concentration is comparatively high, the size of colloidal crystals is too small to be recognized with the naked eye. Monovoukas and Gast[68] observed colloidal single crystals in detail by using polarizing microscopy.

3.3.5 Melting Temperature

Williams et al.[4] have measured melting temperature, T_m, of colloidal crystals for the first time as a function of sphere concentration. I[18,69-71] measured melting temperatures of exhaustively deionized colloidal crystals. Fig. 3.16 shows three sets of melting temperatures of polystyrene spheres (109 nm in diameter) measured by Williams et al.[4] (shown by circles, reported in 1976), by us[18] (crosses in 1991), and by us[71] again for exhaustively deionized samples (triangles in 1994). The sphere samples of the last two papers are mono-dispersed polystyrene spheres purchased from Dow Chemical Company. Williams et al. also reported that the samples had the same sphere size, and were obtained from the same company. Thus the three samples are believed to be the same. Clearly, the melting temperatures increased significantly with years reported, especially in the range of low concentrations of the spheres. This increase in T_m values is undoubtedly the result of the improvement of the deionization of the sample suspensions.

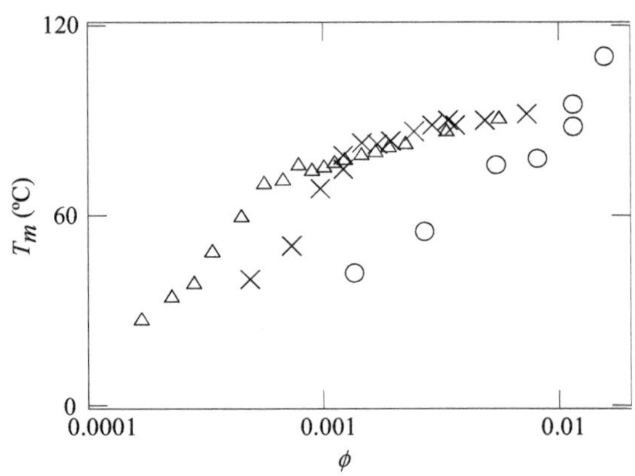

Figure 3.16 Melting temperatures of polystyrene spheres (109 nm in diameter) as a function of sphere concentration. Williams et al.[4] (*circles*); Okubo[18] (*crosses*); Okubo[71] (*triangles*).

Following the theoretical discussion by Williams et al.,[4] T_m of colloidal crystals is given by Equation 14:

$$T_m = C\phi^{1/3}\exp\left(-A\phi^{1/6}\right)\left\{\phi^{-1}\ln\phi^{-1}-\left(\phi^{-1}-1\right)\ln\left(\phi^{-1}-1\right)\right\}^{-1} \quad 14$$

Here, ϕ is the concentration of spheres in volume fraction. C and A are the constants. Equation 14 was fitted to the experimental T_m data by adjusting the values of C and A. Agreement between the observation and theory was excellent. The constants C and A thus estimated are compiled in Table 3.4. Both C and A increased substantially with the year reported proceeds. The ϕ_{350} values were almost insensitive to the sphere size between 90 and 190 nm, and ϕ_{350} values of D1C25 and CS61 seem to be twofold larger than those of the other spheres examined. Here ϕ_{350} means the sphere concentration in volume fraction where T_m becomes 350° K.

Table 3.4 Parameters ϕ_{350}, C, and A on the Melting of Typical Colloidal Crystals

Spheres	d (nm)	ϕ_{350}	$10^{-3}C$	A
Polystyrene Sphere				
D1C25*	85	0.004	156	6.0
D1C25†b	85	0.0034	117	5.2
D1C27*	91	0.002	170	6.1
D1C27†	91	0.001	283	7.4
Williams et al.‡	109	0.005	160	6.0
D1B76*	109	0.0013	211	6.6
D1B76†	109	0.001	300	7.6
D1P30*	109	0.003	207	6.7
D1P30†	109	0.0017	242	7.0
D1K88*	137	0.0025	195	6.5
D1K88†	137	0.0022	257	7.2
D1B72*	173	0.003	172	6.2
D1B72†	173	0.002	241	7.0
VN1001†	192	0.0012	274	7.3
D1B28*	212	0.004	184	6.4
Colloidal Silica Sphere				
CS61†	81.2	0.0035	185	6.4
CS81†	103	0.0018	242	7.0
CS91†	110	0.0014	245	7.0
CS121†	136	0.0017	260	7.2
CS22†	178	0.00083	300	7.5
CS161†	184	0.0026	228	6.9

From Okubo T: J Chem Phys *95: 3690-3697, 1991.*

†*From Okubo T:* Colloid Polymer Sci *270:1018-1026, 1992.*

‡*From Williams R, Crandall RS, Wojtowicz PJ:* Phys Rev Lett *37:348-351, 1976.*

ϕ_{350}: Colloidal concentration in volume fraction where T_m becomes 350° K.

The enthalpy (heat) of melting, ΔH, and entropy of melting, ΔS, are given by Equations 15 and 16, respectively.[4]

$$\Delta H = Nk_BC\phi^{1/3}\exp\left(-A\phi^{1/6}\right) \tag{15}$$

$$\Delta S = Nk_B\left\{\phi^{-1}\ln\phi^{-1} - \left(\phi^{-1}-1\right)\ln\left(\phi^{-1}-1\right)\right\}, \tag{16}$$

where N is the number density of spheres per cm^3, and k_B is Boltzmann's constant. The kind of spheres has no influence on the absolute values of ΔH. In most cases, ΔH decreased slightly from 23 to 20 kJ/mol as ϕ increased from 0.001 to 0.004. The effect of sphere size was also insignificant, except for very small spheres (e.g., D1C25 and CS61). ΔS decreased uniformly in the range of 80 to 40 J/mol K as ϕ increased from 0.001 to 0.02. ΔS values were evaluated to be approximately 20-17 J/mol K. It should be noted here that the colloidal crystallization is accompanied with the decrease in both ΔH and ΔS terms, similar to those of usual crystals. In other words, colloidal crystallization is driven by enthalpy, not by entropy. Colloidal crystallization takes place as the result of some kind of intersphere attraction forces, not only by the intersphere repulsion forces, as may be assumed at first glance. Here we should note that colloidal crystallization always takes place in a closed vessel. Even in an air–suspension interface, the crystals are closed by the interfacial tension of the medium phases. In this closed condition, intersphere repulsion is balanced by the sphere-cell wall repulsion forces. Thus intersphere repulsion is apparently transmitted to the effective intersphere attraction with the help of thermal motion of colloidal spheres. However, of course, the original intersphere forces are repulsion forces exclusively. In summary, the effective intersphere attraction is induced from the intersphere repulsion in a closed vessel with the help of Brownian translational movement of each spheres. Interestingly, ΔS is independent from parameters C and A, and also from the degree of deionization of suspensions. The T_m values were compared with the theory of Robbins et al.,[72] which treats the repulsive Yukawa potential between colloidal spheres. The melting temperature of colloidal silica spheres in aqueous ethanol and aqueous ethylene glycol mixtures were measured by using reflection spectroscopy.[70] The T_m data were analyzed successfully with the theory of Williams et al.[4] T_m decreased by the addition of ethanol or ethylene glycol.

It should be mentioned here that colloidal crystallization in the hard systems has been found to be driven by an increase in entropy.[73-75] As was described in Section 3.2, colloidal crystallization takes place simply by the maximizing free volume. The maximizing packing effects correspond to increase in entropy compared with the entropy of the disordered (liquid) structure, because the local movements of spheres in the crystal state are more free than that of liquid. Thus, for the soft systems of ionic spheres, the enthalpic loss induced by the electrostatic interactions must be considered in addition to the entropy-driven effects.

3.3.6 Blinking Phenomena of Single Crystals

"Blinking" of small, single crystals (or crystallites) was observed for colloidal suspensions very close to the critical concentration of melting, especially in the presence of a small amount of foreign salt or aqueous alcohol mixtures.[76] The blinking of the crystallites continued for more than 24 hours, and resembled the blinking lights on a Christmas tree. At one moment, small crystallites appeared, but in the next moment they disappeared, and new crystallites appeared in other places. The lifespan of the crystallites was several seconds at room temperature, and became shorter with rising temperature. The main cause for the blinking is ascribed to the rotational and translational thermal movement of colloidal single crystals (crystallites), which are surrounded in phase equilibrium with a liquid structure of suspension. Reflection spectra exhibiting the blinking were taken.[76] Every measurement gave slightly different peak profiles; peak wavelengths, intensities of the peaks, and the ratios of reflection intensities from *fcc* to body-centered cubic (*bcc*) structures changed as time elapsed. This is explained by the fact that the locations of the crystallites in suspension change with time, and the sizes or shapes of crystallites are also different depending on the location. It is highly plausible that the character of the Bragg layers of the crystallites formed near the cell wall, where the reflected light is detected, changes with time. The suspension systems exhibiting the blinking are compiled in Table 3.5.[76] Note that *blinking* is different than *flickering*. The crystallites in the stable colloidal crystal suspension often flicker like twinkling stars in the night sky. This occurs because the crystallites are easily distorted or even broken temporarily by the very weak shearing forces in suspension induced by the faint vibration of a desk, for example.

The crystalline sediments of monodisperse silica spheres stearylated formed by gravity.[77] Phase diagram and dynamics of colloidal crystals and colloidal liquids have been discussed experimentally and theoretically by many researchers in detail.[10,16,78-84]

3.4 Lattice Structure of Colloidal Crystals

In general, lattice structures of colloidal crystals of spherical particles of a single component are *fcc* and *bcc* lattices. The other lattices, however, were observed when two plates of a cell were oriented in the wedge geometry.[7] From much data on the colloidal lattice structures, it is concluded that the *fcc* lattices are most stable and they transform into *bcc* lattices when (1)

Table 3.5 Suspensions of CS91 Spheres Exhibiting Blinking*,†

Experiment Number	Conditions	ϕ	T_m (°C)	Temperature Range (°C)
1	With resins	0.00232	60	≈ 11-55
2	Without resins	0.0435	27	≈ 10-27
3	Without resins	0.0464	92.5	≈ 11-91
4	NaCl (2.5×10^{-5}M)	0.0841	92	≈ 10-90
5	EtOH (30 vol%), with resins	0.00725	69	≈ 60-67
6	EtOH (30 vol%), with resins	0.0087	76.2	≈ 60-67
7	EtOH (60 vol%), with resins	0.0131	46	≈ 10-45
8	EtOH (60 vol%), with resins	0.0145	75	≈ 14-74.5
9	EtOH (60 vol%), with resins	0.0174	70.5	≈ 10-69
10	EtOH (60 vol%), with resins	0.0377	92	≈ 87-92
11	EG (30 vol%), with resins	0.00551	31.5	≈ 10-30
12	EG (30 vol%), with resins	0.00638	55	≈ 52-55
13	EG (60 vol%), with resins	0.0102	60	≈ 50-60

**From Okubo T:* J Colloid Interf Sci *153:587-591, 1992.*
†The observation was made within 12 hours after suspension preparation.
EG, ethylene glycol; *EtOH,* ethyl alcohol; *NaCl,* sodium chloride.

particle concentration is lowered, (2) a small amount of salt is added, and (3) suspension temperature is raised.[18,85-89] The *fcc* and *bcc* structures are assigned definitely from the static light-scattering (SLS) measurements. However, the transformation between *fcc* and *bcc* is followed conveniently from the reflection spectroscopy. Reflection intensity profiles as a function of wavelength appear as a single peak, double peaks, or a single peak with a shoulder. The two peak wavelengths are always close together, with a difference of only 1.03 in the ratio of the wavelengths, and the peaks appearing at longer and shorter wavelengths are *fcc* and *bcc* lattices, respectively. The ratios of the nearest neighbored intersphere distances, l_f, and the Bragg diffraction distance, d_f, of the *fcc* crystals against those of *bcc*, l_b and d_b, are always 1.0284, as is shown in Equation 17.[85]

$$l_f/l_b = d_f/d_b = 1.0284 \qquad 17$$

Figs. 3.17 through 3.19 show the typical examples of the transformation between *fcc* and *bcc* crystal subphases as a function of sphere concentration, salt concentration, and suspension temperature, respectively.[85,87] It should be noted here that the transformation from *bcc* to *fcc* took place with rising

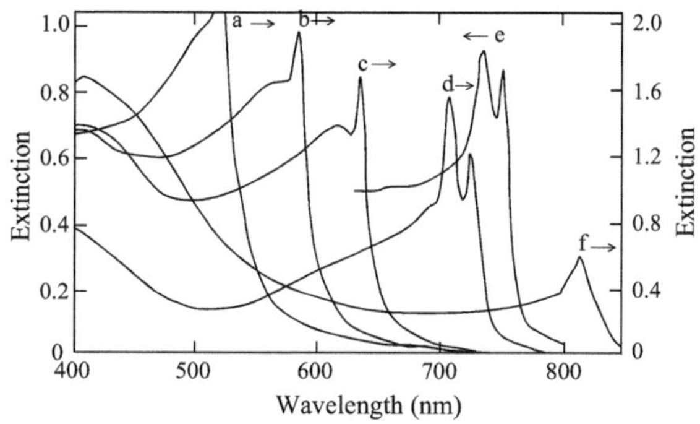

Figure 3.17
Transmission spectra of colloidal crystals of D1B72 spheres (173 nm in diameter). **A,** 0.06 in volume fraction; **B,** 0.0421; **C,** 0.0316; **D,** 0.0210; **E,** 0.0194; **F,** 0.0146. *Redrawn from Okubo T:* J Chem Soc Faraday Trans *1 82:3185-3190, 1986.*

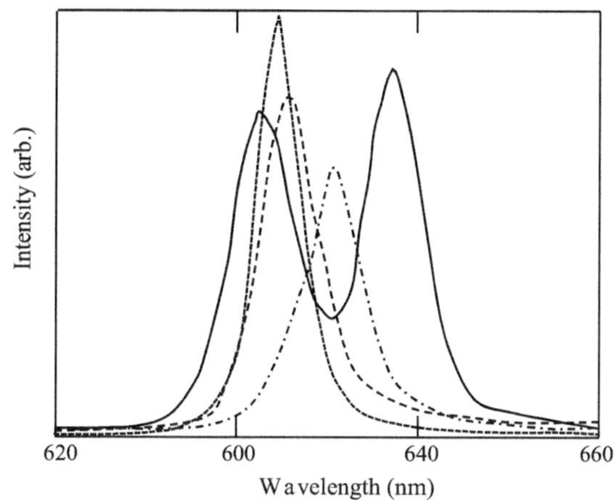

Figure 3.18
Reflection spectra of colloidal crystals of D1B22 spheres with sodium chloride. [D1B22] = 0.0194 in volume fraction; *solid curve:* [KCl] = 0 M; *dotted curve:* 3.33×10^{-6} M; *broken curve:* 6.67×10^{-6} M; *broken-dotted curve:* 1×10^{-5} M. *Redrawn from Okubo T:* J Chem Soc Faraday Trans *1 82:3163-3173, 1986.*

temperature, as opposed to the usual transformation, and has been observed only for colloidal silica spheres at the temperatures close to the crystal melting temperature.[22,69,70] For polystyrene spheres, however, the usual transformation from *fcc* to *bcc* took place near the melting temperature as the suspension temperature was raised.[18] In general, *fcc* lattice is more stable compared with *bcc*, and the *bcc* lattices often appear close to the phase transition toward liquid structures.

Transformation of colloidal crystal lattices from *fcc* to *bcc* often takes place in the elevated pressure within 2000 bar.[90] This tendency is more significant

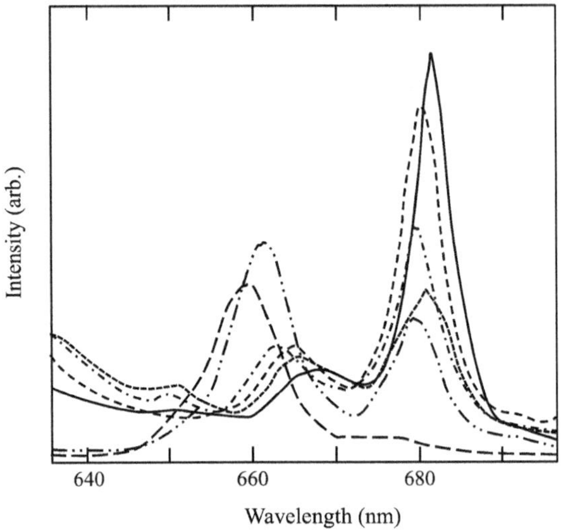

Figure 3.19
Reflection spectra of colloidal crystals of D1B22 spheres (109 nm in diameter) at 15° to 65° C. [D1B22] = 0.0158 in volume fraction; *solid curve:* 15° C; *dotted curve:* 25° C; *broken curve:* 35° C; *dot-broken curve:* 45° C; *dot-dot-broken curve:* 55° C; *long broken curve:* 65° C. *Redrawn from Okubo T:* J Chem Soc Faraday Trans *1 82:3163-3173, 1986.*

for larger spheres. At very high pressures, the melting of the *bcc* structures and the transformation from *bcc* to *fcc* rarely occurred, especially for small spheres. The influence of pressure on the crystal lattice structures is explained by the increase in the effective charge number of spheres and the thinning of the Debye screening length as pressure increases. It should be mentioned here that the (100) plane of *fcc* lattice looks rectangular and similar to the patterns of (110) plane of *bcc* lattice when the colloidal distribution is observed through a microscope in a sedimentation equilibrium.

The lattice spacing of crystal structures of polystyrene spheres at the cell interfaces decreased by approximately 4% to 18% from the mean spacing of the bulk suspension.[91] The lattice spacing at the same sphere concentration was in the order expressed in Equation 18:

$$\text{Quartz glass} > n\text{-hexane} \geqq \text{polystyrene} \approx \text{PMMA} \approx \text{air} \qquad 18$$

This order is explained reasonably by the weak adsorption effect of the spheres around the interfaces of the cell substances induced by the hydrophobic or dipole–dipole interactions. For quartz glass, furthermore, release of the ionic species from the cell wall into the suspension phase is still not neglected, which induces the shortening of the Debye screening length. It should be noted that release of the ionic impurities from the usual glass wall is substantial. Interestingly, the transformation from *fcc* structures to *bcc* lattices in the central area of a rotating disk was proposed in the centrifugal equilibrium state of colloidal crystals.[89] The most important contribution for the *fcc–bcc* transformation is undoubtedly the extended electrical double layers. The charge density of colloidal particles is also an important factor for the transition in relation to the electrical double layers; in other words, a higher charge density results in a higher concentration of the diffusible simple ions and thinner double layers. Coexistence of *fcc* and *bcc* crystal lattices is highly related to the increased free space for particles forming lattice structures (i.e., the *fcc* structure is more dense than the *bcc*). The *fcc* and *bcc* structures are observable visually by a metallurgical microscope, especially in sedimentation equilibrium, and the interparticle distances can be determined from the microscopic observation. Amorphous-solid structures are observed for highly polydispersed colloids, especially in concentrated suspensions. The cationic, charged colloidal crystals form, although the spheres are not so stable chemically in the deionized condition.[92] Furthermore, core shell–type spheres form colloidal crystals in deionized aqueous media.[93,94]

Fig. 3.20 shows close-up color pictures exhibiting the typical hexagonal patterns of diffraction spots for the deionized suspensions of N200 spheres (220 nm in diameter).[57] The suspensions at high sphere concentrations were prepared ($\phi = 0.049$ and 0.098) and deionized with the bag of ion-exchange resins in the cells for the ultramicroscopic observation using a cover glass plate (≈0.12-0.17 mm in thickness) at one end. The hexagonal patterns are

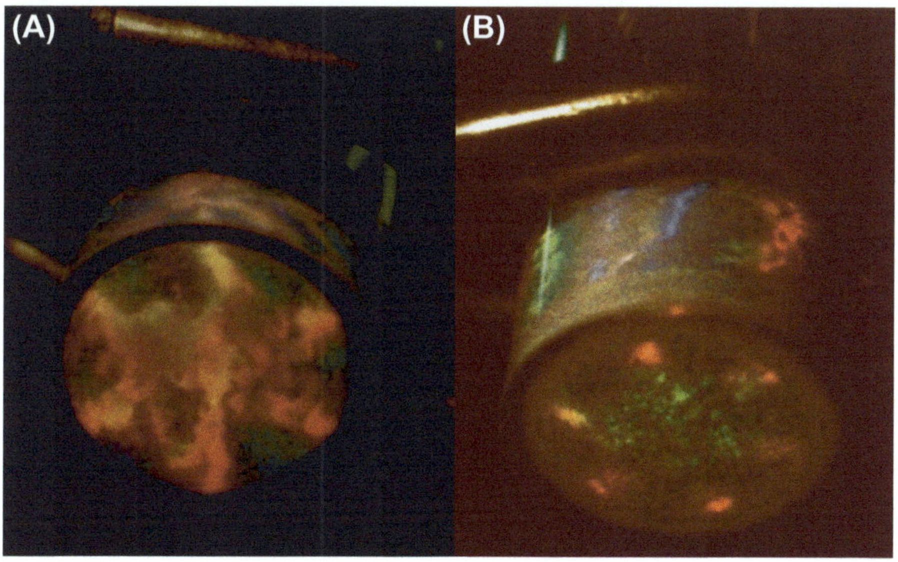

Figure 3.20 Close-up color pictures of N200 spheres at 25° C. **A,** $\phi = 0.0489$; **B,** 0.0978.

originated from arrays of the (111) planes of *fcc* crystals along the cover glass. The similar spot patterns were observed for other colloidal spheres such as D1B33 (269 nm in diameter) and C301 spheres (300 nm), for example. Tomita et al.[95] clarified that size and shape of the spots depend on the sphere size and concentration.

Several types of asymmetrical colloidal crystals have been studied, especially in sedimentation equilibrium in deionized suspension. Cylindrical colloids of nickel dimethylglyoxime (9.6 μm in length, 0.12 μm in diameter) were ordered in a hexagonal array[96] (see Fig. 2-44, *C*). It is now clear that the particles are oriented to reduce the dead space as effectively as possible. Bowl-shaped polystyrene particles were also oriented in a deionized suspension (see Figs. 2-23 and 2-45, *C*).[97-99] Interestingly, other colloidal spheres were observed in the cavities of bowl-shaped particles (see Fig. 2-45, *D*). Both the host and guest particles were not in direct contact, but were kept apart from each other by the electrical double layers formed around these two kinds of particles. The transformation of the particle distribution of ellipsoidal polytetrafluoroethylene from unsymmetrical to symmetrical crystals has been observed, with decreasing ionic concentration of suspension accompanied with the thickening of the electrical double layers (see Fig. 2-23, *B*).[98] Asymmetrically ordered structures, similar to paved bricks, have been formed for rectangular tungstic acid colloids in the sedimentation equilibrium, as is shown in Fig. 3.21. Interestingly, the regions of the single crystals, where all the particles are lying or standing on the cell wall and orient in the same direction, are observable. Furthermore, grain boundaries are recognized between adjacent single crystals. Ordered phases of platelike particles were observed in concentrated dispersions.[97,98] Tungstic acid particles are

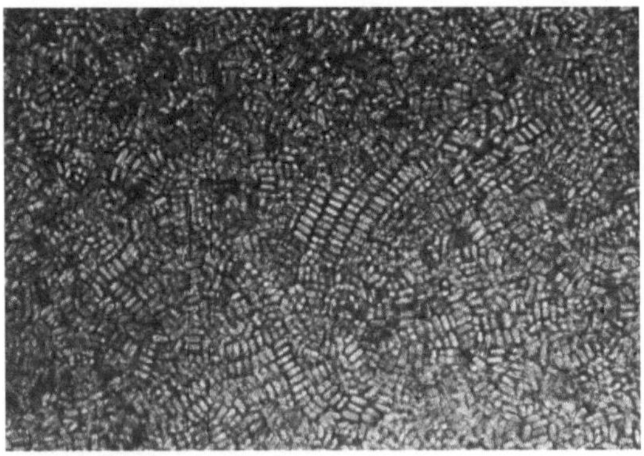

Figure 3.21
Microscopic observation of the unsymmetrically ordered colloidal crystals of tungstic acid particles in sedimentation equilibrium. Initial concentration (ϕ^*) = 0.00135 in volume fraction. Distance between the dashed parallel lines is 1 μm. *From Okubo T, Aotani S:* Colloid Polym Sci *266:1049-1057, 1988.*

ellipsoidal when the particles are freshly prepared. However, they transform to rectangular with time elapsed. The particle is one of the sparingly water-soluble electrolytes. When the suspension is kept with the ion-exchange resins, the particles dissolve slowly and disappear quickly, often in 1 to 2 days. Thus the microscopic observation was made without the coexistence of the resins. The particles appear enlarged and are coated with the electrical double layers, because, as is clearly observed in Fig. 3.21, the particles do not make direct contact. Ion concentration of the diffusible simple ions is fairly high ($\approx$0.1 mM), and then the interparticle distance is not so far compared with that in the deionized suspension.

It is well known that the rodlike and disklike colloidal particles form nematic, smectic (or lamellar), and columnar ordering. Onsager[100] studied the nematic ordering of tobacco mosaic virus and found that the nematic ordering takes place because the loss in orientational entropy is compensated for by the gain in packing entropy. The nematic suspensions of rodlike boehmite (AlOOH),[101,102] Li_2Mo_6 wires,[103] V_2O_5 ribbons,[104] and imogolite nanotubes[105,106] have been studied intensively. A magnetically oriented microcrystal array was obtained for $LiCoPO_4$ microrods, mesocrystals of *L*-alanine, and lysozyme.[107-111]

3.5 Distorted Colloidal Crystal, Ordered Colloidal Liquid, and Colloidal Liquid

In general, several intermediate subphases exist between typical colloidal crystal and liquid (i.e., distorted crystal or ordered amorphous solid and ordered liquid). Distorted colloidal crystals retain regular structures over short ranges and for short periods. In most cases, distorted crystals form for the polydispersed particles in size and shape; standard deviations divided by the mean values (polydispersity indices) are approximately 0.1 to 0.3. When the aqueous suspensions are deionized, for example, the effective polydispersity indices, where the electrical double layers coating the particles are taken into account, decrease significantly to approximately 0.1 or even smaller. In general, the colloidal crystallization (or amorphous solidification) takes place when *g* factor (thermal fluctuation divided by the nearest interparticle distance) is smaller than 0.1 by Lindemann's rule for crystal melting, as was discussed in Section 2.2. The distorted crystals were observed even for the aggregates in addition to the highly dispersed particles. Typical examples of the distorted crystals are shown in Fig. 3.22 for gold particles[98] and in Fig. 3.23 for diamond (*A*),[112] microcrystalline cellulose (MCC; *B*), aggregates of polystyrene spheres of H1001 (*C*)[113] and N1000 (*D*),[113] respectively. Properties of the particles are compiled in Table 3.6. All the particles were negatively charged, and the extended electrical double layers

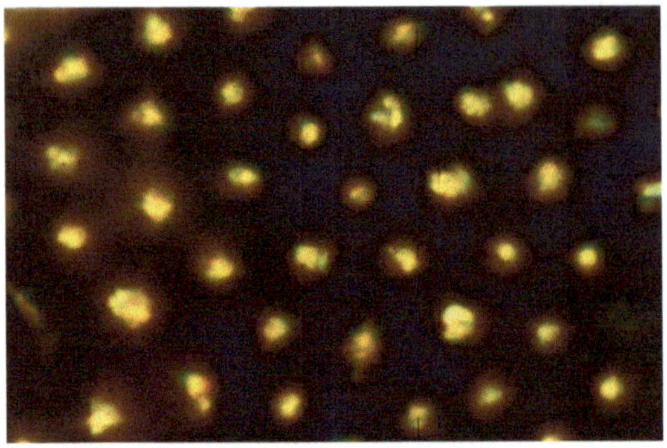

Figure 3.22 Microscopic observation of distorted crystals of polydispersed gold particles in sedimentation equilibrium. $\phi^* = \approx 2 \times 10^{-6}$. Distance between the dashed lines is 1 μm.

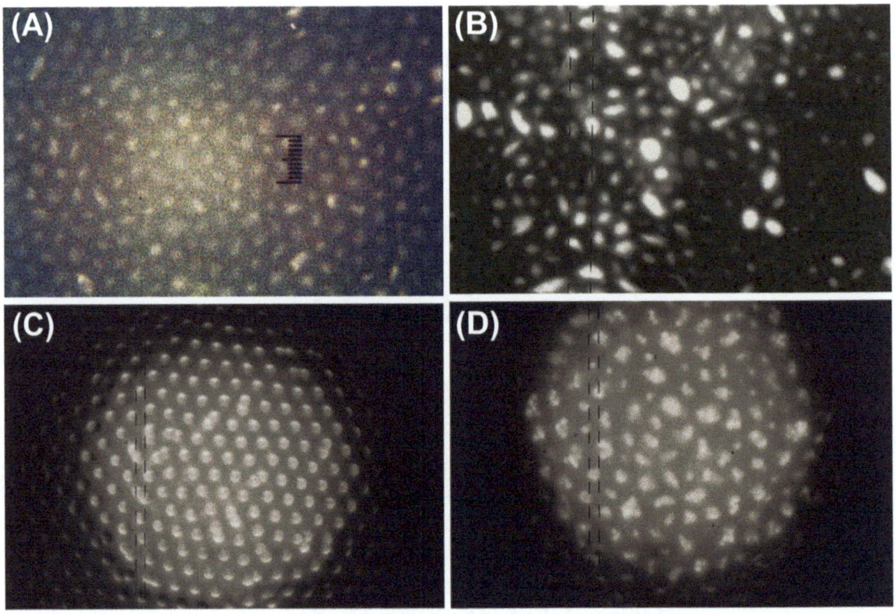

Figure 3.23 Microscopic observation of distorted crystals of **A,** polydispersed diamond particles: $w^* =$ 0.0084 wt%, full scale is 10 mm.[112] **B,** MCC: $\phi^* = 3.1 \times 10^{-5}$, distance between the dashed lines = 5 μm.[113] **C,** H1001 spheres in ethylene glycol: 10 vol% + water mixtures 1.53×10^{-3}, 1 μm. **D,** N1000 spheres: 8×10^{-4}, 4 μm in sedimentation equilibrium.

formed around each particle. Thus the effective size of the particles increased and the sphere behaved as enlarged spheres with the similar thickness as the double layers in the equation: $d_{eff} = d + 2 \times l_D$. Then the effective polydispersity (s/d_{eff}) decreased by approximately 0.1, and the distorted crystal structures formed. The importance of the electrical double layers for the formation of distorted crystals is clear even for the particles of high polydispersity. These pictures were taken in the exhaustively deionized and the sedimentation equilibrium states.

Table 3.6 Properties of Gold,* Diamond,† and Microcrystalline Cellulose Particles‡

Particle	Size (μm)	s/d	Sp. Density (g/cm^3)	Charge Density (μC/cm^2)	ζ-Pot (mV)
Gold particles	≈0.8 ± ≈0.5	0.6	19.3	—	—
Diamond particles	1.5 ± 0.32	0.2	2.87	—	−13.6
MCC	2.7 ± 0.6	0.2	1.5	230	—

*From Okubo T, Aotani S: Colloid Polym Sci *266:1049-1057, 1988.*
†From Okubo T: Colloid Polym Sci *291:1623-1629, 2013.*
‡From Okubo T: Colloid Polym Sci *268:1159-1166, 1990.*
MCC, Microcrystalline cellulose; *s/d,* polydispersity.

Phase transition between colloidal crystal and liquid structures are recognized with the naked eye from the disappearance of the enlarged single crystals (crystallites) with further dilution, although the iridescent colors still remained in the liquid structures. The phase transition was studied also by the transmitted-light (transmission) spectra.[88] The spectra of the crystal suspension showed a sharp peak, whereas those of liquid were like round hills in the spectra. Fig. 3.24 shows the typical example of the transmission spectra of D1C27 polystyrene spheres in the presence of NaCl. The broken curve shows the coexistence of the high but broad shoulder at the shorter wavelengths from the sharp peak. The high but broad shoulder is assigned to the "ordered liquid" structures. By the addition of further salt, the transmission peak disappeared and the weak broad hills assigned to the typical liquid appeared instead. In Fig. 3.24, liquid structures remained in the salt range of 3.33×10^{-5} to 1×10^{-4} M of NaCl.

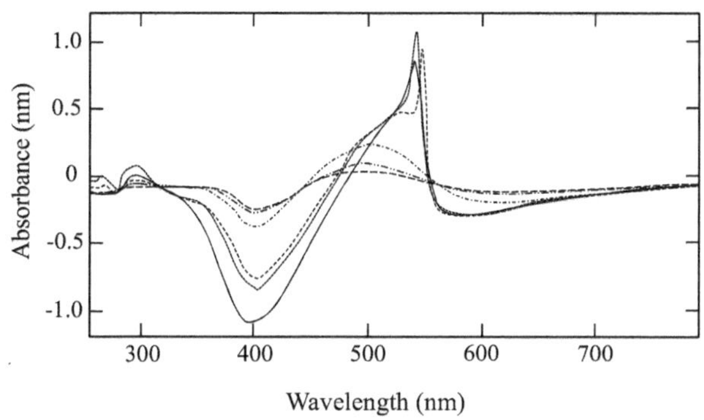

Figure 3.24
Transmission spectra of D1C27 suspensions at 25° C. $\phi = 0.019$, *solid curve:* [NaCl] = 0 M; *dotted:* 3.33×10^{-6}; *broken:* 1×10^{-5}; *broken-dotted:* 3.33×10^{-5}; *broken-dot-dot:* 6.67×10^{-5} M; *solid:* 1×10^{-4} M. Reference suspension contains 3.33×10^{-4} M of NaCl. The path length of quartz optical cell is 1 mm. *From Okubo T:* J Chem Soc Faraday Trans 1 *86:2871-2876, 1990.*

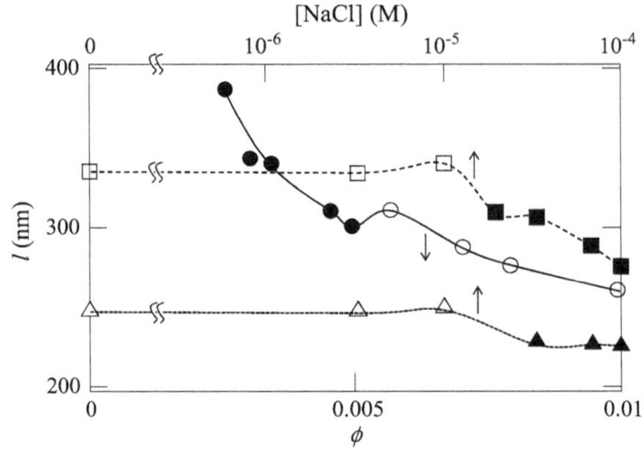

Figure 3.25 Plots of intersphere distance (*l*) of D1C25 (*circles*), D1C27 (*triangles*), and D1B76 (*squares*) against sphere concentration (D1C25) and concentration of NaCl (D1C27 and D1B76). *Open symbols:* crystal; *solid symbols:* liquid.

Fig. 3.25 shows the plots of the nearest interparticle distance; *l* values estimated from the sharp and broad peak wavelengths as a function of sphere concentration; and the concentration of NaCl for D1C25 (circles), D1C27 (triangles), and D1B76 colloidal spheres (squares). Here, open and closed symbols indicate the crystal and liquid structures, respectively. The *l* values in the liquid structures decreased uniformly with sphere concentration, but jumped up by several percent at the critical sphere concentration, where the liquid transformed to the crystal.

The most plausible phase diagram of colloidal crystallization is shown schematically in Fig. 3.26. The diagram is sensitive to the polydispersity index (s/d) and effective polydispersity index (s/d_{eff}) in addition to the colloidal concentration and size. Here, s is the dispersion index of colloidal size and d_{eff} is

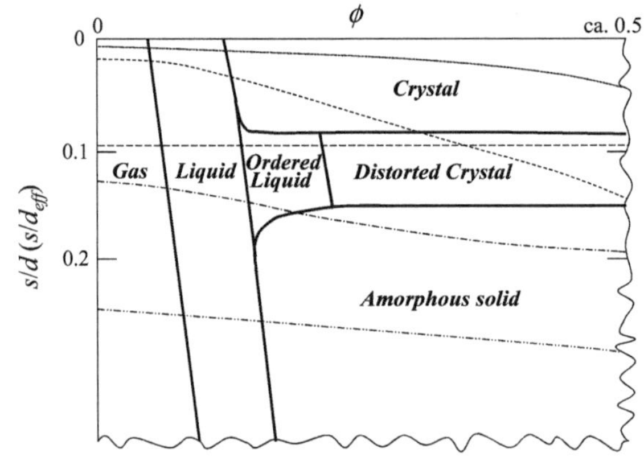

Figure 3.26 Phase diagram of colloidal crystallization as a function of sphere concentration, ϕ and polydispersity index of spheres, s/d (or s/d_{eff}).

the effective diameter of colloidal spheres, including the electrical double layers ($d_{eff} = d + 2l_D$). l_D is given by Equation 1 as described in Section 3.2. In deionized aqueous dispersions, crystallization takes place even for the spheres having an s/d larger than 0.1, because the s/d_{eff} value becomes lower than 0.1, which is the criterion of crystallization from Lindemann's rule. At approximately 0.1 to 0.2 in s/d_{eff} values, ordered liquid and distorted crystal subphases exist at the higher concentration of the colloidal spheres of colloidal liquid subphase, as is shown in Fig. 3.26. When s/d_{eff} value of spheres in the deionized dispersion changes following the dotted curves with sphere concentration, the gas, liquid, and crystal phases will appear in order with increasing sphere concentration. When in the deionized dispersion s/d_{eff} value changes along the short broken line, gas, liquid, crystal, and distorted crystal form in order with concentration, ϕ. When contribution of the electrical double layers is absent and eh s/d value was approximately 0.09, the gas, liquid, ordered liquid, and distorted crystal structures form. Furthermore, if the s/d_{eff} values change following the dot-broken and dot-dot-broken curves, the phases of gas–liquid–ordered liquid–amorphous solid and gas–liquid–amorphous solid phases should appear according the diagram, respectively. Again, Fig. 3.26 demonstrates the important role of the s/d_{eff} values in colloidal crystallization.

3.6 Alloy Crystals of Colloidal Mixtures

3.6.1 Lattice Structures and Morphologic Characteristics of Alloy Crystals

Opal is one of the typical examples of colloidal alloys fixed in solid state, which may form slowly during the course of sedimentation of the mixtures in suspension having different but monodispersed colloidal spheres. Colloidal alloy structures are interesting because most of the structures of metal alloys also appear in colloidal suspensions of binary mixtures of spheres having different sizes. Sanders et al.[62,114,115] clarified that the super lattice structures are self-assembled by the packing effects of the two different sizes of monodispersed colloidal spheres, A ($d_A = 362$ nm in diameter) and B (210 nm), for example. In this case, the diameter ratio, d_B/d_A, is 0.58, and A and B spheres are arranged in AB_2- and AB_{13}-type structures. A schematic picture of an opal is shown in Fig. 3.27. Yoshimura and Hachisu[63,116,117] and We[118] found five different super lattices, including AB_2 and AB_{13} in a binary suspension of polystyrene spheres, as is shown in Fig. 3.28. Bartlett et al.[119] further identified the super lattice structures in detail. The super lattice structures observed hitherto in colloidal systems are of the AlB_2, $NaZn_{13}$, $CaCu_5$, $MgCu_2$, NaCl, and AB_4 types. Note that these alloy structures are explained by changes in the effective size ratio (i.e., effective size of small spheres including the

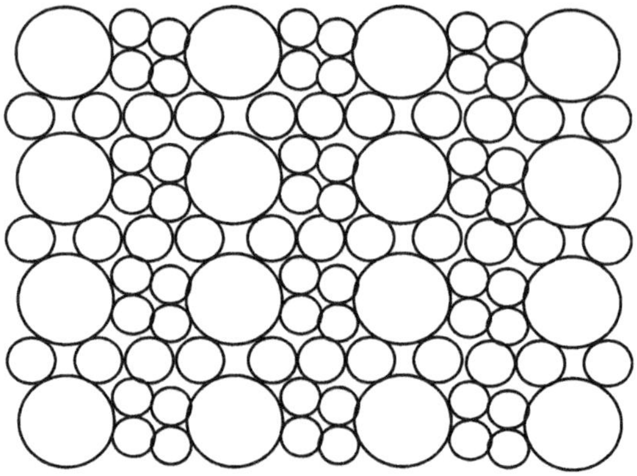

Figure 3.27
Schematic representation of the ordered arrays of Brazilian opal. Large spheres: 362 nm in diameter, small spheres: 210 nm.
Redrawn from Sanders JV: Phil Mag A *42:705-720, 1980, Fig. 1.*

electrical double layers vs. that of large spheres, and further the segregation effect). Constant-volume phase diagram containing three eutectics was determined using spheres of PMMA cores stabilized by thin layers of poly-12-hydroxystearic acid.[119-121] Hachisu observed NaCl-type crystal structure from the mixtures of monodispersed colloidal silica spheres (530 nm in diameter) and gold spheres (80 nm).[122] The stability of binary colloidal spheres was discussed theoretically.[123] Schofield et al.[124,125] obtained the maximum volume fractions of binary colloidal alloys: AB(CsCl), AB(NaCl), AB_2, AB_6(*bcc*), AB_6(sc), AB_{13}, as a function of d_B/d_A.

Colloidal alloy structures in the exhaustively deionized suspension were studied with use of microscopic and spectroscopic techniques in sedimentation equilibrium.[118,126] Substitutional solid solution (*sss*)–type alloy structures formed in mixtures of D1C25 (85 nm in diameter) and D1C27 (91 nm), where the ratio of small spheres to large is 0.93. When D1C25 and D1B76 (109 nm) are mixed, the structures were eutectic mixtures of *sss*-type lattices and Mg_2Cu_2-type super-lattices. The alloy structures of D1C25 and D1B72 (137 nm) spheres were the eutectic mixtures of (1) *sss*-type lattices, (2) $MgCu_{12}$-type super lattices, (3) *fcc* crystals of D1B72 spheres, and (4) crystals of D1C25 spheres. D1C25 and D1B41 (173 nm) suspensions were also eutectic mixtures of (1) $NaZn_{13}$-type super lattices, (2) D1C25 lattices, and (3) lattices of D1B41 spheres. These dramatic changes in the alloy structure depending on the combination of the spheres were explained reasonably by the changes in the effective size ratio (effective size of small spheres including the electrical double layer thickness against that of large spheres) and the segregation effects. The elastic moduli of the alloy structures were between 18 to 108 Pa.

Very large single crystals formed even for the binary mixtures of colloidal spheres.[61] Typical examples of the color pictures are shown in Fig. 3.29 for D1C25 (85 nm in diameter) + D1C27 (91 nm) and CS81 (103 nm) + CS121

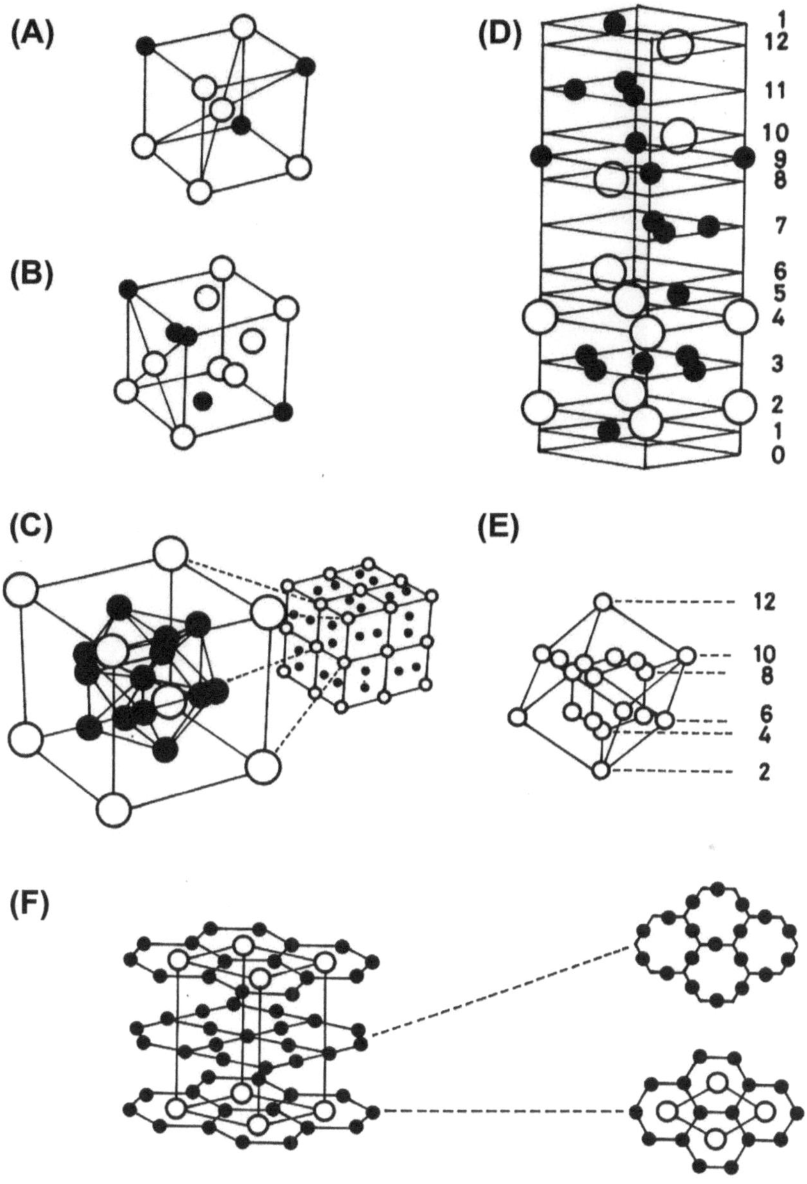

Figure 3.28 Alloy structures appeared in the binary colloidal spheres: **A,** *bcc* lattice; **B,** *fcc* lattice; **C,** $NaZn_{13}$-type super lattice; **D** and **E,** $MgCu_2$-type super lattice; **F,** $CaCu_5$-type super lattice. *From Okubo T:* J Chem Phys *93:8276-8283, 1990.*

(136 nm) mixtures. The alloy crystals look more turbid than the crystals of single-component spheres. In the deionized suspensions, both mixtures gave *sss*-type single crystals. We should note here that the colloidal alloys form also by the packing effects and the intersphere repulsion forces in a closed vessel.

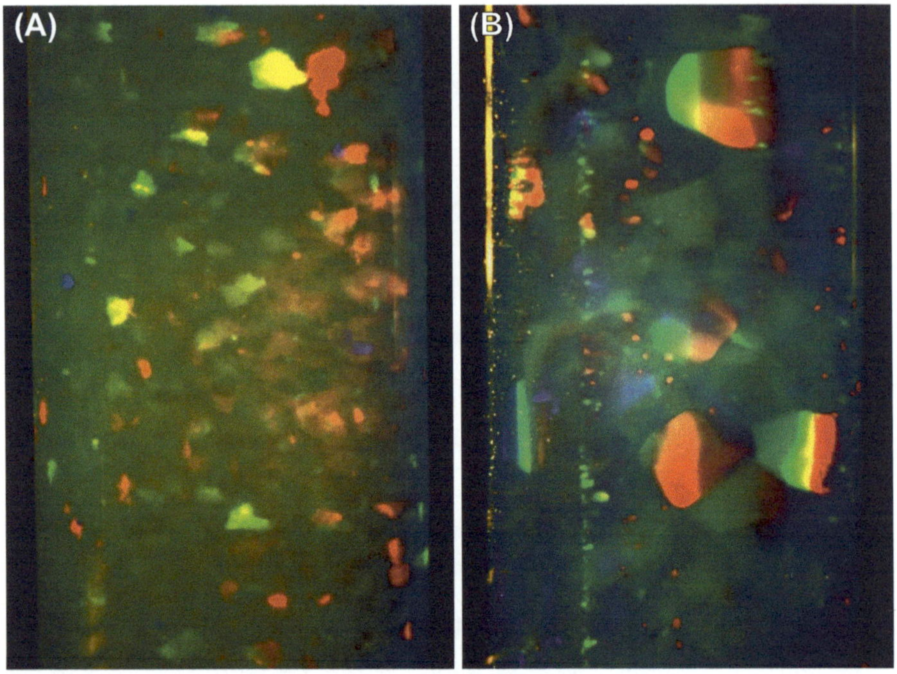

Figure 3.29 Close-up color pictures of alloy crystals: **A,** D1C25 (85 nm in diameter, $\phi = 0.00188$ in volume fraction) + D1C27 (91 nm, 0.000718); **B,** CS81 (103 nm, 0.000288) + CS121 (136 nm, 0.000579).

3.6.2 Phase Diagram

A phase diagram of the alloy crystals were studied in the exhaustively deionized state.[61] Interestingly, crystallization was disturbed more effectively when small spheres were added in the suspension of large spheres. In other words, the colloidal crystals of small spheres are more stable than those of large spheres by the addition of the other spheres. One of the typical examples of the phase diagram of alloy crystals is shown in Fig. 3.30 for D1B76 + D1C25 mixtures. Crystal structure of the alloy crystals is deduced to be the *sss* type, because the effective diameter, including the thickness of the electrical double layers, is very close, within 2%. Triangles show the amorphous-solid structures. The extended graph is given in Fig. 3.30, *B,* to see the crystal melting region. The melting region is shifted toward the axis of D1B76, which means the melting of the alloy crystals takes place more easily with the addition of smaller spheres in the mixtures. It should be noted here that the amorphous-solid structures were not observed for the mixtures of D1C25 and D1C27. The ratios of small size against large ones are larger than 0.93 and close to each other, within 7% when the electrical double layers are taken into account. It is well known that the crystals melt when the fluctuation parameter comes to be larger than 0.1 (i.e., Lindemann's rule).

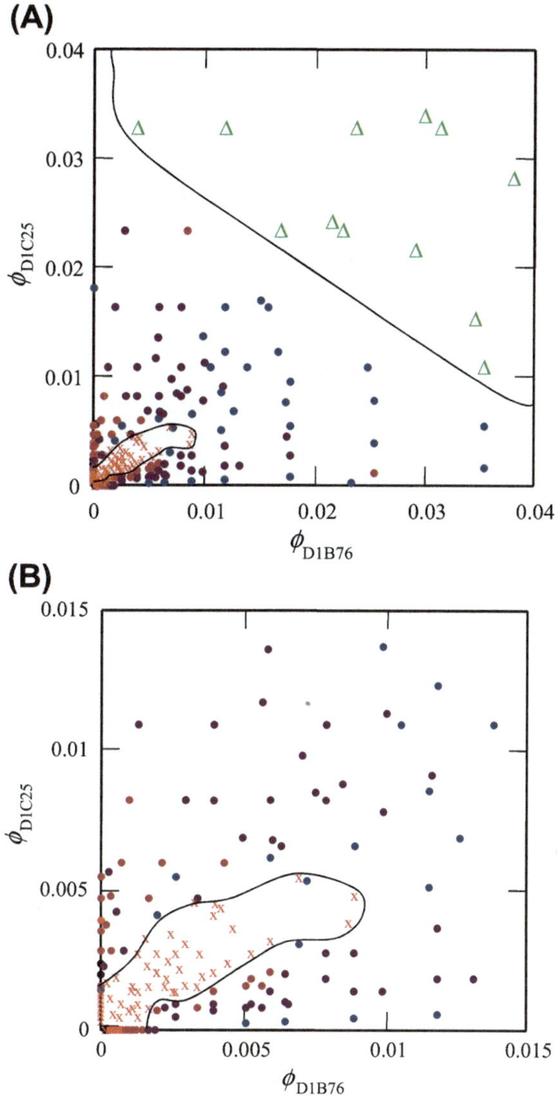

Figure 3.30
A, Phase diagram of D1C25 + D1B76 mixtures at 23° C. Solid *circles* are crystal; the size of single crystals is roughly shown by the change from reddish (*large*) to bluish (*small*) colors. *Crosses* are liquid. *Triangles* are amorphous solid. **B,** Extended diagram. ***A** redrawn from Okubo T, Fujita H:* Colloid Polymer Sci *274:368-374, 1996.*

3.6.3 Segregation Effect

Segregation effect means (1) the repelling of aggregates or outsized particles from the ordered region of the major particles or (2) the separation effects of the different-sized or different-shaped particles toward the different phases of likely particles. The cause for the segregation is not always clear.

However, the thermal motion of all the particles in an external field (e.g., gravity and flow) is most plausible. Hachisu et al.[127] observed several kinds of the segregation phenomena microscopically. It is well known that separation takes place into the layers of the large and small spheres into upper and lower parts, respectively, in a gravity field. These segregation effects were discussed theoretically.[128-130] We should note that the segregation effects are explained also by one of the interface phenomena: When large and small monodispersed colloidal particles are mixed, the different sized spheres are apt to join together in groups of like-sized particles and phases.

Tsuchida et al.[131,132] made microgravity experiments on the crystal growth rates of a colloidal alloy and found the segregation effects disappear in microgravity. Fig. 3.31 shows the rate coefficient ratios k_{0g}/k_{1g} in the growth process of the alloy crystals of D1W52 (polystyrene spheres, 88 nm in diameter) + D1B76 (polystyrene, 109 nm; circles), CS91 (colloidal silica, 110 nm) + CS121 (colloidal silica, 136 nm; crosses), D1B76 + CS91 (triangles) and D1B76 + CS121 mixtures. The rate coefficient k was obtained as the reciprocal period where the initial linear line in the peak intensities intersects two horizontal lines, giving initial and final intensities in the reflection spectroscopy at $t = t_i$ and t_f, respectively: $k = 1/(t_f - t_i)$. The t_i value corresponds to the induction time for crystallization. The rate coefficients at 0 g and 1 g are k_{0g} and k_{1g} at the same experimental conditions except gravity. Clearly, large k_{0G}/k_{1G} values were obtained for CS91 + CS121 and D1B76 + CS91 mixtures. The former system is a combination of the same sphere density but different sphere sizes, and the latter is a mixture of spheres having the same size but different specific densities. In normal gravity (1 g), small spheres sediment faster than large ones by the segregation effect, and heavy spheres sediment faster than light ones. The maximum value of k_{0g}/k_{1g} was approximately 1.7, and the microgravity effect was large. In microgravity, the segregation effect should disappear and totally homogeneous mixing must occur, which leads to faster alloy structure formation (Fig. 3.32). The absence of convectional flow of spheres induced by gravity must also be favorable for alloy distribution. The significant acceleration effect of microgravity supports the alloy ordering of colloidal spheres, which is determined by the mixing packing density for a given ratio of the sphere diameters, including the thickness of the electrical double layers.

For the binary crystals of D1B76 + CS121 and D1W52 + D1B76, k_{0g}/k_{1g} was close to unity irrespective of the mixing ratio. The former system is a combination of small, less dense spheres and large, dense spheres, where the segregation effects from the size and density differences cancel each other out at 1 g. For the mixture of D1W52 + D1B76, the densities of the spheres (1.05) are close to that of water, and furthermore both spheres are so small that the effective sizes including the double layers are similar to each other. These are the main reasons why the microgravity effect was not observed.

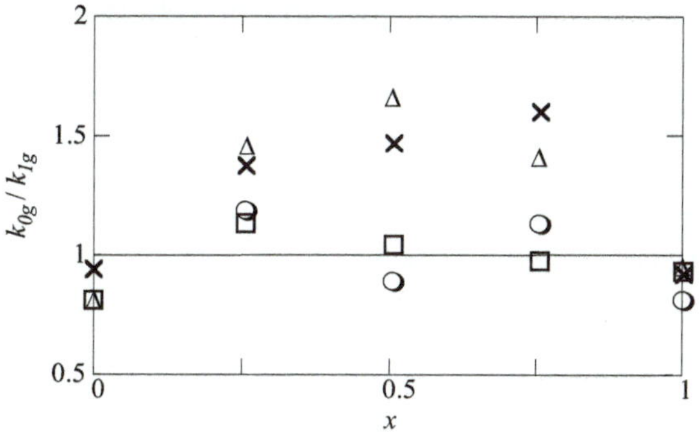

Figure 3.31
Rate coefficient ratios in microgravity experiments as a function of sphere's volume fraction in $[D1W52]_{1-x}$ + $[D1B76]_x$ (*circles*), $[CS91]_{1-x}$ + $[CS121]_x$ (*crosses*), $[D1B76]_{1-x}$ + $[CS91]_x$ (*triangles*) and $[D1B76]_{1-x}$ + $[CS121]_x$ (*squares*) mixtures at 25° C. $\phi_{total} = 0.005$. *Data from Okubo T, Tsuchida A, Takahashi K, Taguchi K, Ishikawa M:* Colloid Polym Sci *278:202-210, 2000.*

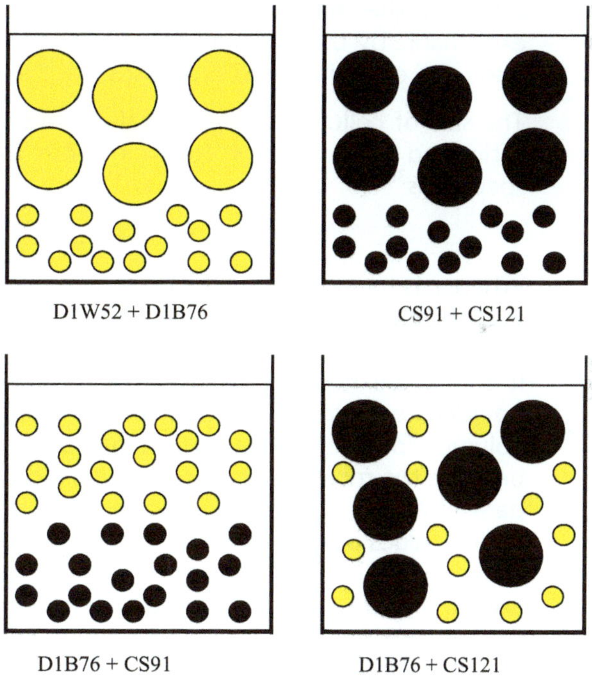

Figure 3.32
Schematic representation of the segregation effects in the mixtures of large and small spheres with the same density (D1W52 + D1B76 and CS91 + CS121), light and heavy spheres with the same size (D1B76 + CS91), and light small and heavy large spheres (D1B76 + CS121).

Here it should be further noted that the acceleration effect in microgravity was largest at mixing ratio (x) = 0.5. The values of k_{0G}/k_{1G} for the single-component systems were always less than unity (i.e., the colloidal crystallization were retarded in microgravity).[131,132] One of the main causes is the lack of the extra downward diffusion of spheres, which enhances intersphere collision in microgravity. The other reason is the lack of convection in microgravity, which is not favorable for the crystallization of single-component spheres. Thus the segregation effects play an important role for the colloidal alloy structures. It should be noted here that the significant segregation effects were often observed in the drying patterns for the mixtures of large and small spheres, which are described in Chapter 5.

3.7 Kinetics of Colloidal Crystallization

The kinetic analyses of colloidal crystallization have been made by the reflection spectroscopy,[133-136] light-scattering,[36,137-141] and microscopy and charge-coupled device (CCD) video observation.[9] Most kinetic measurements on colloidal crystallization showed an induction period, after which the crystal growth starts, especially in diluted suspensions. This observation supports the finding that the kinetics of colloidal crystallization are explainable by the classical diffusive crystallization theory, including nucleation and crystal growth processes.[133-136]

The size of colloidal single crystals from the homogeneous nucleation, L, is estimated by Scherrer's equation from the half-width of the reflection spectra as[36,85,142]:

$$L \propto n\Delta q^{-1} \tag{19}$$

Where n is the number of the order in the Bragg equation, q is the scattering vector, and Δq denotes q_l-q_s, with q_l and q_s referring to the largest and smallest wavelengths corresponding to the half-width of the reflection peak. It should be noted here that Equation 19 is derivable by assuming the powderlike random distribution of a large number of crystals in the reflection volume. However, this assumption is difficult to hold for colloidal crystals, because the most dense planes of colloidal single crystals are apt to orient parallel to the cell plane. Furthermore, it is quite difficult to obtain the accurate values of L from the reflection spectroscopy, because the prove area detecting reflection light is not small enough and smaller values than real crystal size are always obtained experimentally.

On the other hand, the peak intensity, I, of the reflection spectroscopy is approximated by Equation 20[36]:

$$I \propto N_{cryst} L^3 \propto L^3 \tag{20}$$

Here, N_{cryst} is the number of single crystals in the reflecting volume, which is directly proportional to the number of crystals in the final stages of the

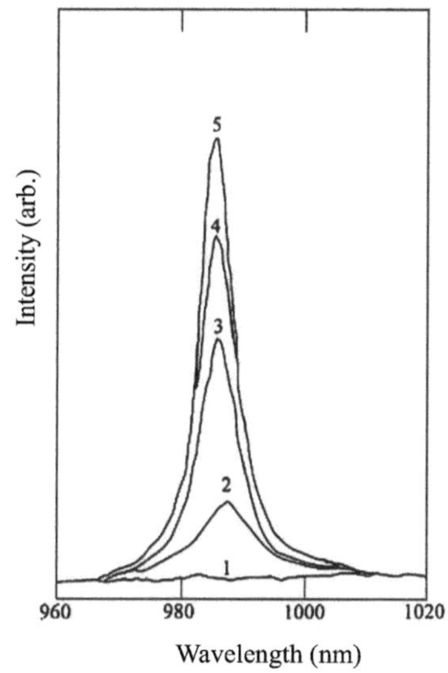

Figure 3.33 Reflection spectra during crystallization of CS91 spheres at 24° C. $\phi = 0.00179$, 17 days after suspension preparation with the resins. Curve 1: 5 seconds after mixing; 2: 15 seconds; 3: 20 seconds; 4: 25 seconds; 5: 600 seconds. *Modified from Okubo T, Okada S:* J Colloid Interf Sci *192:490-496, 1997.*

crystallization process, being equal to the total number of nuclei formed in the whole course of crystallization. Fig. 3.33 shows a typical example of the reflection spectra as a function of time elapsed after the colloidal crystals are melted out by the stopped-flow mixing. Fig. 3.34 shows the size versus time plots evaluated using Equation 20 and the average crystal size at the final stage with the naked eyes.

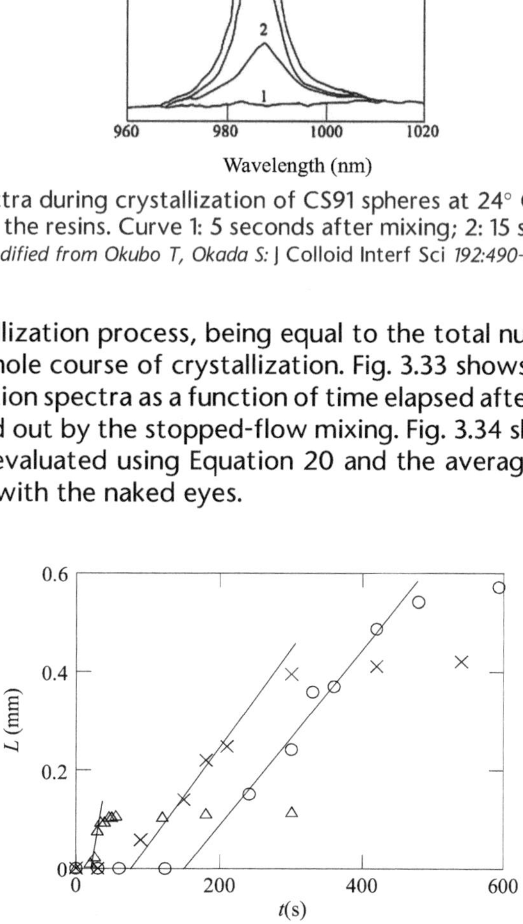

Figure 3.34
Crystal size estimated from the half-width during the course of crystallization of CS91 spheres at 23° C. *Circles:* $\phi = 0.00143$; *crosses:* 0.00179; *triangles:* 0.00262. *Modified from Okubo T, Okada S:* J Colloid Interf Sci *192:490-496, 1997.*

The number of nuclei that germinate per unit of time, the nucleation rate, v_n, is given by

$$v_n = \phi l^3 \Big/ \left[(4\sqrt{2}/3)\pi(d_o/2)^3 L^3 t_i\right], \qquad 21$$

where ϕ is the volume fraction of sphere particles, and L and l are the mean size of single crystals formed and the nearest-neighbor intersphere distance, respectively. d_o is the diameter of sphere particles; t_i is the induction period and corresponds to the time of nucleation process. A typical example of the v_n values is shown in Fig. 3.35. Clearly, nucleation rates range from 10^{-4} to 10^4 nuclei mm^{-3}s^{-1} and increased very sharply as sphere concentration increased (see Fig. 3.34). The induction periods were from 1 to 500 seconds and became larger with decreasing sphere concentration for CS91 spheres, as is shown in Fig. 3.34.[134] An increase in sphere concentration is followed by an increase in v_n and number of nuclei, and resulted in a decrease in size of the single crystals formed, as discussed by Dhont et al.[36] It should be mentioned here that size of the single crystals, observed with the naked eye, is polydisperse (e.g., see Figs. 3.3, 3.4, 3.9, 3.11, 3.13, and 3.29). Therefore the clear-cut separation of the nucleation step from the crystallization process is difficult, and the nucleation reaction may remain operative even in the crystal growth period.

Colloidal crystallization processes are composed of the fast growth of a metastable crystal (step 1) and the slow reorientation toward the stable crystal (step 2). The first crystal growth rate of a metastable crystal, v_1, is given by[134]:

$$v_1 = v_{1\infty}[1-\exp(-\sigma)]. \qquad 22$$

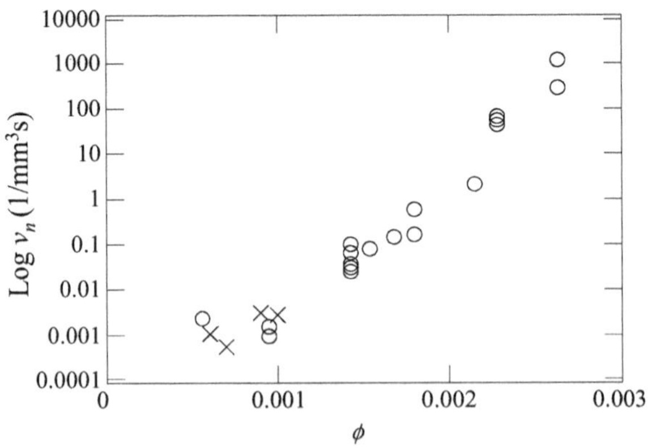

Figure 3.35 Nucleation rates of CS91 spheres at 23° C. Circles[130]; crosses.[129] *Redrawn from Okubo T, Okada S: J Colloid Interf Sci 192:490-496, 1997.*

Here, σ is the relative supersaturation given by $(\phi - \phi_c)/\phi_c$. The critical concentration of melting is represented by ϕ_c. The maximum crystal growth rate is represented by $v_{1\infty}$. Equation 22 is further simplified as:

$$v_1 = v_{1\infty} - v_{1\infty}\phi_c/\phi \qquad 23$$

Equation 23 means that the growth rate decreases linearly as the reciprocal sphere concentration increases, which was supported experimentally, but only in a very low sphere concentration range.[134] Fig. 3.36 shows v_1 thus evaluated versus $1/\phi$ plots. The data are scattered. And v_1 were 15.3 μm/s.

The peak profile in the reflection spectroscopy became sharp with time during the course of crystallization, which corresponds to an increase in crystal size. The background intensity decreased during the course of crystallization, which is ascribed soundly to the diminution in the multiple scattering of the suspension. The nearest-neighbor intersphere distance, l is observed from the peak wavelength.[85] The l value is also calculated from the sphere size and concentration.[85]

The fast and slow crystal growth rates, v_1 and v_2, were evaluated. The v_1 values in the diluted suspensions decrease linearly as the reciprocal concentration of spheres increases.[134] It should be noted that the importance of the dynamic phase transition has been reported for colloidal crystals and for other crystals such as metals and polymers. The slow step of crystallization decreases as sphere concentration increases. The slow crystallization step is ascribed to the "reorientation" of the metastable crystals formed in the first step toward the stable ones matched with the other crystals mediated by the grain boundaries and the cell wall. When sphere concentration is high, the number of the metastable crystals formed is huge and their size becomes quite small. Then a long time is required for the crystals to be matched with

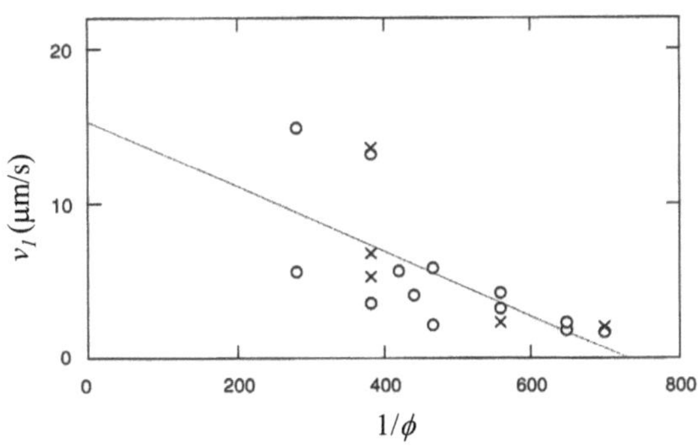

Figure 3.36
Crystal growth rates as a function of reciprocal sphere concentrations estimated from the cube-root method (*circles*) and half-width method (*crosses*) for CS91 spheres at 23° C. *Modified from Okubo T, Okada S: J Colloid Interf Sci 192:490-496, 1997.*

other crystals and the cell wall. The Ostwald ripening is also highly plausible during the slow process, although direct observation was not successful until Wong et al.[143] observed the ripening for colloidal crystallization processes.

We assume here that the maximum growth rate, $v_{1\infty}$, is determined by the diffusion of single spheres near the crystal surface. Then:

$$v_{1\infty} = 4D/\xi \tag{24}$$

Here ξ is the mean diffusion length and D is the diffusion coefficient of spheres in the crystal lattices when ξ and D are assumed to be the mean intersphere distance (l, 1.18 μm) and the diffusion coefficient of spheres in the colloidal gas is calculated by the Stokes-Einstein equation (4.36×10^{-12} m^2/s). Then $v_{1\infty}$ will be given by:

$$v_{1\infty} = 4k_BT/3\pi\eta dl \tag{25}$$

where d is the sphere diameter and $v_{1\infty}$ is calculated to be 14.8 μm/s, which agrees surprisingly well with the observation, 15.3 μm/s. We should note here that D should be the diffusion coefficient in the supersaturated liquids, for example, 1×10^{-12} μm/s in the previous work on dynamic light scattering.[133] Furthermore, the mean diffusion length (ξ) is clearly overestimated when $\xi \cong l$ is taken, because colloidal spheres in supersaturated liquids must move in a cooperative manner by the electrostatic intersphere repulsive forces. In this sense, colloidal crystal systems are quite similar to fused metal systems. Slight movement of the effectively enlarged spheres, including electrical double layers, is enough to allow crystallization. Then the agreement of $v_{1\infty}$ between the calculation and observation is explained by the balancing of two factors in the decrease of D and ξ.

Critical concentration of melting (ϕ_c) of the colloidal crystals in the deionized state is very low, around 0.0003 in volume fraction, as is shown in Table 3.2 and Fig. 3.8, for example. However, the effective concentration of spheres, including the extended electrical double layers, is substantially high, close to 0.74 in volume fraction corresponding to the close packing. Each sphere in the colloidal crystals interacts strongly and in a cooperative manner through the electrostatic repulsion via the double layers. Schatzel and Ackerson[37] proposed the importance of the dynamic phase transition by the density fluctuation in the colloidal crystallization. The important role of the dynamic phase transition has been reported for other crystals such as metals[144] and polymers.[145]

Kinetic analyses of colloidal crystallization were made in a sinusoidal electric field with use of reflection spectroscopy.[136] Colloidal spheres of CS91 were used. The sample suspension in the cell coexisted with mixture of the resins to attain the exhaustively deionized state. The colloidal growth rates, v of the *bcc* lattices have been determined from the increase in the cube-root of the intensity in the sharpened reflection peaks using Equation 20. The value without electric field was 20 μm/s and decreased to 8 μm/s as the voltage applied increased from 0 to 10 V at 1 Hz. Furthermore, v decreased from 30 to 15 μm/s when frequency increased from 0.01 to 10 Hz at $E = 6$ V, and then remained constant irrespective of frequencies higher than 10 Hz up to 10 kHz. The main causes for the retardation of colloidal crystallization are that (1) the additional translational fluctuation of the respective colloidal spheres, including the surrounding electrical double layers, are induced by the low frequency electric field; and (2) the partial melting effect of the forming crystals by the shearing forces of the flow of medium molecules. Interestingly, crystallization of *bcc* lattices was enhanced at low frequencies between 0.01 and 0.5 Hz. At the low frequencies, the reflection spectra were not very smooth and the peak shifted shorter wavelengths was significant during crystallization. On the other hand, (3) at the high frequencies above 1 kHz, the spectra became smooth, and (4) a reduction of the background intensities was clear during crystallization. Furthermore, (5) peak shifts in the reflection spectra at high frequencies were moderate like those in the absence of the current, and (6) *fcc* lattices were not formed at high frequencies; only *bcc* crystals formed. These observations provide strong support for the finding that the lattice spacing of the crystals formed is sensitive to the electric field strength at low frequencies. Two or three relaxation times of the colloidal crystals were observed by the stopped-flow technique and dynamic light-scattering (DLS) measurements, respectively, as will be described in Sections 3.8 and 3.9, respectively. The slow relaxation times were in the order of 10 milliseconds or longer. This means that lattice spacing cannot follow the change in the currents in the high-frequency field, whereas the spacing follows the field of low frequencies. It should be noted that *bcc* crystals always form first under unstable experimental conditions such as at high suspension temperature, in salt-containing suspensions, and under high pressure.

Morphologic close-up observation of crystallites during crystallization was made independently in a sinusoidal electric field. The cell used is shown in Fig. 3.37. Platinum disk electrodes were set perpendicularly in the cell. Mixtures of ion-exchange resins ($\approx$5 ml) were introduced into the cell to maintain the exhaustively deionized state of the sample suspension. Fig. 3.38 shows the typical close-up pictures showing the crystallization process under an electric field. Two kinds of single crystals from homogeneous and heterogeneous nucleation mechanisms formed and growth seems completed after 25.5 minutes. At first glance, the growing processes resemble those without an electric field. Furthermore, the times needed for the complete growth were approximately 20 to 25 minutes irrespective of the applied frequencies between 1 Hz to 1 kHz, and were also similar to those in the absence of the electric field. It should be mentioned here that the close-up

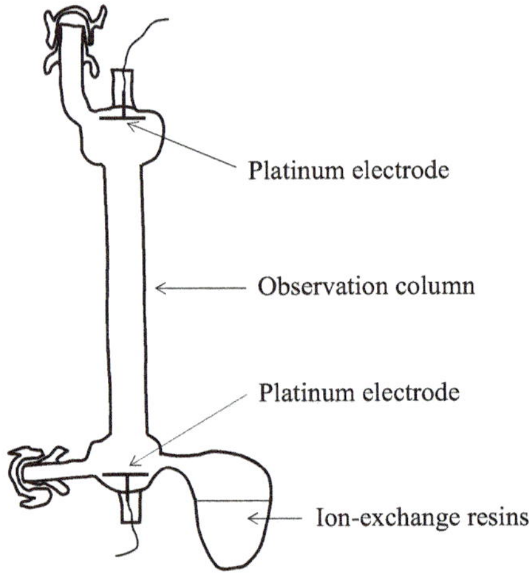

Figure 3.37
An observation cell for close-up color pictures and reflection spectroscopy as a function of height in the electric fields.

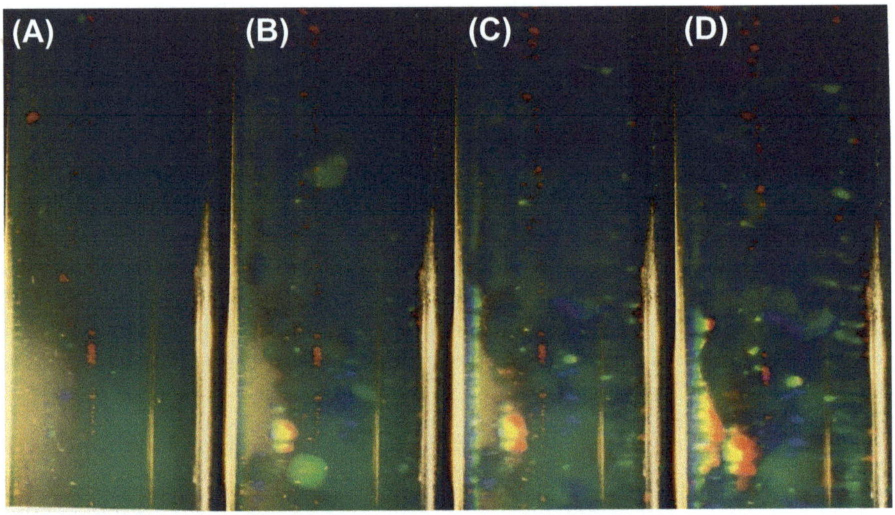

Figure 3.38
Close-up color pictures showing crystal growth processes of CS82 spheres in a sinusoidal electric fields at 25° C. $\phi = 0.00124$; $E = 3$ V, 1 Hz; **A,** 2 minutes after inverted mixing; **B,** 4.5 minutes; **C,** 6 minutes; **D,** 25.5 minutes.

color pictures during the crystal growth processes were not always clearly taken. This is because the Bragg diffraction conditions always change for each observation. However, a clear-cut picture of a single crystal is taken every so often, when the incident angle of light is changed slightly after a series of observation.

Fig. 3.39 shows the morphologic characteristics at the final stage 30 minutes after crystallization starts. From the left, the pictures without electric field

(*A*); at 10 Hz, 3 V (*B*); 100 Hz, 3 V (*C*); and 1 kHz, 3 V (*D*) are shown. Clear-cut differences in the morphologic characteristics were not observed among the pictures. Fig. 3.40 shows the close-up color pictures of the upper parts of the cells approximately 30 minutes after crystallization starts. They look similar at first glance. However, when the three pictures are compared carefully, the crystal facets are observed most clearly for *C* and appear more vague for *B*. The frequency of 1 kHz is too high to influence the crystallization processes, because the fastest structural relaxation time of colloidal crystals is in the order of 10 milliseconds.

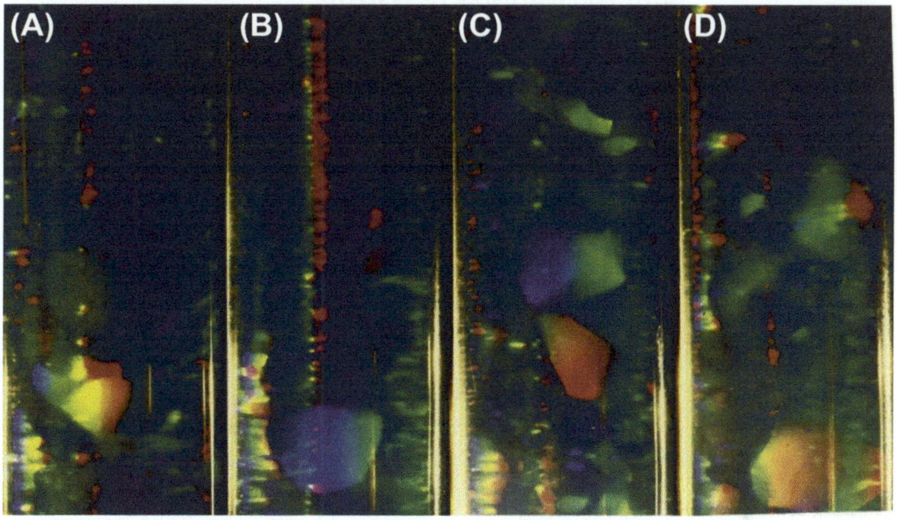

Figure 3.39 Close-up color pictures of colloidal crystals at the final stage of crystal growth at 25° C. CS82 spheres, $\phi =$ 0.00124; **A,** without electric field; **B,** 10 Hz, 3 V; **C,** 100 Hz, 3 V; **D,** 1 kHz, 3 V.

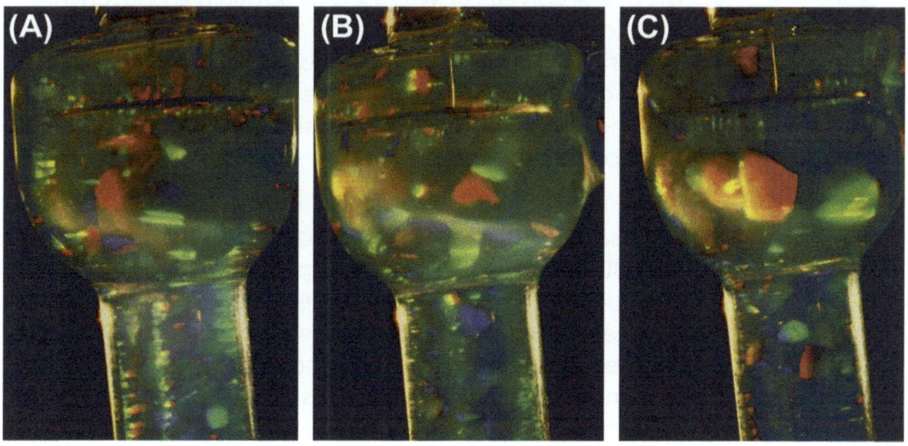

Figure 3.40 Close-up color pictures of single crystals at 25° C approximately 30 minutes after inverted mixing. **A,** Without electric field; **B,** 1 Hz, 3 V; **C,** 1 kHz, 3 V.

Kinetic analyses of colloidal crystallization in alcoholic organic solvents and their aqueous mixtures have been studied by reflection spectroscopy.[146] The kinetic study of colloidal crystallization in acetonitrile has been made for the silica spheres modified with polymers on their surfaces.[147,148] Kinetic study of the crystallization of ellipsoidal particles under an alternating electric field was studied.[149] Our group studied the kinetics of crystallization in the shear flow.[150] Kinetic analyses of Aastuen et al.[17,151] for the colloidal crystallization are very important. Dynamics of crystallization in hard-sphere suspensions has been studied intensively by Ackerson et al. and others.[37,138,139,152-154] Crystallization kinetics has been studied and reviewed by Palberg for the soft- and hard-sphere systems.[38,141,155]

3.8 Static and Dynamic Light-Scattering Measurements

Critical concentrations of melting of colloidal crystals in the deionized suspensions are low, approximately 0.0002 to 0.0003 in volume fraction. Furthermore, turbidity of the colloidal crystals and liquids is fairly low compared with that of gas suspensions, because of the greatly diminished intensity of the multiple-scattering for visible light. For the colloidal crystals exhibiting large (or giant) single crystals, turbidity is fairly low. Therefore reliable and accurate SLS and DLS data are available for aqueous and diluted colloidal crystals[133,156-158] compared with the data obtained from the roughly deionized and then rather high sphere concentrations. The SLS and DLS measurements are made on a DLS spectrophotometer. The goniometer spans from 15 to 160 degrees in the scattering angle (θ) and a helium-neon laser was used as a light source at 633 nm. The light source is vertically polarized after passing through two pinpoint holes, and then focused on the center of the cell. A correlator was used to obtain the photon intensity autocorrelation function. In our laboratory, the stock suspensions, approximately 0.1 in volume fraction of colloidal spheres in a Pyrex laboratory bottle with screw cap (1 L, for resisting high pressure and temperature), are deionized with the cation- and anion-exchange resins mixtures (Bio-Rad) that have coexisted for more than 10 years.

In our laboratory, the sample of 5 ml was prepared by filtering with a disposable syringe filter unit into a Pyrex cuvette (12 mm in outside diameter and 130 mm in length) with a cap. These cuvette cells are grounded carefully in a spectroscopically guaranteed level. SLS and DLS measurements for the samples without resins have been made within 5 hours after the suspension was set in the cell. However, measurements for the exhaustively deionized suspensions with the Bio-Rad resins were made more than 3 to 4 weeks after the suspensions were set. The resins used were rinsed several times carefully by the decantation method with pure water to remove the fragments of the resins, although they are not recognized with the naked eye. The amount of resins added was always much in excess compared with the ionic impurities in suspension. Adsorption of the colloidal spheres on the resin surfaces was safely prevented by the gentle inverted mixing of the sample suspension several times every day. Note that it takes a long time to attain the exhaustively deionized suspension with the coexistence of the ion-exchange resins.

The autocorrelation function of the scattering field $g^{(1)}(\tau)$ is actually found to be a single exponential decaying function only for the suspension containing foreign salt or for gas structures. For liquid and crystal suspensions, the intersphere repulsive interaction is so strong that $g^{(1)}(\tau)$ deviates significantly from a single exponent. In these cases the histogram method of the nonnegative least square (NNLS) analysis was applied in our work to obtain the distribution of the decay rates and the diffusion coefficients or hydrodynamic diameter.[133,156-158] The homodyne photocorrelation spectroscopy measures the normalized intensity autocorrelation function $g^{(2)}(\tau)$, which is related to the autocorrelation function of the scattered field, $g^{(1)}(\tau)$ following the Gauss central-limit theorem.

$$g(2)(\tau) = 1 + \beta|g(1)(\tau)|2 \qquad 26$$

where β is the amplitude. When the suspension of monodispersed spheres is sufficiently diluted so that the intersphere interaction may be negligible, the decay of concentration fluctuation only occurs by the translational diffusion motion of a free particle. The $g^{(1)}(\tau)$ is represented by a single exponential type of decay function, Equation 27 from which the translational diffusion coefficient, D_o is estimated.

$$g^{(1)}(\tau) = \exp\left(-q^2 D_o \tau\right) \qquad 27$$

Here, q is the scattering vector whose magnitude is equal to $(4\pi/\lambda)\sin(\theta/2)$, where λ is the wavelength of the incident beam in medium. From D_o, we may calculate the hydrodynamic diameter, d_H, of the sphere using the Stokes-Einstein equation.

$$d_H = k_B T / 3\pi\eta_o D_o \qquad 28$$

Here η is the solvent viscosity.

In the liquid state where the colloidal particles are locally ordered to be reflected as the maximum of the static structure factor $S(q)$ at q_m, $g^{(1)}(\tau)$ is no more fitted by Equation 27. In this case, the first cumulant $\Gamma_e = K_1$ defined by Equation 29 is conveniently used for characterization of dynamical behavior of the spheres in the liquid state.

$$\ln g^{(1)}(\tau) = \Sigma K_n(-\tau)^n / n! \qquad 29$$

The Γ_e was found proportional to $S(q)^{-1}$ as predicted by the theory. For $q \gg q_m$ where $S(q) \cong 1$, on the other hand, $g^{(1)}(\tau)$ shows the bimodal distribution of the decay rate, where the short-time diffusion coefficient is close to D_o and the long-time diffusion coefficient D_l is smaller than D_o by a factor of three.[159] The short-time behavior represents the Brownian motion of the particle in a cage formed by neighboring particles. The behavior at the long-time end is much longer than the lifetime of the cage, τ_l, according to the notation by Pusey, and represents diffusion through several nearest-neighbor cages.

When the suspension is exhaustively deionized, the colloidal crystal forms, and DLS proves the overdamped motion of the particle on the lattice, which has been investigated by Hurd et al.,[160] Cotter and Clark,[161] and van Megen et al.[162] For example, Cotter and Clark showed that Γ_e from thermal fluctuation with q along lattice symmetry direction was given by Equation 30:

$$\Gamma_e(q) = \omega(q)^2/\lambda(q), \tag{30}$$

Where $\omega(q)$ is the undamped mode frequency of harmonic oscillator and $\lambda(q)$ the damping coefficient. Equation 30 predicts that for single scattering in the harmonic approximation, Γ_e is periodic in reciprocal space, being in agreement with experiments. For DLS measurements on colloidal crystals that are not always randomly oriented in space, the time profile of $g^{(1)}(\tau)$ is not precisely known.

For the crystal suspensions of colloidal silica spheres, the difference $\Delta g^{(2)}(\tau)$ between $g^{(2)}(\tau)$ observed and calculated from cumulant analysis was significant, as is shown in Fig. 3.41, *A*, which supports the existence of the

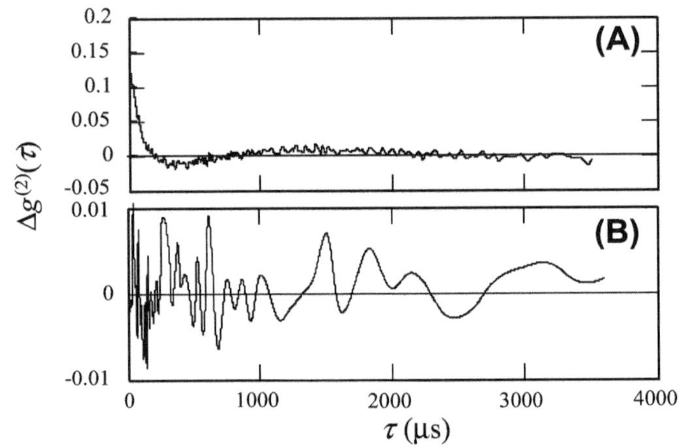

Figure 3.41
The difference in the autocorrelation function from the calculation for colloidal crystals of CS81 spheres in the deionized suspension at 24° C. $\phi = 0.00122$, $\theta = 90$ degrees. **A,** Cumulant analysis; **B,** NNLS analysis. *From Okubo T, Kiriyama K, Nemoto N, Hashimoto H:* Colloid Polym Sci *274:93-104, 1996.*

multimodes for the colloidal crystal. Separation into multimodes from the autocorrelation functions was made by the histogram analyses of the NNLS and Marquardt methods.[163] The data fitting by the NNLS analysis was made successfully, as is seen in Fig. 3.41, *B*. In most cases, three modes have been detected for crystal suspensions. We adapted the NNLS analysis mainly and occasionally used the Marquardt method when we needed to ascertain the validity of the peak separation with the former analysis.

Fig. 3.42, *A* and *B*, show the SLS intensities, $I(q)$ and effective diffusion coefficients of colloidal crystals of CS91 as a function of scattering vector, q.[157] The colloidal crystal samples are the exhaustively deionized suspensions exhibiting the strong iridescent colors and large single crystals. The colloidal liquid samples are also prepared by diluting the stock suspensions with water and measured within 5 hours without further deionization with the resins. Surprisingly, sharp double peaks appeared in I versus q plots for the crystals. The measurements have been made at intervals of half the degree, especially around the peaks. However, the interval was not always narrow enough for the determination of the most reliable values of the peak scattering vector, q_m. These double peaks were assigned to be (111) and (110) reflections of *fcc* and *bcc* from the left. The difference in the peak angles is

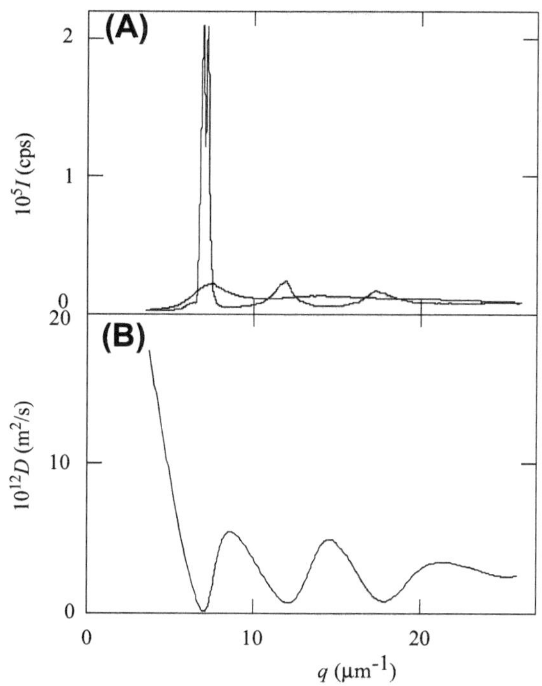

Figure 3.42 Light-scattering intensities and diffusion coefficients of colloidal crystals of CS91 spheres from SLS (*A*) and DLS (*B*) measurements at 24° C. ϕ = 0.00080, curve with sharp peaks: with resins; broad curve: without resins. *Data from Okubo T, Kiriyama K:* J Mol Liquids *72:347-364, 1997.*

approximately 1 degree only. Assignment of the respective peaks, q_m to the crystal planes is made rather easily from comparison of the observed peak vectors with the calculation from Equation 31:

$$q_{hkl} = 2\pi/a(h^2 + k^2 + l^2)^{1/2} \tag{31}$$

Here, h, k, and l are the Miller's indices and a the unit cell size. The Bragg distances (d_B) are determined from the q_m values in the $S(q)$ curves using the Bragg equation, $d_B = 2\pi/q_m$. Several SLS measurements of colloidal crystals, liquids, and gases were made in the highly deionized and diluted concentrations, validating the work of pioneering researchers.[104,161,164-166]

Colloidal crystal lattices transformed from *fcc* to *bcc* subphase when the sphere concentration decreased. It is now clear that the crystal structures of colloidal crystals are *fcc* (or closed packed) or *bcc* when the external fields are not applied. The *fcc* lattice transforms to *bcc* subphases when (1) sphere concentration decreases, (2) salt concentration is raised, (3) suspension temperature is raised within the usual range, (4) charge density of spheres is high, (5) pressure is applied, or (6) time has elapsed after suspension preparation without ion-exchange resins.[3,4,6,9,11,20,68,72,81,167-170]

It should be noted here that the intensity peaks of colloidal crystals, $I(q)$, are so sharp and the background intensity is so low in the exhaustively deionized suspension that there should be no significant difference with the profiles of structural factor $S(q)$. Equation 32 holds true.

$$I(q) \propto P(q)S(q) \tag{32}$$

Here, $P(q)$ is the single particle form (particle) factor and can be estimated experimentally from the $I(q)$ measurements on the same suspension, but in the presence of salt, for example. For spherical particles, $P(q)$ has been given as Equation 33 theoretically when the Rayleigh approximation is valid.[171]

$$P(q) = \left\{3[\sin(qd/2) - (qd/2)\cos(qd/2)]/(qd/2)^3\right\}^2 \tag{33}$$

Using the observed $I(q)$ and Equation 33, $S(q)$ versus q curves were estimated.

Fig. 3.42 shows the concentration dependency of $S(q)$ for ordered liquid (A) and crystals (B–D) of CS91 spheres in the exhaustively deionized suspensions with resins. Sphere concentrations range from 0.000159 to 0.0053 in volume fraction. $S(q)$ value of the first peak in Fig. 3.43, A, is approximately 2.5, which supports the ordered liquid nature of the suspension. The sharpest peak in Fig. 3.43, B, was assigned to be (110) reflection of *bcc* lattice, and other peaks were also from the *bcc* lattices. The suspension corresponding to the curve, B, at $\phi = 0.000531$ showed the giant single crystals and the sedimentation of these crystals occurred when SLS measurements were made 12 hours after the suspension was set. Fig. 3.42, C,

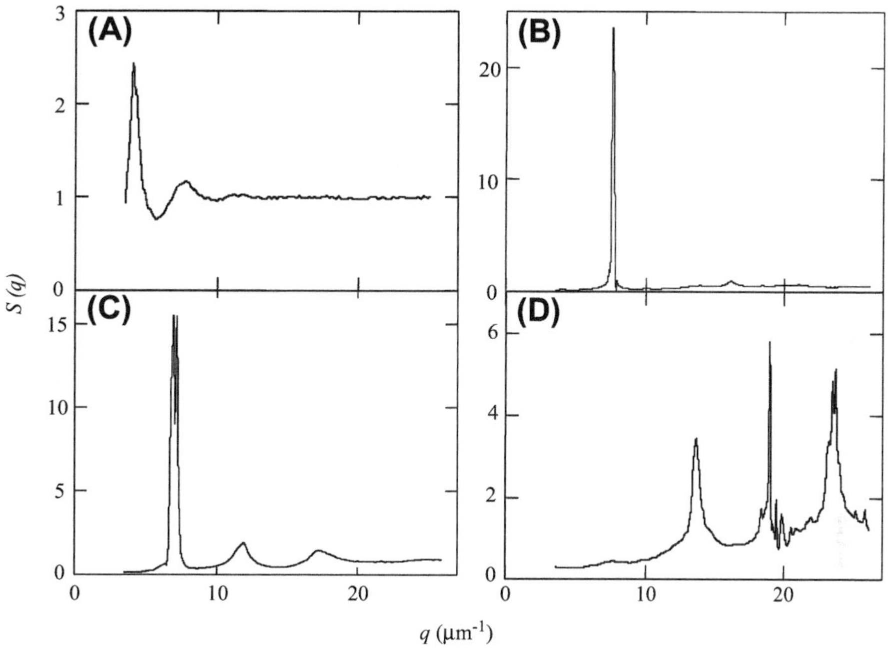

Figure 3.43
S(*q*) versus *q* plots for CS91 crystals with resins. **A,** $\phi =$ 0.000159; **B,** 0.000531; **C,** 0.000796; **D,** 0.00531. *Data from Okubo T, Kiriyama K:* J Mol Liquids *72:347-364, 1997.*

shows the $S(q)$ curve corresponding to the $I(q)$ curve shown in Fig. 3.42, *A*. The lattice structures were assigned to be mixtures of *fcc* and *bcc* for curve *D*. The primary sharp peak must be fused from (111) plane of *fcc* lattice and (110) of *bcc*, where these two peaks were so close. The Bragg distances observed closely agreed with the calculation from the sphere size and concentration. It should be noted here that the magnitudes of $S(q_m)$ were large, approximately 23 and 16 for curves shown in Fig. 3.43, *B* and *C*, respectively. These magnitudes may be too large, when the magnitudes indicating the coordination number of crystal is taken into account. The experimental errors of the magnitudes are large, because the measurements around the peak were made at intervals of half the degree, very short and at the limit of the instrument. However, this interval was not narrow enough to get reliable $S(q_m)$ values.

Fig. 3.43 shows the typical $S(q)$ curves for colloidal gas and liquid of CS91 spheres for colloidal suspensions without ion-exchange resins. Fig. 3.44, *C*, shows the disordered liquid. Curves shown in Fig. 3.44, *D*, support the ordered (or super-cooled) liquids. Sphere concentrations range from 0.000080 to 0.0027 without resins. The curve *A* is a typical for colloidal gas, because the $S(q)$ values kept unity irrespective of *q* values. The $S(q)$ curves of the ordered liquid have been studied in detail by Hayter et al.,[172-174] Hartl

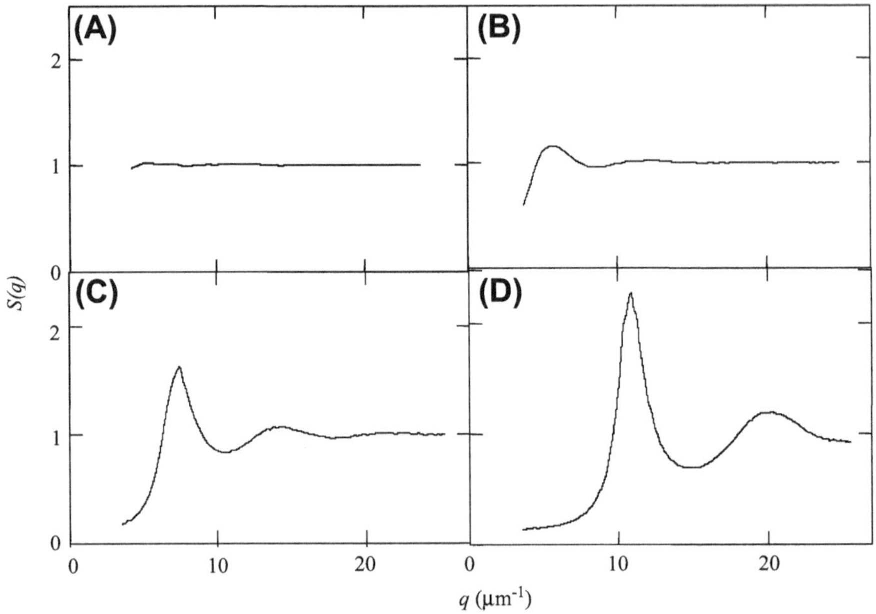

Figure 3.44 $S(q)$ versus q plots of gases and liquids of CS91 spheres without resins. **A,** $\phi = 0.000080$; **B,** 0.00027; **C,** 0.00080; **D,** 0.0027. *Data from Okubo T, Kiriyama K: J Mol Liquids 72:347-364, 1997.*

and Versmold,[175,176] van Megen et al.,[177] Ottewill,[8] and Klein et al.[178] The magnitudes of $S(q_m)$ of liquids range from approximately 1.5 (disordered liquids) to approximately 2.5 (ordered liquids).

The diffusion coefficients evaluated from the cumulant analysis are shown in Fig. 3.41, *B*. The diffusion coefficients for the suspensions showing peaks in the light-scattering curve were substantially low. According to Pusey and Tough,[179] the first cumulant K_1 in Equation 29 is given by:

$$K_1 = D_o q^2 / S(q) \qquad 34$$

where D_o is the diffusion coefficient for a system of noninteracting particles. The effective diffusion coefficient, D_{eff} is given by:

$$D_{eff} = D_o / S(q) \qquad 35$$

Constancy of $D_{eff} \times S(q)$ has been satisfactory for the colloidal liquid within experimental errors.[179-181] Figs. 3.42, *A* and *B*, support the validity of Equation 35 for the colloidal crystal system. It should be noted that the primary peaks were not always highest among the peaks, which is different from the expectation for the powder of crystals. These observations support the finding that the colloidal single crystals are large and the direction of their crystal planes are not randomized.

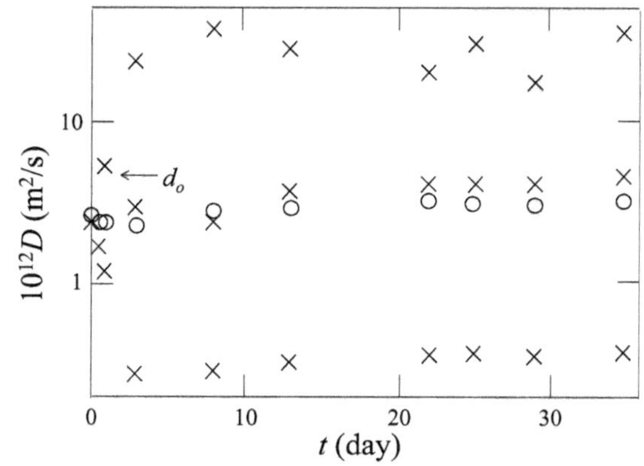

Figure 3.45
Diffusion coefficients of CS81 spheres as a function of time elapsed with resins at 24° C. $\phi = 0.00049$; *circles:* cumulant analysis; *crosses:* Marquardt analysis; $\theta = 90$ degrees. *From Okubo T, Kiriyama K, Nemoto N, Hashimoto H:* Colloid Polym Sci *274:93-104, 1996.*

When the suspension concentration of CS81 is 0.00049 in volume fraction with ion-exchange resins (Fig. 3.45), the structure changed from gas to liquid within 1 day after the suspension was set, and to crystal after 3 days.

Fig. 3.46 shows the sphere concentration dependence of the diffusion coefficients without resins obtained from data analysis with the cumulant method (circles) and also with the NNLS method (crosses). When the suspension concentration is lower than 0.0012 in volume fraction without resins for CS81 spheres, the structure was gas and changed from gas to liquid, and further to crystal at higher concentrations than 0.002 and 0.005 in volume fraction. The liquid structure was characterized by the short (D_s) and long

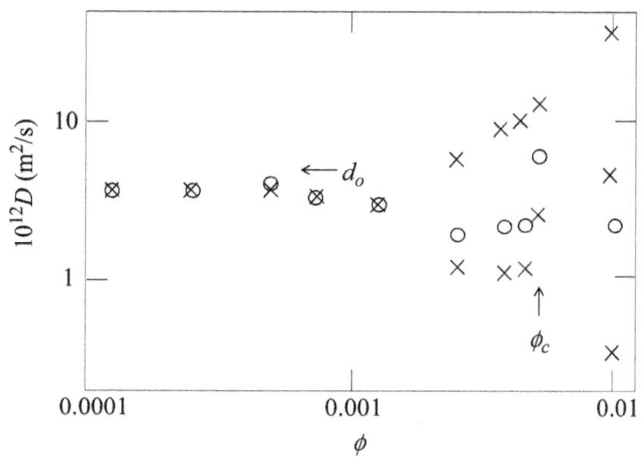

Figure 3.46
Diffusion coefficients of CS81 spheres as a function of sphere concentration without resins at 24° C. Diameter and critical concentration of melting are indicated by d_o and ϕ_c, respectively. *Circles:* cumulant analysis; *crosses:* Marquardt analysis; $\theta = 90$ degrees. *From Okubo T, Kiriyama K, Nemoto N, Hashimoto H:* Colloid Polym Sci *274:93-104, 1996.*

diffusion coefficients (D_l) with $D_s = D_o$ and $D_l \cong D_s/3$. On the other hand, we found that time profiles of $g^{(2)}(t)$ of the colloidal crystal were better represented by curves fitted from respective modes, only for simplicity. Owing to our experimental setting for DLS measurements, we could neither confirm the periodic nature of the first cumulant, although the cumulant analysis resulted in a poor fit, nor calculate the decay rate from elastic constants of the colloidal crystals. Three apparent diffusion coefficients ($D_f > D_s > D_l$) calculated from respective rates are different from one another by one order of magnitude. Values of the fastest and the slowest modes, D_f and D_l, did not change with time once the crystal structure was formed, whereas D_s of the medium mode very gradually increased with time and became close to the intrinsic D_o value. The medium mode, strongest in amplitude, is assigned as the Brownian oscillation of the sphere around the equilibrium lattice point, taking into account the results for the liquid suspension. Visual observation of the crystal structures using a metallurgical microscope have been the synchronous movement of colloidal spheres.[3,182] The magnitude of Brownian oscillation, $<\delta^2>^{1/2}$ was determined by the repulsive potential valley of crystal lattice. According to the Lindemann's rule of crystal melting, the non dimensional parameter, g given by $<\delta^2>^{1/2}/l$ (l: intersphere distance), is less than 0.1 for a stable crystal. When g equals unity, the sphere distribution is ideal gas. The order of magnitude of the elastic modulus, G, is written in terms of the magnitude of the synchronous fluctuation, d, as in Equation 36.[5,11,20]

$$G \cong f/l \cong (k_B T/<\delta^2>)/l \cong N k_B T/g^2 \qquad 36$$

Here, f is the elastic force constant because the magnitude of g has been estimated from the rigidities of colloidal crystals to be 0.02 to 0.1, $<\delta^2>$ of spheres ranges from 24 to 120 nm for $\phi = 0.00049$ for CS81, for example. The order of magnitude is comparable with $q^{-1} = 53.5$ nm in which our DLS experiments were performed but smaller than Debye-length (l_D). The slowest mode cannot be assigned at the transitional diffusional motion of the spheres through several nearest-neighbor shells, because such a motion is prohibited in the stable crystal lattice. The slowest mode may be related to the overdamped oscillation mode of the colloidal crystal lattice. It should be further noted here that the slowest mode is correlated to or corresponds to the oscillating shear waves of the elastic crystals, which has been clarified for the colloidal crystals.[6,7,183,184]

The D values as a function of sphere concentration were also observed with resins systems, although the graph was omitted to save space (see Fig. 10 in Okubo et al. [1996][156]). Concerning the fastest mode, the easiest assignment is to relate to the effect of multiple scattering. D_f was, roughly speaking, independent of ϕ, and ϕ-independent amplitude to this mode may support the previous idea in this relatively dilute region. However, DLS measurements on 10× concentrated suspensions with the liquid structure did not show any symptom of the multiple scattering. For aqueous solutions of poly-*L*-lysine hydrobromide with very low concentration (c_s) of added NaBr salt, the apparent D increased with decreasing c_s from 2×10^{-11} m^2/s at $c_s =$ approximately 10^{-1} mol/L to 4×10^{-10} m^2/s at $c_s = 10^{-4}$ mol/L.[185] This pronounced salt effect was interpreted by a theory of the electrically neutral fluctuation of local permittivity in the polyelectrolyte solutions.[186] The rigid silica spheres have only 3 degrees of freedom in comparison with a high number of internal degrees of freedom to the flexible macromolecules. The previously described theory may not be literally applicable for explanation of the fastest mode. However, we may not deny the possibility that the enhanced diffusion caused by the interaction between charged sphere and the counter-ion clouds might be in the completely deionized state.

The D_s looked to increase gradually with ϕ and exceeds D_0. The d_{eff} becomes equal to l at approximately $\phi = 0.001$, above which the electrical double layers start to overlap each other. Such overlapping may affect the diffusive motion of the sphere around the equilibrium position to some extent. An increase of D_l with ϕ was noticeable and may be represented as $D_l \propto \phi^{1/3}$. The lattice dynamics predicts that the undamped frequency ω_0 with the sinusoidal varying term in Equation 30 is proportional to the elastic constant of the lattice. Then Γ_e or D_l at constant q should be proportional to $\phi^{2/3}$ from Equation 34, when g, the index of stability of the crystal, is independent of ϕ. More accurate measurements on ϕ dependence of G seem necessary for dilute colloidal crystals. Validity of the prediction by Cotter and Clark may have to be discussed in more detail for the suspension of the extremely low concentrations of colloidal crystals.[161]

Colloidal single crystals are broken easily, even by the shear stress caused by the suspension flow occurring in small displacement of the cell. This occurs because the elastic modulus of the single crystals is extremely low.[11,16] When the suspension displaying colloidal crystals is shaken and left to stand, the suspension shows iridescent colors in time. Within several minutes, very small single crystals form and sometimes they show the blinking phenomena by the Brownian movement of the single crystals.[187] The growing process of the single crystals was observed from 10 to 40 minutes after shaking, and seemed, to the naked eye, to be completed after approximately 60 minutes.

Formation processes of the liquid structure were also pursued for the suspension with Bio-Rad resins coexisted in the observation cuvette by repeated $g^{(2)}(\tau)$ measurements. The $g^{(2)}(\tau)$ were analyzed with both the cumulant method and the Marquardt method to obtain the first cumulant, Γ_e, and also two decay rates, from which the apparent diffusion coefficients D were evaluated. In the gas suspension, the cumulant and Marquardt analyses gave identical single D values corresponding to the translational diffusion of free particles, which monotonically decreased with time and became constant. This means clearly that the Debye screening length (l_D) increases during the course of deionization process with the resins, and that the hydrodynamic diameter of spheres increases as the electrical double layers expands. The forced fit by the cumulant analysis for the colloidal liquids gave D values similar to those found in the gas structure. However, the Marquardt analysis gives the presence of two modes to time profiles of $g^{(2)}(\tau)$. From the decay rate of the gas suspension the translational diffusion coefficient, D_o, was estimated soundly.

The diffusion coefficients and the hydrodynamic diameters of CS91 spheres are displayed in Fig. 3.47 as a function of salt concentration.[157] The most impressive feature in this figure is that D_o increased significantly by 20% as concentration of NaCl increased. This change is explained in terms of thinning of the electrical double layers with increasing ionic concentration of suspension. The dotted curve shows the calculated value of d_{eff} ($= d + 2 \times l_D$) using Equation 1. Good agreement between the experiment and the calculation is attained only at high concentrations of NaCl. Calculated salt concentration dependency is much more substantial compared with the observation. The larger difference at lower ionic concentrations is ascribed to the fact that the electrical double layers are soft and easily distorted. The electrical double layers play an essential role for crystal formation at higher sphere concentration, but dynamically only a fraction of water inside the double layer moves with the sphere. Nevertheless, the hydrodynamic diameter, d_H, in the deionized state is much larger than the true diameter, d (110 nm), of the sphere and the greatest amount of water is contained in the shell of the thickness, $(d_H - d)/2 = 15$ nm. At any rate, an important contribution of

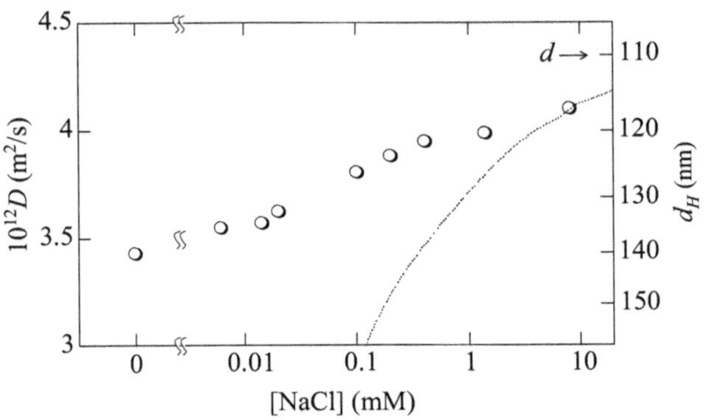

Figure 3.47
Diffusion coefficients and hydrodynamic diameters of CS91 spheres as a function of NaCl concentration at 24° C. $\phi = 0.000796$; *circles:* cumulant analysis; $\theta = 90$ degrees; *dotted curve* shows the calculated values of d_{eff}. *Okubo T, Kiriyama K:* J Mol Liquids *72:347-364, 1997.*

the electrical double layers is not denied at all by this difference. This salt influence on the diffusion behavior strongly supports the finding that the colloidal spheres are always accompanied by the electrical double layers and the translational movement of the former is retarded as the latter expands. It should be noted that the observed D_0 values at high concentrations of NaCl are still lower than that calculated from the true diameter, *d*, as is shown in Fig. 3.47. This may be ascribed to the hydration effect around the colloidal spheres. The importance of the electrical double layers on the translational Brownian movement of colloidal spheres in gas suspension has been supported by other experiments[188] as described in Chapter 2 (see Figs. 2-21 and 2-22).

SLS and DLS measurements of very small colloidal suspensions (22 nm in diameter) have been made,[189] where crystallization did not take place in any concentration, even in the exhaustively deionized state. However, colloidal liquid structures appeared. Back scattering of light measurements on the several colloidal crystal systems were reported.[166] The reader is advised to review several important works on the interparticle structural factor, $S(q)$, from video-tape observation, light-scattering, X-ray, and neutron-scattering measurements for the deionized and salt-containing colloidal suspensions.[8,182,190-199] DLS measurements have been made in detail for colloidal crystals.[200-204] Particle motion was studied by tracer diffusion method for concentrated colloidal suspensions.[205] Effects of polydispersity of colloidal spheres on the structural factor was discussed in detail.[206]

3.9 Structural Relaxation Times of Colloidal Crystals

The structural relaxation time (τ), the time of fluctuation of spheres in both crystal and liquid suspensions, is one of the most fundamental dynamic properties. I observed the τ values of colloidal crystals and liquids for the first time by spectrophotometric and conductance-stopped-flow (CSF) measurements.[207,208] The ordered structure should be broken down during the rapid continuous flow because of the strong shearing forces arising from the velocity gradient. When the suspension flow is stopped, the colloidal spheres revert to moving freely and ordering reappears. The appearance of the ordering is observed by the changes in the absorbance of transmitted light by the flow birefringence effects. Suspension conductance is also expected to change with time in the CSF cell, because the electrical anisotropy of the electrical double layers occurs as a result of the shear rate flow and relaxes toward an equilibrium state after the flow is stopped.[209-217] The reaction systems studied by CSF method hitherto were micelle equilibria of ionic surfactants,[218] polyelectrolyte complexation,[219-221] inclusion reactions,[222-224] enzymatic reactions,[223,225] rotational relaxation times,[217,226-229] and

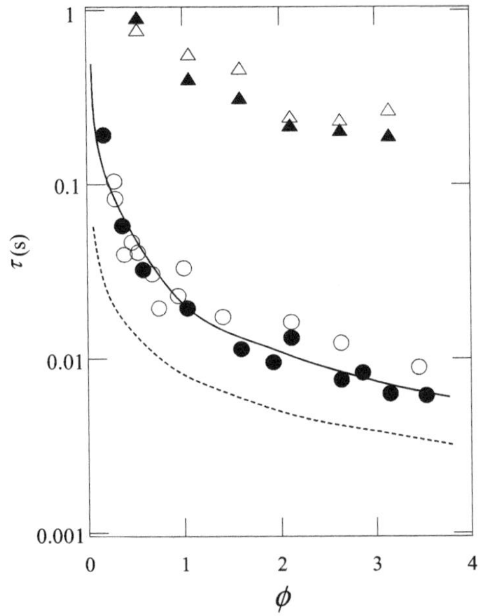

Figure 3.48 Structural relaxation times of colloidal crystals and liquids of D1B22 spheres observed by SPSF (*circles:* τ_f; *triangles:* τ_s) and CSF measurements (*solid circles:* τ_f; *solid triangles:* τ_s) at 25° C. *Solid curve:* τ_f calculated using d_{eff}; *broken curve:* using real diameter. *Okubo T:* J Chem Phys *87:3022-3028, 1987.*

others.[230-232] Fig. 3.48 shows the typical results of τ values obtained for colloidal crystals and liquids of polystyrene spheres (D1B22). The influence of colloidal size and the addition of potassium chloride on the τ values are demonstrated in Fig. 3.49 and 3.50, respectively. Two kinds of τ values, fast (τ_f) and slow (τ_s) steps are observed by the stopped-flow technique. The former is given by Equation 37.

$$\tau_f = l^2/6D_t, \qquad 37$$

where l is the mean intersphere distance, and D_t is the cooperative translational diffusion constant in strongly interacting structure. The D_t value of spheres is given by the Stokes-Einstein equation for self-diffusion in Equation 2.

The essential point in the fluctuation of spheres is that the D_{tr} and τ_f values observed in the colloidal crystals agree closely with the calculated values from Equations 37 and 2 when d_{eff} is used instead of d (true diameter of colloidal sphere in Equation 2),[207,208] as is shown in Fig. 3.48. Agreement between the calculation and observation in the deionized suspensions of monodisperse polystyrene spheres was excellent. It should be noted here that d_{eff} value, which includes Debye-screening length, does not correspond to the hydrodynamic length. Ottewill et al. observed particle motion directly with a microscopy and concluded that the observed translational diffusion

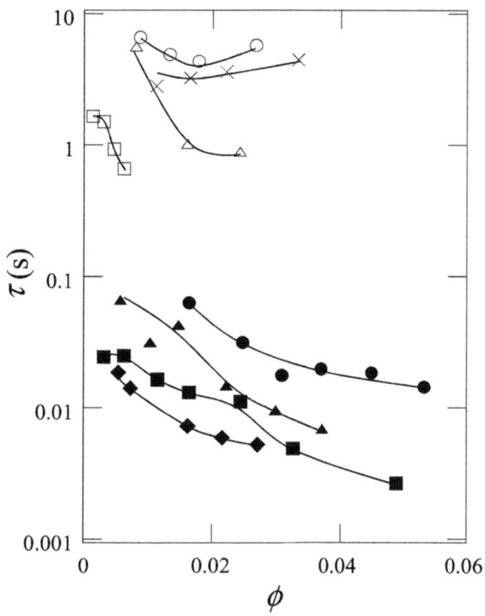

Figure 3.49 Structural relaxation times of colloidal crystals and liquids of polystyrene spheres observed by CSF measurements at 25° C. *Circles:* D1A35; *crosses:* D1A12; *triangles:* N1000; *squares:* D1C11; *solid circles:* G5301; *solid triangles:* D1B72; *solid squares:* D1B76; *diamonds:* D1C27. *Okubo T:* J Colloid Interf Sci *117:165-171, 1987.*

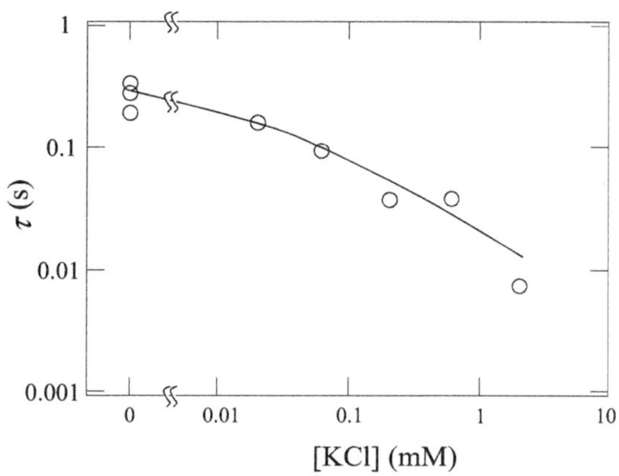

Figure 3.50 Structural relaxation times of colloidal crystals and liquids of LSN spheres in the presence of potassium chloride observed by CSF measurements at 25° C. *Okubo T:* J Colloid Interf Sci *117:165-171, 1987.*

constants agreed with the value calculated from Equation 2 using the real value of sphere diameter.[233] However, the tendency of $D_{obs} < D_{calc}$ is not denied in their paper, especially for small spheres (i.e., $D_{obs} = 3.6 \times 10^{-9}$ cm^2s^{-1} and $D_{calc} = 4.0 \times 10^{-9}$ cm^2s^{-1} for $d = 0.458$ μm). The importance of the electrical double layers on the diffusion properties was clarified, as was described in Chapter 2 using Figs. 2-21 and 2-22, and further Fig. 3.47 in this chapter. Fig. 3.49 shows that τ values increase as sphere size increases,[207] which is consistent with the dependences of τ_f on D_t and then d_{eff} in Equation 37. Interestingly, the τ values decreased sharply with salt addition, as is shown in Fig. 3.50. This result is also explained by the decrease in d_{eff} values with salt addition.

The τ_f value is given by the simple relation shown in Equation 38.[234,235]

$$\tau_f \approx \eta / G \qquad 38$$

where G denotes the rigidity or dynamic (local shear) elastic modulus of the suspension. G ranges from 0.01 Pa to 10 Pa, depending on the particle size and concentration of colloidal crystals, and these values are not far from those of previous reports by the dynamic method.[6,236-239]

The effect of the electrical double layers on the translational diffusion constant of colloidal spheres has been discussed using photon correlation spectroscopy[240,241] and theoretically.[242] It should be noted further that the diffusion constant showed a minimum when the double layer thickness was comparable to the radius of the particle.[10]

3.10 Rigidity of Colloidal Crystals

The elastic modulus is one of the most fundamental properties of colloidal crystals and has been measured by various techniques. The static elastic moduli of crystals of monodisperse polystyrene spheres were evaluated by Crandall and Williams from reflection spectrum measurements at sedimentation equilibrium.[243] Static moduli of colloidal crystals have been measured, especially in the diluted and deionized suspension state, by several methods; these are microscopic observation[95,97,112,182,244-249] and reflection spectrum methods in sedimentation and diffusion equilibrium,[118,250-252] reflection spectroscopy under centrifugal compression,[253] and a direct current (*dc*) electric field method.[254] On the other hand, the "dynamic" (shear) elastic modulus has been measured especially for the concentrated suspension state and discussed theoretically by Goodwin and Smith[255] by shear wave propagation experiments. They measured elastic moduli of flocculated suspensions of colloids in the presence of foreign salt.[256-258] The dynamic moduli of colloidal crystals, especially in the deionized suspensions, have been studied by Mitaku et al.[5,64,237,238] Joany,[259] Fujita and Ametani,[260] van Megen et al.,[261] Benzing and Russel,[262] Lindsay and Chaikin,[6,235] and Pieranski et al.[239] Rough values of dynamic rigidities of colloidal crystals were evaluated from the structural relaxation measurements using Equation 38.[207,208] I have measured the dynamic moduli from rotatory viscometry.[263-265]

The principles of the measurements of the static rigidities are described briefly for four main techniques: reflection spectroscopy in the sedimentation, diffusion, centrifugal, and in the DC electric field methods. According to Crandall and Williams,[243,250] the intersphere distance, l, or peak wavelength, λ_p, is shown by Equations 39 and 40.

$$l - l_m = (\rho_{eff} g_o l_m \phi_m / G)(h - h_m) \quad 39$$

$$\lambda_p - \lambda_{pm} = (\rho_{eff} g_o \lambda_{pm} \phi_m / G)(h - h_m) \quad 40$$

Here h and h_m are the heights of the horizontal plane examined and the horizontal midplane of the crystal suspension phase in the cell. The lattice spacing, the peak wavelength, and the sphere concentration in volume fraction at the midplane are represented by l_m, λ_{pm}, and ϕ_m, respectively. The initial sphere concentration is equal to ϕ_m. The effective density given by the specific gravity of sphere, ρ_{eff}, minus that of the solvent, g_o, is the gravitational constant. The rigidity (elastic modulus), G, is obtained from the slope of l (or λ_p) versus h plots at the sedimentation or diffusion equilibrium state. In the sedimentation experiments, the initial sample suspensions are set in the cell with a small bag of the ion-exchange resin's mixtures. And then the reflection spectra are taken several times as a function of heights until a state of equilibrium is reached. In the diffusion experiments,[252] on the other hand, the stock suspension is introduced carefully into the bottom of the observation cell containing solvent with or without an ion-exchange bag. The initial ascending velocity of the interface increases sharply as the ionic concentration of suspension decrease. The translational diffusion coefficients, D_t, of the colloidal spheres were surprisingly large, approximately 700-fold compared with that calculated using the Stokes-Einstein equation. The important role of the intersphere electrostatic repulsion forces was demonstrated for the diffusion processes.

A new and convenient technique for measuring elastic modulus of colloidal crystal was developed from the reflection spectroscopy at the centrifugal fields.[253] Beautiful color bands appeared on the observation cell surface, as is shown in Fig. 3.51. Centrifugal compression takes place for the lattice spacing (l) of the crystal suspension away from the center. The repulsion force, F_1, resulting from the deformation of the elastic crystals at the distance from the center of the cell, r(I), is given by Equation 41, where G is the rigidity and l(I) is the lattice spacing at r(I).

$$F_1 = -G[(l(\mathrm{N}) - l(\mathrm{I}))/l(1)]r(\mathrm{N})\theta L \quad 41$$

The innermost point and an arbitrary point of the suspension in the cell are denoted by I = 1 and N. An arbitrary small central angle is represented by θ; L is

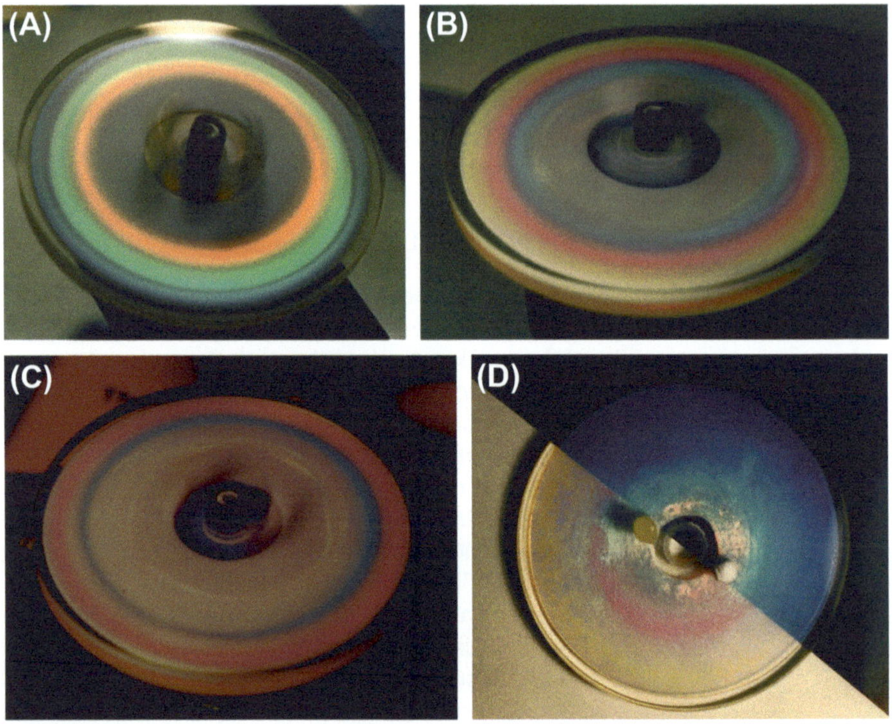

Figure 3.51 Close-up color pictures of colloidal single crystals of D1K88 (*A, B*) and D1B72 spheres (*C, D*) in a rotating cell. $\phi_{D1K88} = 0.0617$, $\phi_{D1B72} = 0.0820$. *Okubo T:* J Am Chem Soc *112:5420-5424, 1990.*

the width of the rotational observation cell. On the other hand, the centrifugal force applied on the suspension (F_2) is given by Equation 42, where w is the angular velocity, and ρ and ρ_o are the specific gravities of spheres and solvent.

$$F_2 = w^2(\rho-\rho_o)I\theta\sum\left\{\phi(\mathrm{I})I(\mathrm{I})^2[I(\mathrm{I})-I(\mathrm{I}-1)]\right\} \quad 42$$

The concentration of the spheres in volume fraction at point $I(I)$ is represented by $\phi(\mathrm{I})$ and is given by Equation 43.

$$\phi = (9.136d_o/\lambda_p)^3 \quad 43$$

From the equality, $F_1 = F_2$ at the centrifugal equilibrium, $G(\mathrm{N})$ is given as:

$$G(\mathrm{N}) = \left\{\left[w^2(\rho-\rho_o)I(1)\right]/[r(\mathrm{N})\times[I(\mathrm{N})-I(1)]]\right\}\sum\left\{\phi(\mathrm{I})r(\mathrm{I})^2[r(\mathrm{I})-r(\mathrm{I}-1)]\right\} \quad 44$$

Rigidity of colloidal crystals is evaluated in a *dc* electric field. When we combine the equation proposed by Crandall and Williams[243] and by Tomita and van de Ven,[266] the interparticle distance at a distance z from the anode plate, $a(z)$, is given by:

$$[a(z)-a(z_m)]/a(z_m) = 18\varepsilon\zeta Eg(\tau,z)\phi^*(z-z_m)/d^2G \quad 45$$

Here $a(z_m)$ is the interparticle distance at the midplane in the electrode cell at z_m, d is the sphere's diameter, ε is the permittivity of the solvent, z is the zeta potential of a sphere, E is the electric field strength, and $g(\tau,\zeta)$ is the relaxation effect discussed by Wiersma et al.,[267] and ϕ^* is the initial concentration of spheres. When $z = 0$, Equation 45 is modified to:

$$I/I_o = 1-[9\varepsilon\zeta g(\tau,\zeta)\phi^* LE]/d^2G \qquad 46$$

Here $a(0)$, $a(z_m)$, and z_m have been replaced by I, I_o, and $L/2$, respectively. L is the optical path length of the electrode cell. G is evaluated from the slope of I versus LE plots.

The experimental results on the rigidity of colloidal crystals are summarized as follows. First, the logarithms of rigidities (log G) of colloidal crystals, including amorphous solids, increased linearly with a slope of unity as log N (sphere number per unit volume) increased. Fig. 3.52 shows the plots of log G against log N for some typical colloidal crystals of monodispersed polystyrene spheres with diameter ranging from 91 to 109 nm. G values are denoted by open circles; crosses and open triangles are static, which were evaluated by us at the sedimentation and diffusion equilibrium. Diamonds symbols show the G values of D1B76 polystyrene sphere crystals observed

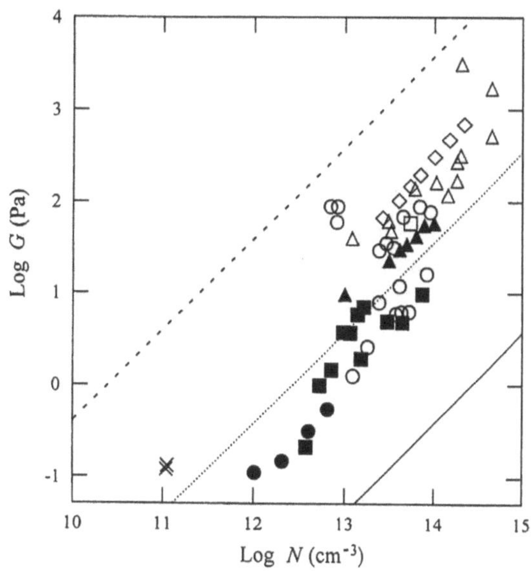

Figure 3.52
Rigidities of colloidal crystals as a function of number density of colloidal spheres. *Open circles:* reflection spectroscopy[250]; *crosses:* microscopy[247]; *open triangles:* rotator viscometer[263]; *open squares: dc* electric fields[254]; *diamonds:* centrifugal field[253]; *solid circles:* Crandall et al.[243]; *solid triangles:* Mitaku et al.[64]; *solid squares:* Lindsay et al.[6]; *solid line:* $g = 1$; *dotted line:* 0.1; broken line: 0.01.

from the centrifugal techniques.[248] The filled circles, filled squares, and filled triangles denote dynamic rigidities.[6,64,243]

The order of magnitude of the modulus is written in terms of the magnitude of the thermal fluctuation δ of a sphere as[3]:

$$G \approx f/l \approx (k_B T/ <\delta^2>)/l \approx N k_B T/g^2 \quad 47$$

where f is the force constant, l is the intersphere distance, and δ is the thermal fluctuation of a sphere in the effective potential valley. k_B and T are the Boltzmann constant and the temperature. Introducing a nondimensional parameter g for $<\delta^2>^{1/2}/l$, the modulus is obtained as a linear function of N. When $g = 1$, Equation 47 gives the rigidity of an ideal gas having the same sphere concentration. Lindemann's law of crystal melting tells us that $g < 0.1$ holds for a stable crystal. As is clear in Fig. 3.51, the g factors of the exhaustively deionized colloidal crystals especially shown in the open symbols are located around 0.03. It is interesting to note here that Equation 47 holds for all kinds of substances such as metals, proteins, ice, and colloidal crystals, for example. Rough values of the substances are estimated using Equation 47. It should be mentioned here that the order of magnitude of G is estimated by Equation 48:

$$G \cong NU \quad 48$$

where U is the electrostatic repulsion energy between neighboring spheres.[6,243,268] If the intersphere distance and the effective charges on a sphere are taken to be l_o (μm) and α, G is given roughly as $10^{-5}\alpha^2/\varepsilon l_o^4$ Pa. Increase of G with decreasing l_o and decreasing ε is understandable. The values of G for D1C25, D1C27, and D1B22 crystals, for example, were estimated to be 6, 14, and 15 Pa, respectively. The observed elastic moduli were larger than this rough estimation.[264]

Second, colloidal crystallization took place for very large polystyrene spheres (≈2-42 μm in diameter), which was observed by microscopy in the sedimentation equilibrium.[245] The particles floated above the cover glass apart from each other, with moving translational mode around their equilibrium positions, as is shown in Fig. 3.53, although the movement of the spheres is not recognized in the still picture. The G values and the g factors were compiled in Table 3.7. The crystals were highly compressed by the gravitational field and g factors decreased sharply from 0.03 to 0.0004 with increasing sphere sizes from 2 μm to 42 μm.[245]

Third, the rigidities of amorphous solids of anisotropic shaped colloidal particles of bentonite and MCC were evaluated from the sedimentation equilibrium method.[113] The characteristics and the results of the rigidity measurements are shown in Table 3.8. There exist no differences in the elastic properties between the spherical and nonspherical particles.

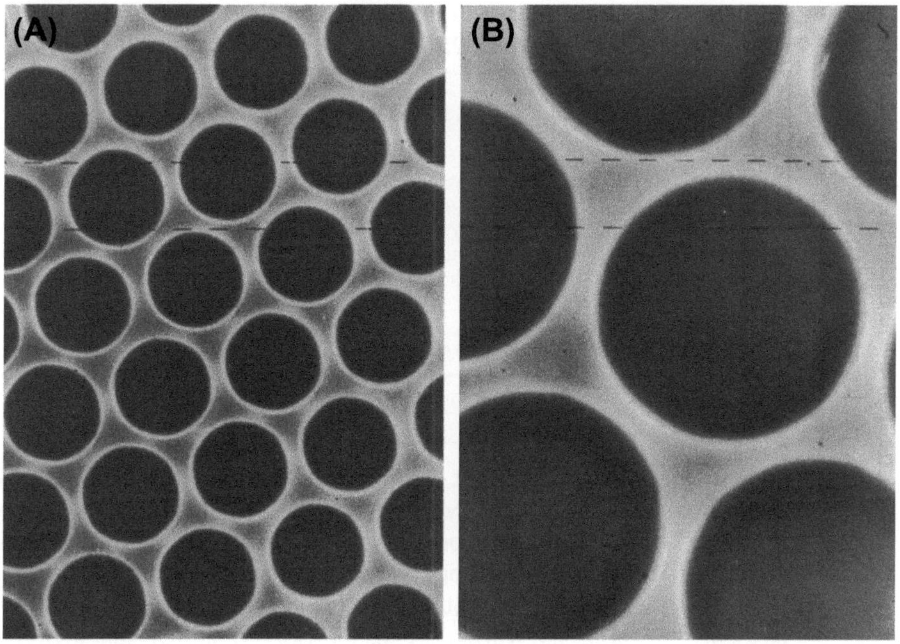

Figure 3.53
Microscopic pictures of SS151 (*A*) and SS411 (*B*) spheres in the sedimentation equilibrium in 20 vol% methanol aqueous and deionized suspension. **A,** 0.0238 vol%; **B,** 0.0591 vol%. *From Okubo T, Aotani S:* Colloid Polym Sci *266:1042-1048, 1988.*

Table 3.7 Rigidities of Colloidal Crystals of Large Polystyrene Spheres

Spheres	Diameter (μm)	*s/d*	*N* (1/cm³)	*G* (Pa)	*g*
SS021	2.06 ± 0.06	0.029	2.1×10^{10}	0.085	0.03
SS051	4.99 ± 0.15	0.030	4.2×10^{9}	0.21	0.01
SS151	15.0 ± 0.5	0.033	2.7×10^{8}	0.37	0.0015
SS201	20.0 ± 0.6	0.030	1.1×10^{8}	0.30	0.0011
SS411	41.7 ± 2.0	0.048	1.6×10^{7}	0.28	0.0004

G, Rigidity; *g*, g-factor; *N*, number density; *s/d*, polydispersity.

Table 3.8 Characteristics and Rigidities of Bentonite and MCC Colloidal Particles

Particles	Size (μm)	Specific Gravity (g/cm³)	Charge Density (μC/cm²)	*N* (1/cm³)	*G* (Pa)	*g*
Bentonite	2.3 ± 1.0	2.8	490	2×10^{10}	0.12	0.03
MCC	2.7 ± 0.6	1.5	230	2×10^{10}	0.09	0.03

G, Rigidity; *g*, g-factor; *MCC*, microcrystalline cellulose; *N*, number density.

Table 3.9 Salt Effects on the Rigidity and Yield Stress of Colloidal Crystals of D1C25 Spheres at 25° C $\phi_{D1C25} = 0.01$.

NaCl (M)	G (Pa)	Yield Stress (Pa)
0	490	2.0
8.2×10^{-7}	230	1.4
2.0×10^{-6}	130	1.7
4.1×10^{-6}	300	1.7
6.7×10^{-6}	130	1.6
8.2×10^{-6}	180	1.3
1.2×10^{-5}	5.3	0.6
1.6×10^{-5}	6.1	0.5 ± 0.2
2.0×10^{-5}	0.94	0.3 ± 0.1

G, Rigidity; *NaCl*, sodium chloride.

Fourth, the influence of salt addition to the dynamic rigidities of colloidal crystals of D1C25 spheres was measured with a rotatory viscometer from the shear stress-shear rate profiles.[263] Interestingly, both *G* and yield stress remain constant when the concentration of NaCl was smaller than 1×10^{-5} M. However, above this concentration, both parameters decreased abruptly, as is shown in Table 3.9. Mitaku et al.[64] reported similar salt effects from the yield stress. The critical concentration of NaCl (i.e., 1×10^{-5} M) corresponds to the melting of the colloidal crystal to the liquid structures.

Fifth, rigidity of the colloidal crystals, including amorphous solid, is larger than that of colloidal liquid when comparisons are made at the same sphere concentration. This is consistent with Equation 47, where g factors of liquid are larger than 0.1 and always larger than the corresponding solid (mean value of g is ≈ 0.03). Of course, the rigidity of colloidal gas is always smaller than the liquid and crystal at the same sphere concentration.

Sixth, the rigidity of colloidal crystals increases substantially as the deionization of the suspension proceeds. It takes a long time to achieve an exhaustively deionized state with use of the mixtures of cation- and anion-exchange resins coexisted in the sample suspension. In our experience, it took several months to several years to achieve the exhaustively deionized state of the stock suspension. Change in the critical concentration of melting after the sample suspension preparation was shown in Table 3.2.

Seventh, the rigidity of amorphous solid increased as the polydispersity of spheres increased, and the *G* value of crystal was smaller than that of amorphous solids when comparison was made at the same particle concentration.

Eighth, rigidity of the structured suspensions increases as the charge density of the particles decreases. This is due to the thickening of the electric double layers formed around colloidal spheres. The ion concentration in suspension is the sum of the concentrations of the diffusible counter-ions, foreign salt, and both H^+ and OH^- from the dissociation of water (see Equation 1 in this chapter). This effect is also consistent with the salt effects described previously. Some researchers may think that stable and hard colloidal crystals form for the colloidal particles of high charge density because of the effects of the stronger interparticle repulsion forces. The truth, however, is just opposite: The thickness of electrical double layers is determined by the concentration of free-state simple ions in the suspension phase, not the charge density of the particles. Colloidal particles of high-charge densities release a lot of counter-ions to the suspension phase, resulting in the thinning of the double layers instead.

Ninth, rigidity in the deionized state at a certain concentration of colloidal particles in volume fraction first increases with increasing particle size, but begins to decrease with further size increases. A maximum value of G was observed around 85 nm in diameter of spheres. The main cause for the increase of G with decreasing size is the fact that the contribution of the electrical double layers increases as the size of the colloidal particles decreases. For very small particles, the stable double layers do not form by their too-violent thermal motion. Furthermore, the size difference between colloidal particles and small ions and solvent molecules becomes less significant and the stable electrical double layers will not form for very small particles.

Tenth, rigidities of colloidal alloys were sensitive to their crystal structures, *sss*, or super-lattices, for example.[269] Rigidities of colloidal alloys of binary mixtures of colloidal silica spheres (CS82) with larger spheres (CS91, CS121, and CS161) were studied by the reflection spectroscopy in the sedimentation equilibrium. The total concentration of spheres in volume fraction was kept constant in the mixing. The *sss* structures formed for CS82 (103 nm in diameter) + CS91 (110 nm) and for CS82 + CS121 (136 nm). A superlattice, probably $MgCu_2$-type, formed for CS82 + CS161 (184 nm) mixtures. Fig. 3.54 shows the plots of the rigidity of the mixtures as a function of the mixing ratios. G values changed depending on the suspension structures and on the mixing ratio. When the rigidities of colloidal alloys were plotted as a function of the total number density of spheres (N_{total}), the rigidities increased as the fraction of other spheres increased. Furthermore, G values decreased sharply when comparatively few large spheres were mixed with the small spheres.

Furthermore, the colloidal crystals in alcoholic organic solvents in the deionized state have been studied by the reflection spectroscopy in sedimentation equilibrium.[251] As is shown in Fig. 3.55, G values first increased and then decreased as the fraction of the organic solvents in the mixtures further

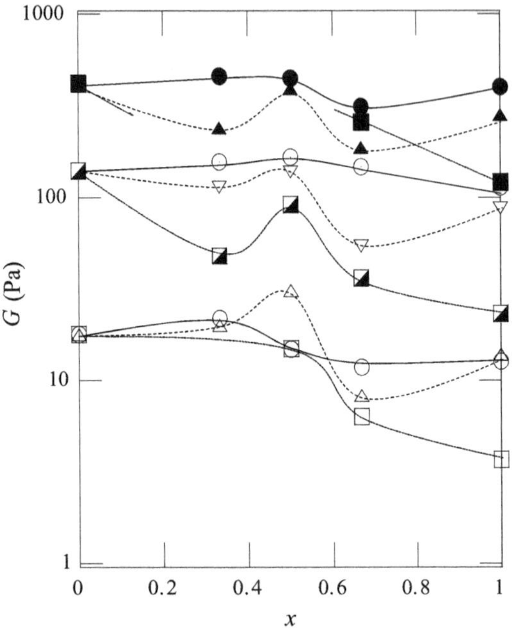

Figure 3.54
Rigidities of colloidal sphere mixtures as a function of fraction, *x*. $\phi_{CS82} + \phi_{CS91} = 0.005$ (*circles*), 0.02 (*dotted circles*), 0.05 (*solid circles*); $\phi_{CS82} + \phi_{CS121} = 0.005$ (*triangles*), 0.02 (*inverted triangles*); and 0.05 (*solid triangles*); and $\phi_{CS82} + \phi_{CS161} = 0.005$ (*squares*), 0.02 (*half solid squares*), and 0.05 (*solid squares*).

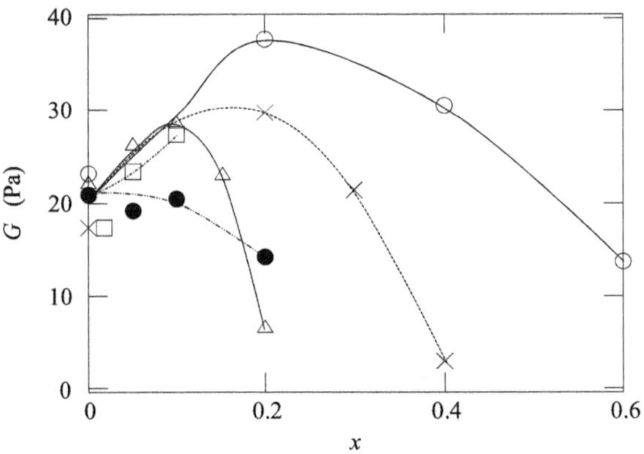

Figure 3.55
G of D1B76 spheres as a function of volume fraction of organic solvents (x). *Circles:* methanol; *crosses:* ethanol; *triangles:* propanol; *squares:* dioxanes; *solid circles:* ethyleneglycol. $\phi_{D1B76} = 0.00214$ (methanol), 0.00232 (others). *From Okubo T:* J Chem Soc Faraday Trans *1 86:151-156, 1990.*

increased. The way in which *G* is influenced by the organic solvents was explained reasonably by the change in the effectively dissociated charge numbers of the colloidal spheres. The fraction of the free-state counter-ions (β) decreased first and then increased by the addition of methanol, ethanol, and propanol.[57] Furthermore, β values also decreased and then increased by the addition of dioxane and ethylene glycol. Decrease in β values corresponds to an increase in the Debye screening length, l_D. It should be noted here that the rigidities of colloidal crystals of very large polystyrene spheres from 2 to 42 μm in diameter observed from the microscopy in the sedimentation equilibrium increased from 17% to 260% by the addition of 20 vol% of methanol.[245]

The dynamic elastic moduli of concentrated colloidal crystals of monodispersed polystyrene spheres, D1C25 and D1B22, were studied by means of a cone-plate type rheometer.[265] In the experiments, the exhaustively deionized state was not always kept during the measurements, because the small amount of the samples in the cell were open with air. Furthermore, the measurements that include the ion-exchange resins are often impossible to obtain without disturbing the rotation of the cell. However, the concentrations of the samples were high, 0.133 and 0.125 in volume fraction for D1C25 and D1B22, and the ion-contamination during measurements were not so serious compared with the measurements of diluted suspensions. The suspensions showed strong iridescent colors resembling those in Fig. 3.1 but were slightly more transparent. Furthermore, marked non-Newtonian flow was observed, which gives an evident yield stress. Dynamic modulus was mostly independent of the angular frequency. As the salt concentration increased, however, the iridescent colors disappeared and the system showed Newtonian flow. It is highly plausible that the crystal-liquid transition begins to occur in the ionic concentration range where the Yukawa potential between colloidal spheres is smaller than the thermal energy k_BT and $l_{eff}\,(= d + 2l_D)$ is close to or less than the distance between neighboring spheres in the crystal structure. Here, d is the diameter of colloidal sphere and l_D is the Debye-screening length. Dynamic rigidity of D1C25 and D1B22 were 150 Pa and 41 Pa at $w = 1\ s^{-1}$, and the latter decreased to 6.3 Pa with 0.3 mM of NaCl. The salt effects on G were quite similar to the data described by Lindsay and Chaikin.[6] G values obtained theoretically by Lindsay and Chaikin[6] and Robbins et al.[270] were close to the experimental values of D1C25 and D1B22. The crystal structures were kept in the nonlinear viscoelastic region of 40% dynamic domain. Furthermore, our work suggested that the flow and the nonlinear deformation of colloidal crystals are due to a slip in a lattice plane at a constant stress.

Rigidity of colloidal crystals in acetonitrile and nitrobenzene has been measured by the reflection spectroscopy in sedimentation equilibrium.[271] The colloidal spheres used were the silica spheres (136 nm in diameter) modified with poly(maleic anhydride-co-styrene), poly(methyl methacrylate), or polystyrene. The fluctuation parameter (g factors) of crystal lattices were between 0.038 and 0.045, being larger than those in the deionized aqueous systems, 0.028 in mean value. These results support the finding that the excluded-volume effects play an important role for the colloidal crystallization in the organic solvents. However, contribution of the electrical double layers was still effective in addition of the excluded-volume effects.

3.11 Viscosity and Viscoelasticity of Colloidal Crystals

Viscosities of colloidal crystals and liquids in exhaustively deionized aqueous suspensions have been measured with an Ostwald-type and a rotational viscometer.[110,263,272-276] The viscosity of the deionized suspension was sensitive to the degree of deionization of the suspension and decreased substantially when the suspension was contaminated with ionic impurities. This was found during the experiments on the viscosity of colloidal crystal samples. The viscosity continued to increase every week and saturated to the reliable values after several years for polystyrene spheres. Typical examples of the reduced viscosity, η_{sp}/c (specific viscosity, η_{sp}, divided by particle concentration, c) of colloidal liquids is shown in Fig. 3.56 for Ludox-SM colloidal silica spheres as a function of sphere concentration in g/ml.[273] Surprisingly, the viscosity of the deionized suspension at the liquid state was very high and the η_{sp}/c decreased sharply as the sphere concentration increased. All the viscosity experiments on colloidal spheres without resins hitherto have shown the increasing tendency of η_{sp}/c with increasing sphere concentration. Colloidal silica spheres are negatively charged and their surface charges do not change throughout the range of concentrations of

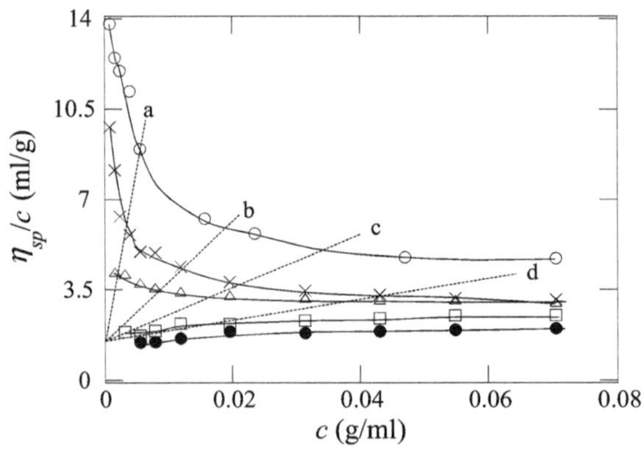

Figure 3.56
Plots of η_{sp}/c against c for Ludox-SM spheres at 25° C. *Circles:* [NaCl] = 0 M; *crosses:* 6.67×10^{-6} M; *triangles:* 2×10^{-5} M; *squares:* 6.67×10^{-4} M; *solid circles:* 2×10^{-3} M. *Dashed lines:* **A,** $n = 1 \times 10^{-4}$ M; **B,** 3×10^{-4} M, **C,** 5×10^{-4} M; **D,** 1×10^{-3} M. *From Okubo T:* Naturwiss *75:91-93, 1988.*

spheres. These extraordinary behaviors of viscosity in the deionized suspension will be explained reasonably with the thinning of the electrical double layers as sphere concentration increases. By the addition of NaCl, these behaviors turned toward normal; in other words, η_{sp}/c increased with c, which is quite similar to the viscosity behavior of polyelectrolyte solution as discussed by Pals and Hermans.[277] It should be noted here that size of Ludox-SM spheres are small, 12 nm in diameter, and the colloidal crystals never form. Colloidal crystals form when the monodispersed spheres are larger than approximately 40 nm in diameter. Furthermore, Nishida et al. explained the decrease in the reduced viscosity with sphere concentration using the electrostatic interaction theory for the charged spheres.[278]

The specific viscosity of colloidal spheres can be given by Equation 49.

$$\eta_{sp} = \eta/\eta_o - 1 = k_1\phi + k_2\phi_2 + \ldots. \quad 49$$

Here η and η_o are the viscosities of the colloidal suspension and the solvent. The concentration of the colloids (volume fraction) is represented by ϕ. The Einstein coefficient, 2.5, is k_1, and k_2 denotes the electrostatic interaction parameter, which is very large for the deionized suspensions. The deviation of the observed η_{sp} values from Einstein's equation was significant in the deionized liquids. In the presence of an excess amount of NaCl, however, the η_{sp} versus ϕ relationship agreed with the Einstein equation. Qualitatively speaking, the large k_2 observed implies that the electrostatic repulsion interaction is significant and the shells of the electrical double layers around the spheres are thick. When η_{sp}/c values were replotted as a function of *c*, but in the condition that the concentration of the diffusible ions in suspension, *n* are kept constant at the certain value, the reduced viscosities increased linearly with *c* as shown in the broken lines in Fig. 3.56, *A* through *D*. This condition means that the thickness of the electrical double layers keeps constant irrespective of sphere concentrations. Then the plotting was made for the same effective size of spheres, including the double layers.

Substantial deviation of the reduced viscosity of colloidal liquids in the deionized suspension from the Stokes-Einstein equation is clearly demonstrated in the η_{sp} versus ϕ plots, as is shown in Fig. 3.57. Broken lines show the Equation 49. In the deionized suspension, the intersphere repulsion interaction prevails far from the sphere surfaces and the shells of the double layers are very thick. The figure shows clearly that the η_{sp} versus ϕ curves come close to the Stokes-Einstein equation (shown by a broken line) in the presence of an excess amount of NaCl. By assuming that the simplified relation, Equation 50, holds for the deionized suspension, the effective volume fraction, ϕ_{eff} of the sphere, including the Debye length, is estimated.

$$\eta_{sp} = k_1\phi_{eff} \quad 50$$

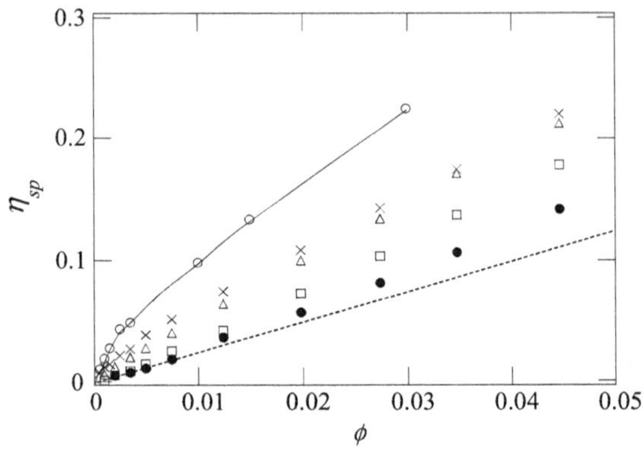

Figure 3.57 Plots of η_{sp} against ϕ for Ludox-SM spheres at 25° C. *Circles:* [NaCl] = 0 M; *crosses:* 6.67 × 10^{-6} M; *triangles:* 2 × 10^{-5} M; *squares:* 6.67 × 10^{-4} M; *solid circles:* 6.67 × 10^{-3} M; *dashed line:* Stokes-Einstein coefficient.

Here, l_D is now calculable using the relation, Equation 51, where d is the real diameter of spheres.

$$\phi_{eff}/\phi = [(d + 2 \times l_D)/d]^3 \quad 51$$

Viscometric properties of colloidal crystals are very peculiar. A peak in the relative viscosity versus concentration curves was observed for the deionized systems (Fig. 3.58).[272] The peak shifted toward higher concentrations by the addition of NaCl. The peak appearance is correlated to the phase transition between colloidal crystals and liquids in the shear flow condition. Suspensions at the concentrations near the peaks showed the beautiful iridescent colors, and the size of the flickering crystallites became smaller as the sphere concentration increased. At the higher concentrations far from the peak position, the crystallites were too small to be recognized with the naked eye.

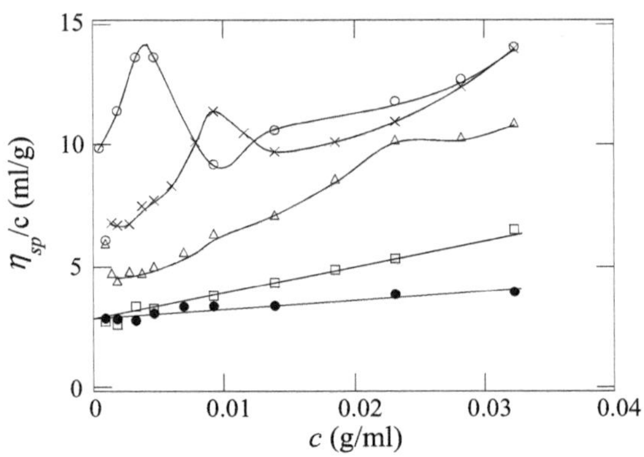

Figure 3.58 Plots of η_{sp}/c against c for LSN spheres at 25° C. *Circles:* [NaCl] = 0 M; *crosses:* 6.67 × 10^{-6} M; triangles: 2 × 10^{-5} M; squares: 6.67 × 10^{-5} M; solid circles: 2 × 10^{-4} M. *From Okubo T:* Naturwiss *75:91-93, 1988.*

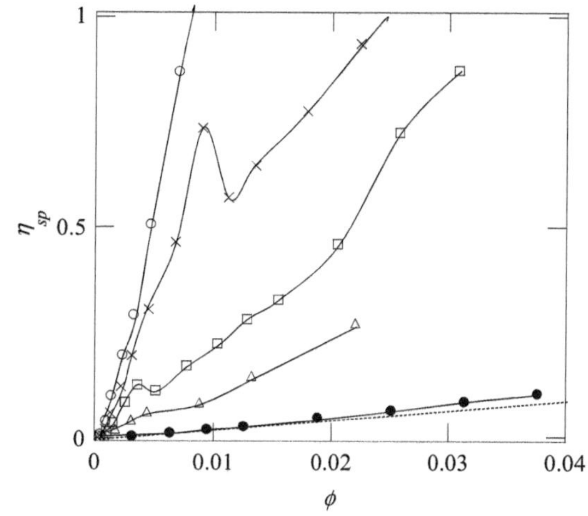

Figure 3.59 Plots of η_{sp} against ϕ for D1C25 (*circles*), D1C27 (*crosses*), LSN (*triangles*), D1B22 (*squares*), and N800 (*solid circles*) spheres at 25° C in the deionized suspension. *Dashed line:* Stokes-Einstein coefficient. *Okubo T:* J Chem Phys *87:6733-6739, 1987.*

The η_{sp} versus ϕ plots for colloidal crystals of various sizes of spheres ranging from 85 nm in diameter to 780 nm are shown in Fig. 3.59.[272] A clear, significant deviation from Stokes-Einstein equation was observed, especially for small spheres such as D1C25 and D1C27. The deviation was less significant as the sphere size increased. For N800 spheres, however, the η_{sp} values were close to Equation 45. This clearly shows the contribution of the electrical double layers for the viscosity of very large spheres being small even in the deionized state. It is clear that the viscosities of colloidal crystals (or amorphous solid) are highest and then decrease in the order of liquid and gas.

A sharp increase in shear stress with increasing shear rate was observed for the colloidal crystals, from which the dynamic rigidity was evaluated as described in the preceding section. Logarithms of the reduced viscosity η_r of colloidal crystals decreased linearly with a slope close to −1 as log [*shear rate*] increased. One of the typical examples is shown in Fig. 3.60.[267] This supports the solid nature of the suspensions. Table 3.10 compiles the slopes of log [*relative viscosity*] versus log [*shear rate*] plots for typical structured colloidal suspensions studied. The plots of log η against log [*shear rate*] for colloidal liquid supported the non-Newtonian flow with the slopes smaller than 0.8 but larger than 0.1 in the range of q values larger than approximately 20 s^{-1}. Viscosities of colloidal crystals were studied in detail by Lindsay et al.,[235] Joanicot et al.,[239] and Jorand et al.[268]

Viscoelastic properties of colloidal liquids and crystals in the deionized aqueous suspension have been studied in detail.[265,279-284] The slopes of the log [*shear stress*] versus log [*shear rate*] or log [*storage modulus*] versus log [*angular frequency*] plots for the colloidal crystals were zero, and increased to

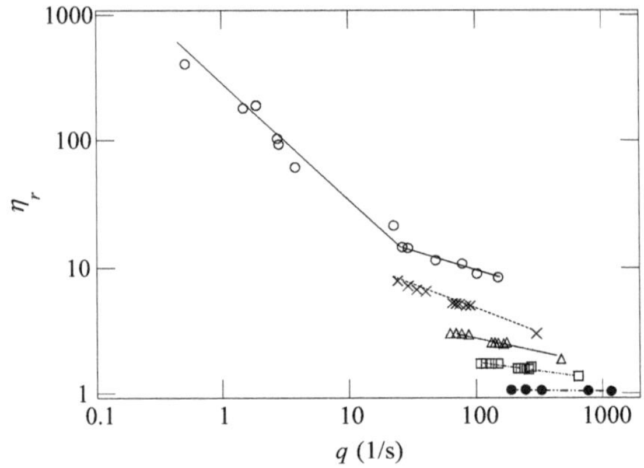

Figure 3.60
η_r versus q plots for D1C25 spheres at 25° C. *Circles:* 0.0275 g/ml; *crosses:* 0.0111 g/ml; *triangles:* 0.00664 g/ml; *squares:* 0.00354 g/ml; *solid circles:* 0.00147 g/ml.
Okubo T: J Chem Phys *87:6733-6739, 1987.*

Table 3.10 Slopes of Log (Relative Viscosity) versus Log (Shear Rate) Plots for Colloidal Crystals, Amorphous Solids, and Liquids in the Deionized Aqueous Suspension

Particles (ϕ)	V or VE	Slope	q (s^{-1})	Structure	Ref
D1C25 (0.0275)	V	−0.93	<10	Crystal	272
D1C25 (0.0632)	VE	−1.0	<50	Crystal	263
D1C27 (0.01∼0.07)	VE	−1.0	<40	Crystal	263
D1B76 (0.098)	VE	−0.93	<20	Crystal	263
D1B76 (0.098)	V	−1.0	<1	Crystal	274
D1B41 (0.061)	VE	−0.93	<40	Crystal	263
C351 (0.021)	VE	−0.84	<70	Distorted crystal	263
S065 (0.081)	V	−1.0	<1	Crystal	274
SS29 (0.103)	V	−1.0	<1	Crystal	274
Bentonite (0.0161)	V	−1.0	<2	Amorphous solid	113
Snowtex-20L	VE	−0.76	<100	Liquid	263

V, Viscometry; *VE,* viscoelasticity.

unity as the structure transformed to liquids. The relationship between rheologic and optical properties of colloidal crystals was studied by simultaneous measurements of reflection spectroscopy and rheologic measurements for colloidal crystals of silica spheres.[279,280] Shear stress of colloidal crystals increased linearly as strain increased, which supports the finding that colloidal crystals are elastic. In the range of large strain, the crystals showed the yielding and flow, resulting in a constant value of shear stress irrespective of strain. The intersphere spacing estimated from the peak wavelength was insensitive to the small strain applied. However, the spacing decreased, approaching a constant value when the strain further increased. These results support the finding that the crystals are deformed slightly but keep their original form of single crystals by the shear at small strains. However, the sliding of the lattice planes may occur accompanied with the deformation of the electrical double layers from spherical to flamelike. Furthermore, partial melt of the crystal structure occurred when the shear rate was high enough, which was confirmed from a decrease in the peak intensity in the reflection spectra.

The rheooptical study on the colloidal alloys of binary mixtures of silica spheres (110 and 300 nm in diameter) were made in an exhaustively deionized suspension.[281] The shear stress showed also the yielding in the stress-strain curves. The intersphere spacings increased first and reached a constant value as strain increased, whereas they decreased for the crystals of single-component spheres. The yielding in the stress-strain profiles were observed for both colloidal crystals of single-component spheres and colloidal alloys.[282-284] The strain dependencies on the peak wavelength and intensity in the reflection spectroscopy differ between the crystal planes and the partial melting, which were supported at the rather high shear stresses and rates.

3.12 Physicochemical Properties

Colloidal particles in suspension state are assumed to be a kind of polyelectrolyte, because a large number of charges always distribute on the colloidal surfaces in the polar media. From this standpoint, physicochemical studies of colloidal suspensions such as mean activity coefficient, osmotic coefficient, heat of dilution, and partial molal volume have been studied. Sugimura et al.[285] measured the mean activity coefficients, osmotic coefficients, and single-ion activities of monodisperse polystyrene spheres in a colloidal crystal state, although the proposal of intersphere attraction in this work is denied later strongly and intersphere repulsion is clarified to be essential for colloidal crystallization.[11] There is no experimental evidence to support the idea of attraction. Rymden[286] measured the osmotic pressure and then the osmotic coefficient of gaslike and liquidlike colloidal dispersions of carboxyl group-containing spheres by a membrane osmometer against pure water. Very small osmotic coefficients ranging from 0.02 to 0.04 have been observed, which supported the significant counter-ion binding in a colloidal sphere. Note here that a simple method to measure osmotic pressure was proposed.[287,288] Takano and Hachisu[288] derived an equation of state for colloidal crystal and liquid by direct osmotic pressure measurements of colloidal dispersions. A theoretical approach to the equation of state has been made mainly by computer simulation and the results supported the effective soft-sphere model.[7,290-294]

Dielectric dispersion may afford important information on the dynamic properties of colloidal crystals. However, most work has been done in the presence of salt, NaCl, for example.[12,294-296] Thus the systems were always gaslike or liquidlike. I measured conductivity dispersion of colloidal crystals.[20] The characteristic frequency increased sharply as the sphere size decreased, sphere concentration increased, or salt concentration increased.

Refractive indices of colloidal dispersions and films, which are important for the study of physicochemical properties such as light-scattering analyses, were measured.[297-303] The preliminary values of refractive indices of colloidal crystals were measured.[304] The reduced refractive index increments of colloidal liquid were higher than those of corresponding colloidal gas in the presence of NaCl. A sharp peak was observed in the reduced refractive index increment of colloidal crystals versus a sphere concentration curve. The peak corresponds to the phase transition between colloidal crystals and liquids. The problem of how to separate the Bragg diffraction term from the observed increments still remains.

3.13 External Field Effects on Colloidal Crystals

The rigidity of colloidal crystals is very small, on the order of 10^{-2} to 10^{3} Pa, as was discussed in Section 3.10. On the other hand, those of metals are from 10^{10} to 10^{12} Pa. Thus colloidal crystal is distorted easily by the weak external fields, such as the gravitational field, shearing forces, an elevated pressure, an electric field, centrifugal compression, and so on.

One of the most important characteristics of colloidal crystals is optical properties represented by Bragg diffraction of light. Iridescent colors change as a function of sphere concentration, because the lattice spacing is in the range of light wavelength and increases as sphere concentration decreases. Fig. 3.61 is one of the close-up color pictures of colloidal single crystals of CS81 spheres in a plane cell. Sphere concentration is quite low at 0.00095 in volume fraction for the four pictures. The suspension was exhaustively deionized. Red, orange, yellow, green, and bluish crystals in the bulk phase far from the cell wall are formed from the homogeneous nucleation mechanism. On the other hand, green to orange pillarlike crystals grow up from the heterogeneous nucleation mechanism along the cell wall. Interestingly, color changes when the incident angles of light source or close-up camera is changed slightly. Coloring of the colloidal crystals is due to the Bragg diffraction of white light by the lattices of colloidal crystals.

The gravitational compression of colloidal crystals occurs quite easily, especially for the diluted crystal suspensions Fig. 3.62. In most cases, colloidal crystals sedimented by gravity reach the sedimentary equilibrium state after several weeks. From the top, the layers of solvent only, colloidal gas, liquid, and crystal states form separately. From the height dependence

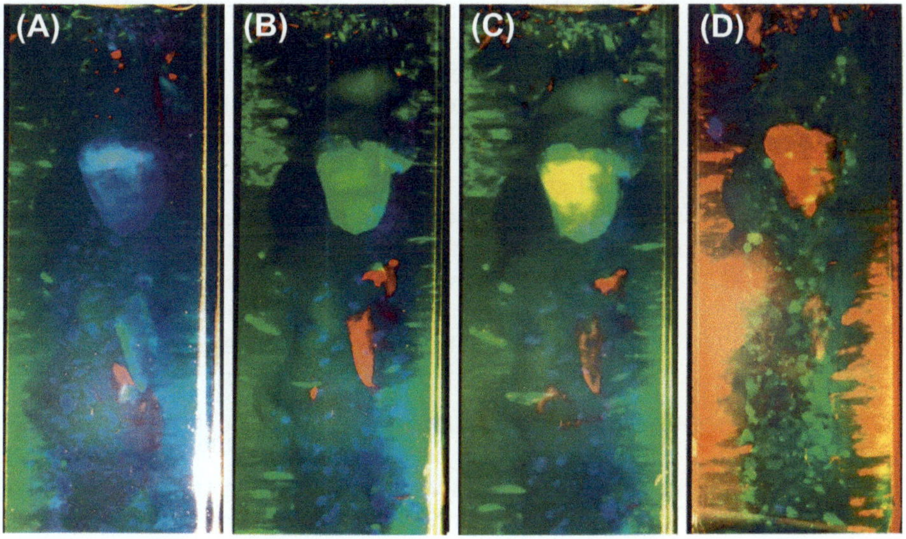

Figure 3.61 Close-up color pictures of colloidal single crystals of CS81 spheres in a plane cell with different angles of incident light against the cell plane. $\phi = 0.000952$; with resins 7 days after suspension preparation and 20 minutes after mixing.

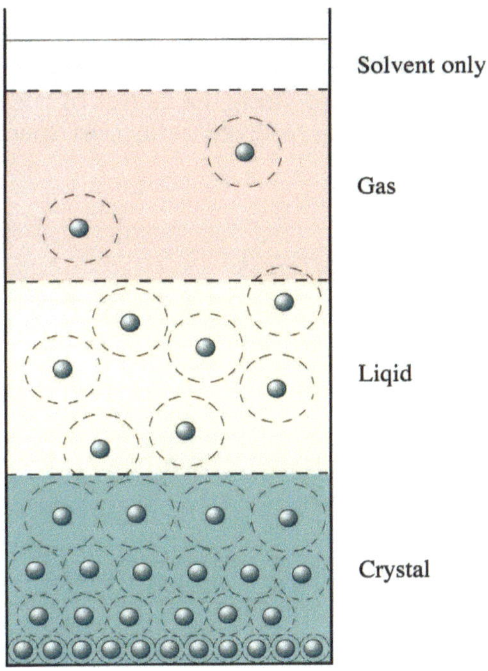

Figure 3.62 Schematic representation of colloidal distribution in the sedimentation equilibrium state.

of the lattice spacing determined from the reflection spectroscopy, the rigidity of the colloidal crystals has been determined. Furthermore, from the concentration dependence of the intersphere distance, at the bottom layer in the observation cell, the static elastic modulus can be determined in the sedimentation equilibrium, as was described in Section 3.10.

Colloidal crystals in a rotating disk are also compressed in the centrifugal field.[253] Beautiful color bands appeared, as is shown in Fig. 3.51. The elastic moduli at various sphere concentrations can be obtained from the change in the reflection peak wavelengths as a function of the distance from the center in the centrifugal equilibrium. These color bands clearly show that the intersphere interactions are "repulsive," and the repulsive forces are counterbalanced with the centrifugal fields.

Electric force–induced change in the lattice structure has been studied intensively by many researchers[11,20,29,305-307] Several electrooptical effects of colloidal crystals were studied. The transformation effect of wave form from square to triangle has been observed, as is shown in Fig. 3.63.[307-317] This effect is explained by the fact that the translational motion of the charged spheres is too slow to follow up changes of the electric field. However, this effect takes place only under certain experimental conditions such as sphere size and voltages or frequencies applied. In most cases, waveforms induced were more complex compared with the effects shown in Fig. 3.63.

We observed second-order harmonics in the electrooptic effects of colloidal crystals (Fig. 3.64).[312] The frequency of the output was just twice that of the input field. To study the causes of the generation of the second-order harmonics, time-resolved reflection spectra were measured in a sine wave

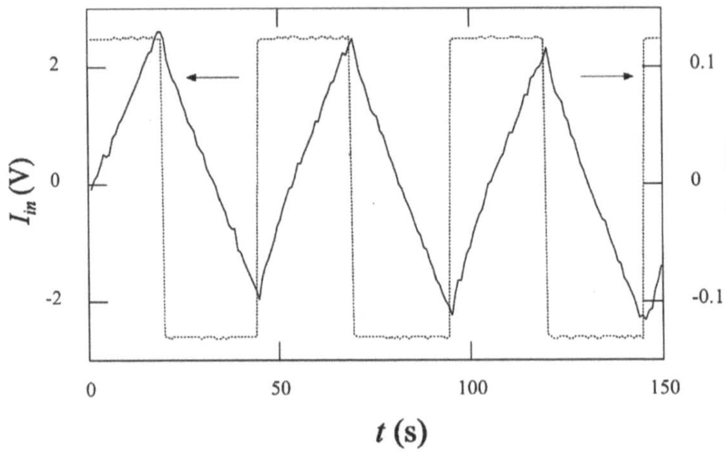

Figure 3.63 Waveforms of the applied square electric fields (*dotted line*) and the reflection responses (*solid line*) of colloidal crystals of CS82 spheres at 25° C. $\phi = 0.048$, $E = 1.9$ V/cm, $f = 0.02$ Hz. *Okubo T, Tsuchida A, Takahashi T, Iwata A:* J Colloid Interf Sci *207:130-136, 1998.*

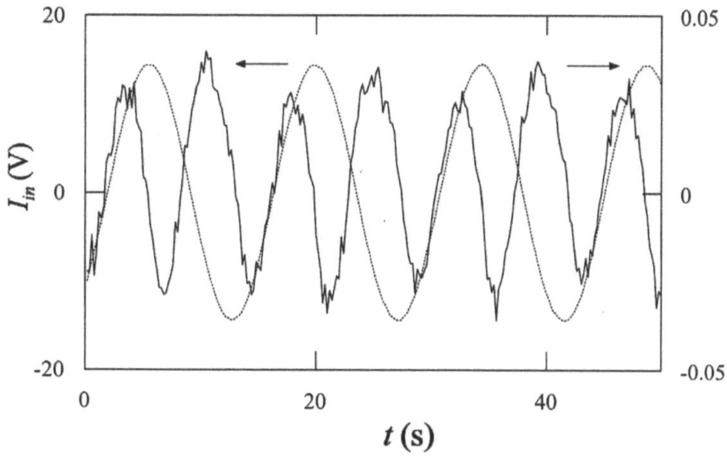

Figure 3.64 Waveforms of applied AC electric fields (*dotted line*) and the reflection responses (*solid line*) of colloidal crystals of CS82 spheres at 25° C. $\phi = 0.048$, $E = 3.7$ V/cm, $f = 0.07$ Hz. *Okubo T, Tsuchida A, Takahashi T, Iwata A:* J Colloid Interf Sci *207:130-136, 1998.*

electric field. The peaks shifted to shorter and then longer wavelengths periodically. Changes in the peak profiles, including the peak intensities and half-widths, were also observed under a sine wave electric field. Plots of the peak wavelengths were plotted against time. The frequency observed was the same as that applied. However, the integrated intensities of the reflection peak profile for the whole change of wavelengths were modulated drastically, and the frequency was just twice that applied. In other words, second-order harmonics were obtained for the integrated intensity. The calculation of the integrated intensities is consistent with observation using the photomultiplier. The observed harmonics were caused by the elastic nature of colloidal crystals themselves. In general, large second-order harmonics were generated at high voltages and low applied frequencies.

The synchronous oscillation was observed even after the sine wave electric current was turned off, as is shown in Fig. 3.65.[310] The measuring cell (of a volume of approximately 10 ml) is shown schematically in the figure. Attached to a sled, the cell is placed in a thermostat metal box. The movement of the sled in the box is restricted in one direction along the incident light beam. Unpolarized white light is incident on the sample parallel to the electrodes. The observation plane is perpendicular to the applied electric field. Light scattering is measured at an observation angle of 90 degrees. The scattered intensity of the unperturbed system (I_o) is compensated and the field-induced light scattering variations ($\Delta I = I_E - I_o$) were observed. The alternating current (*ac*) component of the effect is determined by the amplitude of the steady-state oscillations, and its variations for different volume fractions are shown in the figure by normalized curves. Fig. 3.65 shows ΔI values of the electrooptic response obtained by consequent variation of the distance between the observed spot and the electrodes (points

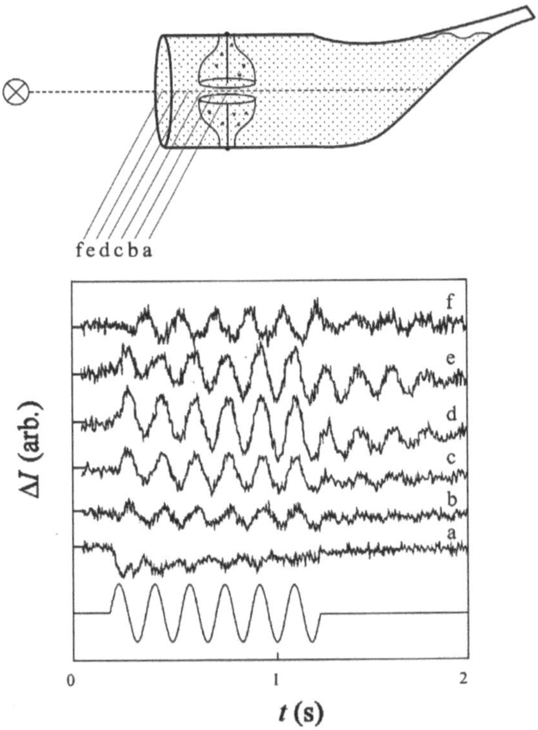

Figure 3.65 Scheme of the electrooptic measuring cell and electrooptic responses as measured by the variations of light intensities at different points (*A–F*). Platinum plates electrodes (area: 1 cm^3; distance: 2.6 mm), *f* = 5.5 Hz, *E* = 154 V/cm, and *f* = 0.007. *Stoimenova M, Dimitrov V, Okubo T:* J Colloid Interf Sci *184:106-111, 1996.*

B through *F* in the cell picture, step 1 mm). Both the amplitude and the phase of the oscillation changed and, despite the evident signs of bad acoustics, they demonstrated the formation of standing waves. Measuring the *ac* amplitude at the maximum, synchronous oscillation was clear even after the sine wave electric current was turned off. This is clearly ascribed to the sheared wave of the colloidal crystals.

Fig. 3.66 shows the amplitudes, *ac* components of the electrically induced shear waves as a function of the frequency applied.[305] The peak shifted to the higher frequencies as sphere concentration increased. The appearance of the peaks demonstrates the existence of some characteristic frequencies for the viscoelastic colloidal crystals. These harmonic oscillation effects were studied using electric light-scattering measurements. At 90 degrees, scattered light is detected from incident white light. Hofmann et al.[318] has reported the ringing gel phenomena on the liquid crystal systems, which is similar to the existence of the self-resonance frequency in the colloidal crystals reported here the photonic-trapping effect of colloidal crystals has been studied.[319,320]

Yoshinaga et al. studied the electrooptics of colloidal crystals in organic solvents and gels.[321] Fujita and Ametani clarified the nonlinear effect for the

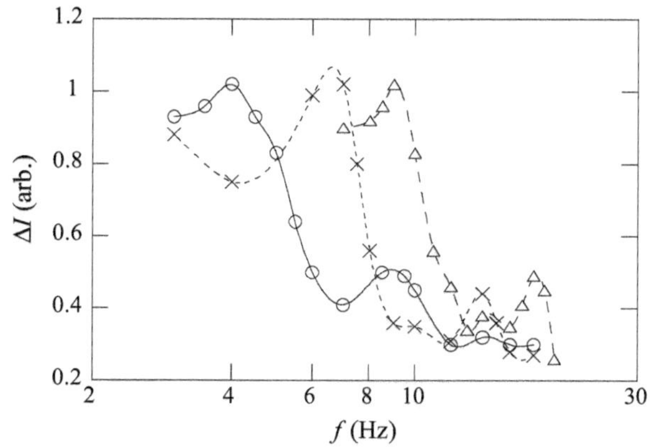

Figure 3.66 Amplitudes of the *ac* component (ΔI) plotted against frequencies of the *ac* electric fields. E =192 V/cm; *circles:* ϕ = 0.007; *crosses:* 0.01; *triangles:* 0.013. *Stoimenova M, Dimitrov V, Okubo T:* J Colloid Interf Sci *184:106-111, 1996.*

modulation signal from the optical measurements in the *ac* electric field.[260,322] Two-dimensional ordering induced by the electric field has been reported by several research groups.[323-325] Magnetic field effects on the orientation of anisotropic-shaped colloidal particles are now well known.[105-111,326] Crystal-liquid phase transition induced by ultraviolet light irradiation was reported for colloidal suspension containing malachite green.[327]

Colloidal crystal structures have been studied at high pressure (up to 200 MPa).[90] Enhancement of *bcc* lattice formation or transformation from *fcc* to *bcc* lattices were observed under elevated pressure. These observations were explained by the increase in the effective charge number on the colloidal surface and the decrease in the Debye screening length as pressure increases.

Shearing deformation of colloidal crystals has been studied by several researchers. Ackerson and Clark reported that the Bragg diffraction spots diffused as the shear rate increased, and turned into diffraction rings characteristic of liquid structures at high shear rates.[14,234,328] Lindsay and Chaikin observed a shear melting at high shear rates.[6] On the other hand, Tomita and van de Ven reported that their suspension did not exhibit disappearance of order at any shear rate, but showed changes in the lattice structure at moderate volume fraction (0.05 to 0.50).[329] Jorand et al. proposed that the shear-induced melting was due to screw dislocation.[268] Furthermore, Ackerson et al. conducted small-angle neutron scattering measurements of colloidal crystals undergoing shear.[194] Small G values have been obtained under shear.[14,328,330-333] The *fcc* and *bcc* lattices coexisted under flow, although the lattices are more or less deformed and melted at high shear rates.[334] The deformation of the electrical double layer from a spherical shape was deduced at low sphere concentrations under shear, and the sliding-layer model proposed by Tomita and van de Ven[329] was supported, especially at high-sphere concentrations. Structural changes under shearing flow have been further studied by several researchers.[335-339]

The intersphere distances in the crystal of monodisperse polystyrene spheres in the deionized suspension are sensitive to the coexisted water-soluble polymers in the order.[340]

$$\text{PEG} > \text{PVA} > \text{none} > \text{PVP} \cong \text{HPC} > \text{PAA}_\text{m} \tag{52}$$

Here, *PEG, PVA, PVP, HPC,* and PAA_m indicate *polyethylene glycol, polyvinyl alcohol, polyvinyl pyrrolidone, hydroxypropyl cellulose,* and *polyacrylamide,* respectively. On the other hand, the order in the intersphere distances is from the microscopy observation in the sedimentation equilibrium.[339]

$$\text{PVA} > \text{PEG} > \text{PVP} > \text{HPC} > \text{PAAm} \tag{53}$$

These orders are explained by the strength of the intersphere repulsion forces through the electrical double layers, which is sensitive to the polymer adsorption on the colloidal surfaces by the hydrophobic or dipole–dipole interactions. The scheme of the adsorption mechanism has been proposed by Russel et al.[12] Furthermore, crystal structures in organic solvents and water-organic solvent mixtures have been studied in several papers.[2,47,150,246,247,249,341]

3.14 Colloidal Crystallization in Microgravity

To compensate the sedimentation effects of colloidal dispersions on the ground, some researchers have studied colloidal dispersions, at which the densities of the particles and medium molecules are matched, using sucrose-water mixtures instead of water, for example. The other convenient method is the application of *dc* electric field perpendicularly for the anionic- (or cationic-) charged colloidal particles. The preliminary pseudomicrogravity experiments of crystal growth were made. The cell used was the same as Fig. 3.37 for study the crystal growth under the *ac* electric fields. Fig. 3.67 shows the influences of the *dc* electric field on the colloidal single crystals. Fig. 3.67, *A*, shows that the crystals move upward slowly by application of 2 V (upper electrode is positive) *dc* current for 10 minutes. Then 6 V was applied for 4 more minutes (Fig. 3.67, *B*). Clearly, blocklike single crystals from the homogeneous nucleation look to move upward, keeping their shapes and sizes. The pilelike crystals along the cell wall formed from the heterogeneous nucleation mechanism inclined upward, as is shown clearly in Fig. 3.67, *B*. Next, the electric field applied was turned down (−6 V) for 2 minutes. The inclined crystals along the cell wall looked back to the perpendicular to the cell wall (*C*). When the application of the current continued for 4 more minutes, blocklike crystals moved downward. Furthermore, the arrayed

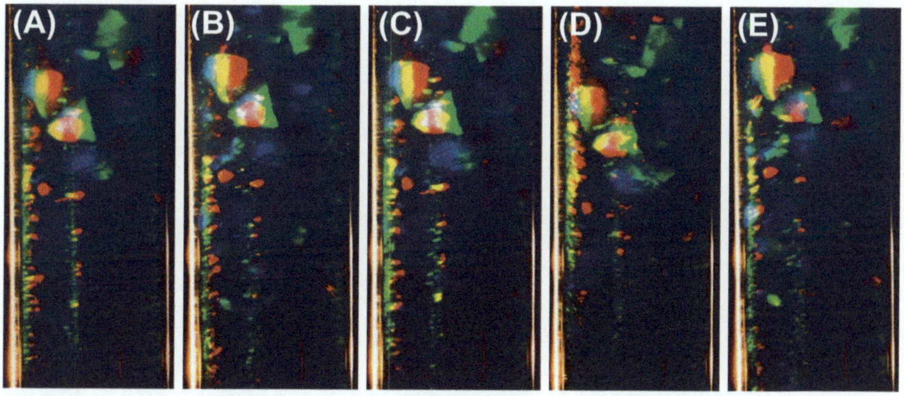

Figure 3.67
Electric field effect on the colloidal single crystals of CS82 spheres formed at 25° C. Type III cell; $\phi = 0.00124$; *dc* electric field; **A,** $E = 2$ V for 10 minutes; **B,** 6 V for 4 minutes; **C,** −6 V for 2 min; **D,** −6 V for 4 minutes; **E,** 6 V for 21.5 min; exposure = 0.7 seconds, iris = 22.

crystals along the wall were also inclined downward significantly. Finally, the electric field was turned to +6 V and maintained for 21.5 minutes. Then, the pilelike crystals inclined again upward, but broke in part. The blocklike crystals in the bulk phase kept their size and shape. Clearly, the whole suspension flew continuously during the application of *dc* electric current upward or downward depending on the sign of the *dc* current. On the ground, the convectional flow of suspension should take place. Thus complex flow of the suspension should occur in the pseudomicrogravity condition under the electric field.

Fig. 3.68 is the typical example showing the crystal growth processes when 6 volts of *dc* current are applied. Under the strong *dc* currents, crystals from both homogeneous and heterogeneous nucleation mechanisms grew. However, most of large crystals were broken to small pieces of crystallites after 20 minutes. Fig. 3.69 demonstrates that the single crystals finally formed are stable without *dc* electric current (see *A*). However, both blocklike and pillarlike crystals are broken apart by the higher *dc* electric fields.

Fig. 3.70 shows the reflection peaks as a function of height under *dc* electric current and in its absence. When the *dc* current at +0.3 V was applied, pseudomicrogravity state is achieved. However, the other factors like convectional flow are not eliminated. We should note further that the pseudomicrogravity state is attained after applying the *dc* electric field continuously for 1 to 2 months. In conclusion, it was found that the actual experiments in microgravity are most important. In microgravity, gravitational convectional flow of particles and solvent should stop definitely. It should be noted here that Marangoni convection induced by the temperature distribution of suspension never disappear even in microgravity.

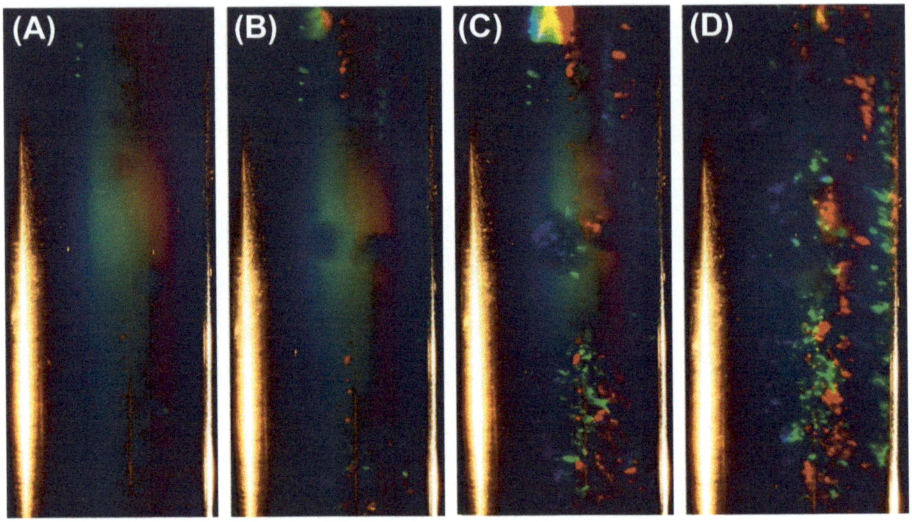

Figure 3.68
Electric field effect on the colloidal single crystals of CS82 spheres formed at 25° C. Type III cell; $\phi = 0.00124$; *dc* electric field; $E = 6$ V; **A,** after 4 minutes; **B,** after 6 minutes; **C,** after 9 minutes; **D,** after 20 minutes; exposure = 1 second, iris = 22.

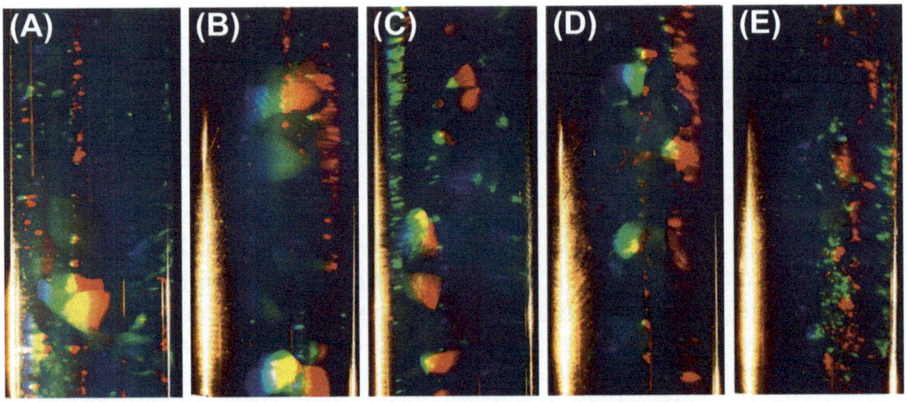

Figure 3.69
Electric field effect on the colloidal single crystals of CS82 spheres formed at 25° C. Type III cell; $\phi = 0.00124$; *dc* electric field; **A,** $E = 0$ V after 31 minutes; **B,** 2 V after 25 minutes; **C,** 3 V after 31 minutes; **D,** 4 V after 25 minutes; **E,** 6 V after 20 min; exposure = 1.5 seconds (*A-D*), 1 second (*E*); iris = 22.

Zhu et al.[342] reported mainly morphologic properties of the colloidal crystals of the hard-sphere systems in microgravity using experiments performed on the space shuttle Columbia. Colloidal crystallization has been studied kinetically in microgravity, produced by parabolic flights of an aircraft.[131] Fig. 3.71 shows the scheme of the measuring apparatus. A peristaltic pump circulates the colloidal suspension first to the ion-exchange resin column and then the observation cell to deionize the sample

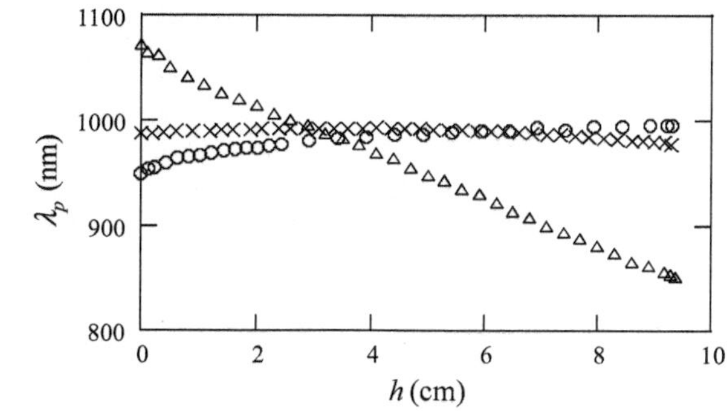

Figure 3.70 Peak wavelengths in the reflection spectra of CS82 spheres as a function of height at 25° C. $\phi = 0.0012$, *circles:* 0 V after 72 days; *crosses:* +0.3 V after 57 days; *triangles:* +1 V after 30 days.

suspension continuously. A light beam from a halogen lamp hits the cell wall using a Y-type optical fiber cable and the reflection light is measured on a photo-multichannel analyzer system. Close-up CCD observation was also made. A Mitsubishi MU-300 type jet aircraft was used. The measuring apparatus set in the aircraft is shown in Fig. 3.72. Four types of measurements were made for each flight: (1) in a plane on the ground before flight, (2) during flight to the test area, (3) in microgravity, and (4) in a plane on the ground after flight. The first and second measurements were made to check the ionic contamination of the sample during the flight. The third and fourth measurements were adopted to examine the effects of the slight vibration of the airplane on the crystal growth rates.

Fig. 3.73 shows the plots of the rate coefficients for *fcc* lattice growth against $1/\phi$. Clearly, data at 0 *g* and 1 *g* gave good linearity. The magnitude of the slope at 0 *g* was smaller than that at 1 *g*. The limiting rate coefficients

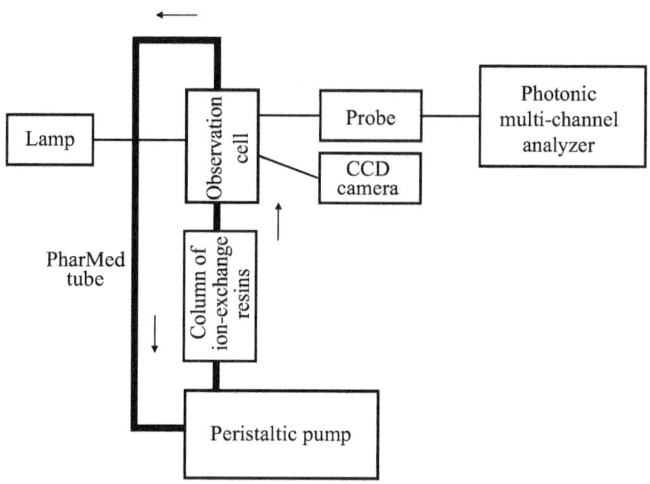

Figure 3.71 Schematic representation of circulation-type measuring apparatus. *Okubo T, Tsuchida A, Okuda T, Fujitsuna K, Ishikawa M, Morita T, Tada T:* Colloids Surf *A153:515-524; A160:311-320, 1999.*

Figure 3.72 Circulation-type measuring apparatus set in the aircraft.

at $1/\phi = 0$ were 0.94 s^{-1} and 1.26 s^{-1} at 0 *g* and 1 *g*, respectively. The mean values of the rate coefficients were smaller by 25% in microgravity compared with those in gravity. The limiting rate coefficients for *bcc* lattice growth were 0.38 s^{-1} and 0.43 s^{-1} at 0 *g* and 1 *g*, respectively. The magnitude of the decrease in microgravity was approximately 10% compared with that in gravity. One of the main causes for the retardation in microgravity was suggested to be elimination of the downward diffusion of spheres, which may enhance the intersphere collisions. No convection of the suspensions in microgravity was also or much more important than the elimination effect of the downward diffusion. Microgravity experiments of colloidal crystallization of silica spheres were also made in the presence of NaCl.[343] Time-resolved reflection spectroscopy was made with a continuous circulating-type stopped-flow cell system (see Figs. 3.71 and 3.72). Crystal growth rate coefficients decreased as the salt concentration increased both at 0 *g* and 1 *g*, and those in microgravity were smaller than those in normal gravity by 16% (maximum), especially in water and in the presence of a small amount of the salt, less than 2×10^{-6} M. Furthermore, the rates in flight at 1 *g* were larger by 15% (maximum) compared with those at 1 *g* on the ground.

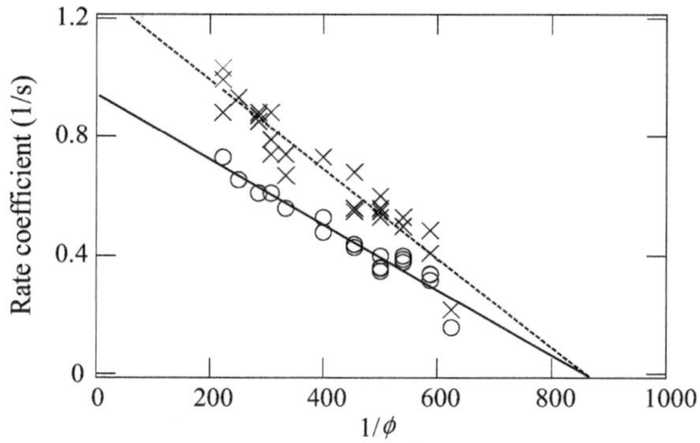

Figure 3.73
Rate coefficient of colloidal crystallization plotted against $1/\phi$ for CS91 spheres at 0 *g* (*circles*) and 1 *g* (*crosses*) at 25° C. For *fcc* crystal lattice. *Okubo T, Tsuchida A, Okuda T, Fujitsuna K, Ishikawa M, Morita T, Tada T:* Colloids Surf *A153:515-524; A160:311-320, 1999.*

Disappearance of the downward diffusion of spheres and no convection of the suspensions are important for the retardation in microgravity. Ishikawa et al.[344] studied microgravity effects on the crystal growth dynamics of polystyrene spheres in the deionized aqueous suspensions. Results showed a power-law relationship between size and time, $L \propto t^{\alpha}$, where L is the size of crystallites and t is time. The α values were 0.33 ± 0.03 in microgravity and 0.25 ± 0.02 in normal gravity, respectively.

Crystal growth rates of the colloidal alloys of binary mixtures of monodispersed polystyrene and silica spheres have been studied in microgravity.[132] The rates increased substantially up to approximately 1.7-fold in microgravity compared with those in normal gravity, as is shown in Fig. 3.30. We should note that the segregation effect is familiar for the binary mixtures of colloid and powder systems; large spheres are segregated upward and small ones downward in normal gravity (see Fig. 3.32 and Section 3.6.3 in this chapter). In microgravity such segregation should disappear and the homogeneous mixing takes place favorably, which leads to fast alloy crystallization. Gravitational effects on the alloy crystallization from bimodal colloidal dispersions were discussed by Bartlett and Ottewill.[345] Microgravity experiments on the other properties of colloidal dispersions were described in Section 2.13.

3.15 Two-Dimensional Colloidal Crystals

Two-dimensional distributions of polystyrene spheres at the air–water interface has been studied by Schuller[346] and Sheppard and Tcheurekdjian[347] from the measurement of surface pressure–surface area isotherms.

Two-dimensional ordering of polystyrene spheres was also studied by Aveyard et al.[348-351] Pieranski noted that electrostatic repulsion between charged particles is enhanced at an interface between an aqueous phase and air, especially at large particle separation.[7] At the interface a particle has an asymmetric counter-ion distribution that results in a dipole moment normal to the water interface. It has also been shown that the interaction near a water–air interface became repulsive and long-range.[352,353] The interaction energy decayed inversely proportional to the salt concentration. Aveyard et al.[351] observed a transition from the original hexagon to a rhombohedral when the monolayers at the air–water surfaces are compressed on a Langmuir trough. Furthermore, they observed that the particle structure at air–water interfaces was very sensitive to the salt concentration in the aqueous phase. At low salt concentration, an ordered structure formed resulting from the interparticle repulsion, while at high salt concentration the spheres formed 2D clusters.[351] Two-dimensional colloidal crystals also form by the cooperative intercolloidal electrostatic repulsion like 3D colloidal crystal. The melting of 2D crystals and solids have been studied in detail.[354-360] Two-dimensional colloidal crystals were studied at the electrode interface. ABC-type triblock terpolymer films formed highly oriented and ordered double-helical rod structures by solvent evaporation.[361] Two-dimensional phase transition of colloidal crystals has been discussed theoretically.[362] Synthetic studies on the fixed 2D colloidal crystals have been performed recently.[363-369] Fujii et al. synthesized soft Janus-type 2D colloidal crystal films using an air–water interface as a place to introduce softness, asymmetry, and regularity. One of the typical examples is shown in Fig. 3.74, where scanning electron microscope (SEM) images of 2D colloidal crystal films composed of polystyrene spheres inserted in polypyrrole film.[370] Fig. 3.74 is courtesy Prof. S. Fujii.

3.16 Biological Colloidal Crystals

In general, crystal-like ordering of colloidal particles in a suspension state is called a *colloidal crystal,* as described in the introduction. Colloidal-crystal ordering in solid state is often observed in biological systems such as insects, birds, and fish. Sea mice are polychaete *Aphroditidae* and one of marine worms, found along oceanic continental shelves. They have a brightly iridescent, furry appearance. The fur consists of hairs that are packed with tubes like a hexagonal honey comb. A transmission electron microscope (TEM) image of the ordered structure within the spines looks just like hexagonal 2D colloidal crystals.[371,372]

The brightly colored tail feathers of a peacock (*Pavo* spp.) is famous for its structural colors. Yoshioka and Kinoshita clarified the structures theoretically and with SEM and TEM measurements.[373] Structural colors of the wings of

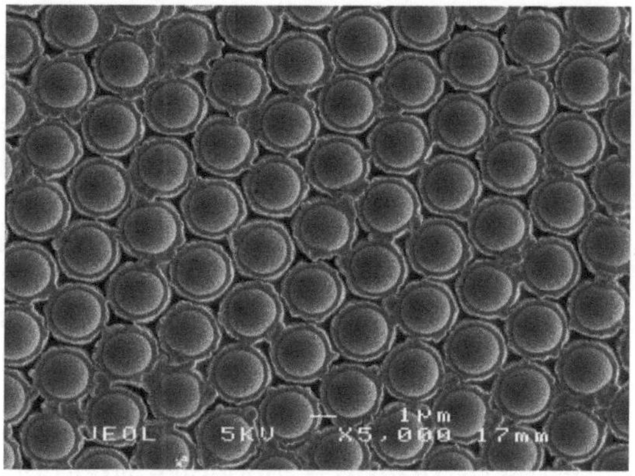

Figure 3.74 SEM image of a 2D colloidal crystal film composed of polystyrene spheres and polypyrrole. *Courtesy Professor S. Fujii.*

the Morpho butterflies were also studied in detail by Kinoshita and Yoshioka.[374,375] The cooperation of regularity and irregularity in the structures was the main cause for the characteristic blue color of the Morpho wings. Furthermore, it was found that the pigments close to the iridescent structure enhance the blue coloring by reducing background light. The Japanese jewel beetle, *Chysochroa fulguidissima,* emits beautiful iridescent colors. Optical analyses of the colors of the dorsal surface were investigated in detail.[376] The color change of fish *Motile iridophores,* living in tropical areas is due to the reflection of light from the multilayered thin-film interference.[377,378]

3.17 New Colloidal Crystal Systems

Typical colloidal spheres are hard, solid spheres. Recently new types of colloidal crystals have been developed. Gel spheres develop in between linear-type polymers and hard colloidal spheres from the point of view of colloidal chemistry. Most gel spheres are white and turbid in aqueous media, which means the interfaces have refractive indices different than that of solvent molecules. Thus fundamental structures and properties of most gel crystals are quite similar to those of colloidal crystals. Gel crystals are reviewed and discussed in Chapter 4.

Recently, Fujii et al.[379] studied colloidal crystallization of poly-*n*-butyl acrylate (PBA) colloidal spheres (320 nm $\pm$ 50 nm in diameter) in exhaustively deionized aqueous suspensions. The glass transition temperature of this polymer is $-54°$ C and the rubberlike liquid state occurs at room temperature.[380,381] The colors are due to the emulsion crystal formation. A close-up color picture of stock suspension of PBA spheres is shown in Fig. 3.75. Strong iridescent and metallic colors were observed with the ion-exchange resins in a bottle. The critical concentration of melting of the crystal structures were approximately 0.01 and 0.03 in volume fraction (at approximately 1.2 and 3.5 in wt%) with and without resins, respectively. The sharp and broad reflection peaks coexisted in the reflection spectrum, which supports the coexistence of the crystal and distorted crystal structures. The main cause for the formation of distorted crystals is due to the fact that the polydispersity index given by the dispersity divided by the mean diameter (30/320 = 0.094) is a bit smaller than but close to 0.1, the criteria between crystals and distorted crystals. Another cause is the formation of the associates, dimers, and trimers, for example, of PBA spheres especially under agitation of the sample bottle. During the course of drying suspension, the iridescent colors were strengthened. However, the colors disappeared at the final stage of drying, because the suspension becomes unstable as water content decreases. The final state of the particles is a rubberlike amorphous-solid structure.

Strongly colorful and angle-independent ordered liquid (at lower sphere concentration) and distorted crystal structures (at higher concentration) formed in ionic liquids[382-386] were reported by Prof. M. Watanabe et al.

Figure 3.75 Close-up color picture of colloidal crystals of PBA at 25° C. Stock suspension, 6.2 wt%, with ion-exchange resins. *From Fujii S, Nakamura Y, Tsuchida A, Okubo T:* Colloid Polym Sci *292:2303-2310, 2014.*

PMMA-grafted colloidal silica spheres were dispersed in the ionic liquid, 1-ethyl-3-methyl imidazolium bis(trifluoro methane sulfonyl) amide ([C2mim] [NTf_2]) at room temperature. The iridescent colors changed with particle concentration, as is shown in Fig. 3.76. Phases of the ordered liquid and the distorted crystal structure are called *colloidal glass* by some researchers.

Grzybowski et al. reported a new type of diamondlike colloidal crystals formed from oppositely charged and equally sized gold and silver particles.[387] Furthermore, core-shell type clusters, where the shell is formed by the particles in excess, were prepared from oppositely charged silver and gold particles.[388,389] These regulated aggregates are a highly interesting finding, because most researchers thought that only polydisperse aggregates could form from cationic and anionic colloidal spheres by the too-strong electrostatic interparticle attraction. The important role of the electrical double layers was also proposed for crystallization.

Oil-in-water type emulsions containing 10% surfactant showed the iridescent colors ascribed to the lyotropic liquid-crystal structures.[390] J. Watanabe et al.[391] prepared brightly colored emulsion crystals composed of the mixtures of aqueous $Na_2S_2O_3$ solution and tetrahydrofuran solution of hydroxylpropyl cellulose. Color changed from deep purple to pink as the concentration of $Na_2S_2O_3$ changed from 45 wt% to 50 wt%. The color change was explained by Mie's scattering theory in combination with the Christiansen effect.

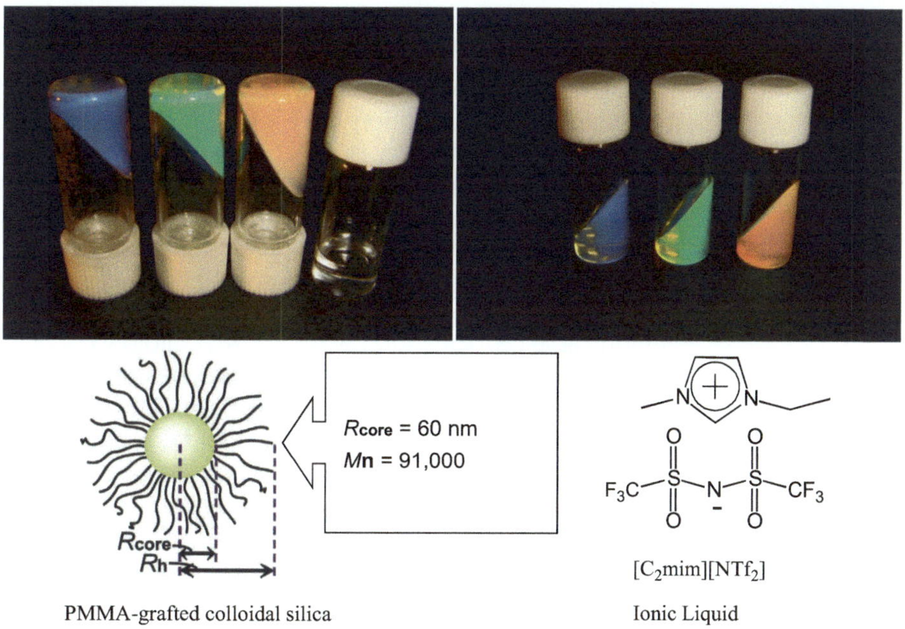

Figure 3.76 Distorted colloidal crystals of PMMA-grafted silica spheres in ionic liquid. *Courtesy Professor M. Watanabe.*

As was described in Chapter 2, spherical plasma with positive charges formed crystal structures.[49] This plasma crystal supports the important role of the interparticle electrostatic repulsion for colloidal crystallization. Recently, a keen attention has been paid to the gaslike, liquidlike, and crystal-like distributions of dusty plasmas in space, (i.e., low-temperature plasmas containing charged macroparticles).[392] The main cause for the plasma crystallization is the interparticle electrostatic repulsions, which is similar to colloidal crystallization.

Leunissen and et al.[393,394] found stable micron-sized water droplets in polar oils without any additives. Under appropriate conditions the oil-dispersed water droplets formed a *bcc* lattice with the lattice spacing of the order of 10 μm, and some of the crystals were stable for approximately 2 years without droplet coalescence. This crystallization has been explained by the electrostatic repulsion between charged water droplets.[393-398]

3.18 Application of Colloidal Crystals

Let's summarize first the characteristics of colloidal crystals before describing their application. Table 3.11 compiles the representative characteristics of colloidal crystals, especially in deionized suspension, which are grouped into preparation, structure, and property. All the characters are believed to be important for application studies.

3.18.1 Preparation

The higher the monodispersity in size and shape of the colloidal particles, the more easily crystallization takes place. However, crystallization occurs whenever monodispersity is lower than 0.05 in general. All kinds of colloidal particles are charged negatively in most cases, in the polar medium phase. However, nonionizable particles do not repel the counterionic species so much into an aqueous phase compared with the ionizable colloids. When the ionic concentrations of suspensions are low, the widths of the electrical double layers extend. Thus colloidal crystallization occurs more easily for nonionizable particles such as Au and Ag colloidal particles than the ionizable ones such as sulfonate- and carboxyl-group–containing spheres. It should be noted that the electrical double layers extend more significantly for nonionizable particles. Deionization of colloidal suspensions with coexistence of the mixed ion-exchange resins is also very important for crystallization. Enlarged l_D values of the suspension by the coexistence of ion-exchange resins are favorable for the colloidal crystallization. Plane cell is more favorable for crystallization than the curved wall cell such as a test tube, vial, or capillary tube.

Table 3.11 Characteristics of Colloidal Crystals

Preparation	
1. Monodispersity effects	Size and shape, $d_{eff} = d + 2 \times l_D$ Particles with monodispersities within 0.1 crystallize
2. Nonionic particles	Best for crystallization (Au, Ag...); Enlarged l_D
3. Deionization	Coexistence of ion-exchange resins; Enlarged l_D Decrease in ionic impurities, shielding of cells, etc.
4. Wall effects	Most dense crystal plane orient along the cell wall Homo- and heteronuclei form far from and along wall
Structures	
5. *fcc* or *bcc*	*fcc* is more stable in low temperature, low ionic concentration, etc.
6. Homo- and heteronucleations	Far from and along the wall
7. Lattice spacing tunable	Colors (λ_p) are regulated with particle concentration
Properties	
8. "Suspension" but "crystal"	Appears to be a suspension, but is crystal in particle structure
9. Reversible melting	By shearing forces, raised temperature, other external fields Off in the external fields resulted in re-crystallization
10. Soft crystal	Very soft elastic and viscoelastic crystal, $G = Nk_BT/g^2$
11. Iridescent colors	Structural colors by Bragg diffraction; Beautiful! Blinking crystals, color change with sphere concentration
12. External field responsible	Electrooptic effects (harmonics, oscillation, etc.) Rheooptics, centrifugal forces, pressure, etc.

bcc, Body-centered cubic; *fcc*, face-centered cubic.

3.18.2 Structures

Colloidal lattice structures are influenced greatly by the cell wall. Most dense crystal planes (1,1,1) and (1,1,0) for *fcc* and *bcc* lattices are often oriented along the cell wall. The cell wall substantially influences crystal growth from the heterogeneous nucleation mechanisms. Colloidal crystal structures are simple. Most crystals are *fcc* or *bcc*. They form in the conditions of stable and unstable atmospheres, respectively; *fcc* crystals form more easily in the concentrated and deionized suspensions. In the presence of foreign salt or elevated temperatures, transition from *fcc* to *bcc* subphases takes place. Single crystals from the homogeneous and heterogeneous nucleation mechanisms form always far from the cell wall and close to and along the cell wall, respectively. Furthermore, single crystals from the heterogeneous nucleation form first along the cell wall, and then the blocklike and randomly directed single crystals from homogeneous nucleation mechanisms form. The wall effects are very important for colloidal crystallization. Lattice spacing of the usual crystals such as metals (atomic crystal), proteins (molecular crystal), and ice (molecular crystal) does not change so much in any experimental conditions.

However, lattice spacing of colloidal crystals changes by the sphere concentration or sphere size exclusively. Thus the iridescent colors and single crystal size changes systematically by changing the sphere concentration or sphere size. Usual crystals form by the interparticle attraction forces, whereas colloidal crystals form by the intersphere repulsion forces.

3.18.3 Properties

Let's consider the specific properties of colloidal crystals. First, colloidal crystals look like suspension at first glance, because the medium molecules are liquid. However, three states of colloidal dispersions are determined by the distribution of colloidal particles, not by the medium molecules. Thus colloidal crystals are really crystal in the strictest sense. Most properties of crystals are expected to exist also for colloidal crystals. Lattice spacing of colloidal crystals are often in the range of light wavelengths, which is in contrast to the fact that the spacing of usual crystals are in the range of an X-ray's wavelength, for example. Furthermore, colloidal crystals are distorted and then melt to the liquids by the external fields (e.g., shearing forces, elevated temperatures, and high pressures). However, when the external fields are cut off, crystallization takes place again every time. The crystallization and melting are reversible. One of the merits of colloidal crystals is the fact that the reversible melting occurs in the weaker external fields more easily than with the usual crystals.

The structural relaxation times of colloidal crystals are much larger than those of usual crystals. Colloidal crystals are very soft; their rigidities are low compared with those of usual crystals. Approximate rigidity of substances (G) is given by the equation $G = Nk_BT/g^2$, where N is the number density of colloidal spheres, k_BT is the thermal energy per each sphere, and g is the fluctuation parameter given by the ratio of the thermal fluctuation of particles against the nearest-neighbored interparticle distance. This equation of G holds to both colloidal and usual crystals equally. Colloidal crystals emit the light following the Bragg diffraction when white light hit the sample. The iridescent colors are regulated freely by changing the sphere concentration or colloidal size. Colloidal crystals are highly sensitive to many kinds of external fields. Number density of colloidal crystals is extremely low and the particles are always charged, especially negatively, and are the main cause for the high responsiveness to the external fields. In conclusion, colloidal crystals are very soft ionic crystals.

3.18.4 Fixation of Crystal Structures

Fixation of colloidal crystal structures in solid-state has been the first step of most application studies of colloidal crystals hitherto. Polymethyl acrylate films containing crystal arrays of colloidal silica spheres were prepared successfully by Ford et al.[399] and Yoshinaga et al.[400] The initial studies in most work on colloidal lithography and photonic crystal described here has focused on the fixation of colloidal crystal structure. However, I think colloidal crystal in suspension state is also highly interesting, because all the dynamic modes of characters of colloidal crystals are available in application. We should note that crystal structures remain firm, but most of the dynamic characters of colloidal crystals are frozen by the fixation of crystal structures.

3.18.5 Colloidal Lithography

In general, the lithography process is divided into two steps. The first is making the master for providing a replicated pattern. The second is the replication, where the pattern transfer is made quickly and precisely. In the standard techniques of lithography, replication shorter than several tens nanometer scale was difficult.[401] For the fabrication of nanometer-scale structures, electron- and ion-beam lithography techniques are appropriate. However, these methods need longer processing time and have higher costs.[402] To overcome these difficulties, colloidal lithography has been developed.[401] Fischer and Zingsheim[403] evaporated colloidal suspension (300 nm spheres in diameter) on a glass plate and obtained the regions of hexagonally packed spheres. After platinum deposition, they obtained ordered arrays of triangular areas. Later the deposition over the entire surface of a macroscopic substrate was achieved.[404] Furthermore, the coating techniques such as spin-coating and Langmuir-Blodget method were developed.[405,406] Recent developments on colloidal lithography are amazing.[407,408]

3.18.6 Photonic Crystals

Photonic crystal refers to the structured solid materials composed of two components having different refractive indices and having periodically ordered structures in the scale of light wavelength. By the materials, photonic band-gap (PBG) filter is formed. This technique is interesting for building the new type of devices. The first report on the PBG was made by Yablonovitch in 1991.[409] Structure exhibiting full PBG in the microwave, millimeter, and submillimeter regimens were fabricated more than 10 years ago. Now, submicron length-scale photonic crystal has been developed rapidly using colloidal crystals as the starting material.[410-422] Asher et al.[423] synthesized monodispersed hollow polymer spheres and studied the photonic crystal function, including magnetic field effects. Theoretical predictions were also shown based on energy considerations. Two-dimensional crystal of monodispersed magnetic particles was formed as a liquid surface.[424] Saado et al.[425] found 2D and 3D self-assembly of magnetic particles by the interparticle magnetic repulsion forces on the liquid surface and layer-by-layer technique, respectively. The possibility of the spheres as PBG applications was also reported. Furthermore, tunable paramagnetic colloidal photonic crystals were developed.[426] Fudouzi et al.[427] prepared photonic rubber sheets with tunable color by elastic deformation, and studied also the tunable structural color in colloidal photonic crystals.[428] To obtain large single crystals, Blaaderen et al. studied the template-directed colloidal crystallization.[429] Semiconductor crystals smaller than the bulk exciton Bohr radius shows the highly effective band-gap of the material. ZnS growth on CdSe crystals was prepared to have very high fluorescent quantum yield of 50%.[425,426,429,430] There are still many difficulties to solve before practical use can be made of colloidal crystals.

References

1. Vanderhoff W, van de Hul HJ, Tausk RJM. Overbeek JThG. In: Goldfinger G, editor. *Clean surfaces: their preparation and characterization for interfacial studies.* New York: Marcel Dekker; 1970. pp. 15–44.
2. Hiltner PA, Papir YS, Krieger IM. *J Phys Chem* 1971;**75**:1881–6.
3. Kose A, Ozaki M, Takano K, Kobayashi Y, Hachisu S. *J Colloid Interf Sci* 1973;**44**:330–8.
4. Williams R, Crandall RS, Wojtowicz PJ. *Phys Rev Lett* 1976;**37**:348–51.
5. Mitaku S, Ohtsuki T, Kishimoto A, Okano K. *Biophys Chem* 1980;**11**:411–6.
6. Lindsay HM, Chaikin PM. *J Chem Phys* 1982;**76**:3774–81.
7. Pieranski P. *Contemp Phys* 1983;**24**:25–73.
8. Ottewill RH. *Ber Bunsenges Phys Chem* 1985;**89**:517–25.
9. Aastuen DJW, Clark NA, Cotter LK, Ackerson BJ. *Phys Rev Lett* 1986;**57**:1733–6.
10. Pusey PN, van Megen W. *Nature* (London) 1986;**320**:340–2.
11. Okubo T. *Acc Chem Res* 1988;**21**:281–6.
12. Russel WB, Saville DA, Schowalter WR. *Colloidal dispersions.* Cambridge: Cambridge University Press; 1989. pp. 329–365.
13. Sood AK. *Solid State Phys* 1991;**45**:1–73.
14. Ackerson BJ, Clark NA. *Phys Rev Lett* 1981;**46**:123–6.
15. Okubo T. *J Chem Soc Faraday Trans* 1 1988;**84**:1163–9.
16. Monovoukas Y, Gast AP. *J Colloid Interface Sci* 1989;**128**:533–48.
17. Aastuen DJW, Clark NA, Swindall JC, Muzny CD. *Phase Transitions* 1990;**21**:139–55.
18. Okubo T. *J Chem Phys* 1991;**95**:3690–7.
19. Okubo T. *Naturwiss* 1992;**79**:317–20.
20. Okubo T. *Prog Polym Sci* 1993;**18**:481–517.
21. Simon R, Palberg T, Leiderer P. *J Chem Phys* 1993;**99**:3030–6.
22. Okubo T. *Colloid Polym Sci* 1993;**271**:190–6.
23. Okubo T. *Langmuir* 1994;**10**:1695–702.
24. Okubo T. *Langmuir* 1994;**10**:3529–35.
25. Fujita H, Ametani K. *Jpn J Appl Phys* 1977;**16**:1091–6.
26. Tobochnik J, Chapin PM. *J Chem Phys* 1988;**88**:5824–30.
27. Thirumalai D. *J Phys Chem* 1989;**93**:5637–44.
28. Okubo T. In: Schmitz KS, editor. *Macro-ion characterization. From dilute solutions to complex fluids.* Washington DC: American Chemical Society; 1994. pp. 364–80.
29. Okubo T. *Colloids Surf A* 1996;**109**:77–88.
30. Okubo T, Tsuchida A. *Forma* 2002;**17**:141–53.
31. Okubo T. In: Kinoshita S, Yoshioka S, editors. *Structural colors in biological systems.* Osaka, Japan: Osaka University Press; 2005. pp. 267–86. 2005.
32. Okubo T, Yoshimi H, Shimizu T, Ottewill RH. *Colloid Polym Sci* 2000;**278**:469–74.
33. Okubo T. *Polym J* 2008;**40**:882–90.

34. Russel WB. *Phase Trans* 1990;**21**:127–37.
35. Harkless CR, Singh MA, Nagler SE, Stephenson GB, Jordan-Sweet JL. *Phys Rev Lett* 1990;**64**:2285–8.
36. Dhont JKG, Smits C, Lekkerkerker HNW. *J Colloid Interf Sci* 1992;**152**:386–401.
37. Schatzel K, Ackerson B. *J Phys Rev E* 1993;**48**:3766–77.
38. Wurth N, Schwarz J, Culis F, Leidener P, Palberg T. *Phys Rev E* 1995;**52**:6415–23.
39. van Megen W. *Transport Theor Stat Phys* 1995;**24**:1017–51.
40. Conway JH, Sloane NJA. *Sphere packing, lattices and groups.* 3rd ed. New York: Springer-Verlag; 1998.
41. Hales TC. *Discr Comput Geom* 1997;**17**:1–51.
42. Hales TC. *Discr Comput Geom* 1997;**18**:135–49.
43. Fejes Toth L. *Lagerungen in der Ebene, auf der Kugel und im Raum.* 2nd ed. Heidelberg: Springer; 1972.
44. Thue A. *Christiania Vid Selsk Skr* 1910;**1**:3–9.
45. Tanemura M, Matsumoto T. *Z Kristallographie* 1997;**212**:637–47.
46. Aoyama K, Ogawa K, Kimura Y, Fujiyoshi Y. *Ultramicroscopy* 1995;**57**:345–54.
47. Kose A, Hachisu S. *J Colloid Interf Sci* 1974;**46**:460–9.
48. Ottewill RH. *Langmuir* 1989;**5**:4–9.
49. Itano WM, Bolinger JJ, Tan JN, Jelenkovic B, Huang XP, Wineland DJ. *Science* 1998;**279**:686–9.
50. Overbeek JThG. *J Chem Phys* 1987;**87**:4406–8.
51. Dickinson E. *Chem Phys Lett* 1978;**57**:148–50.
52. Dickinson E. *Faraday Discuss Chem Soc* 1978;**65**:127–9.
53. Dickinson E. *J Chem Soc Faraday Trans* 2 1979;**75**:466–73.
54. Dickinson E, Parker R, Lai M. *Chem Phys Lett* 1981;**79**:578–82.
55. Dickinson E, Parker R. *J Phys Lett* (Paris) 1985;**46**:L229–32.
56. Barat JUL, Hansen JP. *J Phys* 1986;**47**:1547–53.
57. Okubo T. *Ber Bunsenges Phys Chem* 1987;**91**:516–9.
58. Okubo T. *Colloid Polymer Sci* 2006;**285**:245–50.
59. Okubo T. In: Hubbard A, editor. *Encyclopedia of surface and colloid science.* New York: Marcel Dekker; 2002. pp. 1300–9.
60. Okubo T. In: Salamone JC, editor. *The polymeric molecules encyclopedia. Synthesis, properties and applications*, vol 2C. Boca Raton, FL: CRC Press; 1996. pp. 1290–8.
61. Okubo T, Fujita H. *Colloid Polymer Sci* 1996;**274**:368–74.
62. Murray MJ, Sanders JV. *Phil Mag A* 1980;**42**:721–40.
63. Yoshimura S, Hachisu S. *J Phys* (Paris) 1985;**46**(C3):115–26.
64. Mitaku S, Ohtsuki T, Okano K. *Jap J Appl Phys* 1980;**19**:439–48.
65. Okubo T, Fujita H, Kiriyama K, Yamaoka H. *Colloid Polym Sci* 1996;**274**:73–80.
66. Yoshinaga K, Chiyoda M, Ishiki H, Okubo T. *Colloids Surf A* 2002;**204**:285–93.
67. Boistelle R, Astiev JP. *J Crystal Growth* 1988;**90**:14–30.
68. Monovoukas Y, Gast AP. *Langmuir* 1991;**7**:460–8.
69. Okubo T. *J Chem Phys* 1991;**96**:2261–8.
70. Okubo T. *Colloid Polymer Sci* 1992;**270**:1018–26.
71. Okubo T. *Colloid Polymer Sci* 1994;**272**:440–6.
72. Robbins MO, Kremer K, Grest GS. *J Chem Phys* 1988;**88**:3286–312.
73. Frenkel D. *Physica A* 1999;**263**:26–38.
74. Auer S, Frenkel D. *Nature* 2001;**409**:1020–3.
75. Gasser U, Weeks ER, Schofield A, Pusey PN, Weitz DA. *Science* 2001;**292**:258–62.
76. Okubo T. *J Colloid Interf Sci* 1992;**153**:587–91.
77. Emmett S, Lubetkin SD, Vincent B. *Colloids Surf* 1989;**42**:139–53.
78. Voegtli LP, Zukoski IV CF. *J Colloid Interf Sci* 1991;**141**:79–91.
79. Shih WY, Aksay IA, Kikuchi R. *J Chem Phys* 1987;**86**:5127–32.

80. Kremer K, Robbins MO, Pincus P. *J Chem Phys* 1983;**79**:1474–9.
81. Hone D, Alexander S, Chaikin PM, Pincus P. *J Chem Phys* 1983;**79**:1474–9.
82. Chaikin PM, Pincus P, Alexander S, Hone D. *J Colloid Interf Sci* 1982;**89**:555–63.
83. Barnes CJ, Chan DYC, Everett DH, Yates DE. *J Chem Soc Faraday* 1978;**2**:136–48.
84. Kaldasch J, Laven J, Stein HN. *Langmuir* 1996;**12**:6197–201.
85. Okubo T. *J Chem Soc Faraday Trans* 1 1986;**82**:3163–73.
86. Okubo T. *J Chem Soc Faraday Trans* 1 1986;**82**:3175–83.
87. Okubo T. *J Chem Soc Faraday Trans* 1 1986;**82**:3185–90.
88. Okubo T. *J Chem Soc Faraday Trans* 1 1990;**86**:2871–6.
89. Okubo T. *J Am Chem Soc* 1990;**112**:5420–4.
90. Okubo T. *J Chem Soc Faraday Trans* 1 1988;**84**:1949–59.
91. Okubo T. *J Chem Soc Faraday Trans* 1 1988;**84**:4161–8.
92. Okubo T, Kimura H, Hase H, Yamaguchi K, Nagai K. *Colloid Polym Sci* 2004;**282**:250–5.
93. Okubo T, Kimura H, Hase H, Lovell PA, Erington N, Thongnoi P. *Colloid Polym Sci* 2005;**283**:393–401.
94. Okamoto J, Kimura H, Tsuchida A, Okubo T, Ito K. *Colloids Surf B Biointerf* 2007;**56**:231–5.
95. Tomita M, Takano K, van de Ven TGM. *J Colloid Interf Sci* 1983;**92**:367–82.
96. Okubo T. *Angew Chem Intern Ed Engl* 1987;**26**:765–7.
97. Okubo T, Aotani S. *Naturwiss* 1988;**75**:145–6.
98. Okubo T, Aotani S. *Colloid Polym Sci* 1988;**266**:1049–57.
99. Brown ABD, Clarke SM, Rennie AR. *Langmuir* 1998;**14**:3129–32.
100. Onsager L. *Ann NY Acad Sci* 1949;**51**:627–59.
101. Buining PA, Philipse AP, Lekkerkerker HNW. *Langmuir* 1994;**10**:2106–14.
102. van Bruggen MPB, Dhont JKG, Lekerkerker HNW. *Macromolecules* 1999;**32**:2256–64.
103. Davidson P, Gabriel JC, Levelu AM, Batail P. *Adv Mater* 1993;**5**:665–8.
104. Davidson P, Gabriel JCP, Pileni MP, editors. *Nanocrystals forming mesoscopic structures.* Weinheim: Wiley-VCH; 2005. pp. 173–212.
105. Kajiwara K, Donkai N, Hiragi Y, Inagaki H. *Makromol Chem* 1986;**187**:2883–94.
106. Donkai N, Hoshino H, Kajiwara K, Miyamoto T. *Makromol Chem* 1993;**194**:559–80.
107. Genoud JY, Staines M, Mawdsley A, Manojlovic V, Quinton W. *Supercond Sci Technol* 1999;**12**:663–71.
108. Kimura T, Kimura F, Yoshino M. *Langmuir* 2006;**22**:3464–6.
109. Kimura T, Chang C, Kimura F, Maeyama M. *J Appl Crystallogr* 2009;**42**:535–7.
110. Kimura F, Kimura T, Oshima W, Maeyama M, Aburaya K. *J Appl Crystallogr* 2010;**43**:151–3.
111. Kimura F, Mizutani K, Mikami B, Kimura T. *Cryst Growth Des* 2011;**11**:12–5.
112. Okubo T. *Colloid Polym Sci* 2013;**291**:1623–9.
113. Okubo T. *Colloid Polym Sci* 1990;**268**:1159–66.
114. Sanders JV. *Phil Mag A* 1980;**42**:705–20.
115. Sanders JV, Murray MJ. *Nature* 1978;**275**:201–3.
116. Hachisu S, Yoshimura S. *Nature* 1980;**283**:188–9.
117. Yoshimura S, Hachisu S. *Prog Colloid Polym Sci* 1983;**68**:59–70.
118. Okubo T. *J Chem Phys* 1990;**93**:8276–83.
119. Bartlett P, Ottewill RH, Pusey PN. *Phys Rev Lett* 1992;**68**:3801–4.
120. Bartlett P, Ottewill RH. *J Chem Phys* 1992;**96**:3306–18.
121. Bartlett P, Ottewill RH, Pusey PN. *Phys Rev Lett* 1992;**68**:3801–4.
122. Hachisu S. *Phase Transition* 1990;**21**:243–9.
123. Shih WY, Smith WH, Aksay IA. *J Chem Phys* 1989;**90**:4506–12.
124. Schofield AB. *Phys Rev E* 2001;**6405**:1–4. 051403.
125. Schofield AB, Pusey PN, Radcliffe P. *Phys Rev E* 2005;**72**:031407.

126. Okubo T. *J Chem Phys* 1987;**87**:5528–33.
127. Hachisu S, Kose A, Kobayashi Y, Takano K. *J Colloid Interf Sci* 1976;**55**:499–509.
128. Williams JC. *Powder Technol* 1976;**15**:245–51.
129. Rosato A, Stradburg KJ, Prinz F, Swendsen RH. *Phys Rev Lett* 1987;**58**:1038–40.
130. Vanell LLM, Rosato A, Dave RN. *Phys Rev Lett* 1997;**78**:1255–8.
131. Okubo T, Tsuchida A, Okuda T, Fujitsuna K, Ishikawa M, Morita T, Tada T. *Colloids Surf* 1999;**A153**:515–24 (misprinting in part); A160:311–320.
132. Okubo T, Tsuchida A, Takahashi K, Taguchi K, Ishikawa M. *Colloid Polym Sci* 2000;**278**:202–10.
133. Okubo T, Okada S, Tsuchida A. *J Colloid Interf Sci* 1997;**189**:337–47.
134. Okubo T, Okada S. *J Colloid Interf Sci* 1997;**192**:490–6.
135. Okubo T, Tsuchida A, Kato T. *Colloid Polym Sci* 1999;**277**:191–6.
136. Okubo T, Ishiki H. *J Colloid Interf Sci* 2000;**228**:151–6.
137. Th Palberg, Simon R, Wurth M, Leiderer P. *Prog Colloid Polym Sci* 1994;**96**:62–71.
138. He Y, Ackerson BJ, van Megen W, Underwood SM, Schatzel K. *Phys Rev E* 1996;**54**:5286–97.
139. Harland JL, van Megen W. *Phys Rev E* 1997;**55**:3054–67.
140. van Megen W, Ackerson BJ. *Prog Colloid Polym Sci* 1997;**104**:97–103.
141. Th Palberg. *J Phys Condens Matter* 1999;**11**:R323–60.
142. James RW. In: Bragg L, editor. *The optical principles of the diffraction of X-rays: the crystal state*, vol. 2. Ithaca, NY: Cornell University Press; 1965.
143. Wong EM, Bonevich JE, Pearson PC. *J Phys Chem* 1998;**B102**:7770–5.
144. Fujita H: Proceedings of the Fourth Asia-Pacific Conference Workshop on electron microscopy, p. 215.
145. Imai M, Mori K, Mizukami T, Kaji K, Kanaya T. *Polymer* 1992;**33**:4451–6.
146. Okubo T, Okada S. *J Colloid Interf Sci* 1998;**204**:198–204.
147. Okubo T, Ishiki H, Kimura H, Chiyoda M, Yoshinaga K. *Colloid Polym Sci* 2002;**280**:290–5.
148. Yoshinaga K, Chiyoda M, Ishiki H, Okubo T. *Colloids Surf* 2002;**A204**:285–93.
149. van de Ven TGM, Baloch MK. *J Colloid Interf Sci* 1990;**136**:494–508.
150. Tsuchida A, Takyo E, Taguchi K, Okubo T. *Colloid Polym Sci* 2004;**282**:1105–10.
151. Aastuen DJW, Clark NA, Cotter LK, Ackerson BJ. *Phys Rev Lett* 1986;**57**:1733–6.
152. Ackerson BJ, Schatzel K. *Phys Rev E* 1995;**52**:6448–60.
153. Schatzel K, Ackerson BJ. *Phys Scripta* 1993;**T49**:70–3.
154. Henderson SI, Mortensen TC, Underwood SM, van Megen W. *Physica A* 1996;**233**:102–16.
155. Palberg T. *J Phys Condens Matter* 2014;**26**:333101.
156. Okubo T, Kiriyama K, Nemoto N, Hashimoto H. *Colloid Polym Sci* 1996;**274**:93–104.
157. Okubo T, Kiriyama K. *J Mol Liquids* 1997;**72**:347–64.
158. Okubo T, Kiriyama K. *Ber Bunseng Phys Chem* 1996;**100**:849–56.
159. Brenner SL. *J Phys Chem* 1976;**80**:1473–7.
160. Hurd AJ, Clark NA, Mockler RC, O'Sullivan WJ. *Phys Rev A* 1982;**26**:2869–81.
161. Cotter LK, Clark NA. *J Chem Phys* 1987;**86**:6616–21.
162. van Megen W, Underwood SM, Pusey PN. *J Chem Soc Faraday Trans* 1991;**87**:395–401.
163. Lawson CL, Hansen RJ. *Solving least squares problems*. Upper Saddle River, NJ: Prentice Hall; 1974.
164. Fujita H, Ametani K. *J Appl Phys* 1977;**16**:1907–17.
165. Pusey PN, van Megen W, Bartlett P, Ackerson BJ, Parity JG, Underwood SM. *Phys Rev Lett* 1989;**63**:2753–6.
166. von Berlepsch H, Strey R. *Colloid Polym Sci* 1994;**272**:577–83.
167. Luck W, Klier M, Wesslau H. *Naturwiss* 1963;**50**:485–94.
168. Hiltner PA, Krieger IM. *J Phys Chem* 1969;**73**:2386–9.

169. Williams R, Crandall RS. *Phys Lett* 1974;**A48**:225–6.
170. Schefer DW, Ackerson BJ. *Phys Rev Lett* 1975;**35**:1448–51.
171. Berne B, Pecora R. *Dynamic light scattering*. New York: Plenum; 1980.
172. Hayter JB, Penfold J. *Mol Phys* 1981;**42**:109–18.
173. Hansen P, Hayter JB. *Mol Phys* 1982;**46**. 651–485.
174. Snook IK, Hayter JB. *Langmuir* 1992;**8**:2889–97.
175. Hartl W, Versmold H. *Z Phys Chem Neue Folge* 1987;**153**:1.
176. Hartl W, Versmold H. *J Chem Phys* 1988;**88**:7157–61.
177. van Megen W, Underwood SM, Pusey PN. *Phys Rev Lett* 1991;**67**:1586–9.
178. D'Aguanno B, Klein R. *J Chem Soc Faraday Trans* 1991;**87**:379–90.
179. Pusey PN, Tough RJA. In: Pecora R, editor. *Dynamic light scattering*. London: Plenum; 1985.
180. Dalberg PS, Boe A, Strand KA, Sikkeland T. *J Chem Phys* 1978;**69**:5473.
181. Grunner F, Lehmann W, Degiorgio V, Corti M, Giglio M. *Light scattering in fluids and macromolecular solutions*. New York: Plenum; 1980.
182. Okubo T. *J Chem Phys* 1989;**90**:2408–15.
183. Dubois-Violette E, Pieranski P, Rothen F, Strzelecki L. *J Phys* (Paris) 1980;**41**:369–76.
184. Stoimenova M, Okubo T. *J Colloid Interf Sci* 1995;**176**:267–71.
185. Nemoto N, Matsuda H, Tsunashima Y, Kurata M. *Macromolecules* 1984;**17**:1731–5.
186. Lin SC, Lee WI, Schurr JM. *Biopolymers* 1978;**17**:1041–64.
187. Okubo T. *J Colloid Interf Sci* 1992;**153**:587–91.
188. Okubo T. *J Phys Chem* 1989;**93**:4352–4.
189. Okubo T, Tsuchida A. *Colloid Polym Sci* 2002;**280**:438–45.
190. Brown JC, Pusey PN, Goodwin JW, Ottewill RH. *J Phys* (Paris) 1975;**A8**:664–82.
191. Ackerson BJ, Clark NA. *Faraday Discuss Chem Soc* 1983;**76**:219–28.
192. Cebula DJ, Goodwin JW, Jeferey GC, Ottewill RH, Parentich A, Richardson RA. *Faraday Discuss Chem Soc* 1983;**76**:37–52.
193. Philipse AP, Vrij A. *J Chem Phys* 1988;**88**:6459–70.
194. Ackerson BJ, Hayter JB, Clark NA, Cotter L. *J Chem Phys* 1986;**84**:2344–9.
195. Goodwin JW, Ottewill RH. *J Chem Soc Faraday Trans* 1991;**87**:357–69.
196. Bartlet P, Ottewill RH, Pusey PN. *J Chem Phys* 1990;**93**:1299–312.
197. Smith GS, Sirota EB, Satinya CR, Plano RJ, Clark NA. *J Chem Phys* 1990;**92**:4519–29.
198. Ashdown S, Markovic I, Ottewill RH, Lindner P, Obertur RC, Rennie AR. *Langmuir* 1990;**6**:303–7.
199. Reus V, Belloni L, Zemb T, Lutterbach N, Versmold H. *J Phys II France* 1997;**7**:603–26.
200. van Veluwen A, Lekerkerker HNW, de Kruif LG, Vrij A. *J Chem Phys* 1987;**87**:4873–80.
201. Gruner F, Lehmann W. *J Phys* (Paris) 1979;**A12**:L303–7.
202. Hess W, Klein R. *Adv Phys* 1983;**32**:173–283.
203. Klein R, Hess W. *Faraday Discuss Chem Soc* 1983;**76**:137–50.
204. van Megen W, Snook I. *Faraday Discuss Chem Soc* 1983;**76**:151–63.
205. Ottewill RH. Williams NStJ. *Nature* (London) 1987;**325**:232–4.
206. D'Aguanno B, Klein R. *J Chem Soc Faraday Trans* 1991;**87**:379–90.
207. Okubo T. *J Colloid Interf Sci* 1987;**117**:165–71.
208. Okubo T. *J Chem Phys* 1987;**87**:3022–8.
209. Shindewolf U. *Z Elektrochem* 1954;**58**:697.
210. Heckmann K. *Z Phys Chem NF* 1956;**9**:318–37.
211. Eigen M, Schwarz G. *J Colloid Sci* 1957;**12**:181–94.
212. Heckmann K, Gotz KG. *Elektrochem* 1958;**62**:281–8.
213. Gotz KG, Heckmann K. *J Colloid Sci* 1958;**13**:266–72.
214. Schwarz G. *Z Phys Chem* 1959;**19**:286–314.
215. Gotz KG. *J Colloid Sci* 1965;**20**:289–390.
216. Kern EE, Anderson DK. *J Polym Sci* Part A 1968;**6**:2765–72.

217. Okubo T. *J Am Chem Soc* 1987;**109**:1913–6.
218. Okubo T, Kitano H, Ishiwatari T, Ise N. *Proc Roy Soc London Ser A* 1979;**366**:81–90.
219. Okubo T. *Biophys Chem* 1980;**11**:425–31.
220. Okubo T, Enokida A. *J Chem Soc Faraday Trans* 1 1983;**79**:1639–48.
221. Okubo T, Hongyo K, Enokida A. *J Chem Soc Faraday Trans* 1 1984;**80**:2097–8.
222. Kitano H, Hasegawa J, Iwai S, Okubo T. *J Phys Chem* 1986;**90**:6281–4.
223. Kitano H, Hasegawa J, Iwai S, Okubo T. *Polym Bull* 1986;**16**:89–93.
224. Okubo T, Maeda Y, Kitano H. *J Phys Chem* 1989;**93**:3721–3.
225. Kitano H, Maeda Y, Okubo T. *Biophys Chem* 1989;**33**:47–54.
226. Okubo T. *Makromol Chem Suppl* 1985;**14**:161–77.
227. Okubo T. *J Chem Soc Faraday Trans* 1 1988;**84**:703–12.
228. Okubo T. *Macromolecules* 1989;**22**:1818–21.
229. Okubo T, Shimizu T. *J Colloid Interf Sci* 1990;**135**:300–3.
230. Okubo T. In: Oosawa F, editor. *Dynamic aspects of polyelectrolytes and biomembranes.* Japan: Kodansha; 1982. pp. 111–7.
231. Okubo T, Ise N. 1:109–114, 1978.
232. Sawamoto M, Higashimura T, Enokida A, Okubo T. *Polym Bull* 1980;**2**:309–13.
233. Cornel RM, Goodwin JW, Ottewill RH. *J Colloid Interf Sci* 1979;**71**:254–66.
234. Clark NA, Ackerson BJ. *Phys Rev Lett* 1980;**44**:1005–8.
235. Lindsay HM, Dozier PWD, Chaikin M, Klein P. *J Phys* 1986;**A19**:2583–9.
236. Mitaku S, Ohtsuki T, Enari K, Kishimoto A, Okano K. *Jap J Appl Phys* 1978;**17**:305–13.
237. Ohtsuki T, Mitaku S, Okano K. *Jap J Appl Phys* 1978;**17**:627–35.
238. Ohtsuki T, Kishimoto A, Mitaku S, Okano K. *Jap J Appl Phys* 1981;**20**:509–18.
239. Joanicot HM, Jorand M, Pieranski P, Rothen F. *J Phys* (Paris) 1984;**45**:1413–21.
240. Schmacher GA, van de Ven TGM. *Faraday Discuss Chem Soc* 1987;**83**:75–81.
241. Schmacher GA, van de Ven TGM. *J Chem Soc Faraday Trans* 1991;**87**:971–6.
242. Ohshima H, Healy TW, White LR, O'Brien RW. *J Chem Soc Faraday Trans* 2 1984;**80**:1299–317.
243. Crandall RS, Williams R. *Science* 1977;**198**:293–5.
244. Okubo T. *Colloid Polym Sci* 1987;**265**:522–7.
245. Okubo T, Aotani S. *Colloid Polym Sci* 1988;**266**:1042–8.
246. Okubo T. *J Chem Phys* 1987;**86**:2394–9.
247. Okubo T. *J Chem Phys* 1987;**86**:5182–7.
248. Okubo T. *J Chem Phys* 1988;**88**:6581–4.
249. Okubo T. *J Chem Phys* 1988;**88**:2083–6.
250. Okubo T. *J Chem Soc Faraday Trans* 1 1989;**85**:455–66.
251. Okubo T. *J Chem Soc Faraday Trans* 1 1990;**86**:151–6.
252. Okubo T. *J Chem Phys* 1995;**102**:7721–7.
253. Okubo T. *J Am Chem Soc* 1990;**112**:5420–4.
254. Okubo T. *J Chem Soc Faraday Trans* 1 1987;**83**:2487–95.
255. Goodwin JW, Smith RW. *Faraday Discuss Chem Soc* 1974;**57**:126–9.
256. Buscal R, Goodwin JW, Hawkins MW, Ottewill RH. *J Chem Soc Faraday Trans* 1 1982;**78**:2873–87.
257. Goodwin JW, Gregory T, Miles JA, Warren BCH. *J Colloid Interf Sci* 1984;**97**:488–95.
258. Goodwin JW, Hughes RW, Patridge SJ, Zukoski CF. *J Chem Phys* 1986;**85**:559–66.
259. Joanny JF. *J Colloid Interf Sci* 1979;**71**:622–4.
260. Fujita H, Ametani K. *Jap J Appl Phys* 1979;**18**:753–8.
261. van Megen WJ, Snook IK, Watts RO. *J Colloid Interf Sci* 1980;**77**:131–7.
262. Benzing DW, Russel WB. *J Colloid Interf Sci* 1981;**83**:178–90.
263. Okubo T. *Ber Bunsenges Phys Chem* 1988;**92**:504–11.
264. Okubo T. *J Colloid Interf Sci* 1990;**135**:259–62.
265. Matsumoto T, Okubo T. *J Rheol* 1991;**35**:135–48.

266. Tomita M, van de Ven TGM. *J Phys Chem* 1985;**89**:1291–6.
267. Wiersma PH, Loeb AL. Overbeek JThG. *J Colloid Interf Sci* 1966;**22**:78.
268. Jorand M, Koch AJ, Rothen F. *J Phys* 1986;**47**:217–27.
269. Okubo T, Ishiki H. *Colloid Polym Sci* 2001;**279**:571–8.
270. Robbins MO, Kremer K, Grest GS. *J Chem Phys* 1988;**88**:3286–312.
271. Okubo T, Ishiki H, Chiyoda M, Yoshinaga K. *Colloid Polym Sci* 2002;**280**:446–53.
272. Okubo T. *J Chem Phys* 1987;**87**:6733–9.
273. Okubo T. *Naturwiss* 1988;**75**:91–3.
274. Okubo T. *Polym Bull* 1988;**20**:269–76.
275. Okubo T. *J Phys Chem* 1990;**94**:1962–6.
276. Okubo T, Takezawa K, Kimura H. *Colloid Polym Sci* 2000;**278**:571–5.
277. Pals DTF, Hermans JJ. *Rec Trav Chim Pay-Bas* 1952;**71**:433.
278. Nishida K, Kiriyama K, Kanaya T, Kaji K, Okubo T. *J Polym Sci Part B Polym Phys* 2004;**42**:1068–74.
279. Okubo T, Kimura H, Hatta T, Kawai T. *Phys Chem Chem Phys* 2002;**4**:2260–3.
280. Kimura H, Okubo T. *Colloid Polym Sci* 2002;**280**:579–83.
281. Okubo T, Kimura H, Kawai T, Niimi H. *Langmuir* 2003;**19**:6014–8.
282. Kimura H, Niimi H, Tsuchida A, Okubo T. *Colloid Polym Sci* 2005;**283**:1079–84.
283. Kimura H, Nakayama Y, Tsuchida A, Okubo T. *Colloids Surf B Biointerf* 2007;**56**:236–40.
284. Kimura H, Sakurai M, Sugiyama T, Tsuchida A, Okubo T, Masuko T. *Rheol Acta* 2011;**50**:159–68.
285. Sugimura M, Okubo T, Ise N, Yokoyama S. *J Am Chem Soc* 1984;**106**:5069–72.
286. Rymden R. *J Colloid Interf Sci* 1988;**124**:396–402.
287. Rohrsetzer S, Kovacs P, Nagy M. *Colloid Polym Sci* 1986;**264**:812–6.
288. Costello BA, Kim IT, Luckham PF. *J Chem Soc Faraday Trans* 1990;**86**:3693–4.
289. Takano K, Hachisu S. *J Chem Phys* 1977;**67**:2604–8.
290. Alder BJ, Wainwright TE. *J Chem Phys* 1957;**27**:1208–9.
291. Alder BJ, Hoover WG, Toung DAY. *J Chem Phys* 1968;**49**:3688–96.
292. Hoover WG, Gray SC, Johnson KW. *J Chem Phys* 1971;**55**:1128–36.
293. Hopkins AJ, Woodcook LV. *J Chem Soc Faraday Trans* 1990;**86**:3419–28.
294. Schwan HP, Schwarz G, Macznk J, Pauly H. *J Phys Chem* 1962;**66**:2626–35.
295. De Lacey EHB, White LR. *J Chem Soc Faraday Trans* 2 1981;**77**:2007–39.
296. Rosen LR, Saville DA. *Langmuir* 1991;**7**:36–42.
297. Nakagaki M, Heller W. *J Appl Phys* 1956;**27**:975–9.
298. Robillard F, Patistsas AJ. *Can J Chem* 1973;**51**:2395–401.
299. Robillard F, Patistsas AJ. *Can J Chem* 1974;**52**:1571–82.
300. Champion JV, Meeten GH, Senior MJ. *J Chem Soc Faraday Trans* 2 1978;**74**:1319–29.
301. Champion JV, Meeten GH, Senior MJ. *J Colloid Interf Sci* 1979;**72**:471–82.
302. Meeten GH. *J Colloid Interf Sci* 1980;**77**:1–5.
303. Sano Y, Nakagaki M. *J Phys Chem* 1984;**88**:95–9.
304. Okubo T. *J Colloid Interf Sci* 1990;**135**:294–6.
305. Okubo T. *Prog Colloid Polym Sci* 2003;**124**:112–5.
306. Tsuchida A, Okubo T. In: Stoylov S, Stoimenova M, editors. *Molecular and colloidal electro-optics*. New York: CRC Press; 2006. pp. 447–56.
307. Okubo T, Tsuchida A, Stoimenova M. *Adv Colloid Interf Sci* 2011;**162**:80–6.
308. Okubo T. *J Chem Soc Faraday Trans* 1 1988;**84**:3377–88.
309. Stoimenova M, Okubo T. *J Colloid Interf Sci* 1995;**176**:267–71.
310. Stoimenova M, Dimitrov V, Okubo T. *J Colloid Interf Sci* 1996;**184**:106–11.
311. Okubo T, Tsuchida A, Okada S, Kobata S. *J Colloid Interf Sci* 1998;**199**:83–91.
312. Okubo T, Tsuchida A, Takahashi T, Iwata A. *J Colloid Interf Sci* 1998;**207**:130–6.
313. Okubo T, Tsuchida A, Iwata A, Takahashi T. *Colloids Surf* 1999;**A148**:87–94.

314. Stoimenova M, Alekov A, Okubo T. *Colloids Surf* 1999;**A148**:83–6.
315. Stoimenova M, Okubo T. In: Schwarz JA, Contescu CI, editors. *Surfaces of nanoparticles and porous materials*. New York: Marcel Dekker; 1999. pp. 103–24.
316. Okubo T, Tsuchida A, Takahashi T, Iwata A, Okada S, Kobata S, Kobayashi K, Okubo T. *Langmuir* 1999;**15**:4203–7.
317. Tsuchida A, Shibata K, Okubo T. *Colloid Polym Sci* 2003;**281**:1104–8.
318. Gradzielski M, Hoffmann H. In: Bloor DM, Wyn-Johns E, editors. *Dynamic and equilibrium properties of colloidal systems*. Dordrecht: Kluwer; 1990. pp. 427.
319. Shibata K, Yamamoto T, Kurita M, Kimura H, Tsuchida A, Okubo T. *Colloid Polym Sci* 2006;**284**:688–93.
320. Shibata K, Kimura H, Tsuchida A, Okubo T. *Colloid Polym Sci* 2006;**285**:127–33.
321. Shibata K, Kimura H, Tsuchida A, Okubo T, Sato S, Yoshinaga K. *Colloid Polym Sci* 2006;**284**:372–9.
322. Fujita H, Ametani K. *Jpn J Appl Phys* 1979;**11**:1907–17.
323. Trau M, Saville DA, Aksay IA. *Science* 1996;**272**:706–9.
324. van der Zande BMI, Koper GJM, Lekkerkerker HNW. *J Phys Chem B* 1999;**103**:5754–60.
325. Gong T, Marr DWM. *Langmuir* 2001;**17**:2301–4.
326. Fujiwara M, Oki E, Hamada M, Tanimoto Y, Mukuda I, Shimomura Y. *J Phys Chem A* 2001;**105**:4383–6.
327. Gu ZZ, Fujishima A, Sato O. *J Am Chem Soc* 2000;**122**:12387–8.
328. Clark NA, Hurd AJ, Ackerson BJ. *Nature* (London) 1979;**281**:57–60.
329. Tomita M, van de Ven TGK. *J Colloid Interf Sci* 1984;**99**:374–86.
330. Weitz DA, Dozier WD, Chaikin PM. *J Phys Colloq* 1985;**46**:257–68.
331. DI Meglio JM, Weitz DA. *Chaikin PM: Phys Rev Lett* 1987;**58**:136–9.
332. Ackerson BJ, Clark NA. *Physica* 1983;**A118**:221–49.
333. Ackerson BJ, Clark NA. *Phys Rev* 1984;**A30**:906–18.
334. Okubo T. *J Chem Soc Faraday Trans* 1 1988;**84**:1171–80.
335. Hoffman RL. *Future directions in polymer colloids*. Dordrecht: Martinus Nijhoff; 1987. pp. 151–165.
336. Mathis C, Bossis G, Brady JF. *Prog Colloid Polym Sci* 1988;**76**:16–9.
337. Lindner P, Markovic I, Oberthur RC, Ottewill RH, Rennie AR. *Prog Colloid Polym Sci* 1988;**76**:47–50.
338. Bossis G, Brady JF, Mathis C. *J Colloid Interf Sci* 1988;**126**:1–15.
339. Mathis C, Bossis G, Brady JF. *J Colloid Interf Sci* 1988;**126**:16–25.
340. Okubo T. *J Chem Soc Farady Trans* 1 1987;**83**:2497–504.
341. Cornell RM, Goodwin JW, Ottewill RH. *J Colloid Interf Sci* 1979;**71**:254–66.
342. Zhu J, Li M, Reger R, Meyer W, Ottewill RH. STS-73 Space Shuttle Crew, Russell WB, Chaikin PM. *Nature* (London) 1997;**387**:883–5.
343. Tsuchida A, Taguchi K, Takyo E, Yoshimi H, Kiriyama S, Okubo T, Ishikawa M. *Colloid Polym Sci* 2000;**278**:872–7.
344. Ishikawa M, Morimoto H, Okubo T. *Intern J Modern Phys* B 2002;**16**:338–45.
345. Bartlett P, Ottewill RH. *Adv Colloid Interf Sci* 1994;**50**:39–50.
346. Schuller H. *Kolloid Z Z Polym* 1967;**216/217**:380–3.
347. Sheppard E, Tcheurekdjian N. *J Colloid Interf Sci* 1968;**28**:481–6.
348. Aveyard R, Clint JH. *J Chem Soc Faraday Trans* 1995;**91**:2681–97.
349. Aveyard R, Binks BP, Fletcher PDI, Rutherford CE. *J Dispersion Sci Technol* 1994;**15**:251–71.
350. Aveyard R, Cooper P, Fletcher PDI, Rutherford CE. *Langmuir* 1993;**9**:604–13.
351. Aveyard R, Clint JH, Nees D, Paunov N. *Langmuir* 2000;**16**:1969–79.
352. Hurd AJ. *J Phys A* 1985;**8**:L1055–60.
353. Goulding D, Hansen JP. *Mol Phys* 1998;**95**:649–55.
354. Murray CA, Wenk RA. *Phys Rev Lett* 1989;**62**:1643–6.

355. Glaser MA, Clark NA. *Phys Rev* A 1990;**41**:4585–8.
356. Weider T, Glaser MA, Hanley HJM, Clark NA. *Phys Rev* B 1993;**47**:5622–8.
357. Pouligny B, Malzbender R, Ryan P, Clark NA. *Phys Rev* B 1990;**42**:988–91.
358. Murray CA, van Winkle DH. *Phys Rev Lett* 1987;**58**:1200–3.
359. Trau M, Saville DA, Aksay IA. *Langumuir* 1997;**13**:6375–81.
360. Solomentsev Y, Bohmer M, Anderson JL. *Langmuir* 1997;**13**:6058–68.
361. Hong S, Higuchi T, Sugimori H, Kaneko T, Abetz V, Takahara A, Jinnai H. *Polym J* 2012;**44**:567–72.
362. Naidoo KJ, Schnitker J. *J Chem Phys* 1994;**100**:3114–21.
363. Hayashi S, Kumamoto Y, Suzuki T, Hirai T. *J Colloid Interf Sci* 1991;**144**:538–47.
364. Hulteen JC, Treichel DA, Smith MT, Duval ML, Jensen TR, van Duyne RP. *J Phys Chem B* 1999;**103**:3854–63.
365. Lenzmann F, Li K, Kitai AH, Stover HDH. *Chem Mater* 1994;**6**:156–9.
366. Wood MA. *J Res Soc Interf* 2007;**4**:1–17.
367. Li L, Zhai T, Zeng H, Fang X, Bando Y, Golberg D. *J Mater Chem* 2011;**21**:40–56.
368. Xu H, Goedel WA. *Langmuir* 2002;**18**:2363–7.
369. Marczewski D, Goedel WA. *Nano Lett* 2005;**5**:295–9.
370. Fujii S, Kappl M, Butt HJ, Sugimoto T, Nakamura Y. *Angew Chem Int Ed* 2012; **51**:9809–13.
371. Mc Phedran RC, Nicorovici NA, Mc Kenzie DR, Botten LC, Parker AR, Rouse GW. *Aust J Chem* 2001;**54**:241–4.
372. Welch VC. In: Kinoshita S, Yoshioka S, editors. *Structural colors in biological systems. Principles and applications.* Osaka, Japan: Osaka University Press; 2005. pp. 53–71.
373. Yoshioka Y, Kinoshita S. *Forma* 2002;**17**. 169–153.
374. Kinoshita S, Yoshioka S. In: Kinoshita S, Yoshioka S, editors. *Structural colors in biological systems. Principles and applications.* Osaka, Japan: Osaka University Press; 2005. pp. 113–40.
375. Kinoshita S, Yoshioka S. In: Kinoshita S, Yoshioka S, editors. *Structural colors in biological systems. Principles and applications.* Osaka, Japan: Osaka University Press; 2005. pp. 141–52.
376. Hariyama T, Hironaka M, Horiguchi H, Stavenga DG. In: Kinoshita S, Yoshioka S, editors. *Structural colors in biological systems. Principles and applications.* Osaka, Japan: Osaka University Press; 2005. pp. 153–76.
377. Kasukawa H, Oshima N. *Pigment Cell Res* 1987;**1**:152–7.
378. Fujii R, Kasukawa H, Miyaji K, Oshima N. *Zool Sci* 1989;**6**:477–86.
379. Fujii S, Nakamura Y, Tsuchida A, Okubo T. *Colloid Polym Sci* 2014;**292**:2303–10.
380. Gierenz G, Karmann W, editors. *Adhesives and adhesive tapes.* Weinheim: Wiley-VCH; 2008.
381. Possart W, editor. *Adhesion: current research and applications.* Weinheim: Wiley-VCH; 2005.
382. Ueno K, Inaba A, Sano Y, Kondoh M, Watanabe M. *Chem Commun* 2009:3603–5.
383. Ueno K, Sano Y, Inaba A, Kondoh M, Watanabe M. *J Phys Chem B* 2010;**114**:13095–103.
384. Ueno K, Inaba A, Ueki T, Kondoh M, Watanabe M. *Langmuir* 2010;**26**:18031–8.
385. Ueno K, Watanabe M. *Langmuir* 2011;**27**:9105–15.
386. Ueno K, Watanabe M. *Langmuir* 2011;**27**:9105–15.
387. Kalsin AM, Fialkowski M, Paszewski M, Smoukov SK, Bishop KJM, Grzybowski BA, Schatz GC, Grzybowski BA. *Science* 2006;**312**:420–4.
388. Kalsin AM, Pinchuk AO, Smoukov SK, Paszewski M, Schatz GC, Grzybowski BA. *Nano Lett* 2006;**6**:1896–903.
389. Kalsin AM, Kowalczyk B, Smoukov SK, Klajn R, Grzybowski BA. *J Am Chem Soc* 2006;**128**:15046–7.
390. Friberg SE, Al-Bawab A, Bozeya A, Aikens PA. *J Colloid Interf Sci* 2009;**336**:345–51.

391. Watanabe J, Okoshi K, Sano N, Edo S. In: Kinoshita S, Yoshioka S, editors. *Structural colors in biological systems. Principles and applications.* Osaka, Japan: Osaka University Press; 2005. pp. 319–27.
392. Fortov VE, Khrapak AG, Khrapak SA, Molotkov VI, Petrov OF. *Phys Uspekhi* 2004;**47**:447–92.
393. Leunissen ME, van Blaaderen A, Hollingworth AD, Sullivan MT. *Proc Natl Acad Sci USA* 2007;**104**:2585–90.
394. Leunissen ME, Zwanikken J, van Roij R, Chaikin PM, van Blaaderen A. *Phys Chem Chem Phys* 2007;**9**:6405–14.
395. Zwanikken J, van Roij R. *Phys Rev Lett* 2007;**99**:178301.
396. Born M. *Z Phys* 1920;**1**:45–8.
397. Verwey EJW, Niessen KF. *Phil Mag* 1939;**28**:435–46.
398. Israelachvili JN. *Intermolecular and surface forces.* 2nd ed. London: Academic Press; 1992.
399. Jethmalani JM, Ford WT, Beaucage G. *Langmuir* 1997;**13**:3338–44.
400. Yoshinaga K, Mouri E, Ogawa J, Nakai A, Ishii M, Nakamura H. *Colloid Polym Sci* 2004;**283**:340–2.
401. Xia Y, Rogers JA, Paul KE, Whitesides GM. *Chem Rev* 1999;**99**:1823–48.
402. Smith HI. *Physica E* 2001;**11**:104–9.
403. Fischer UC, Zingsheim HP. *J Vac Sci Technol* 1981;**19**:881–5.
404. Deckman HW, Dunsmuir JH. *J Vac Sci Technol* B 1983;**1**:1109–12.
405. Deckman HW, Dunsmuir JH, Garoff S, McHenry JA, Pfeiffer DG. *J Vac Sci Technol B* 1988;**6**:333–6.
406. Hayness CL, Van Duyne RP. *J Phys Chem B* 2001;**105**:5599–611.
407. Ingert D, Pileni MP. In: Pileni MP, editor. *Nanocrystals forming mesoscopic structures.* Weinheim: Wiley-VCH; 2005. pp. 295–305.
408. Choi FG, Jang SG, Kim S, Lee E, Han CS, Yang SM. *Adv Funct Mater* 2006;**16**:33–40.
409. Yablonovitch E, Gmitler TJ, Meade RD, Rapper AM, Brammer KD, Joannopoulos JD. *Phys Rev Lett* 1991;**67**:3380–3.
410. Pradhan RD, Bloodgood JA, Watson GH. *Phys Rev* 1997;**B55**:9503–7.
411. Wijnhoven JEG, Vos WL. *Science* 1998;**281**:802–4.
412. Zakhidov AA, Baughman RH, Igbal Z, Culi C, Khayrullin I, Dantas SO, Marti J, Ralchenko VG. *Science* 1998;**282**:897–901.
413. Levi BG. *Phys Today* 1999;**52**:17–8.
414. Megen M, Wijnhoven JEGJ, Lagendijk A, Vos WL. *J Opt Soc Am* 1999;**B16**:1403–8.
415. Park SH, Xia Y. *Langmuir* 1999;**15**:266–73.
416. Debord JD, Lyon LA. *J Phys Chem* 2000;**B104**:6327–31.
417. Freemantle M. *Chem Eng News* 1998;**76**:32–7.
418. Shim M, Wang C, Guyot-Sionnest P. *J Phys Chem* 2001;**B105**:2369–73.
419. Yang SM, Ozin GA. *Chem Commun* 2000:2507–8.
420. Yin Y, Lu Y, Gates B, Xia Y. *J Am Chem Soc* 2001;**123**:8718–29.
421. Yin Y, Xia Y. *Adv Mater* 2002;**14**:605–8.
422. Matsuo S, Fujine T, Fukuda K, Juodkazis S, Misawa H. *Appl Phys Lett* 2003;**82**:4283–5.
423. Xu X, Asher SA. *J Am Chem Soc* 2004;**126**:7940–5.
424. Wen W, Zhang L, Sheng P. *Phys Rev Lett* 2000;**85**:5464–7.
425. Saad Y, Golosovsky T, Ji M, Davidov D, Avni Y, Frenkel A. *Opt Mater* 2001;**17**:1–6.
426. Davidson P, Gabriel JCP. In: Pileni MP, editor. *Nanocrystals forming mesoscopic structures.* Weinheim: Wiley-VCH; 2005. pp. 173–212. chapt 7.
427. Fudouzi H, Sawada T. *Langmuir* 2006;**22**:1365–8.
428. Furumi S, Fudouzi H, Sawada T. *J Mater Chem* 2012;**22**:21519–28.
429. van Blaaderen A, Ruel R, Wiltzius P. *Nature* 1997;**385**:321–4.
430. Hines MA, Guyot-Sionnest P. *J Phys Chem* 1996;**100**:468–71.

4 Gel Crystallization

4.1 Gel Crystals

Blocklike and colloidal gels are well known for their phase transition of swelling and shrinking under the atmospheric conditions such as temperature, salt concentrations, and solvent. The thermodynamic volume effects in the gel systems are now familiar among researchers.[1-13] Physicochemical properties of gels have been studied intensively.[14-24] Keen attention has been paid to the gel crystals, because the coupling effects of the optical, dynamic, and electrical properties have interesting applications in many fields of science.

Gel crystals are grouped into three systems: (Type 1) crystal-like ordering of gel spheres (particles) in a suspension state, (Type 2) gels including interior crystal arrays of colloidal particles (including hollow of liquid or gas state), and (Type 3) crystal-like arrays of gel spheres formed at the plane or curved interfaces (Fig. 4.1). All of these systems are interesting because their structure and function are similar to biological assemblies such as skin, muscle, and crystalline lens. Furthermore, most gels are highly sensitive to external fields and other atmospheres. Type-1 gel crystals are similar to the typical colloidal crystals of hard spheres, although the interfaces of gels with medium phase are more or less vague and fuggy compared with colloidal crystals. Typical gel spheres in of the first type are cross-linked linear-type polymers. Recently, many kinds of core-shell–type gel spheres (gel shell or gel core) have been prepared. Mixtures of the gel spheres and colloidal hard spheres are also interesting.

Type 2 gel crystal spheres are composed of colloidal arrays in the gel state medium. Properties of type 2 gel crystals are similar to those of gels themselves, but optical properties originated from the crystal arrays are further added in most cases. Many kinds of type 3 gel crystals are available. Drying patterns of monodispersed gel spheres are also categorized as type 3 gel crystals. Typical gel substances are composed of synthetic and biological polymers. Typical polymer gels studied for crystallization are compiled in Table 4.1 for three types of gel crystals. In this chapter, type 1 and 3 gel crystals are mainly discussed. However, type 2 gel crystals are attractive for

Colloidal Organization. http://dx.doi.org/10.1016/B978-0-12-802163-7.00004-0

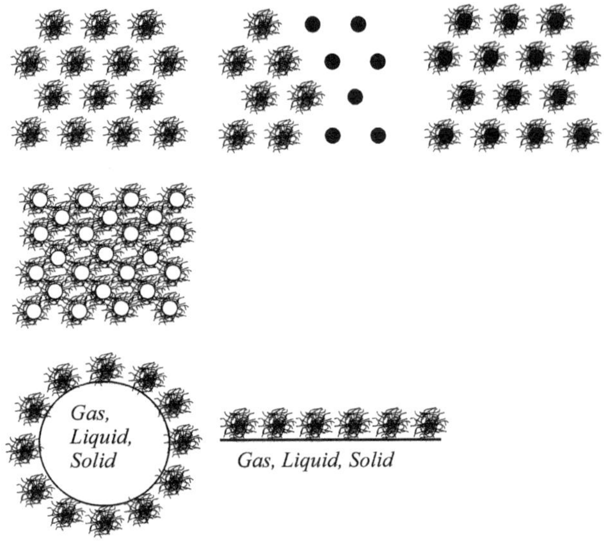

Figure 4.1
Types 1, 2, and 3 gel crystals.

Table 4.1 Typical Polymers Forming Type 1, 2, and 3 Gel Crystals

Anionic Gels (in water medium)
Poly(*N*-isopropylacrylamide) (pNIPAm) Poly(*N*-ethylacrylamide) Poly(*N*-methylacrylamide) Poly(*N*-isopropylacrylamide-co-ethylene glycol) Poly(*N*-isopropylacrylamide-co-acrylic acid) Poly(*N*-isopropylacrylamide-co-3-acrylamidephenylboronic acid) Poly(*N*-methylacryloylglycinemethylether) Poly(ethylene glycol-co-methylether methacrylate) Poly(methacrylic acid-co-methylether methacrylate) Poly(methacrylic acid-co-ethylene glycol) Poly(methyl methacrylate)
Cationic Gels
Poly(2-vinylpyridine) (P2VP)
Biological Gels
Gelatin (polypeptide) Agar-agar (vegetable gelatin, polysaccharide) Arrow-root starch (polysaccharide) Potato starch (polysaccharide) *Nata de coco* (polysaccharide [cellulose derivative])

preparing optical and mechanical devices. We should note that type I crystals are really gel crystals. But elementary substances of type 2 and 3 crystals are fixed and restricted from free thermal fluctuation. In other words, the translational and rotational movements of gel particles are highly restricted in the gels of types 2 and 3. Thus they should be called *frozen crystals.* Furthermore, future study of the crystallization of monodispersed anisotropic gel particles to identify the special functions and properties of unisotropic gel crystals is exciting.

Let's consider type 1 gel crystals in more detail. In general, polymer solution, which is composed of linear-type polymers and their solvents, is transparent. The main cause for the transparency is that the unit's size of dynamic thermal motion of the typical polymers, which is close to the size of their monomer units, is very small compared with the light wavelengths and close to the size of solvents (Fig. 4.2). No interfaces exist between the polymer solutes and solvents (medium molecules), because the monomer units and solvent molecules exchange their location freely by their thermal motion. On the other hand, colloidal dispersions of hard spheres are composed of large, solid particles and small, medium molecules. The size of the typical colloidal particles is close to or larger than the light wavelengths, and 100- to 1000-fold larger than that of the medium molecules. Thus colloidal dispersion is turbid except in the refractive index matched condition between particles and medium molecules. In other words, turbid dispersion supports the existence of the interfaces in a sense of colloid chemistry.

Most colloidal particles in water, on the other hand, have interfaces with the medium molecules, and get negative charges on their interfaces by one of three mechanisms: (1) dissociation of ionizable groups introduced to the interfaces from the initiators during polymerization of monomers, (2) the preferential adsorption of ions from suspension to reduce the high free-energy level at the particle–water interfaces, and (3) the differences in the

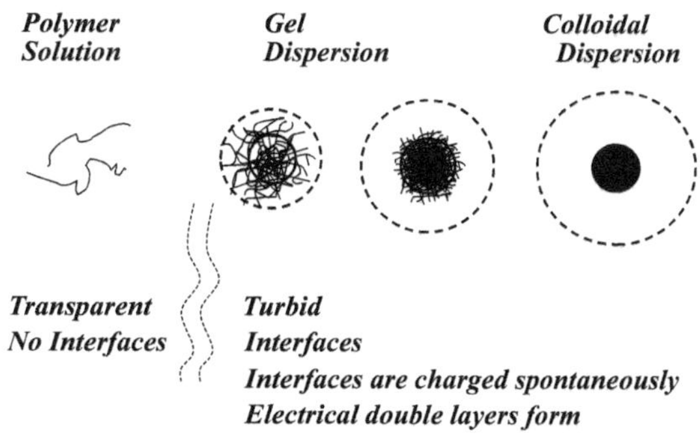

Figure 4.2
Schematic picture showing the comparative characteristics of polymer solution, gel dispersion, and colloidal dispersion. *Edited from Okubo T, Suzuki D, Yamagata T, Horigome K, Shibata K, Tsuchida A:* Colloid Polym Sci *289:1273–1281, 2011.*

dielectric constants of colloidal particles and water. The definite cause is not clarified yet. Typical examples of negatively charged particles are the colloidal crystallization of monodispersed gold and silver particles. They acquire negative charges spontaneously when they disperse in water, and then they always show negative ζ potential in suspension state. However, these particles should have no charges on the sphere surfaces when the spheres do not come in contact with water.[25] We should further note here that the charges on a colloidal sphere are introduced spontaneously when amount of the charges is not sufficient for the stabilization of dispersion, whereas the charges are bound with their counter-ions when the charge amount on a colloidal sphere is excessive. This charge renormalization effect has been discussed by Alexander et al. and others (see Sections 2.5 and 2.6).[26-29]

The ionic groups at the interfaces leave their counter-ions in the medium phase, and these excess charges accumulate near the interface, forming an electrical double layer. The counter-ions and co-ions in the diffuse region are distributed according to a balance between the thermal diffusive forces and the forces of electrical interaction with colloidal particles. The importance of the electrical double layers on the colloidal crystallization has been clarified by many researchers.[25-52] Giant colloidal single crystals have formed in the exhaustively deionized and diluted suspension of colloidal spheres.[44,53-55] The interparticle distance is just in the range of light wavelengths and each single crystal reflects light in different colors by the Bragg diffraction, which gives colloidal crystals beautiful and fantastic strong iridescent structural colors. It should be noted here that in our experience, the colloidal spheres smaller than approximately 40 nm in diameter hardly form crystal structures even though the spheres are monodisperse and the suspensions are deionized exhaustively. This is because of the very vigorous Brownian thermal motion of such small spheres and difficulty in the formation of the electrical double layers. Colloidal crystals were not formed for the monodisperse and 29-nm spheres at any sphere concentrations, although the typical liquid-like structures, ordered liquid, formed instead.[56] The electrical double layers might not be formed firmly when the sphere size is not large enough compared with that of the simple electrolytes or medium molecules, because the interfaces between the similar-sized spheres and the solvent molecules are not formed firmly.

Most gel dispersions are also turbid, and their nature looks to be just in between linear-type polymer solution and colloidal dispersion, as is shown in Fig. 4.2 schematically. When the degree of cross-linking is low, the gels will behave like polymer solution, whereas the gels will behave as colloidal particles as their degree of cross-linking increases. Thus study of the contribution of the electrical double layers on type 1 gel crystallization is interesting.

We should note here that type 1 gel crystals are really crystal; the gel spheres move with the translational and rotational modes by the thermal energy $k_B T$, but are keeping their equilibrium positions. On the other hand, thermal movement of type 2 gel crystals is greatly restricted. Type 2 gel crystals, however, may be much more useful practically to build up the external-field responsible devices, because the contraction and extension of gels is significant compared with other types of gel crystals. Type 1 gel crystals are discussed primarily in sections 4.3 through 4.8. Type 2 gel crystals are described in sections 4.9 and 4.10, and type 3 in sections 4.11 and 4.12.

4.2 Preparation and Characterization of Gel Spheres

Thermosensitive poly(*N*-isopropylacrylamide) (pNIPAm) gel spheres were prepared by Pelton et al., for the first time.[57,58] New types of pNIPAm gel spheres such as copolymer and core-shell types have been developed intensively.[59-64] Important investigations have been performed on the characteristic properties of gels, including blocklike gels (e.g., phase transition, density, cross-linking density, shearing flow, rheology, light-scattering, X-ray and neutron scattering, nuclear magnetic resonance, electron microscopy, fluorescence, electrophoresis).[22,65-101] Several research groups have performed most of the type 1 gel crystallization studies in aqueous media using the thermosensitive gel spheres of pNIPAm and their mixtures with other colloidal spheres.[102-118] The important role of the electrical double layers on the gel crystallization has been demonstrated using the monodisperse thermo-sensitive gel spheres of pNIPAm in the exhaustively deionized aqueous suspension. Type 1 crystallization of the cationic gel spheres of poly(2-vinylpyridine) (P2VP) pyridine partially cross-linked has been also studied in detail.[119-121] pH-responsible phase transition effects were reported for lightly cross-linked sterically stabilized P2VP spheres by Fujii et al.[122] These cationic gel spheres were very hard compared with pNIPAm gel spheres. However, small cationic gel spheres contained approximately 70 vol% of water inside gel spheres in an aqueous suspension state.

Table 4.2 shows several characteristics of a series of pNIPAm gel spheres: pNIPAm(70-5), pNIPAm(200-0.5), pNIPAm(200-2), pNIPAm(200-5), pNIPAm (200-10), pNIPAm(400-5), pNIPAm(600-5) and pNIPAm(1500-5). Here, the first and second numbers in a parenthesis indicate the rough values of sphere size at 20° C (low temperature) in nanometer and the degree of cross-linking in mol%, respectively. The chemical structure of pNIPAm is shown in Fig. 4.3. The sphere size, *d*, is estimated from the dynamic light-scattering (DLS) measurements and decreased sharply and transitionally as the suspension temperature increased, as is shown in Fig. 4.4. The temperatures of the

Table 4.2 Characteristics of colloidal crystals of pNIPAm gel systems.

Gel spheres	d (nm)		ρ (wt%/vol%)		f_w		Critical conc. of crystallization (wt% (ϕ_c)) With resins			Critical conc. of crystallization (wt% (ϕ_c)) without		T_t	I_D/d at 1wt%		ζ-pot. (mV)
	25°C	45°C	25°C	45°C	25°C	45°C	20°C	33°C	45°C	20°C	33°C		25°C	45°C	25°C
	[20°C]	(50°C)	[20°C]		[20°C]		[25°C]			[25°C]					[20°C]
pNIPAm(70-5)	[72]	(28)	——	——	——	——	—— (——)	——	—— (——)	——	——	35	——	——	——
pNIPAm(200-0.5)	320	120	0.030	0.61	0.978	0.56	0.015 (0.005)	0.17	0.3 (0.0049)	0.8	2	32	0.15	1.29	-10.1
pNIPAm(200-2)	260	100	0.035	0.59	0.975	0.57	0.015 (0.0043)	0.05	0.3 (0.0051)	0.4	0.3	33	0.19	1.26	-14.8
pNIPAm(200-5)	225	110	0.088	——	0.937	——	[0.01] (0.0011)	——	——	[0.8]	——	33	0.43	——	-13.5
pNIPAm(200-10)	200	120	0.122	0.62	0.912	0.55	0.01 (0.00082)	0.03	0.06 (0.00097)	0.15	0.6	35	0.54	1.28	-16.7
pNIPAm(400-5)	[410]	220	[0.11]	0.68	[0.921]	0.51	0.02 (0.0018)	0.04	0.07 (0.00103)	0.3	——	33	[0.51]	1.34	-14.4
pNIPAm(600-5)	520	290	0.127	0.72	0.908	0.48	0.03 (0.0024)	0.12	0.35 (0.0049)	0.8	2	33	0.56	1.38	-17.9
pNIPAm(1500-5)	[1500]	505	——	0.80	——	0.58	—— (——)	——	—— (——)	——	——	31	——	——	-13.4

pNIPAm, thermo-sensitive gel spheres of poly(*N*-isopropylacrylamide).

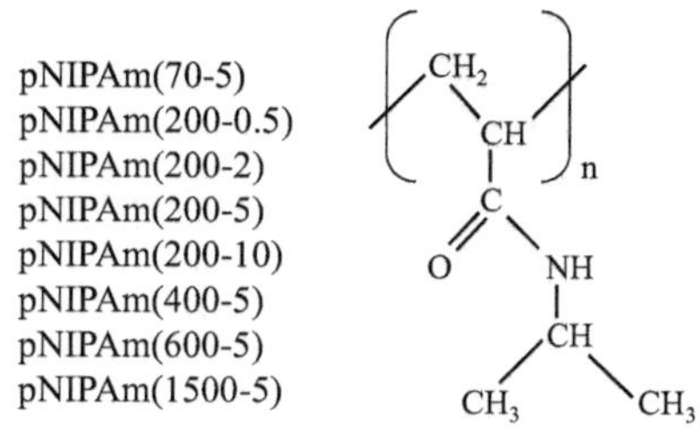

Figure 4.3 Chemical structures of pNIPAm spheres.

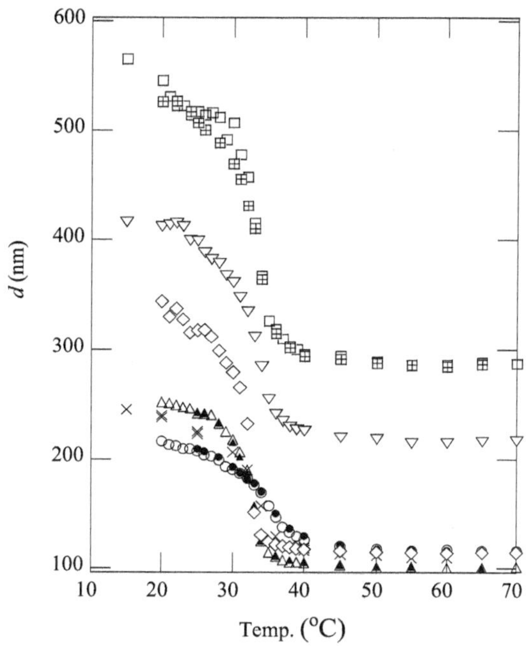

Figure 4.4 Diameters of pNIPAm gel spheres as a function of temperature. *Circles:* pNIPAm (200-10); *crosses:* (200-5); *triangles:* (200-2); *diamonds:* (200-0.5); *reversed triangles:* (400-5); *squares:* (600-5).

transition were approximately 33° C for each sphere. Deionized stock suspensions with concentrations ranging from approximately 2 to 4 wt% emitted strong iridescent colors at any temperatures between 10° C and 55° C.

We estimated the effective density of a gel sphere in the suspension state (ρ) from the ratio of the gel concentration in weight-percent divided by volume-percent as evaluated from the peak wavelength in the reflection spectroscopy. The density increased sharply as the suspension temperature increased five to twentyfold, as is shown in Table 4.2. At an elevated temperature, the gel spheres must be compacted with the polymer chains and the colloidal interfaces with a water medium to form firmly. Many researchers believe this sharp temperature dependence of gel conformation is due to the enhanced intersegmental hydrophobic interactions with elevated temperature. The ρ values of the gel spheres increased as the degree of cross-linking or size of the spheres increased. Equations 1 and 2 hold for the ρ values.

$$\text{pNIPAm(200-0.5)} < \text{pNIPAm(200-2)} < \text{pNIPAm(200-5)} < \text{pNIPAm(200-10) at } 25^\circ\text{ C} \tag{1}$$

$$\text{pNIPAm(200-5)} < \text{pNIPAm(400-5)} < \text{pNIPAm(600-5)} < \text{pNIPAm(1500-5) at } 25^\circ\text{ C} \tag{2}$$

The ρ value is the mean density (i.e., the ratio of the total weight of gel spheres against the total volume of the spheres, including volume of the solvents within the sphere regions in a suspension state). It is highly plausible that the local density in the center of a sphere is much higher than the density of the exterior when the hydrophobic intersegment affinitive interactions are taken into account in a water medium. We should note here that the volume fraction of gel spheres, ϕ value, is estimated from the nearest-neighbored intersphere distance; l values are estimated from the peak wavelength of reflection spectroscopy using Equation 3.[123]

$$\phi = (\pi/6)(d/l)^3 \tag{3}$$

Here, d is the diameter of gel spheres and is assumed to be close to the hydrodynamic size estimated from the DLS measurements, although the discussion on the validity of the method remains.

The ζ potentials measured from electrophoresis spectroscopy were negative and decreased slightly as the degree of the cross-linking or sphere size increased. These results are consistent with the colloidal nature of the gel spheres; in other words, the water–gel interfaces form more firmly as the sphere density increases. The negative charges on the gel interfaces are originated mainly from the negative charges from the initiator, potassium persulfate used in the polymerization processes, but also from the spontaneous charging effects at the interfaces between the gel spheres and water, as was discussed in section 4.1.

Gel sizes of pNIPAm were estimated as the hydrodynamic diameter (d) from the DLS measurements of the diluted suspensions of gel spheres. The adequacy of evaluation of the size in suspension state from the DLS method remains to be determined. However, I believe that the DLS is the best of the various techniques, which include electron microscopy (including the cryo-method), optical microscopy, and spectrophotometries such as laser diffraction particle size (LDPS) analysis measurements. Richtering et al. discussed the crystal lattice structures and the fuzzy nature of the interfaces of gel crystals of pNIPAm-co-acrylic acid).[124]

Characteristic properties of cationic gel spheres, lightly cross-linked poly (2-vinylpyridine) (P2VP) stabilized with polyethylene glycol (PEGMA) (PEGMA-P2VP[120] and μPEGMA-P2VP[125]) and lightly-cross-linked P2VP spheres (AIBA-P2VP[120] and CAIBA-P2VP[121]) are given in Table 4.3. Syntheses of these cationic gel spheres are shown schematically in Fig. 4.5. The diameters were measured by various techniques, including scanning electron microscopy, LDPS analyzer method, electrophoretic light-scattering (ELS) measurements, and DLS measurements. In Table 4.3 describing the size of CAIBA-P2VP and PEGMA-P2VP gel spheres, *d* decreased slightly when the degree of cross-linking increased from 0.1 to 1 mol%, which is consistent with the dependency of the size of pNIPAm gel spheres. Zeta-potentials of AIBA-P2VP decreased as the degree of cross-linking increased, whereas those of CAIBA-P2VP, PEGMA-P2VP, and μPEGMA-P2VP increased. The cause for the different tendencies is not clear yet. In general, tendencies of PEGMA-P2VP and CAIBA-P2VP are understandable, because the compactness of the gel increases as the degree of cross-linking and the charge density increases.

Table 4.3 Characteristics of Cationic Gel Spheres P2VP at 25° C

Gel Spheres	d (nm)	p	f_w	Critical Concentration of Melting: wt% With Resins	Critical Concentration of Melting: ϕ_c With Resins	Critical Concentration of Melting: wt% Without	l_D/d	ζ-Pot. (mV)
CAIBA-P2VP(0.1)	113	0.28	0.75	0.5	(0.0179)	1.0	0.63 (at 2.17 vol%)	+36
CAIBA-P2VP(0.5)	112	0.33	0.70	0.5	(0.0152)	1.4	0.85 (1.69)	+41
CAIBA-P2VP(1)	107	0.26	0.76	0.7	(0.027)	1.5	0.50 (2.50)	+41
AIBA-P2VP(0.1)	170	0.51	0.54	0.2	(0.0039)	0.8	0.26 (10.6)	+60
AIBA-P2VP(0.5)	170	0.70	0.36	0.18	(0.0026)	0.5	0.38 (9.65)	+58
AIBA-P2VP(1)	180	0.81	0.26	0.7	(0.0086)	1.5	0.45 (8.63)	+47
PEGMA-P2VP(0.1)	400	0.83	0.25	1.1	(0.0133)	1.2	0.65 (5.14)	+16
PEGMA-P2VP(0.5)	390	0.73	0.34	0.7	(0.0096)	1.0	0.59 (5.16)	+22
PEGMA-P2VP(1)	385	0.93	0.16	0.7	(0.0075)	1.0	0.68 (5.15)	+23
μPEGMA-P2VP(0.1)	1170	≈0.9*	≈0.2	—	(—)	—	—	+11
μPEGMA-P2VP(0.5)	1210	≈0.9*	≈0.2	—	(—)	—	—	+20
μPEGMA-P2VP(1)	1120	≈0.9*	≈0.2	—	(—)	—	—	+29

Values not measured, but deduced.

AIBA-P2VP, CAIBA-P2VP, PEGMA-P2VP and mPEGMA-P2VP: cationic gel spheres of poly(2-vinylpyridine).

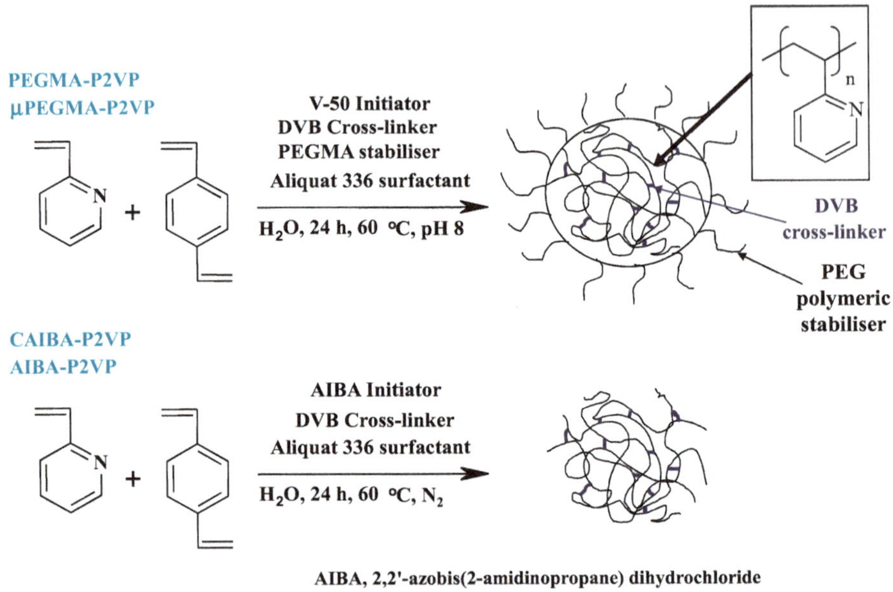

Figure 4.5 Scheme of cationic gel sphere of lightly cross-linked P2VP stabilized with PEGMA; PEGMA-P2VP and μPEGMA-P2VP, and lightly cross-linked P2VP; CAIBA-P2VP and AIBA-P2VP.

The ρ values of a series of P2VP gel spheres increased sharply as sphere size increased. The ρ values of all the cationic P2VP spheres appear to be insensitive to the degree of cross-linking of spheres when large experimental errors are taken into account, as is shown in Table 4.3. Furthermore, the ρ values of pNIPAm gel crystals were insensitive to the sphere concentration. It should be mentioned here that the density of AIBA-P2VP in the dried state should be close to that of PEGMA-P2VP, because the most prevalent component of both gel spheres is P2VP polymer. The specific density of PEGMA-P2VP was measured to be 1.1 in mean value.[119] Thus the volume fractions of the AIBA-P2VP gel polymers and water in the suspension state are evaluated to be 0.74 (=0.81/1.1) and 0.26, 0.64 (=0.7/1.1) and 0.36, and 0.45 (=0.51/1.1) and 0.55, for AIBA-P2VP(1), AIBA-P2VP(0.5), and AIBA-P2VP(0.1), respectively.[120] Furthermore, the volume fractions of the PEGMA-P2VP gel polymers and water in a suspension state are estimated to be 0.75 (=0.83/1.1) and 0.25, 0.66 (=0.73/1.1) and 0.34, and 0.85 (0.93/1.1) and 0.15, respectively, for PEGMA-P2VP(0.1), PEGMA-P2VP(0.5), and PEGMA-P2VP(1).[119] For PEGMA-P2VP(0.1), the quarter volume of a gel sphere is filled with water in suspension state. Clearly, AIBA-P2VP and PEGMA-P2VP are hard and close to the typical colloidal hard spheres. The ρ values of μPEGMA-P2VP were not observed by the reflection spectroscopy, because the iridescent colors from the higher order reflection were not strong enough to take the spectra. However, the ρ value was deduced to be approximately 0.9 from the sphere size dependency of the ρ values of a series of other cationic gel spheres. The ρ values were in the order of Equation 4, irrespective of the degree of cross-linking.

$$\text{CAIBA-P2VP} < \text{AIBA-P2VP} < \text{PEGMA-P2VP} < \mu\text{PEGMA-P2VP} \qquad 4$$

Equation 4 supports the finding that the small size of CAIBA-P2VP gel spheres are softest among the P2VP spheres examined, and μPEGMA-P2VP spheres are hardest and close to the hard colloidal spheres such as polystyrene and colloidal silica spheres. It is plausible that the main cause for the increasing compactness of P2VP spheres with increasing sphere size is due to the strong hydrophobic interpolymer segment's attractive forces in water media. The forces become stronger with an increasing amount of P2VP moieties within a gel sphere. The volume fractions of the gel polymer and water within μPEGMA-P2VP sphere are, therefore, evaluated to be approximately 0.8 (=0.9/1.1) and 0.2, irrespective of the degree of cross-linking. This means that only one fifth of the volume of a gel sphere is filled with water in the deionized aqueous suspension state.

4.3 Morphologic Characteristics and Phase Diagram of Type 1 Gel Crystals

Close-up color pictures of typical gel single crystals of pNIPAm(200-0.5), pNIPAm(200-2), pNIPAm(200-5), pNIPAm(200-10), pNIPAm(400-5), and pNIPAm(600-5) spheres in a test tube are shown in Figs. 4.6 to 4.11, respectively. Fig. 4.6, *A* through *D,* are taken with ion-exchange resins at 20° C; Fig. 4.6, *E* and *F,* are taken without ion-exchange resins at 20° C.[113] The bluish and brownish spherical particles distributed in the suspension in *A* are the ion-exchange resins coexisted. Fig. 4.6, *A,* supports the finding that the crystal lattices of pNIPAm(200-0.5) are rigid enough to suspend the large, heavy resin particles. The inner surfaces of the test tube in Fig. 4.6, *A,* emit brownish to bluish iridescent colors. The size of the single crystals was very small, but can be partially recognized with the naked eye. When gel concentration decreased from 0.39 wt% to 0.039 wt%, crystal size increased, as is clear in Fig. 4.6, *A* through *C*. When the gel concentration decreased further, 0.018 wt% size in turn decreased (see Fig. 4.6, *D*). The orange and greenish particles seen in Fig. 4.6, *B,* are the gel single crystals. The large single crystals of approximately 2 mm from the homogeneous nucleation were observed at 0.039 wt% (Fig. 4.6, *C*). In general, the gel crystal size of pNIPAm with a low degree of cross-linking was not so large compared with the size of spheres with a higher degree of cross-linking, and the former suspensions were more turbid. Two kinds of single crystals exist, as is shown in Fig. 4.6, *C* and *D*. Red, blue, yellow, and blocklike crystals formed from the homogeneous nucleation mechanism in the bulk phase far from the cell wall. The color of the single crystals changed depending on their location. The

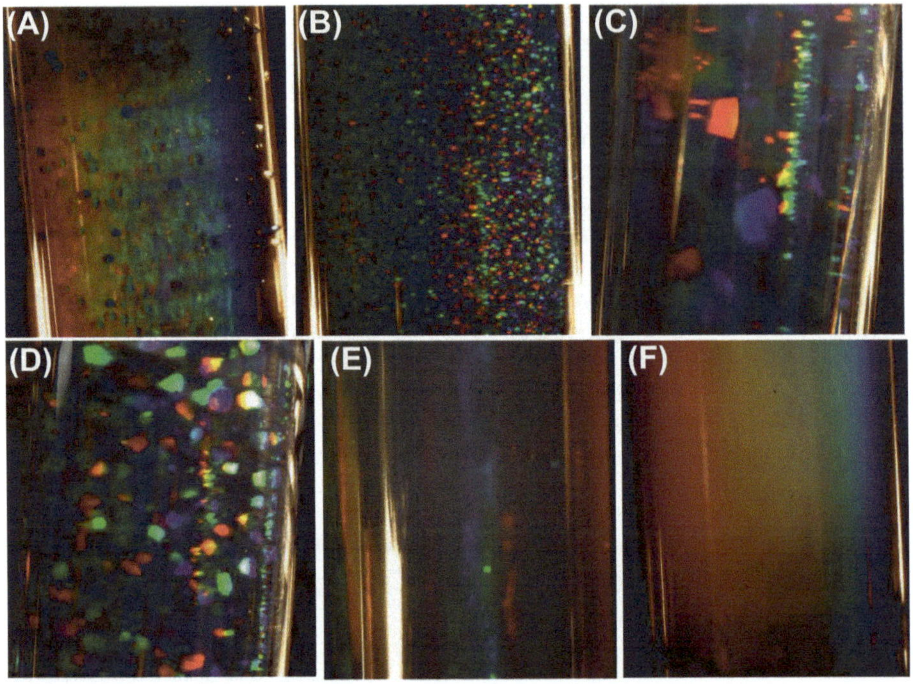

Figure 4.6
Close-up color pictures of pNIPAm(200-0.5) with resins (*A-D*) and without resins (*E-F*) in a test tube at 20° C. **A,** 0.39 wt%; **B,** 0.078; **C,** 0.039 wt%; **D,** 0.0181 wt%; **E,** 0.90 wt%; **F,** 0.39 wt%. *Edited from Okubo T, Suzuki D, Yamagata T, Horigome K, Shibata K, Tsuchida A:* Colloid Polym Sci *289:1273–1281, 2011.*

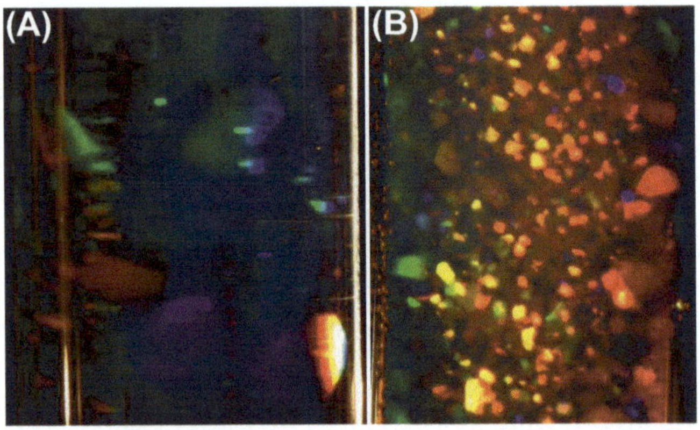

Figure 4.7
Close-up color pictures of pNIPAm(200-2) with resins in a test tube. **A,** 0.052 wt% at 20° C; **B,** 0.197 wt% at 33° C. *Edited from Suzuki D, Horigome K, Yamagata T, Shibata K, Tsuchida A, Okubo T:* Colloid Polym Sci *289:1799–1808, 2011.*

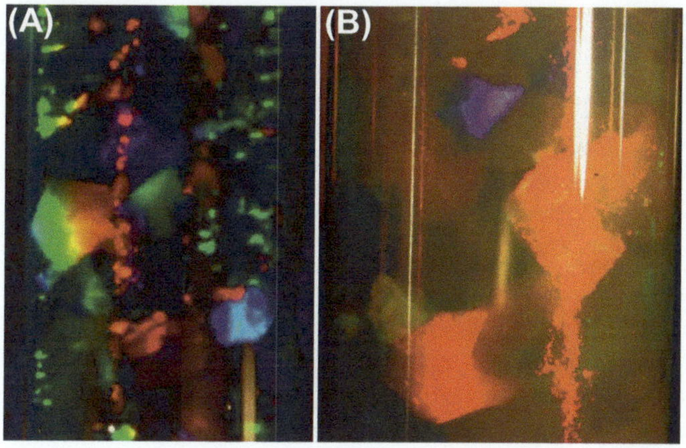

Figure 4.8
Close-up color pictures of pNIPAm(200-5) in a test tube at 25° C. **A,** 0.116 wt% with resins; **B,** 1.74 wt%, [NaCl] = 3.5×10^{-5} M. *Edited from Okubo T, Suzuki D, Yamagata T, Katsuno A, Sakurai M, Kimura H, Tsuchida A:* Colloid Polym Sci *289:291–299, 2011.*

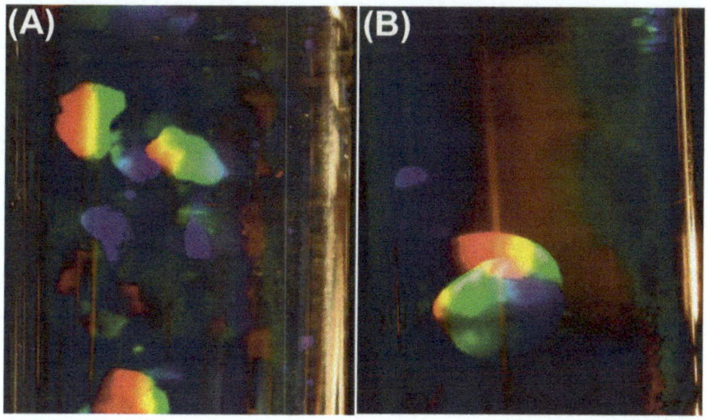

Figure 4.9
Close-up color pictures of pNIPAm(200-10) in a test tube at 20° C. **A,** 0.062 wt% with resins; **B,** 0.219 wt% without resins. *Edited from Suzuki D, Horigome K, Yamagata T, Shibata K, Tsuchida A, Okubo T:* Colloid Polym Sci *289:1799–1808, 2011.*

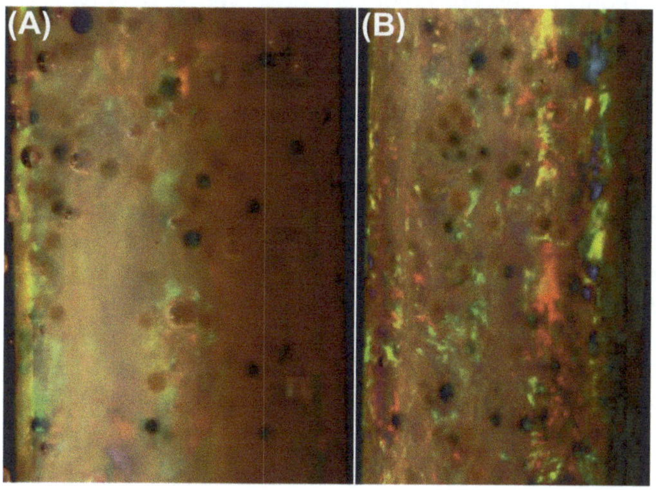

Figure 4.10
Close-up color pictures of pNIPAm(400-5) with resins in a test tube; 3.89 wt%, **A,** 20° C, **B,** 45° C. *Suzuki D, Yamagata T, Horigome K, Shibata K, Tsuchida A, Okubo T:* Colloid Polym Sci *290:107–117, 2012.*

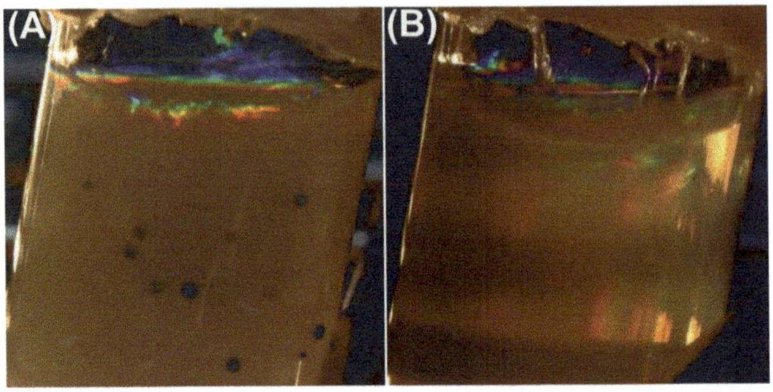

Figure 4.11
Close-up color pictures of pNIPAm(600-5) with resins in a test tube at 20° C. **A,** 0.90 wt%; **B,** 0.179 wt%.
Suzuki D, Yamagata T, Horigome K, Shibata K, Tsuchida A, Okubo T: Colloid Polym Sci *290:107–117, 2012.*

color change within the single crystals in Fig. 4.6, *C* and *D,* is due to the change in the direction of the crystal planes and the incident angles of light through the curved wall of the test tube. When a test tube was rotated slowly, the single crystals exhibiting the strong iridescent colors disappeared suddenly, but the other single crystals appeared in the other places. On the other hand, the pillarlike crystals observed in Fig. 4.6, *C* and *D,* grew from the heterogeneous nucleation mechanism along the cell wall. It should be noted here that each single crystal of the typical colloidal sphere is surrounded by the grain boundaries and packed densely.[40,44,55-53,126,127] The morphologic characteristics of the single crystals of the gel spheres is similar to the characteristics of the typical colloidal spheres.[27,40,44-47,53-55,127-132] These iridescent colors are clearly ascribed to the Bragg diffraction of the incident light through the crystals. The single crystals are packed densely, and the appearance and disappearance of the crystals takes place depending on the Bragg diffraction conditions. Without ion-exchange resins, small single crystals are recognized when sphere concentration is 0.90 wt% (Fig. 4.6, *E*), and only iridescent colors are observed at 0.39 wt% (Fig. 4.6, *F*).

Fig. 4.7 shows typical examples of the macroscopic pictures of pNIPAm (200-2).[114] Critical concentration of crystallization (or melting) increased sharply as suspension temperature increased, as is clear in Table 4.2. Large single crystals were observed at 0.052 wt% and at 20° C (see Fig. 4.7, *A*). Furthermore, the critical concentration decreased slightly as the degree of cross-linking increased. Fig. 4.8 shows the close-up pictures of pNIPAm (200-5) gel spheres with ion-exchange resins (Fig. 4.7, *A,* 0.116 wt%) and with 3.5×10^{-5} M of sodium chloride (NaCl) coexisted (Fig. 4.7, *B*) at 25° C.[112] With resins, the suspension became transparent with reduced multiple scattering. At 0.116 wt% ($\phi = 0.0132$) (Fig. 4.7, *A*), colorful giant single crystals formed from the homogeneous nucleation mechanism. Greenish or reddish pillarlike crystal arrays along the cell wall are grown from the heterogeneous nucleation. Interestingly, giant single crystals formed even in the presence of rather high salt concentration, 3.5×10^{-5} M of NaCl. Formation of such large single crystals have been extremely difficult to observe for the typical colloidal hard spheres. The morphologic characteristics of single crystals are seen in Fig. 4.9 for pNIPAm(200-10), which look most transparent and clear among the gels that include lower degrees of cross-linking, and are similar to the typical colloidal crystals of hard spheres at first glance.[114] Morphologic observations of the four samples support that the interfaces between gel spheres and medium water form more stably as the degree of cross-linking increases from 0.5 to 10 mol%.

Figs. 4.10 and 4.11 demonstrate that observation of iridescent colors interior to the suspension phase is difficult, and only the liquid surface areas are observed by the enhanced strength of the multiple-scattering.[115] However, basic morphologic characteristics are the same as those of small gel spheres: Crystals from the homogeneous and heterogeneous nucleation mechanisms coexist. It should be noted here that the tone of the iridescent colors does not change, irrespective of the suspension temperatures between 20° C to 45° C (compare Fig. 4.10, *A* and *B*). Dark bluish and brownish spheres observed in Fig. 4.10 are again the cation- and anion-exchange resin particles. In Fig. 4.11, *B*, giant single crystals are recognized even for very large gel spheres, 600 nm in diameter at 20° C (see Fig. 4.10, *B*), although the shape and size are vague, because the multiple-scattering is substantial.

The critical concentration of melting (or crystallization) (ϕ_c) of pNIPAm gel crystals in wt% (and in volume fraction) are compiled in Table 4.2. Values of ϕ_c are more meaningful when gel crystals are compared with the typical hard-sphere crystals. Fig. 4.12 shows the ϕ_c values of pNIPAm gel crystals as a function of fraction of cross-linking of the spheres. The solid circle at the unity in the horizontal axis is the mean ϕ_c value of the colloidal crystals of hard spheres. pNIPAm gel crystal ϕ_c values were significantly high at 10 to 100 times the value of the typical colloidal crystals ($\phi_c = 0.0002$). pNIPAm(200-x) series ϕ_c values decreased toward that of colloidal crystals as the degree of cross-linking increased. Interestingly, the ϕ_c values at 25° C and 45° C were similar when the experimental errors were taken into account. It is highly plausible that the interfaces of pNIPAm(200-x) gel spheres with water at 25° C are vague and the thickness of the electrical double layers (l_D) are small, but the spheres themselves are highly extended. At high temperatures, on the other hand, sphere size decreased sharply, but the electrical double layers were extended. Thus the two effects of the extension of gel spheres and the thinning of the double layers should be compensated to each other. In Fig. 4.12, ϕ_c values of the gel crystals seem to increase as sphere size increases.

Fig. 4.13, *A* through *F*, show the size distributions of the gel single crystals as a function of gel sphere concentration in volume fraction. Giant crystals of

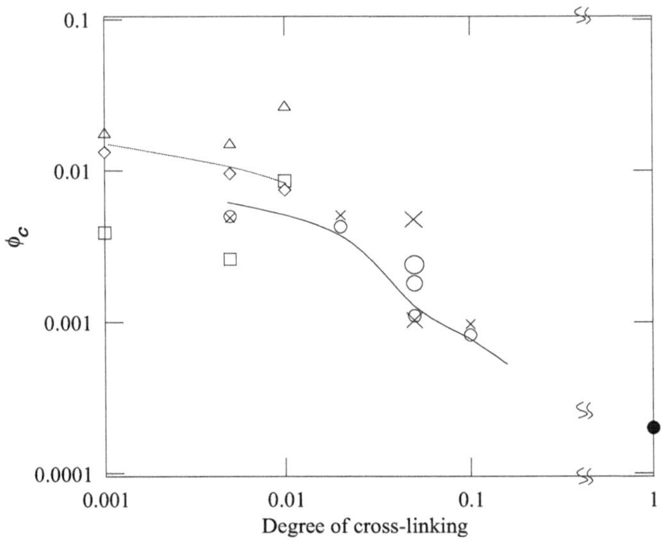

Figure 4.12
Plots of ϕ_c against degree of cross-linking for pNIPAm gel crystals and P2VP crystals. At 20° C, *small circles:* pNIPAm(200-x); *medium circles:* pNIPAm(400-5); *large circles:* pNIPAm(600-5). At 45° C, *small crosses:* pNIPAm(200-x); *medium crosses:* pNIPAm(400-5); *large crosses:* pNIPAm(600-5). At 25° C, *triangles:* CAIBA-P2VP; *squares:* AIBA-P2VP; *diamonds:* PEGMA-P2VP.

approximately 3 mm formed from the homogeneous nucleation mechanisms at $\phi = 0.001$ to 0.011, and decreased slightly as the degree of cross-linking increased. Furthermore, the maximum size of single crystals formed at two- to threefold higher gel concentrations compared with the corresponding ϕ_c values. For the large gel spheres, differences between the largest crystal forming concentration and ϕ_c value became significant; the former was larger than the latter by tenfold or more. These findings in size profiles of gel crystals are characteristic compared with typical colloidal crystals, where the largest crystal forms at a slightly higher concentration than the ϕ_c value.

Phase diagram of pNIPAm(200-5) gel crystals showing the relationship between the ϕ_c values and the concentrations of foreign salt, NaCl, is shown in Fig. 4.14. The solid curve shows the ϕ_c values as a function of the salt concentration. The dotted curve shows the mean ϕ_c values versus salt concentration plots of typical colloidal crystals for comparison. Clearly, the ϕ_c values of gel crystals are higher than those of the typical colloidal crystals by more than tenfold at the same salt concentration. Gel crystal ϕ_c values increased sharply as salt concentration increased and reached very high gel concentrations, 0.2 to 0.3 in volume fraction, not far from the closed-packing concentration of spheres at 0.72. These results clearly support the finding that the interfaces of gel spheres are more or less vague and fuzzy.

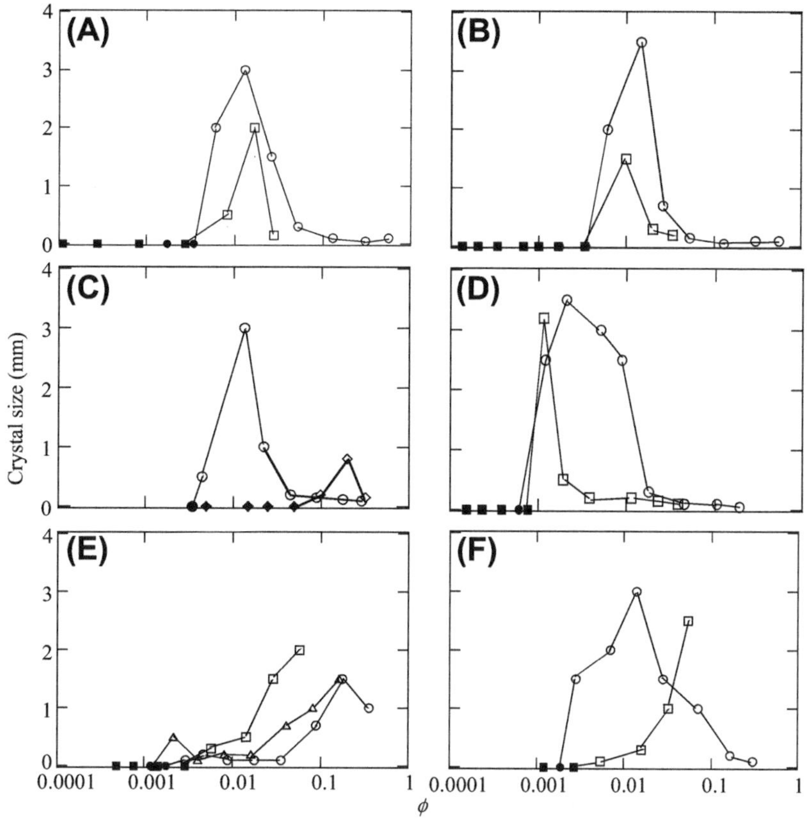

Figure 4.13 Crystal size versus gel sphere concentration (ϕ) plots for pNIPAm(200-0.5) (*A*), pNIPAm (200-2) (*B*), pNIPAm(200-5) (*C*), pNIPAm (200-10) (*D*), pNIPAm(400-5) (*E*), and pNIPAm (600-5); (*F*) gel spheres at 20° C (*circles*), 33° C (*triangles*), and 45° C (*squares*).

We now turn to the crystallization of cationic gel spheres of P2VP. Close-up color pictures of gel crystals in a test tube are shown in Fig. 4.15, *A* through *C*, for CAIBA-P2VP[121]; Fig. 4.15, *D* through *F*, for AIBA-P2VP[120]; and Fig. 4.16 for PEGMA-P2VP.[119] All the suspensions of P2VP gel spheres were coexisted with the ion-exchange resins and showed iridescent colors. AIBA-P2VP emitted metallic and brilliant colors. The colors appear to change from reddish to greenish with increasing sphere concentration. However, height dependencies of the colors are not recognized with the naked eye, although these suspensions are at the sedimentation equilibrium state (see Fig. 4.15, *D* through *F*). These findings indicate that the rigidity of AIBA-P2VP gel crystals is high, resulting in the insensitive height dependence of iridescent colors. All the suspensions were strongly turbid, and only the crystals near the cell wall were observed with the naked eye. Critical concentrations of crystallization (or melting) are also shown in Fig. 4.12 as a function of degree of cross-linking (triangles, squares, and diamonds indicate CAIBA-P2VP, AIBA-P2VP, and PEGMA-P2VP, respectively). The ϕ_c values of cationic P2VP gel spheres shown

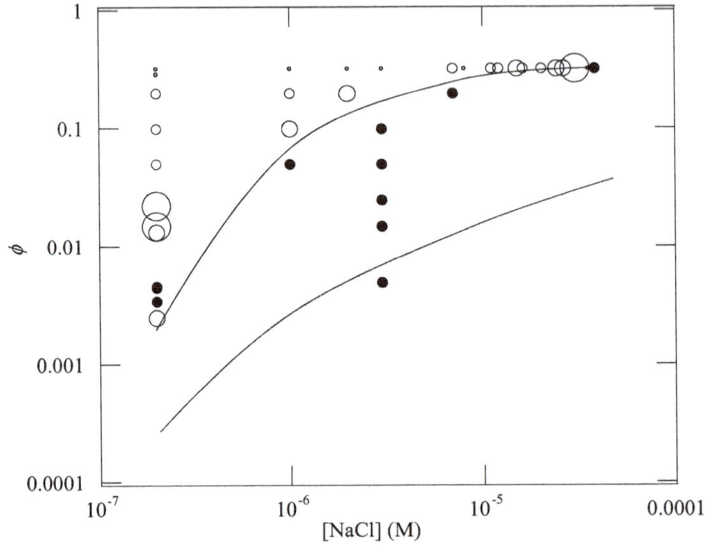

Figure 4.14
Phase diagram of pNIPAm(200-5) gel spheres at 25° C. *Open circles:* crystal, the size of the circles shows the approximately mean crystal size; *solid circles:* liquid.

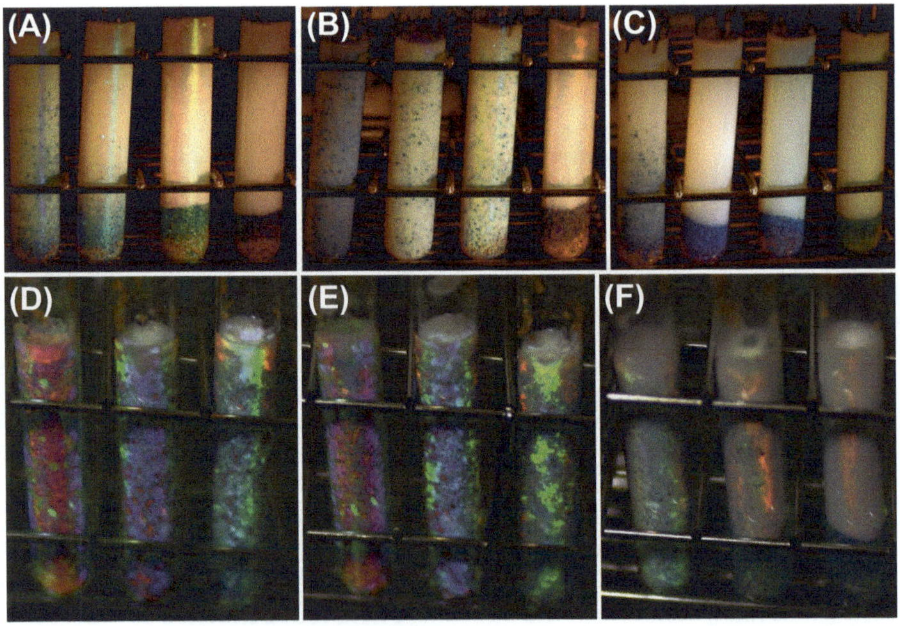

Figure 4.15
Close-up color pictures of cationic gel crystals of CAIBA-P2VP(0.1) (**A,** 4.34, 3.26, 2.17, and 1.09 wt% from the left); CAIBA-P2VP(0.5) (**B,** 6.74, 5.06, 3.37, and 1.69 wt%); CAIBA-P2VP(1) (**C,** 5.01, 3.76, 2.50, and 1.25 wt%); AIBA-P2VP(0.1) (**D,** 8.63, 5.18, and 3.45 wt%); AIBA-P2VP(0.5) (**E,** 9.65, 5.79, and 3.86 wt%); and AIBA-P2VP(1) (**F,** 10.6, 6.36, and 4.24 wt%) in the deionized and in the sedimentation equilibrium at 25° C and 22° C, respectively. *Edited from Okubo T, Fujii S, Aono K, Nakamura Y:* Colloid Polym Sci *291:2569–2577, 2013 and Fujii S, Yamashita Y, Nakamura Y, Tsuchida A, Okubo T:* Colloid Polym Sci *292:1627–1637, 2014.*

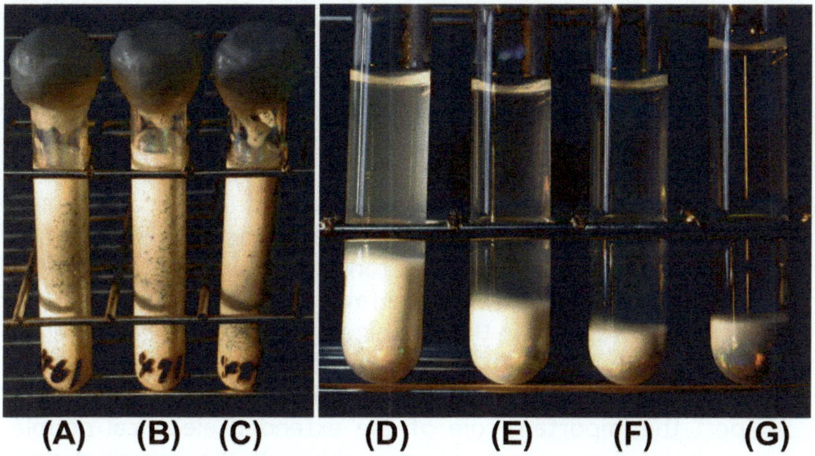

Figure 4.16
Close-up color picture of gel crystals, PEGMA-P2VP(0.1) (**A**, 5.14 wt%); PEGMA-P2VP(0.5) (**B**, 5.16 wt%); and PEGMA-P2VP(1) (**C**, 5.15 wt%); coexisted with ion-exchange resins at 25° C 5 days after setting. Gel crystals of PEGMA-P2VP(0.5) (**D-F**, 3.10 wt%) and PEGMA-P2VP(0.1) (**G**, 3.08 wt%) coexisted with NaCl (**D**, 0 M; **E**, 0.0002 M; **F**, 0.002 M; **G**, 0.006 M) at 25° C 5 days after setting. *Taken from Okubo T, Fujii S, Aono K, Nakamura Y, Tsuchida A:* Colloid Polym Sci *291:1201–1210.*

by the dotted curve are close to or slightly larger than those of pNIPAm anionic gel spheres.

Fig. 4.16 shows gel crystals of PEGMA-P2VP(0.1) (*A*), PEGMA-P2VP(0.5) (*B*), and PEGMA-P2VP(1) (*C*) in the deionized aqueous suspensions. Reddish iridescent colors are observed. Bluish and reddish small single crystals are recognized at the air–liquid interfaces of the suspensions (*A* through *C*). However, the suspensions were totally turbid, because the sphere size is large, approximately 400 nm in diameter, and the multiple-scattering contribution was substantial. The crystallization took place in the sedimentation equilibrium state for the suspensions at higher sphere concentrations shown in Fig. 4.16 (*A* through *C*). Fig. 4.16, *D* through *G*, show the gel crystals of PEGMA-P2VP(0.5) (*D* through *F*) and PEGMA-P2VP(0.1) (*G*) in the presence of NaCl (*D*, 0 M; *E*, 0.0002 M; *F*, 0.002 M; *G*, 0.006 M). Even in the presence of a large amount of salt, gel crystallization was observed in the sedimentation equilibrium. This supports the high monodispersity of the cationic spheres and also high stability of the gel spheres with the addition of salt.

Gottwald et al.[133] proposed the reentrant order–disorder transition from their theoretical study as a function of gel sphere concentration. The static light scattering (SLS) and DLS of gel sphere suspensions. I believe that the polydispersity index of the gel spheres is most important for study of the phase transitions. Phase transition of gel crystals of pNIPAm spheres was also studied by Richtering et al.[124]

4.4 Lattice Structures of Type 1 Gel Crystals

SLS measurements of deionized suspensions of the thermosensitive gel spheres of pNIPAm with various degrees of cross-linking and sizes were made at 20° C and 40° C.[117] Sharp scattering peaks were observed in the scattering curve and they were assigned to the face-centered cubic (*fcc*) or body-centered cubic (*bcc*) lattices in the distribution of gel spheres. The *fcc* and *bcc* crystal structures formed in the stable and unstable conditions, respectively; *fcc* formed more favorably at high sphere concentrations, low temperatures, and in the deionized state. The closest intersphere distances were much longer than the hydrodynamic diameters of the gel spheres, especially at low sphere concentrations. These experimental results support the important role of the extended electrical double layers in the crystallization of gel spheres, although the contribution of the double layers in gel systems is weak compared with that in the typical colloidal spheres.

Fig. 4.17 shows the typical examples of the static light-scattering intensity, $I(q)$, as a function of scattering vector, q, for pNIPAm(200-5) at 0.3 wt% and at 20° C. Blue and red curves indicate the data for the suspensions with and without ion-exchange resins. The suspensions are crystal and liquid, respectively. Sharp peaks appeared for the crystals. The light-scattering intensity measurements were made at intervals of half the degree. However, this interval was not narrow enough to determine the reliable values of peak scattering vector, q_m. The similar profiles in the intensity of scattered light

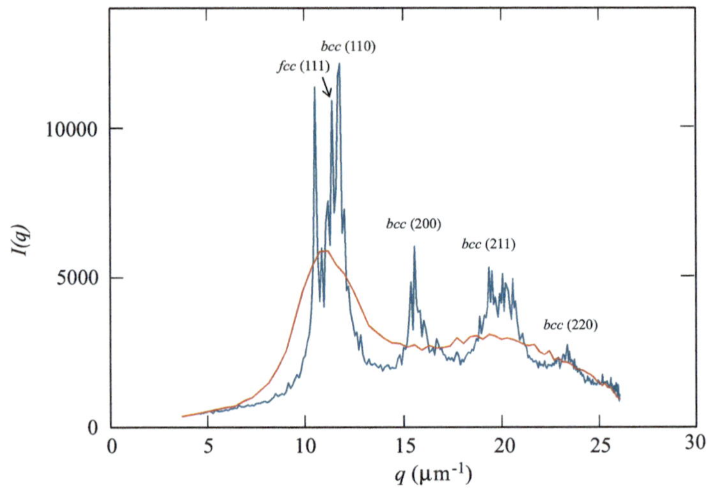

Figure 4.17 Scattering intensity of pNIPAm(200-5) as a function of scattering vector at 20° C; 0.30 wt%; *blue curve:* with resins; *red curve:* without resins. *From Okubo T, Suzuki D, Tsuchida A:* Colloid Polym Sci *291:661–668, 2013.*

showing the peaks have been reported for the typical colloidal crystals in the deionized aqueous suspensions. However, most of the experiments for the colloidal crystals were made for the much lower sphere concentrations than this work on gel crystals.[134-137]

Assignment of the respective peaks, q_m, to the crystal planes was made from comparison of the observed peak vectors with the calculation from $q_{h,k,l} = 2\pi/a(h^2 + k^2 + l^2)^{1/2}$, where *h*, *k*, and *l* are Miller's indices and *a* the unit cell size. As is shown in Fig. 4.17, several extra sharp peaks appeared coupled with the main highest peak. The scattering curves of pNIPAm(200-5) at 40° C showed, however, quite a few of the artifact peaks and the assignment of the crystal lattices was easy, as is shown in Fig. 4.18. At the high temperature of 40° C, crystal growth rate should be much higher than that at 20° C. The incomplete crystallization and aging of the crystals is the main cause for the formation of the artifact peaks at 20° C. During crystallization and aging, it is highly plausible that several kinds of single crystals of similar lattice structures and lattice spacing values will form. My team recently found low rates of crystallization of pNIPAm gels compared with those of typical colloidal spheres such as colloidal silica spheres.[116] Only the highest and main peaks were assigned to be the reflections of *fcc* and *bcc* lattices, as is shown in Fig. 4.18. The Bragg distances (d_B) were evaluated from the q_m values using Bragg equations, $d_B = 2\pi/q_m$. The nearest-neighbored inter-sphere distance (l), which is given by $(3/2)^{1/2} d_B$, agreed satisfactorily with the corresponding l_o values calculated from the sphere concentration, as is shown in Table 4.4. It should be noted here that the peaks are sharp and the background intensity is low. Thus there should not be a significant difference

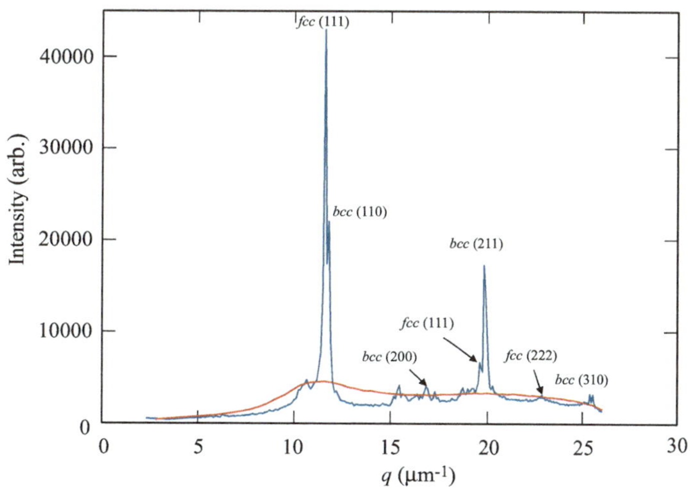

Figure 4.18 Scattering intensity of pNIPAm(200-5) as a function of scattering vector at 40° C; 0.30 wt%; *blue curve:* with resins; *red curve:* without resins. *From Okubo T, Suzuki D, Tsuchida A:* Colloid Polym Sci *291:661–668, 2013.*

Table 4.4 Summary of SLS Measurements of pNIPAm Gel Spheres

Gel Spheres	Temp (°C)	*w* (wt%)	ϕ	l_o (nm)	*l* (nm)			l_D/d_o
					fcc	*bcc*	Liquid	
pNIPAm	20	0.2	0.0164	758	710	700		1.2
(200-10)	40	0.2	0.00322	777	780	760		2.5
pNIPAm	20	0.3	0.0341	663	(710)	650;680	700*	0.9
(200-5)	40	0.3	0.005	619	660	650	680*	2.3
pNIPAm	20	0.4	0.114	485	490	480		0.42
(200-2)	40	0.4	0.0068	511	—	480		1.8
pNIPAm	20	0.4	0.133	609	600	580	570	0.35
(200-0.5)	40	0.4	0.0066	570	—	—		—
pNIPAm	20	0.2	0.0182	1420	1460	1430		1.26
(400-5)	40	0.2	0.00294	1420	1490	1430		2.8

*Without resins.

bcc, Body-centered cubic; *fcc*, face-centered cubic; *pNIPAm*, thermo-sensitive gel spheres of poly(*N*-isopropylacrylamide); *SLS*, static light scattering.

between the profiles in the structure factor, $S(q)$, and the intensity of scattered light, $I(q)$.

$$I(q) = P(q)S(q) \qquad 5$$

Here, $P(q)$ is the single-particle form factor, and can be estimated experimentally from the $I(q)$ measurements on the same suspension but in the presence of salt, for example.[138,139]

The first and second broad peaks in red in Fig. 4.17 show the scattering curve of the gel suspensions without resins. When the red curve is examined carefully, the main peak is composed with double peaks supporting the suspension being *liquid,* but, close to *crystal.* In other words, the suspension shown by the red curve is ordered liquid, which has been discussed by Hartl and Versmold[140] in detail.

At 40° C the artifact peaks almost disappeared in contrast to the curve at 20° C, as is shown in Fig. 4.18. This supports the finding that crystal growth at 20° C does not complete and the peaks corresponding to the several intermediate crystals appear. The *fcc* and *bcc* crystals coexisted. Clearly, sharp double peaks appeared at low *q* in the Fig. 4.18, and were assigned to the (111) and (110) reflections of *fcc* and *bcc* lattices, respectively. Most other peaks were then assigned accurately, as is shown in Fig. 4.18. The difference in the peak scattering vectors between the (111) reflection of *fcc* and (110) of *bcc* was only 1 to 2 degrees in scattering angle. The agreement between the observed distance and the calculation from the sphere concentration and size is satisfactory when the experimental errors of the SLS measurements are taken into account. The red scattering curve in Fig. 4.18 demonstrates, on the other hand, that the suspension without resins is the typical liquid structure.

Let's discuss here the importance of the electrical double layers in the formation of colloidal crystal structure. Ionic groups either dissociated or adsorbed on the colloidal surfaces (as was described in section 4.1), leave their counter-ions in suspension, and these charges accumulate near the surface, forming an electrical double layer. The thickness of the electrical double layer is approximated by the Debye screening length (l_D).

$$l_D = \left(4\pi e^2 n / \varepsilon k_B T\right)^{-1/2} \qquad 6$$

where e is the electronic charge, ε is the dielectric constant of the solvent, k_B is the Boltzmann constant, and n is the concentration of free-state cations and anions in suspension, and is given by $n = n_c + n_s + n_o$, where n_c is the concentration (number of ions per cm^3) of "diffusible" (or "free-state") counter-ions, n_s is the concentration of foreign salt, and n_o is the concentration of both hydronium and hydroxide ions from dissociation of water. To estimate n_c, the fraction of free counter-ions must be known, because most counter-ions are bound tightly with the ions of the colloidal surface. Note that the maximum values of l_D both observed and calculated are approximately 1 μm in water. According to the effective soft-sphere model,[141-145] which is a simple but convenient assumption especially for the de-ionized colloidal suspension, a crystal structure forms when the effective diameter of spheres (d_{eff}) including the Debye screening length is close to or larger than the intersphere distance (l) (i.e., d_{eff} [= diameter (d) + 2 × l_D] ≥ l). Note that the observed intersphere spacing (l) is always close to the calculated mean intersphere distance (l_o), because the colloidal crystallization always takes place by the intersphere electrostatic repulsion forces. Table 4.4 shows that the l_o values of pNIPAm(200-5) systems are close to the l values irrespective of the distribution of gel spheres (i.e., *fcc*, *bcc*, and liquid). It should be noted here that the l_o values did not change irrespective of the suspension temperature. This is consistent with the finding that the crystallization of gel spheres takes place by the electrostatic intersphere repulsion forces. The last column of Table 4.4 shows the l_D/d. For pNIPAm(200-5) gel spheres, l_D/d values were 0.9 and 2.3 at 20° C and 40° C, respectively. These values are clearly close to or larger than the gel size especially at high suspension temperature. This result clearly supports that the gel spheres in the crystal structure do not contact each other and the electrical double layers play an important role for crystallization of gel spheres in addition to the excluded volume effect of the sphere themselves.

Lattice structures of Type 1 gel crystals are also studied by the reflection spectroscopy conveniently and accurately.[112-115] Figs. 4.19 and 4.20 show the typical reflection spectra of the gel crystals of two series of pNIPAm and P2VP gel spheres with coexistent ion-exchange resins (Figs. 4.19, *A-F*, and 4.20, *A-F*). Sharp peaks are ascribed to the Bragg diffraction through the gel's

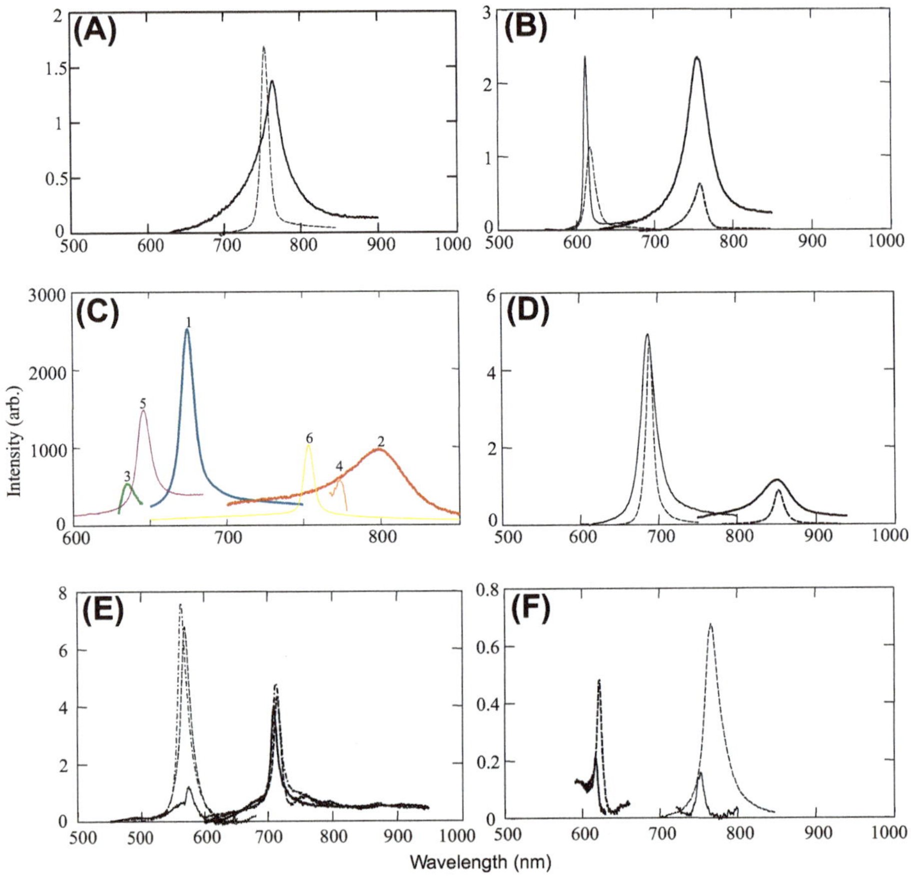

Figure 4.19
Reflection spectra of **A,** pNIPAm(200-0.5), 1.68 wt%, *solid curve:* 25° C, *broken curve:* 45° C. **B,** pNIPAm(200-2), 1.97 wt% (*fine curves*), 1.06 wt% (*broad curves*), *solid curves:* 25° C, *broken curves:* 45° C. **C,** pNIPAm(200-5), *curve 1:* with resins, 2.52 wt%, *curve 2:* 1.55 wt%, *curve 3:* 0.388 wt%, *curve 4:* 0.194 wt%, *curve 5:* without resins, 2.82 wt%, *curve 6:* 1.74 wt%. **D,** pNIPAm(200-10), 2.37 wt% (*fine curves*), 1.28 wt% (*broad curves*), *solid curves:* 25° C, *broken curves:* 45° C. **E,** pNIPAm(400-5), 3.89 wt% (*fine curves*), 1.95 wt% (*broken curves*), *solid curves:* 20° C; *broken curves:* 33° C, *dotted curves:* 45° C. **F,** pNIPAm(600-5), 3.88 wt% (*broad curves*), 2.09 wt% (*fine curves*), *solid curves:* 25° C, *broken curves:* 45° C.

single crystals. In general, the nearest-neighbored intersphere distance (l) is estimated from the primary peak wavelengths (λ_p) in the reflection spectroscopy at the scattering angle of 90 degrees ($q = 90$ degrees). The l values of the *fcc* or *bcc* lattices, which are familiar lattices in both colloidal and gel crystals,[123] are given by Equation 7.

$$l = (3/8)^{1/2}\left\{\lambda_p / \left(n_s^2 - n_a^2\cos^2\theta\right)\right\} \quad 7$$

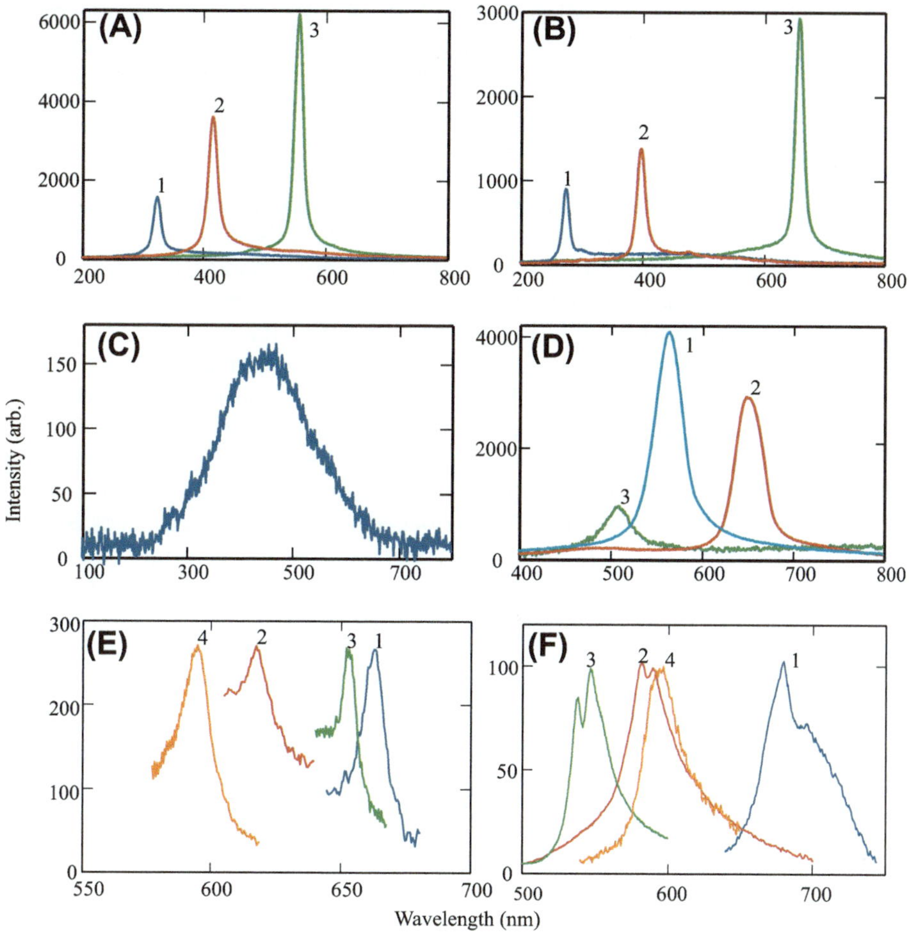

Figure 4.20
Reflection spectra of P2VP spheres at the bottom of the cells at 25° C. **A,** CAIBA-P2VP(0.1) *curve 1:* 4.34 wt%; *curve 2:* 3.26 wt%; *curve 3:* 2.17 wt%. **B,** CAIBA-P2VP(0.5); *curve 1:* 5.06 wt%; *curve 2:* 3.37 wt%; *curve 3:* 1.69 wt%. **C,** CAIBA-P2VP(1), 2.50 wt%, 6 days after setting. **D,** *Curve 1:* AIBA-P2VP(0.1), 10.6 wt%; *curve 2:* AIBA-P2VP(0.5), 9.65 wt%; *curve 3:* 3.45 wt%. **E,** *Curve 1:* PEGMA-P2VP(0.1), 5.14 wt%; *curve 2:* PEGMA-P2VP(0.5), 5.16 wt%; *curve 3:* PEGMA-P2VP(1), 5.15 wt%; *curve 4:* 3.09 wt%; **F,** PEGMA-P2VP(0.5), 3.10 wt%, *curves 1-3:* PEGMA-P2VP(0.1), 3.08 wt%; *curve 4.* Coexisted with NaCl; *curve 1:* 0 M, *curve 2:* 0.0002 M, *curve 3:* 0.0006 M, *curve 4:* 0.002 M.

Thus the nearest-neighbored intergel sphere distance (l) has a linear relationship with the reflection peak wavelengths (λ_p) by Equation 7. The refractive index of the liquid (n) is given by Equation 8.

$$n = \phi \times [\text{refractive index of gel sphere}] + (1 - \phi) \times [\text{refractive index of water}] \quad 8$$

Furthermore, Equation 3 holds between the l values and volume fraction of gel spheres. Then, the ρ values will be insensitive to the degree of cross-linking and the sphere concentration when the experimental errors are taken into account. It should be recalled that the ρ values increased as the degree of cross-linking increased, and were insensitive to the sphere concentration for the pNIPAm[112-115] and P2VP gel crystals.[119-121] Fig. 4.19, *A*, *B*, and *D* through *F*, show the temperature dependency of the reflection spectra at the same sphere concentration. We should note here that the lattice spacing of gel sphere crystals does not change by changing suspension temperature, because the lattice spacing is determined by sphere concentration and not only by gel size. Insensitivity of the lattice spacing of the thermosensitive gel sphere crystals on suspension temperature was supported experimentally by Asher et al.[146] It should be noted here that temperature dependence of the lattice spacing of the type 2 gel crystals is substantial,[147] because the lattice spacing changes following changes of the gel size. Fig. 4.19, *C*, shows the concentration dependency of the reflection spectrum of pNIPAm(200-5) gel crystals with resins (curves 1-4) and without resins.[5,6] The reflection peak shifted toward a longer wavelength as gel concentration decreased. It should be noted that peaks 1, 2, 5, and 6 are the primary reflection peaks. On the other hand, peaks 3 and 4 are the secondary reflection peaks and twice the peak wavelengths are the real λ_p values. The ρ values were 0.088 ± 0.004 irrespective of gel concentration. Reflection spectra of pNIPAm(400-5) and pNIPAm(600-5) were also sharp, and the peak wavelengths were quite insensitive to the suspension temperature (see Fig. 4.19, *E* and *F*).

Fig. 4.20, *A* and *B*, show the reflection spectra at three different concentrations of CAIBA-P2VP(0.1) and CAIBA-P2VP(0.5). Sharp peaks appeared supporting the gel crystal formation. On the other hand, the curve of Fig. 4.20, *C*, for CAIBA-P2VP(1) was a broad peak, which supports the fact that the suspension structure is liquid. However, the suspension emitted strong iridescent colors, and the suspension structure was an ordered liquid state. From the broad peak wavelength, intersphere distance is estimated using Equation 7. Fig. 4.20, *D*, shows the reflection spectra of AIBA-P2VP(0.1) (curve 1), AIBA-P2VP(0.5) (curve 2), and AIBA-P2VP(1) (3), respectively. Fig. 4.20, *E*, shows the typical reflection spectra of the gel crystals of PEGMA-P2VP gel spheres with ion-exchange resins coexisted. Sharp peaks are ascribed to the Bragg diffraction through the single gel crystals. Here, the λ_p values were estimated to be the third, third, third, and ternary reflection peaks, respectively. The wavelengths corresponding to the primary peaks should be three, three, three, and four times of the observed λ_p values. The nearest-neighbored intersphere distances l are estimated to be 914 nm, 852 nm, 901 nm, and 1094 nm for the curves 1, 2, 3, and 4, respectively. Interestingly, Fig. 4.20, *F* where NaCl is introduced, shows the double peaks, which are separated by approximately 3% in the peak wavelengths. As was reported previously,[123,148-150] longer and shorter peaks support the coexistence of the *fcc* and *bcc* lattices, respectively, and further *fcc* structures are more stable than the *bcc* lattices. In the homogeneous suspension state (not sedimentation equilibrium state), the transformation from the *fcc* to the *bcc* lattices took place at high salt concentration, elevated temperature, high pressure, and diluted sphere concentration. Thus the peak shift from the *bcc* to *fcc* structures by the addition of the salt in Fig. 4.20, *F*, means that the crystals become more stable with the introduction of NaCl in the sedimentation state, where the sphere concentration becomes high.

4.5 Kinetics of Type 1 Gel Crystallization

Kinetic studies on Type 1 gel crystals were made mainly with reflection spectroscopy.

The size of the colloidal single crystals from the homogeneous nucleation, L, is estimated by Scherrer's equation from the half-width of the reflection spectrum.[151,152]

$$L \propto n(\Delta q)^{-1} \quad 9$$

where n is the number of the order in the Bragg equation, q is the scattering vector, and Δq denotes $q_l - q_s$, with q_l and q_s referring the largest and smallest wavelengths corresponding to the half-width of the reflection peak. It should be noted here that Equation 9 is derivable by assuming the powderlike random distribution of a large number of colloidal crystals in the reflection volume. However, this assumption is difficult to be maintain, because the most dense planes of the colloidal single crystals are apt to orient parallel to the cell plane.

The peak intensity, I, of the reflection spectra is approximated by Equation 10[50]:

$$I \propto N_{cryst} L^3 \propto L^3 \quad 10$$

where N_{cryst} is the number of single crystals in the reflection volume, which is directly proportional to the number concentration of crystals in the final stages of crystallization processes, being equal to the total number of nuclei that were formed during the course of crystallization. The crystal growth rate, v, is evaluated from the cube-root plot of the reflection intensity using Equation 10 or the half-width of the reflection peak using Equation 9.

Most kinetic measurements on colloidal crystallization, including gel sphere crystallization, have observed an induction period, after which the crystal growth started especially in the diluted suspensions. This observation supports that the kinetics of colloidal and gel crystallization is explainable by the classical diffusive crystallization theory, including nucleation and crystal growth processes. The number of nuclei that germinate per unit time, the nucleation rate, v_n is given by:

$$v_n = N_n/t_i \tag{11}$$

where N_n is the total number of nuclei formed during the nucleation process and t_i is the induction period. Here, we assume that the number of nuclei equals the number of single crystals formed. The number of spheres per mean size of single crystals, N_c, is given for the cubic lattice by:

$$N_c = 2^{1/2}L^3/l^3 \tag{12}$$

where L and l are the mean size of single crystals formed and the closest intersphere distance, respectively. Total number of colloidal spheres (N_T) in an unit volume is $\phi/[(4/3)\pi(d_o/2)^3]$. Then v_n is given by Equation 13.

$$v_n = N_T/N_c t_i \tag{13}$$

It should be mentioned here that the size of a single crystal is highly polydispersed, as has been observed hitherto.[44] Therefore clear-cut separation of the nucleation step from the crystallization process is difficult, and the nucleation reaction may remain even in the crystal growth period. Furthermore, Ostwald ripening of the crystals (i.e., further growth of large crystals and the disappearance of small ones) has not been observed for many colloidal crystallization systems.

According to Wilson[151] and Frenkel,[152] the crystal growth rate, v, is given by:

$$v = v_\infty[1 - \exp(-\Delta\mu/k_B T)] \tag{14}$$

Here, $\Delta\mu$ is the chemical potential difference between the crystal and liquid phases and is given by Equation 15.

$$\Delta\mu = \mu_{cryst} - \mu_{liq} \tag{15}$$

$\Delta\mu > 0$ means that the crystallization will proceed. We should note that Equation 16 holds.[122]

$$\Delta\mu/k_B T = \ln(\phi/\phi_c) \approx \sigma, \tag{16}$$

where σ is the relative supersaturation given by $(\phi - \phi_c)/\phi_c$. Equation 17 derives from Equations 14 and 16.

$$v = v_\infty[1 - \exp(-\sigma)] \tag{17}$$

Equation 17 is further simplified as:

$$v = v_\infty - v_\infty \phi_c/\phi \tag{18}$$

Equation 18 holds in our experiments for typical colloidal hard spheres hitherto.[137,143-150] Equation 19 gives v_∞.

$$v_\infty = 4D_o/\xi \qquad 19$$

Here, D_o is the diffusion coefficient of spheres and ξ is the mean diffusion length (path) in the crystallization suspension. It should be mentioned here that the unit of gel concentration in the kinetic study should be volume fraction (or volume percent) instead of weight percent, as is shown in Equations 16 and 18, for example.

Fig. 4.21 shows the typical examples of the time-resolved reflection spectra during the course of crystallization of pNIPAm(200-10) by the stopped-flow (SF) method at 45° C (Fig. 4.21, *A*) and by the inverted-mixing method at 33° C (Fig. 4.21, *B*). The spectra for pNIPAm(200-2) by the inverted mixing at

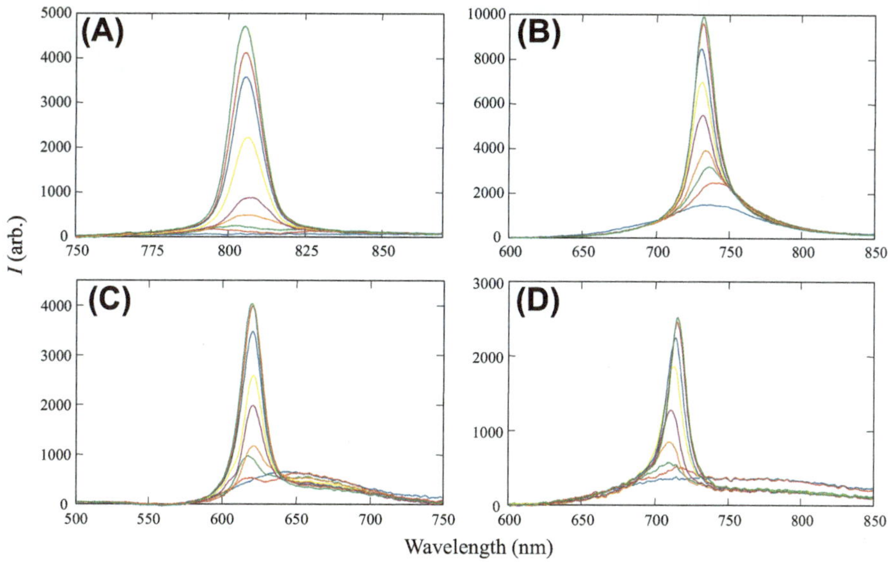

Figure 4.21
Time-resolved reflection spectra during crystallization of pNIPAm gel spheres. **A,** pNIPAm(200-10), SF method, without resins, at 45° C, [pNIPAm(200-10)] = 2 wt%, *blue curve* (lowest one): 0.11 seconds, *red curve:* 0.49 seconds, *green curve:* 0.68 seconds, *orange curve:* 0.99 seconds, *purple curve:* 1.37 seconds, *yellow curve:* 2.93 seconds, *blue curve:* 5.85 seconds, *red curve:* 8.32 seconds, *green curve:* 11.78 seconds. **B,** Inverted mixing method, with resins, at 33° C, [pNIPAm(200-10)] = 2 wt%, *blue curve:* 0 seconds, *red curve:* 0.94 seconds, *green curve:* 1.89 seconds, *orange curve:* 2.84 seconds, *purple curve:* 4.56 seconds, *yellow curve:* 6.08 seconds, *blue curve:* 7.97 seconds, *red curve:* 10.06 seconds, *green curve:* 11.97 seconds. **C,** Inverted mixing method, with resins, at 33° C, [pNIPAm(200-2)] = 1.97 wt%, *blue curve:* 0 seconds, *red curve:* 0.49 seconds, *green curve:* 0.65 seconds, *orange curve:* 0.76 seconds, *purple curve:* 1.03 seconds, *yellow curve:* 1.25 seconds, *blue curve:* 1.71 seconds, *red curve:* 2.36 seconds, *green curve:* 2.77 seconds. **D,** Inverted mixing method, with resins, at 33° C, [pNIPAm(400-5)] = 2 wt%, *blue curve:* 0 seconds, *red curve:* 0.80 seconds, *green curve:* 0.84 seconds, *orange curve:* 0.87 seconds, *purple curve:* 0.91 seconds, *yellow curve:* 1.18 seconds, *blue curve:* 1.41 seconds, *red curve:* 1.67 seconds, *green curve:* 1.82 seconds. *From Okubo T, Suzuki D, Shibata K, Tsuchida A:* Colloid Polym Sci *290:1403–1412, 2012.*

33° C (Fig. 4.21, *C*) and pNIPAm(400-5) at 33° C (Fig. 4.21, *D*) are also given. The SF technique gave the reliable spectra, especially at the initial period after the mixing because the dead time of the mixing in the SF apparatus was very short at approximately 5 ms. However, keeping of the sample suspensions in the deionized state with ion-exchange resins was impossible in the SF cell. Next, the sample suspensions were contaminated continuously during experiments with the foreign ions from the cell wall of the flow mixer, mixing syringe, and the empire tube connecting them. Furthermore, a small amount of the sample suspension in the SF cell was not prevented contamination. As is well known, the crystallization rate decreases substantially as the suspension is contaminated by the foreign salt, because the width of the electrical double layers around the gel spheres decreases with increasing ion concentration in suspension (e.g., see Equation 6). Thus the SF method is not so convenient to obtain reliable kinetic data for the deionized suspension. On the other hand, the dead-time of the inverted mixing method of a test tube cell is rather long, approximately 100 ms to 300 ms even in the shortest cases, because the mixing and sedimentary flow of the suspension and of the ion-exchange resins do not cease so quickly compared with the SF technique. Thus the kinetic data of the initial stage of crystallization, including the induction time, contain large experimental errors. As is shown in Fig. 4.21, *B* through *D*, the reflection spectrum taken at the first time at $t = 0$ s was not a horizontal base, but the broad peak slopes were already detected. Gel crystallization in the rather high sphere concentrations was unexpectedly fast. However, the degree of the deionization is kept in the lowest level during the kinetic measurements in the inverted mixing method, because the sample suspension is always coexisted with the ion-exchange resins in the closed cell. Obtaining the kinetic data of the deionized suspensions is one of the main purposes of this work, and then the inverted mixing method was mainly applied in the kinetic analyses.

As is shown in Equation 18, the rates of crystallization, v, is evaluated from the slopes of the plots of the cube root of the peak intensity with time. However, preliminary analyses of these plots in the inverted mixing method gave large uncertainties especially at the initial stage of crystallization because the dead time of the inverted mixing method is long, as is described previously. Therefore, the rate coefficient, k, was introduced instead of v. The value of $k\,(=v/L_\infty)$ was obtained as the reciprocal period where the initial linear line in the peak intensities (I) intersects two horizontal lines, giving initial and final times at $t = t_i$ and t_f, respectively.

$$k = 1/(t_f - t_i) \qquad 20$$

It should be noted that the t_i values are close to but slightly larger than the real induction times, although the induction time itself should be evaluated from the $I^{1/3}$ versus t plots. Evaluation of k instead of v has been made in some of my kinetic analyses of colloidal crystallization.[153,154]

Fig. 4.22 shows the typical examples of I versus t plots and $I^{1/3}$ versus t plots in the SF mixing methods for the crystallization of pNIPAm(200-10) gel spheres at 45° C. The initial linear lines are also shown for these plots. The kinetic data of k were evaluated from I versus t plots. The typical examples of the I versus t plots of pNIPAm(200-10), pNIPAm(200-2), and pNIPAm(400-5) from the inverted mixing method at 45° C are shown in Fig. 4.23.[116] The linearity of the lines was excellent and reliable k values were obtained.

Fig. 4.24 compares the k values of the gel systems with those of the colloidal silica spheres (flag symbols), where the k values at 20° C and 45° C are denoted by the open and filled symbols, respectively.[116] The gel concentrations in volume fraction were obtained from the ratio of the concentrations in weight percent of gel spheres divided by the ρ values of the corresponding spheres. Several important findings are clear. First, the k values of the gel systems at 20° C stayed small even when sphere concentration increased. This insensitivity of the sphere concentration is notable, because the crystal growth rate of typical colloidal hard spheres increased sharply with increasing sphere concentration. This observation supports the findings that the gel spheres at 20° C are swollen (the ρ values of gel systems are between 0.03 and 0.13), that the interfaces forming between the gel sphere and water phases become vague, and that the electrical double layers do not form as firmly as the colloidal silica spheres. The polymer chains in the outside area of the swollen gel spheres should fluctuate vigorously by the thermal motion,

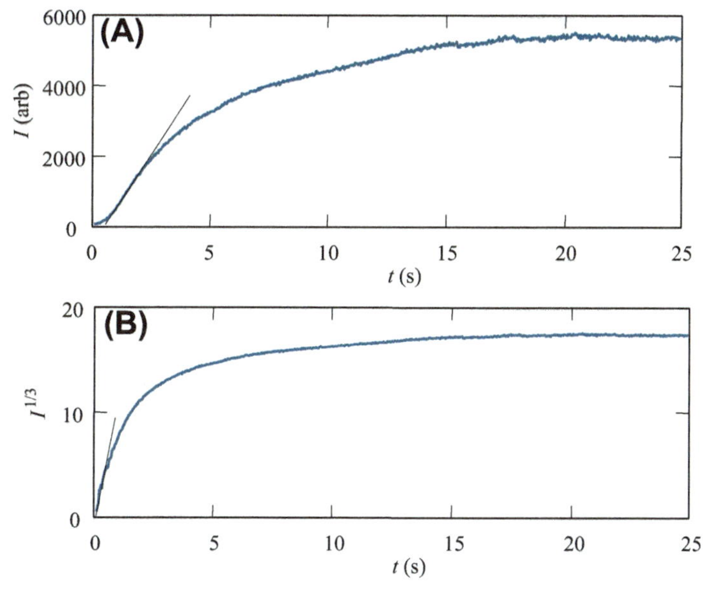

Figure 4.22
A, Reflection peak intensity versus time plots of pNIPAm(200-10) gel spheres at 45° C. SF method, without resins, [pNIPAm(200-10)] = 1.5 wt%. **B,** Cube root of the reflection peak intensity versus time plots of pNIPAm(200-10) gel spheres at 45° C. SF method, with resins, [pNIPAm(200-10)] = 1.5 wt%. *Edited from Okubo T, Suzuki D, Shibata K, Tsuchida A:* Colloid Polym Sci *290:1403–1412, 2012.*

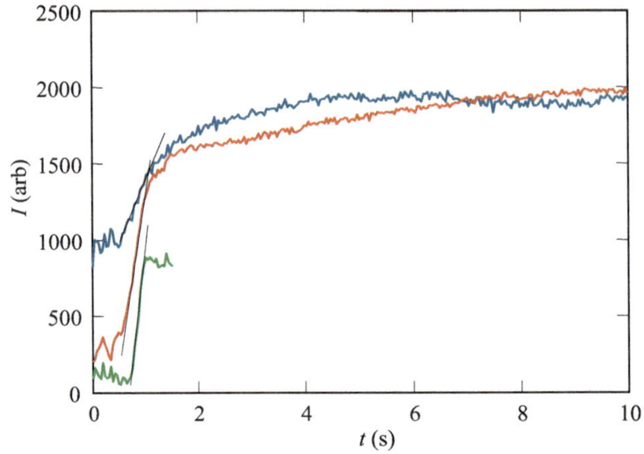

Figure 4.23
Reflection peak intensity versus time plots of pNIPAm(200-10) (*blue curve*), pNIPAm(200-2) (*red curve*) and pNIPAm(400-5) gel spheres (*green curve*) at 45° C. Inverted mixing method, with resins, [pNIPAm(200-10)] = 2 wt%, [pNIPAm(200-2)] = 1.97 wt%, [pNIPAm(400-5)] = 2 wt%. *Edited from Okubo T, Suzuki D, Shibata K, Tsuchida A:* Colloid Polym Sci *290:1403–1412, 2012.*

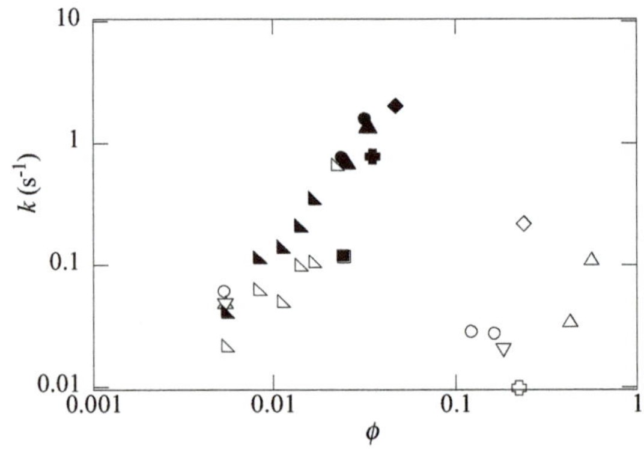

Figure 4.24
Rate coefficients, *k* in the crystallization of pNIPAm and CS83 spheres at 20° C (*open symbols*) and 45° C (*solid symbols*); (200-10) *circles*, (200-5) *broad crosses*, (200-2) *triangles*, (200-0.5) *squares*, (400-5) *reversed triangles*, (600-5) *diamonds*, CS83 triangle flags. *From Okubo T, Suzuki D, Shibata K, Tsuchida A:* Colloid Polym Sci *290:1403–1412, 2012.*

which is one of the causes for the vague interfaces. Second, at 45° C the *k* values of pNIPAm gel systems increased sharply as sphere concentration increased. Third, *k* values of the gel systems at 45° C were substantially large compared with those at 25° C, when the comparison was made at the same concentration in volume fraction. Fourth, the concentration dependencies of the *k* values of silica spheres at 45° C were similar to those of the gel spheres, although the *k* values at 20° C showed substantial variations. These observations are consistent with the fact that the gel spheres at 45° C are highly compact (the ρ values are 0.6 to 0.7). Here, the ρ value of the silica spheres is 2.2, because the specific gravity of silica spheres is 2.2 and the spheres are full of silica (i.e., the dead space without silica in a sphere can be safely neglected), although a silica sphere is composed of small silica spheres of several nanometers. The significant contribution of the electrical double layers formed around the compact gel spheres on the crystallization is clear especially at 45° C in addition of the excluded-volume effect of the gel spheres themselves.

Fig. 4.25 shows the induction times, t_i values of gel and silica spheres as a function of their concentration in volume fraction.[116] The t_i values of gel spheres at 45° C and those of silica spheres at 20° C and 45° C decreased sharply as sphere concentration increased. This tendency is consistent with the finding that the nucleation rates of typical colloidal spheres increase sharply as the sphere concentration increases,[155] as described previously. It should be mentioned here that the t_i values of the gel spheres maintained large values even at the high sphere concentrations. This clearly shows that the nucleation rates of gel spheres were low when the comparison was made at the same sphere concentration in volume fraction. This again demonstrates that the interfaces of the gel spheres are vague and soft compared with the typical colloidal spheres such as colloidal silica spheres.

One of the important findings in the present work is the fact that the crystal growth rates of the swollen gel spheres are small compared with those of typical colloidal hard spheres such as silica spheres. Furthermore, the former often crystallizes at the high sphere concentrations compared with the latter. These experimental results support that the electrical double layers form around the gel spheres, but they are not so stable. At 40° C, however, the ρ values of gel spheres increased sharply up to close unity, and then the interfaces formed stably, similar to those of typical colloidal spheres. In other words, the existence of interfaces between the two phases (i.e., gel and water) is

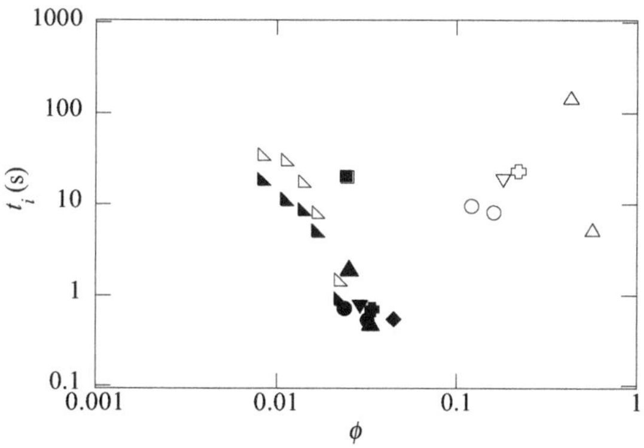

Figure 4.25
Induction times, t_i, in the crystallization of pNIPAm and CS83 spheres at 20° C (*open symbols*) and 45° C (*solid symbols*). (200-10) *circles*, (200-5) *broad crosses*, (200-2) *triangles*, (200-0.5) *squares*, (400-5) *reversed triangles*, CS83 *triangle flags*. *From Okubo T, Suzuki D, Shibata K, Tsuchida A:* Colloid Polym Sci *290:1403–1412, 2012.*

necessary for the formation of the electrical double layers around gel spheres. The rate coefficients, *k*, were discussed here instead of the thermodynamically meaningful crystallization rate, *v*. Further study on the nucleation and crystal growth processes of the crystallization of gel systems is necessary.

4.6 Importance of Electrical Double Layers on Type 1 Gel Crystallization

Colloidal crystallization, including gel crystallization, takes place by the thermal fluctuation of the spheres and the extended electrical double layers in the restricted space of a liquid medium. In other words, electrostatic repulsive intersphere interactions exist. In the deionized state, the electrical double layer is very thick and the intersphere repulsive forces prevail over a long distance as much as micrometers, although the forces become weak. Formation of the crystal-like ordering is nicely explained with the effective soft-sphere model. The effective diameter, d_{eff}, of the spheres includes the Debye screening length, l_D, and is given by the diameter, *d*, plus twice the Debye screening length in Equation 6. Here, *d* of gel spheres is assumed to be close to the hydrodynamic diameter measured by DLS spectroscopy.

$$d_{eff} = d + 2 \times l_D \qquad 21$$

When d_{eff} is similar to or a bit shorter than the intersphere distance, the distribution of spheres is usually liquidlike. When the effective diameter is close to or larger than the observed intersphere distance, crystal-like ordering forms. This effective soft-sphere model has been supported by the systematic comparison between d_{eff} and *l* values. According to the effective soft-sphere model, *l* is close to d_{eff} in crystal suspensions.

The ratio of the Debye screening length (l_D) against the hydrodynamic size in diameter (*d*), l_D/d, is given in Table 4.2 for pNIPAm gel spheres. The ratio is the important parameter to know the relative contribution of the Debye screening effect against the excluded volume effect. The l_D/d values increased sharply as gel concentration decreased or temperature increased. The ratios l_D/d at 1 wt% increased as gel size increased at any temperature. The main cause for these tendencies is that the density of the gel spheres increases with increasing size, as is seen in Table 4.2. The interfaces form more easily for more compact gel spheres. Fig. 4.26 shows the gel crystallization of pNIPAm(200-5) spheres at 1 wt% schematically. Interfaces between the gel spheres and the medium molecules will locate slightly inside the outside edge of the gel spheres. Thus it is summarized that the crystallization of pNIPAm gels takes place by the significant contribution of the extended electrical double layers in addition to the excluded volume effect, especially at low gel concentrations, at high temperatures, at high degrees of cross-linking of gel spheres, and further at large gel spheres. This indicates further that the gel spheres have interfaces with the bulk water phase. In other words, simple ions in a gel sphere

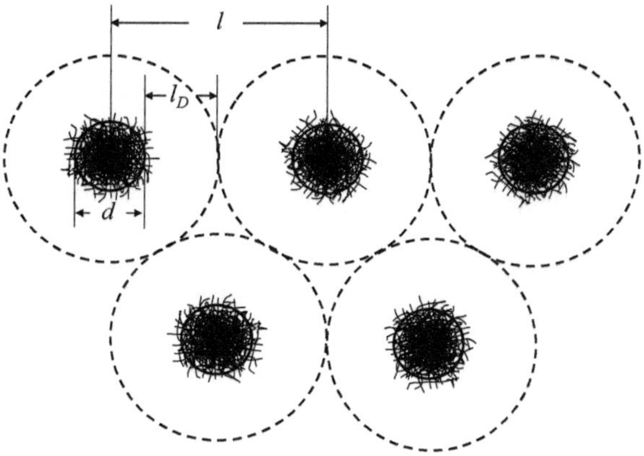

Figure 4.26
Schematic picture of crystal structure of pNIPAm gel spheres surrounded by the electrical double layers. *Solid circle:* hydrodynamic diameter of a gel sphere; *broken circle:* effective size of gel sphere including the electrical double layer ($= d + 2 \times l_D$). *Edited from Okubo T, Suzuki D, Yamagata T, Katsuno A, Sakurai M, Kimura H, Tsuchida A:* Colloid Polym Sci *289:291–299, 2011.*

are placed under restraint from free diffusion in part, although most simple ions and water molecules are able to diffuse rather freely inside the gels.

The ratios of l_D/d values of P2VP gel spheres are listed in Table 4.3. These support the fact that the cationic gel spheres P2VP gels have the interfaces with water phases but the thickness of the electrical double layers are not so thick compared with the anionic charged gel and colloidal hard spheres. The main cause for the small contribution of l_D against d is that the calculation was made at high sphere concentrations compared with those of pNIPAm. Crystallization of the cationic gel spheres takes place by the important contribution of the extended electrical double layers in addition to the excluded volume effect of the gel spheres themselves.

Ballauff et al. analyzed the lattice constant of type 1 gel crystals of pNIPAm spheres with cryo-transmission electron microscopy (cryo-TEM).[109-111] An inequality, $l_{obs} \geq d$, held true. These results of cryo-TEM support the important role of the electrical double layers for the gel crystallization. Eckert and Richtering[124] measured simultaneous static and dynamic three-dimensional cross-correlated light-scattering intensities of concentrated pNIPAm microgels in dimethyl formamide. They found that the microgel structure is different from hard spheres as the cross-link density decreases, leading to a fuzzy particle surface with dangling polymer chains.

4.7 Rigidity of Type 1 Gel Crystals

When colloidal suspensions, including Type 1 gel sphere suspensions, are set in a test tube with coexistence with ion-exchange resins, the spheres start to sediment by the earth gravity and the sedimentation equilibrium state is achieved 1 to 2 months after the setting in most cases. According to the discussion by Crandall and Williams,[156] the nearest-neighbored intersphere distance, l, is given by Equation 22 in the sedimentation equilibrium.

$$(l - l_m)/l_m = (\sigma_{eff} g\, \phi_m/G)(h - h_m) \qquad 22$$

where l_m and ϕ_m indicate the lattice spacing (nearest-neighbored intersphere distance) and the sphere concentration (in volume fraction) at the midplane of the suspension column in the sedimentation equilibrium, respectively. The heights of the arbitrary layer and midplane are represented by h and h_m. The gravitational constant is g. The difference in the specific densities of pNIPAm gel spheres minus that of water in the suspension state is represented by σ_{eff}. G is the rigidity (Young's elastic modulus) of gel crystal.

Fig. 4.27 shows the typical examples of the reflection spectra of the gel crystals of pNIPAm(200-2) with ion-exchange resins coexisted at various heights from 3 mm (blue curve) to 75 mm (orange curve) in the sedimentation equilibrium.[118] Sharp peaks appeared from the Bragg diffraction through the single crystals. In general, the primary peak wavelengths (λ_p) in the reflection spectroscopy of the crystals at the scattering angle of 90° is given from the nearest-neighbored intersphere distance (l) of the *fcc* or *bcc* lattices, which are familiar lattices in colloidal and gel crystals by Equation 7. In the experiments, the concentration of gel spheres is not so high and n is assumed safely to be that of water, 1.33. The l values are further given by Equation 3 as a function of the diameter (d) and volume fraction of gel spheres (ϕ). The ϕ values were evaluated from the l values using Equations 3 and 7. The effective density of the gel spheres in suspension state (ρ) is given by the ratio of the gel concentration in weight percent against that in volume percent (100ϕ), as was described previously.

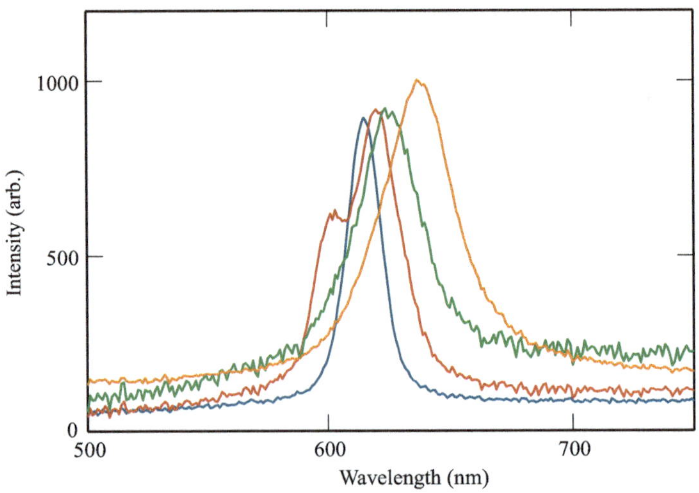

Figure 4.27
Reflection spectra of crystals of pNIPAm(200-2) gel spheres in the sedimentation equilibrium at 20° C. 1.97 wt%, 44 days after the suspensions were set; *blue curve:* height (h) = 3 mm; *red curve:* 27.5 mm; *green curve:* 60 mm; *orange curve:* 75 mm (liquid top). *From Okubo T, Suzuki D:* Colloid Polym Sci *291:1721–1727, 2013.*

Fig. 4.27 shows that most reflection spectra are the single peaks, but double peaks also appeared at $h = 27.5$ mm. The double peaks are often separated approximately 3% in the peak wavelengths. As was reported previously,[123,148-150] longer and shorter peaks support the coexistence of the *fcc* and *bcc* lattices, respectively, in the gel crystals, and further *fcc* structures are more stable than the *bcc* lattices. The transformation from *fcc* to *bcc* lattices took place at high temperature, at high pressure, and at diluted sphere concentration in the homogeneous suspension state (not sedimentary state). It should be mentioned here that the single crystals of *fcc* and *bcc* lattices coexisted so often and the single peak in the reflection spectrum does not always mean the existence of 100% of *fcc* or 100% of *bcc*, but appearance of a broad single peak attributed to the mixture of *fcc* and *bcc* lattices take place often. Furthermore, the reflection spectra having double peaks in Fig. 4.27 support the finding that both *fcc* and *bcc* crystals exist, and the former is rich in the spectroscopy-detected area. Quantitative separation of the contributions from *fcc* and *bcc* crystals is, therefore, not so easy in general. However, Fig. 4.27 supports that the reflection peaks and then lattice spacing of *fcc*-lattices increased as the height increased, because the peaks at the longer wavelength side shifted toward longer wavelengths with increasing height.

Fig. 4.28, *A* and *B,* show typical examples of the peak wavelengths of pNIPAm(200-5) and AIBA-P2VP(0.5) gel crystals as a function of the height from the bottom of the cell in the state of sedimentation equilibrium.[118,120] The gel suspension and ion-exchange resins were set in the test tube cells for 4 days, during which the cells were inverted several times per day to accelerate the deionization processes of the sample suspensions and to prevent the sedimentation of the incompletely deionized gel spheres. After the suspension was thus prepared, the test tube cells were left alone and no longer inverted. The reflection spectrum measurements were made at 17 days, 29 days, and 44 days after the suspensions for pNIPAm(200-5) spheres, for example, were deionized, and the sedimentation equilibrium state was achieved 44 days after the suspensions were left alone. In the sedimentation equilibrium state, increased tendency toward the certain saturated values in the peak wavelengths stopped. Increase in the peak wavelengths with time is clearly due to the slow progress of the deionization of the suspensions with coexistence of the ion-exchange resins.

The peak wavelength observed in Fig. 4.28 increased as the height increased. Interestingly, in the range of small h values, the peak wavelengths that originated from the *fcc* lattices increased linearly as h increased, whereas the peak wavelengths from the *bcc* lattices also increased linearly as h further increased. Furthermore, transformation took place gradually from *fcc* to *bcc* lattices as h increased. We should note here that *fcc* and *bcc* lattices form more favorably at high and low sphere concentrations, respectively.[123,148-150]

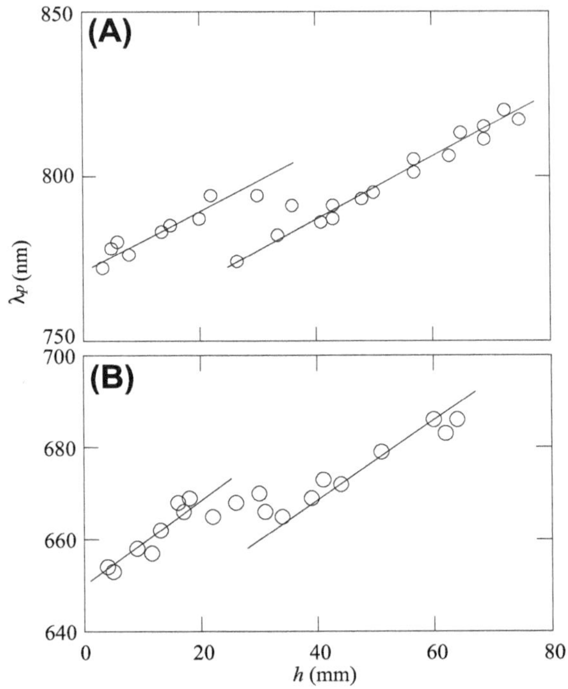

Figure 4.28
A, Peak wavelength of crystals of pNIPAm(200-5) gel spheres in the sedimentation equilibrium as a function of the height at 20° C, 2.78 wt%. **B,** Peak wavelength of gel crystals of AIBA-P2VP(0.5) spheres in the sedimentation equilibrium as a function of the height at 22° C, 9.65 wt%.

The slopes of the linear lines from *fcc* and *bcc* lattices shown by fine, straight lines in Fig. 4.28, *A* and *B,* look to be similar to each other.

Fig. 4.29 shows the typical examples of λ_p versus *h* plots of CAIBA-P2VP(0.1), CAIBA-P2VP(0.5), and CAIBA-P2VP(1) in the sedimentation equilibrium state.[121] The single peaks appeared for these samples, and the linear relationship appeared between the peak wavelength and the height. From the slopes of λ_p versus *h* plots, rigidity *G* of the gel crystals and the ordered liquid (or distorted crystal) was evaluated.

The *G* values estimated are shown in Fig. 4.30 as a function of number density of the gel spheres *N.* As has been discussed by Crandall and Williams[156] and Mitaku et al.,[157,158] the order of magnitude of the modulus *G* is approximated in terms of the magnitude of the thermal fluctuation δ of a sphere as:

$$G \cong f/l \cong \left(k_B T / < \delta^2 > \right)/l \qquad 23$$

where *f* is the force constant, *l* is the intersphere distance, δ is the thermal fluctuation of a sphere in the effective potential valley, k_B is the Boltzmann constant, and *T* is the absolute temperature. Introducing a nondimensional

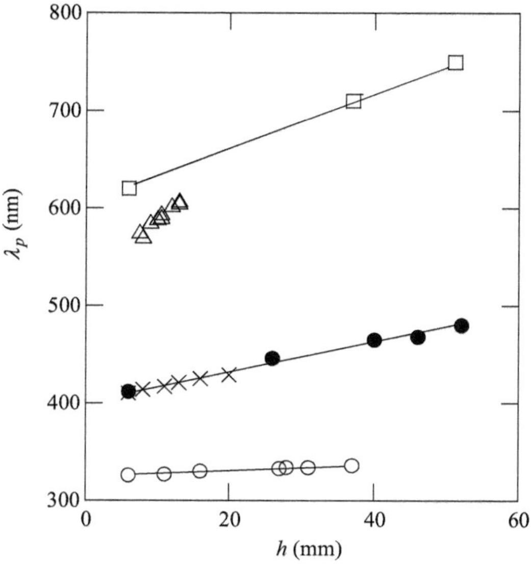

Figure 4.29
Peak wavelengths of CAIBA-P2VP(0.1) (*open circle:* 4.34 wt%, *crosses:* 3.26 wt%, *triangle:* 2.17 wt%), CAIBA-P2VP(0.5) (*square:* 1.69 wt%) and CAIBA-P2VP(1) (*solid circle:* 2.50 wt%) spheres as a function of the height in the sedimentation equilibrium at 25° C. *Edited from Fujii S, Yamashita Y, Nakamura Y, Tsuchida A, Okubo T:* Colloid Polym Sci *292:1627–1637.*

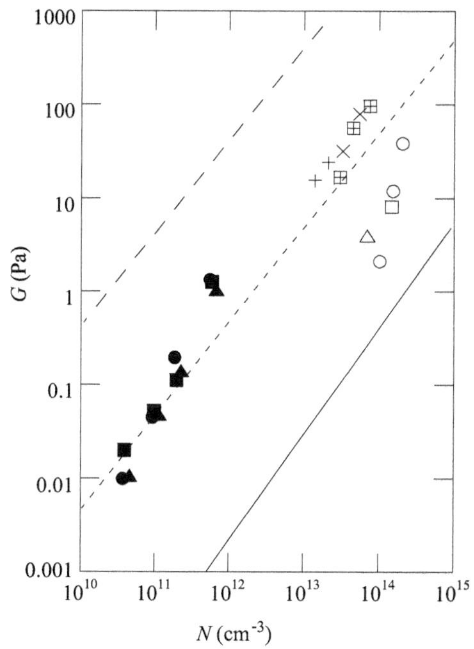

Figure 4.30
Log G versus log N plots of CAIBA-P2VP(0.1) (*open circle,* crystal), CAIBA-P2VP(0.5) (*triangle,* liquid), CAIBA-P2VP(1) (*square,* liquid), PEGMA-P2VP(0.1) (*solid circle*), PEGMA-P2VP(0.5) (*solid triangle*), PEGMA-P2VP(1) (*solid square*), AIBA-P2VP(0.1) (*plus*), AIBA-P2VP (*cross*) and AIBA-P2VP(1) (*plus in square*). *Edited from Fujii S, Yamashita Y, Nakamura Y, Tsuchida A, Okubo T:* Colloid Polym Sci *292:1627–1637.*

fluctuation parameter g for $<\delta^2>^{1/2}/l$, the modulus is obtained as a linear function of N.[157,158]

$$G \cong N\, k_B\, T/g^2 \qquad 24$$

When $g = 1$, Equation 24 gives the elastic modulus of an ideal gas having the same sphere concentration. Lindemann's law of crystal melting tells us that $g < 0.1$ holds for a stable crystal. Short and long dashed lines in Fig. 4.30 represent $g = 0.1$ and 0.01, respectively. It is clear that all the G values of pNIPAm gel spheres shown by plus signs, crosses, and pluses in squares locate around the short dashed line. Open symbols of circles, triangles, and squares are G values of CAIBA-P2VP gel spheres. The solid symbols in Fig. 4.30 show the G values of the cationic gel spheres, AIBA-P2VP and PEGMA-P2VP. Most G values of pNIPAm, AIBA-P2VP, and PEGMA-P2VP spheres are located around short dashed line. The slope of the observed log G values against log N plots of AIBA-P2VP and PEGMA-P2VP looks slightly larger than that of the short dashed line. Thus there exists no difference in the elastic properties between anionic pNIPAm and cationic AIBA-P2VP and PEGMA-P2VP gel spheres. It is clear that the rigidities of CAIBA-P2VP are low compared with other cationic gel spheres and also pNIPAm gel spheres. The g values of CAIBA-P2VP were between 0.15 and 0.28. These results on the very small G and large g values of CAIBA-P2VP gel crystals and amorphous solids support the finding that these gel structures are not so stable compared with AIBA-P2VP, PEGMA-P2VP, and pNIPAm gel systems, or compared with typical colloidal crystals.[157,158] The main cause for the very soft crystals and amorphous solids of CAIBA-P2VP is certainly their high dispersity in the sphere size distribution.

It should be noted here that most G values of the typical colloidal crystals are located between the short dashed and long dashed lines.[45-47] The g values of pNIPAm gel spheres estimated from the G values using Equation 20 were approximately 0.05 through 0.07, 0.05 through 0.09, and 0.05 through 0.15, respectively. These results on G and g values of pNIPAm gel crystals support the finding that the elastic properties of gel crystals is not essentially different from those of typical colloidal crystals or metals, polymers, or other substances.[157,158] These experimental results also emphasize that the gel crystals are soft compared with those of typical hard-sphere systems and the role of the extended electrical double layers for the crystallization of gel spheres is important but weak compared with that of hard colloidal spheres.

4.8 Type 1 Gel Crystals Composed of Gel and Colloidal Hard Spheres

Thermoresponsive Type 1 crystals were available conveniently by mixing colloidal silica spheres (CS82, 103 nm in diameter) and gel spheres of pNIPAm.[107,108] When the small gel spheres were added in the colloidal crystals of silica spheres, the spacing decreased transitionally.[107] On the other hand, the spacing increased with large gel spheres.[108] Thermosensitive gels used were pNIPAm(70-5), 72 nm in diameter at 20° C, and 28 nm at 50° C, which corresponds to a volume change by a factor of 17. The degree of cross-linking was 5%. Thermo-sensitive gel spheres of TG and SVN were also used. The former are triblock copolymers composed of poly(ethylene glycol-b-NIPAm-b-methyl methacrylate) prepared by Tauer and Khrenov.[159] The latter consists of a styrene-vinyl benzyl chloride copolymer as a core surrounded by a hairy layer of pNIPAm chains. Details of preparation of SVN were described in a previous paper.[160] The diameters of TG and SVN decreased from 310 nm at 20° C to 215 nm at 50° C, and from 970 nm at 15° C to 479 nm at 55° C, respectively, corresponding to a volume change by a factor of 3 and 8, respectively.

Changes in the lattice spacing at the temperature between 15° C and 55° C were measured from the reflection spectroscopy using an apparatus shown in Fig. 4.31. A test tube was set in the water bath made of quartz glass. The reflection spectra of the suspensions at an incident angle of 90 degrees through the quartz glass of the bath, the water in the bath, and a test tube (or quartz glass of the long neck optical cell) were recorded on a multichannel photodetector (MCPD-7000G3, Otsuka Electronics) connected to a Y-type optical fiber cable. Temperature in the bath was controlled by the

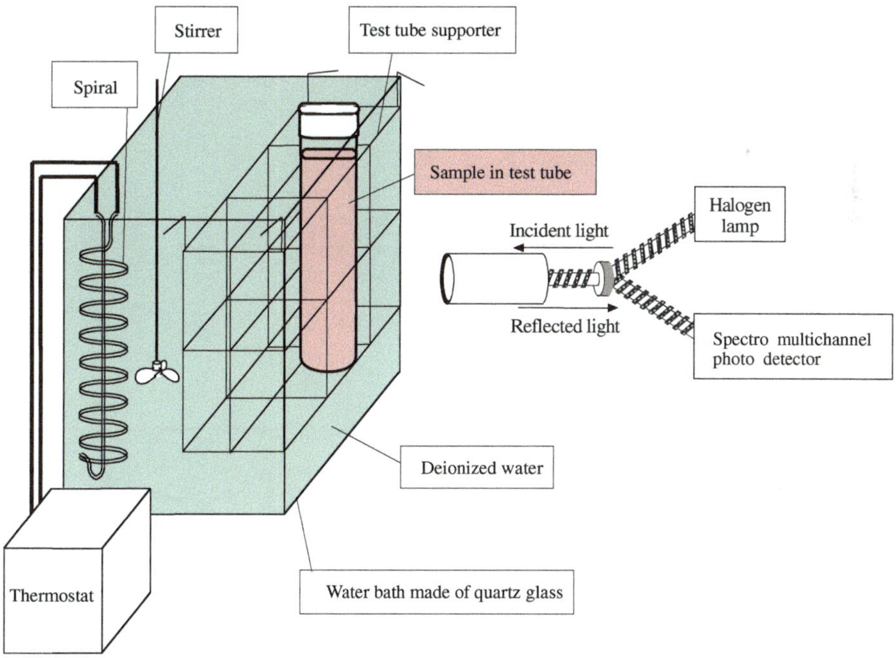

Figure 4.31 Schematic representation of an instrument for the reflection spectrum measurements at regulated temperature.

circulating water from a thermostat. The measuring system is also shown schematically as system B in Fig. 2.33 of Chapter 2.

Typical examples of the reflection spectra of CS82 and pNIPAm(70-5) mixtures at 20° C and 45° C are shown in Fig. 4.32.[107] Interestingly, the peaks at 45° C shifted significantly toward shorter wavelengths and became broad when the concentration of gel spheres increased. On the other hand, the peaks at 20° C shifted to longer wavelengths, although the broadening of the peak took place significantly even at the lower gel concentrations than the experiments at 45° C. These changes in the peak wavelengths indicate clearly that the extension and contraction of the crystal lattice that occur at 20° C and 45° C, respectively, by the addition of the gel spheres.

The lattice spacing, the nearest-neighbor intersphere distance in *fcc* and *bcc* lattices, $l_{obs,f}$, and $l_{obs,b}$ is given by Equations 72 and 73 of Chapter 2. Fig. 4.33 shows the l_{obs} values as a function of suspension temperature at $\phi = 0.032$, where ϕ denotes the concentration of silica spheres in volume fraction. In the absence of gel spheres, l_{obs} did not change within the experimental errors.

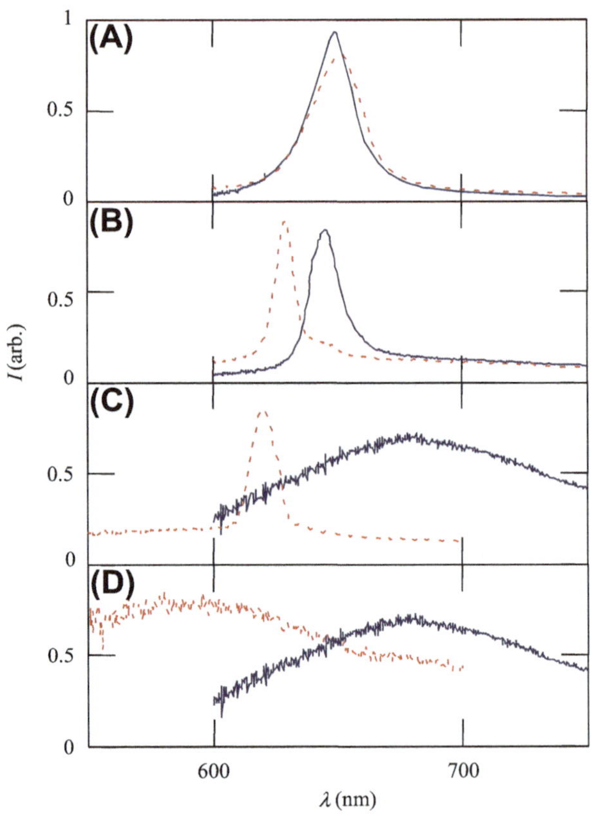

Figure 4.32 Reflection spectra of CS82 + pNIPAm(70-5) gel crystals. $\phi = 0.032$; **A,** $w = 0$ g/ml; **B,** 0.008; **C,** 0.017; **D,** 0.02; *solid curve:* 20° C; *broken curve:* 45° C. *From Okubo T, Hase H, Kimura H, Kokufuta E:* Langmuir *18:6783–6788, 2002.*

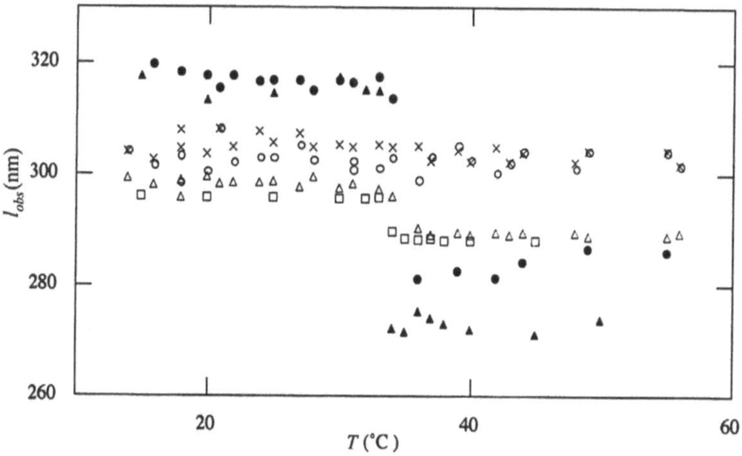

Figure 4.33
l_{obs} versus temperature plots of CS82 + pNIPAm(70-5) gel crystals. $\phi = 0.032$; *circles:* $w = 0$ g/ml; *crosses:* 0.004 g/ml; *triangles:* 0.008 g/ml; *squares:* 0.01 g/ml; *solid circles:* 0.017 g/ml; *solid triangles:* 0.02 g/ml. *Edited from Okubo T, Hase H, Kimura H, Kokufuta E:* Langmuir *18:6783–6788, 2002.*

However, the sharp transitional decrease in l_{obs} was observed clearly when the concentration of pNIPAm(70-5) spheres were higher than 0.008 g/ml. The reversible extension and contraction in the lattice spacing with temperature supports the idea that the weak adsorption of gel spheres takes place with the neighboring colloidal silica spheres in crystal cages. To ascertain this idea, ELS measurements were made for the mixtures of CS82 and pNIPAm(70-5) spheres. The ζ-potential of CS82 spheres, −50 mV increased to a saturated value, approximately −20 mV, as gel concentration increased. This observation is consistent with the weak adsorption between the gel and silica spheres, because the gel sphere layer coated on the silica surfaces must result in the increase in the ζ-potential by the shielding of the charges of silica spheres. The effective diameters of silica spheres, d_{eff}, estimated from the ELS measurements increased significantly from 115 to 200 nm when the gel concentration increased, which also supports weak adsorption of the gels on the silica spheres.

A schematic picture of the thermosensitive colloidal crystal system of CS82 + pNIPAm(70-5) gel mixtures is shown in Fig. 4.34. In the absence of gel spheres (see fig. 4.34, *A*), the crystal structure of silica spheres is kept by the intersphere repulsion interactions intermediated by the electrical double layers surrounding the silica spheres. In the presence of gel spheres, the spheres locate between the neighboring silica spheres and bind weakly with the surfaces. The gel spheres are soft and vague in the outer boundary, as is seen in Fig. 4.34, *B* and *C*. Thus the extension or contraction of the gel spheres with temperature result in the extension or contraction in the lattice spacing of silica spheres. It should be noted here that the electrical double layers

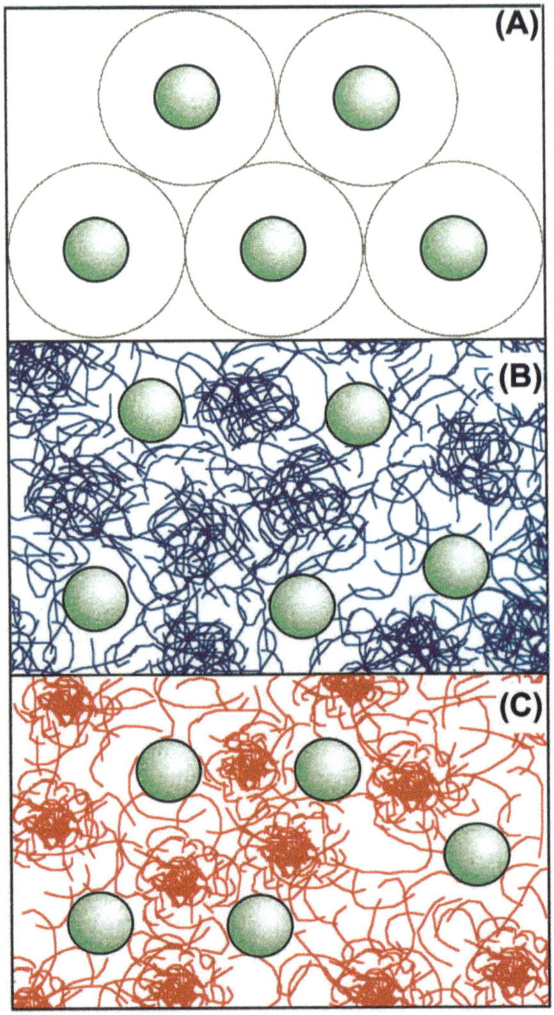

Figure 4.34 Schematic representation of CS82 + pNIPAm(70-5) gel crystals. **A,** Without gel spheres; **B,** with gel spheres at low temperatures; **C,** with gel spheres at high temperatures. *From Okubo T, Hase H, Kimura H, Kokufuta E:* Langmuir *18:6783–6788, 2002.*

always contribute to the crystallization of silica spheres in the coexistence with the gel spheres.

Fig. 4.35 shows the typical examples of the reflection spectra of CS82 and TG gel mixtures.[108] The reflection peaks of CS82 crystals shifted to short wavelengths at 15° C and 35° C, whereas they shifted to longer wavelengths at 50° C. Fig. 4.36 shows the l_{obs} values of CS82 + TG mixtures as a function of suspension temperature and the content of TG. In the absence of gel spheres ($w = 0$ g/ml), l_{obs} was insensitive to suspension temperature. By the addition of gel spheres, l_{obs} increased and then decreased although l_{obs} was insensitive

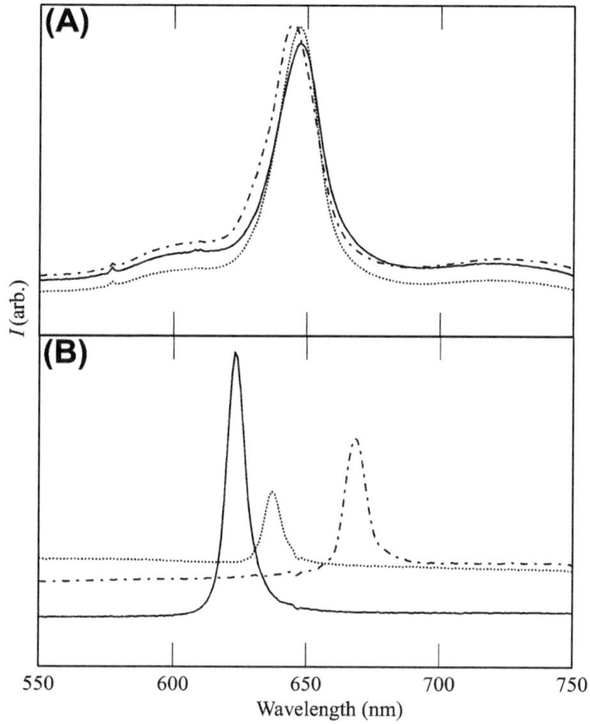

Figure 4.35
Reflection spectra of CS82 (*A*) and CS82 + TG gel crystal (*B*) as a function of temperature. $\phi = 0.032$; **A,** $w = 0$ g/ml; **B,** 0.02 g/ml; *solid curve:* 15° C; *dotted curve:* 35° C; *dotted curve:* 50° C. *Edited from Okubo T, Mizutani T, Okamoto J, Kimura K, Tsuchida A, Tauer K, Khrenov V, Kawaguchi H, Tsuji S:* Colloid Polym Sci *285:351–358, 2006.*

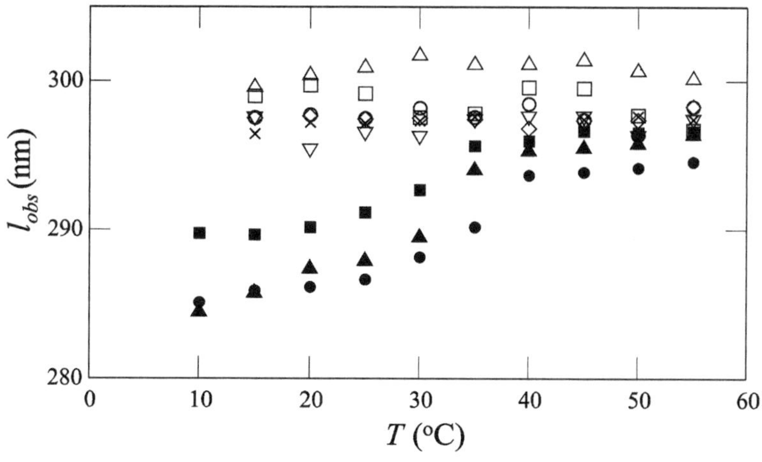

Figure 4.36
Changes in the intersphere spacing (l_{obs}) of CS82 + TG gel crystals. $\phi = 0.032$; *open circles:* $w = 0$ g/ml; *crosses:* 0.0001 g/ml; *open triangles:* 0.0003 g/ml; *open squares:* 0.001 g/ml; *inverse open triangles:* 0.0025 g/ml; *diamonds:* 0.0003 g/ml; *solid circles:* 0.0083 g/ml; *solid triangles:* 0.0167 g/ml; *solid squares:* 0.02 g/ml. *Edited from Okubo T, Mizutani T, Okamoto J, Kimura K, Tsuchida A, Tauer K, Khrenov V, Kawaguchi H, Tsuji S:* Colloid Polym Sci *285:351–358, 2006.*

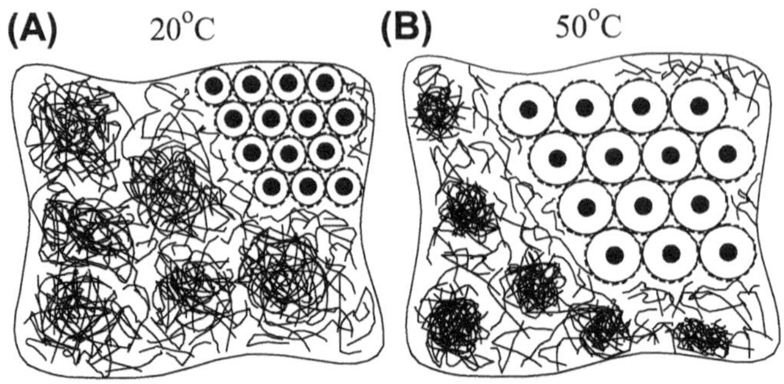

Figure 4.37
Schematic representation of CS82 + TG gel crystals at 25° C (*A*) and 50° C (*B*). *Edited from Okubo T, Mizutani T, Okamoto J, Kimura K, Tsuchida A, Tauer K, Khrenov V, Kawaguchi H, Tsuji S:* Colloid Polym Sci *285:351–358, 2006.*

to temperature when gel concentrations were lower than 0.0038 g/ml. It is surprising to note that by the addition of the gels at concentrations higher than $w = 0.0042$ g/ml, l_{obs} values increased transitionally as suspension temperature increased. The critical temperature of the transition was at 35° C and coincided almost with that of the TG. The l_{obs} values of CS82 + SVN mixtures increased again by the addition of SVN gel spheres, although the graphs showing these are omitted.

The very large gels are difficult to locate between the neighboring silica spheres without breaking the crystal lattices. It is highly plausible that a part of the large gel spheres are segregated from the crystal regions of silica spheres. A schematic picture of the thermosensitive colloidal crystals formed in the mixtures of CS82 spheres and TG gels is shown in Fig. 4.37. In this scheme, the extension or contraction of the gel spheres results in the contraction or extension of the lattice spacing in the crystal regions, when the sum of the gel and crystal regions are kept constant in the vessel.

The SVN gel spheres are composed of polystyrene core and pNIPAm shell parts. The hairy layer is hydrophilic enough to invite silica spheres in it, and parts of silica spheres are likely to be buried in the inner shell of pNIPAm. This situation may result in the decrease of silica spheres in the crystal regions, and then in the increase of the lattice spacing. It should be further mentioned here that the segregation effect of the gel regions from the crystal areas should be incomplete for the hairy, soft nature of gel spheres, especially for the large SVN gels.

4.9 Type 2 Gel Crystals

The porous hydrogels exhibit a rapid change in volume responding to environmental variations, pH, ionic concentration, and temperature, for example. Asher et al.[161] have studied this field for a long time. They prepared crystal

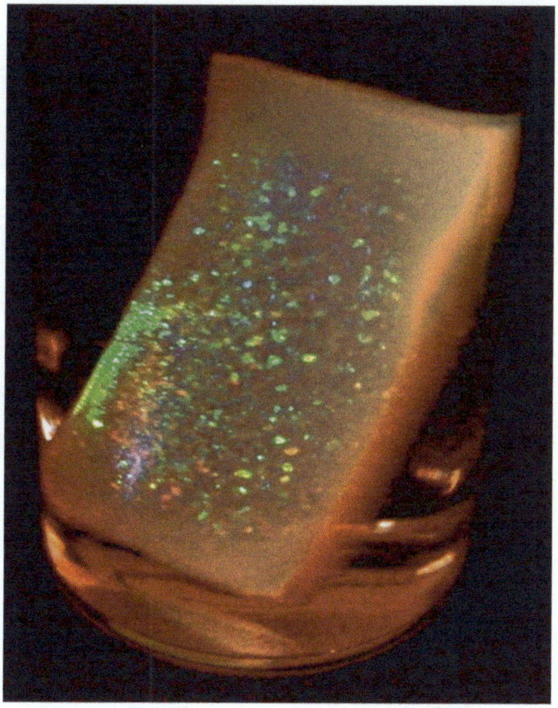

Figure 4.38
Type 2 gel crystals of colloidal silica spheres modified with poly(methyl methacrylate) (PMMA) buried in PMMA gels. Length of the bar is 25 mm. *Courtesy Professor K. Yoshinaga.*

arrays of polystyrene spheres (150 nm in diameter) first in aqueous media and then polymerized the suspension with 50% *N*-vinylpyrrolidone, 17% acrylamide, 4.5% *N,N*-methylene bisacrylamide, and 1% benzoin methylether.[161] The gels keep a crystal structure of spheres and extinction peak wavelength does not change by the gelation. The peak wavelength shifted toward shorter by a uniaxial strain in the plain of the film, and extinction spectra recovered again when stress was relieved. Asher et al. prepared the thermosensitive gels containing crystal arrays of polystyrene spheres in the interior of the gel.[146] Lattice spacing changed sharply with gel temperature. Large gels containing crystal structures of modified silica spheres with polymers emitted strong iridescent colors, as shown in Fig. 4.38, which is kindly donated by Professor Yoshinaga. Type 2 gel crystals in organic solvents were prepared by Yoshinaga et al.[162,163]

An inverse opal structure having a periodically ordered porosity endows Type 2 hydrogels with structural colors based on the Bragg diffraction of light. Asher et al. prepared polyvinyl alcohol hydrogels with crystal arrays as an enabling photonic crystal,[164,165] Keen attention has been paid to the inverse opal-type gel crystals.[166-169] Watanabe and Takeoka prepared a series of Type 2 hydrogel crystals as a glucose sensor.[170] The preparation and

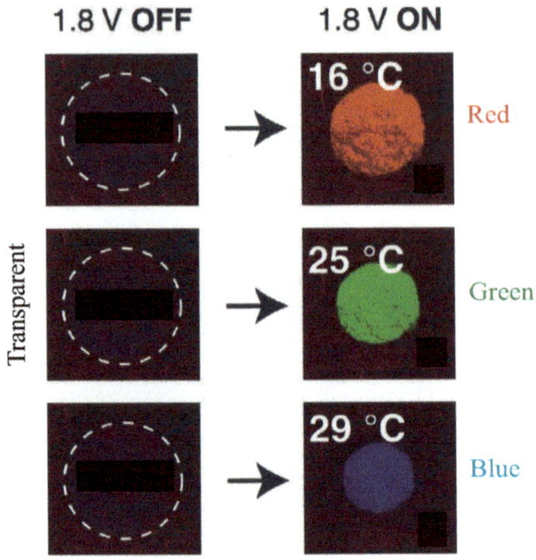

Figure 4.39 Type 2 gel crystals of poly(*N*-isopropylacrylamide-co-sodium methacrylic acid) gel with hollows made from colloidal crystals of silica spheres. *Courtesy Professor M. Watanabe.*

application of the other unique Type 2 gel crystals have been reported.[171-178] A typical example of gel crystals changing their iridescent colors in an electric field is shown in Fig. 4.39. Inverse opal-type copolymer gels of *N*-isopropylacrylamide and sodium salt of methacrylic acid were prepared.[175] This picture was kindly donated by Professor Watanabe. Preparation of multiresponsive (temperature, pH, and ultraviolent–visible spectroscopy) hydrogel photonic crystals was made using poly(NIPAm-co-methacrylic acid).[179] Suzuki et al. synthesized core-shell–type gel crystals containing gold particles.[180] Ikeda et al. synthesized gel crystals containing azobenzene.[181] Other interesting Type 2 gel crystals have been prepared intensively to make the optical application.[182-184] Bimodal swelling of microgel thin films was reported by Sorrell and Lyon.[185]

4.10 External Field Effects on Type 2 Gel Crystals

Electrooptic effects of Type 2 gel crystals were investigated.[186] The alternating current (*ac*) components of the gel crystals (#1 sample) in acetonitrile were smaller than those of crystals without gelation (#2) and in a water system (#3). The peak frequencies were smaller than those of #2 and further #3. Gel crystals are obtained by the addition of gelator Z-L-Ile-C-18 into the dispersion of poly(maleic anhydride-co-styrene)–modified silica spheres, P(MA-ST)/SiO_2 in acetonitrile (Fig. 4.40). Constitutions of the spheres are compiled in Table 4.5. Fig. 4.41 shows the schematic representation of the apparatus used. A light beam from a halogen light source was introduced into a Y-type optical fiber and was focused on the inside cell wall. Intensity change of the reflected light was detected by a photomultiplier and then

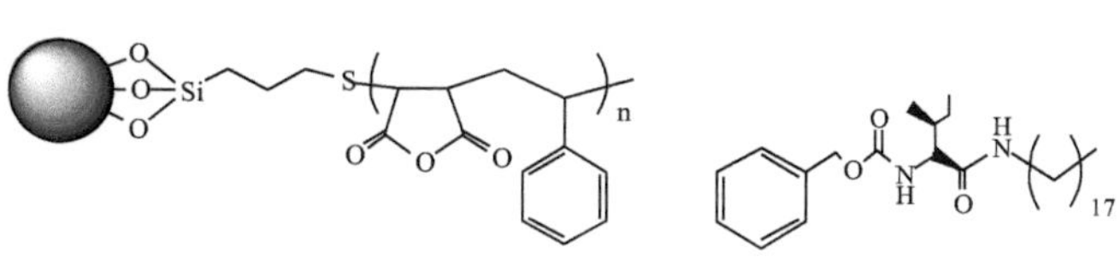

P(MA-ST)/SiO_2 Z-L-Ile-C-18

Figure 4.40
Structure of P(MA-ST)/SiO_2 and the gelator Z-L-Ile-C-18. *From Shibata K, Kimura H, Tsuchida A, Okubo T, Sato S, Yoshinaga K:* Colloid Polym Sci *284:372–379, 2006.*

Table 4.5 Characteristics of #1, #2 and #3 Samples

Code	Spheres	*d* (nm)	ϕ	Solvent	Gelator
#1	P(MA-ST)/SiO_2	135	0.130	acetonitrile	Z-L-Ile-C-18
#2	P(MA-ST)/SiO_2	135	0.133	acetonitrile	—
#3	CS91	110	0.061	water	—

CS91, Colloidal silica spheres ($d = 110 \pm 4.5$ nm, $s/d = 0.041$, charge density: 0.48 $\mu C/cm^2$); *P(MA-ST)/SiO_2:* poly(maleic anhydride-co-styrene)-modified silica spheres.

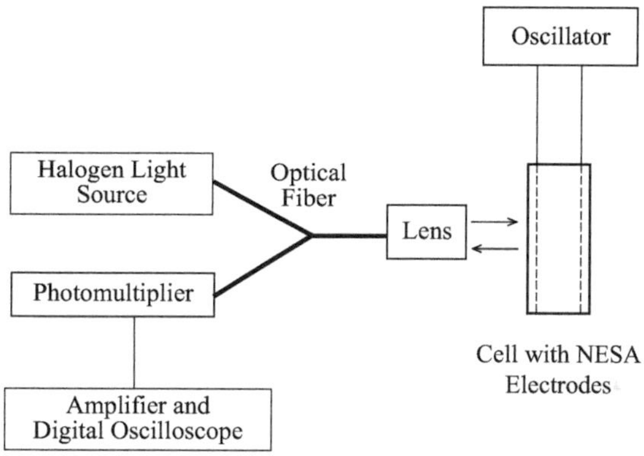

Figure 4.41
Schematic representation of the apparatus for electrooptic measurements of colloidal crystals. *From Shibata K, Kimura H, Tsuchida A, Okubo T, Sato S, Yoshinaga K:* Colloid Polym Sci *284:372–379, 2006.*

recorded on a digital oscilloscope. Time-resolved reflection spectra were taken on a photonic multichannel analyzer.

Fig. 4.42, *A* and *B,* show the frequency (f) dependencies of amplitude (A) and phase difference (φ) of the reflected responses, respectively, at the constant electric field of 2 V/cm. The *A* values of #1 were smaller than those for the ungelated crystals (#2) in all the frequency ranges examined. The 5-mm cell gave larger *A* values than those the for 2-mm cell. These results are reasonable, because the sphere movement of #1 is strongly restricted by gelation. P(MA-ST)/SiO_2 spheres gave the largest *A* values of approximately $f = 0.8$ Hz, regardless of the presence of gelator and the thickness of the cells. Stoimenova et al.[187] pointed out the existence of resonance frequencies for colloidal crystals in aqueous media; in that case the resonance frequencies increase according to the increase in sphere concentration. However, in Fig. 4.42, *A,* the maximum *A* values were the same irrespective of the systems. Therefore in the

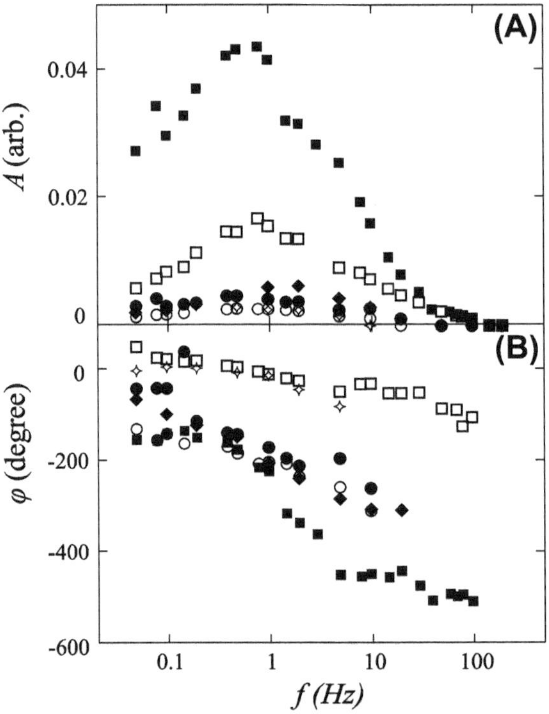

Figure 4.42
Amplitudes (*A*) and phase differences (*B*) in the reflection signals for gel crystals when a sinusoidal electric field is applied at 25° C. $E = 2$ V/cm. P(MA-ST)/SiO_2 with gelator (#1) in 2-mm cell (*open circle*) and 5-mm cell (*filled circle*), P(MA-ST)/SiO_2 without gelator (#2) in 2-mm cell (*open square*) and 5-mm cell (*filled square*), CS91 (#3) in 2-mm cell (*open diamond*), and 5-mm cell (*filled diamond*). *From Shibata K, Kimura H, Tsuchida A, Okubo T, Sato S, Yoshinaga K:* Colloid Polym Sci *284:372–379, 2006.*

present case, the decrease in *A* at the low frequency range ($<$0.8 Hz) must be caused by the partial melting of the colloidal crystals caused by the pseudosteady state application of the low-frequency electric field.[188] On the other hand, decrease of *A* at the high-frequency range ($>$0.8 Hz) is caused by the slow response of the crystals in comparison with the change of applied electric field.[189] The *A* values for the reference CS91 colloidal crystals in water (#3) are also shown in Fig. 4.42, *A*, and they are of a similar order to those for the gel crystals. These results show the higher elastic modulus for the gel crystals than that for ungelated crystals. As for the phase differences (φ), the values decreased substantially when *f* increased for all the systems. These results are consistent with those of the colloidal crystals in aqueous suspension systems.[189-192] *A* increased linearly when *E* increased, irrespective of the thickness of cells and kind of spheres. The higher *E* value causes a larger deformation of the gel crystals. These results are the similar to those reported previously on the colloidal crystals of silica spheres.[191,192]

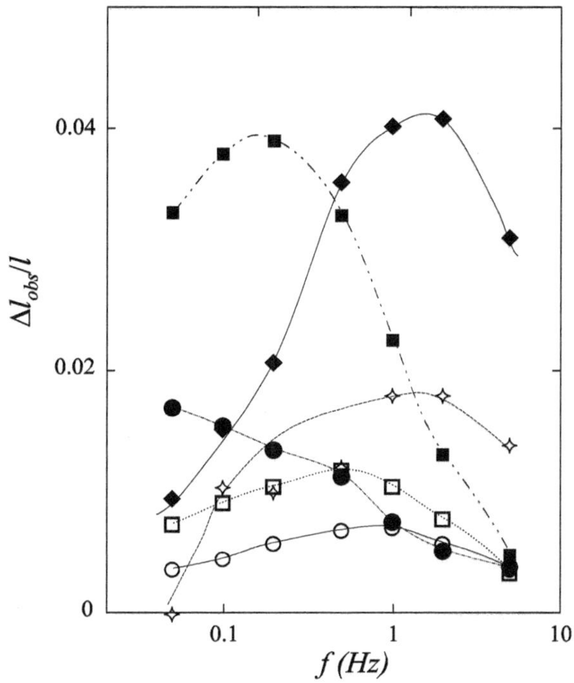

Figure 4.43
Change in the ratios of the intersphere distance divided by that in the absence of electric field for gel crystals when a sinusoidal electric field is applied at 25° C. $E = 2$ V/cm. P(MA-ST)/SiO_2 with gelator (#1) in 2-mm cell (*open circle*) and 5-mm cell (*filled circle*), P(MA-ST)/SiO_2 without gelator (#2) in 2-mm cell (*open square*) and 5-mm cell (*filled square*), CS91 (#3) in 2-mm cell (*open diamond*), and 5-mm cell (*filled diamond*). *From Shibata K, Kimura H, Tsuchida A, Okubo T, Sato S, Yoshinaga K:* Colloid Polym Sci *284:372–379, 2006.*

Fig. 4.43 shows the frequency (f) dependencies of $\Delta l_{obs}/l$, where l_{obs}/l means the change in the intersphere distance divided by that in the absence of electric field. The maximum $\Delta l_{obs}/l$ values in a 2-mm cell appeared at almost the same frequency as approximately 1 Hz. On the other hand, those in a 5-mm cell varied at less than 0.05 Hz for the gelated crystals (#1), approximately 0.1 Hz for the ungelated crystals (#2), and approximately 2 Hz for the reference CS91 crystals (#3). In summary, the optical response for the gel crystals is smaller than that for the ungelated colloidal crystals. This supports the finding that the optical response of the crystals is independent of the sphere mobility in an electric field.

It should be mentioned here that Tanaka et al. studied an electric field collapse effect on gel systems.[193] As described previously, electric field effects on the coloring of the anionic hydrogels were studied by Ueno et al.[175] Suzuki et al.[194] reported autonomously oscillating viscosity changes in microgel dispersions by introducing the ruthenium catalyst for the Belousov-Zhabotinsky reaction into a cross-linked copolymer. Weitz et al.[195] prepared gel crystals with enhanced thermal sensitivity at photonic wavelengths. Takeoka et al. synthesized inverse-opal type photonic gel crystals, which are thermo- and pH-sensitive.[196] Core-shell type hybrid microgels composed of pNIPAm and gold nanoparticles were synthesized to obtain complex, tunable optical materials.[197] In aqueous organic solvent mixtures, gel crystal structures of poly(NIPAm-co-allylamine) were studied.[198] Columnar single crystals were observed. Thermoresponsive photonic gel crystals were studied by Lyon et al.[199] Gel crystals of poly(4-vinylpyridine) microgels in ionic liquid were studied by Lyon et al.[200]

4.11 Type 3 Gel Crystals

Two-dimensional (2D) colloidal crystal arrays were prepared for pNIPAm beads (10 μm in diameter) by the double template polymerization method.[201] Ordered structural adsorption of pNIPAm gel spheres (≈230 nm in diameter) took place on the interface of droplets of hexadecane, toluene, and octanol (≈5-10 mm).[202] The emulsions emitted the iridescent colors. Study on the adsorbed and stabilized emulsions or foams by solid colloidal particles have been reported by Ramsden[203] and Pickering,[204] for example, more than 100 years ago. Silica sols, carbon black, and polystyrene particles were used as emulsifiers to prepare stable oil-in-water and water-in-oil emulsions.[205,206] Studies on stabilizing foams by the mixtures of solid particles with surfactants or surface-active polymers have been made,[205,207,208] but few reports of foams stabilized only by solid particles appeared.[203,205,209-212] Fujii et al.[213,214] found the formation of highly stable foams adsorbed and stabilized gel spheres with ordered polystyrene spheres in water. The foam suspensions emitted the light from the Bragg diffraction of the incident white light and were stable for more than 1 year. The pH-sensitive liquid marbles were prepared by adsorption of P2VP spheres on the liquid water.[215] Adsorption of cross-linked P2VP on the surfaces of gold particles was studied.[216]

Highly ordered 2D microgel arrays were obtained by compression in a Langmuir trough.[217,218] Adsorption of pNIPAm microgels at an oil–water interface was studied also by Richtering et al.[219,220] The tunable swelling and rolling of films assembled via layer-by-layer methods have been prepared from NIPAm-co-acrylic acid microgels and poly(ethylene imine). Depending on pH values, free-standing film or self-rolled scrolls were obtained.[221] Furthermore, microgel multilayers were studied by Lyon et al.[222] Modulation of the deswelling temperature of thermosensitive microgel films were also studied.[223] Peng et al.[224] studied change of the melting kinetics as a function of a layer's number of microgel film. Nozawa et al.[225] studied 2D gel crystal membranes mainly on their phase transition behaviors.

4.12 Type 3 Gel Crystals Observed in Drying Patterns

When any kind of solutions and suspensions are dried on a substrate, macroscopic and microscopic structures of solutes appear in the dried film. In other words, homogeneous dried film is never available. When gel crystal suspensions are dried on a substrate, crystal structures of gel spheres are kept in the dried film in most cases. Drying patterns form by the cooperative processes of (1) wetting of liquid drop on a substrate, (2) evaporation of medium (solvent) molecules, (3) convectional flow of solutes and solvent molecules, (4) sedimentary pattern formation, and (5) solidification. These processes are discussed for many kinds of solutions and suspensions in Chapter 5. We should note here that information about the suspensions and solutions (e.g., shape and size of solute and atmospheric humidity and temperature) is transferred into the drying patterns via the dissipative drying processes. Some information is transferred to the drying patterns accompanied by the selection and amplification effects by the static and dynamic interactions with other solutes or substrate surfaces. Typical convectional and sedimentary patterns observed for gel sphere systems are compiled in Tables 4.6 and 4.7.

Most gel spheres agglomerate on a substrate during the course of drying. The agglomerates are rather monodisperse and several kinds of ordered structures of the agglomerates further form. Drying patterns of gel sphere suspensions of pNIPAm[226-230] were studied on a cover glass, a watch glass, and a Petri glass dish. pNIPAm(70-5) gel spheres did not crystallize even in the

Table 4.6 Typical Convectional Patterns of Gel Spheres

Macroscopic Patterns	
Spoke lines (Terada cell)	pNIPAm, P2VP
Cluster	pNIPAm, P2VP
Bundles	P2VP

P2VP, Cationic gel spheres of poly(2-vinylpyridine); *pNIPAm,* thermosensitive gel spheres of poly(*N*-isopropylacrylamide).

Table 4.7 Typical Sedimentary Patterns of Gel Spheres

Macroscopic patterns	
Clusters	pNIPAm, P2VP
Bundles	P2VP
Inner ring and outside clusters	pNIPAm (e.g., coffee)

P2VP, Cationic gel spheres of poly(2-vinylpyridine); *pNIPAm,* thermosensitive gel spheres of poly(*N*-isopropylacrylamide).

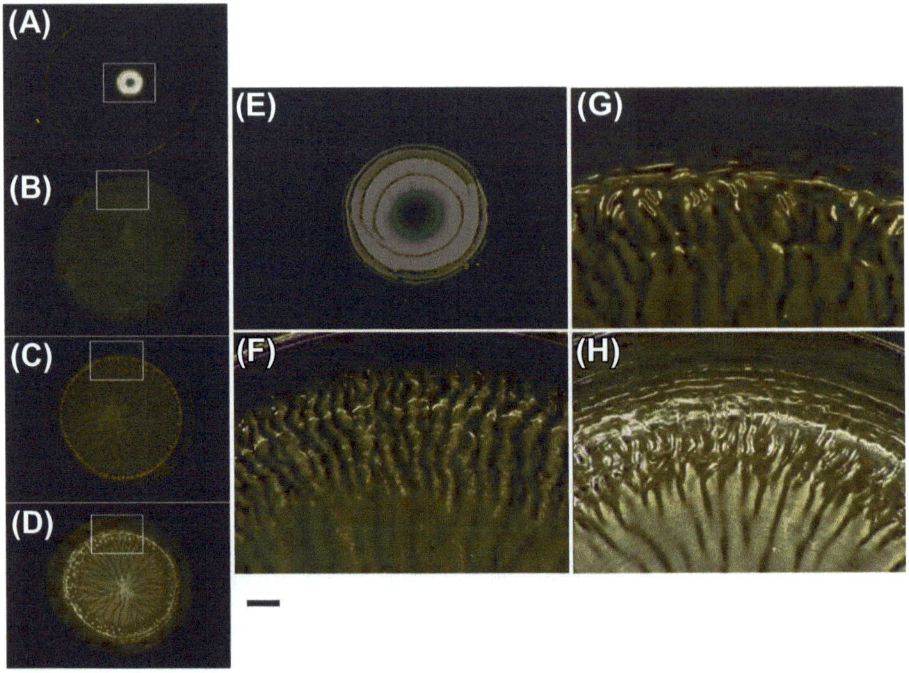

Figure 4.44 Drying patterns of pNPAm(70-5) at 25° C. In water, 0.05 ml; $w = 0.01$ g/ml; **A, E,** 25° C; **B, F,** 35° C; **C, G,** 40° C; **D, H,** 45° C; length of the bar is 2 μm (*A-D*); 0.35 μm (*E-H*). *Edited from Okubo T, Itoh E, Tsuchida A, Kokufuta E:* Colloid Polym Sci *285:339–349, 2006.*

deionized condition, because that size of pNIPAm(70-5) gel spheres is small and close to the limiting size of crystallization, and because their monodispersity is not high enough for crystallization.[226] However, drying patterns of pNIPAm(70-5) are highly interesting. As is shown in Fig. 4.44, drying patterns at 25° C extended to the beautiful flamelike spoke lines transitionally as temperature rose to 35° C, 40° C, and 45° C. The principal macroscopic patterns at 25° C were the single or multiple broad rings of the hill accumulated with the gels. At 45° C, on the other hand, the flamelike spoke-line patterns were observed at the inner area of the broad ring, especially at the gel concentrations higher than 1×10^{-3} g/ml. These observations support the finding that the extended gels at low temperatures are apt to associate weakly to each other, whereas the gels at high temperatures shrink and move freely with the convectional flow diffusion, although the very weakly affinitive intergel interactions still remain. The gravitational and Marangoni convectional flow of the gel spheres and the weak interactions between the gels and substrate (cover glass) are also important for the flamelike spoke-line pattern formation. In the presence of NaCl at high temperatures, the cooperative patterns formed between the gel spheres and the salt.

The most familiar macroscopic drying patterns of any kinds of solutions and suspensions are broad rings and spoke lines, as is described in Chapter 5.

Table 4.8 Six Groups of Macroscopic Drying Patterns and Their Relations with d_f/d_i and S Values

Group	Broad Ring	d_f/d_i	S	Affinitive Interaction	Typical Solutes
A	1BR (inner)	≈0-1, up with concentration	High	Weak intersolute, solute-substrate affinity	Polystyrene, silica spheres
B	2BR (outside) (inner)	1 ≈0-1, up with concentration	High Low	Strong solute-substrate affinity and weak intersolute affinity weak intersolute affinity	pNIPAm, lpNIPAm, PEGMA-P2VP, μPEGMA-P2VP spheres
C	2BR (outside) (inner)	1 ≈0-1, down with concentration	High Low	Strong solute-substrate affinity and strong intersolute affinity	pNIPAm(70-5), CAIBA-P2VP, AIBA-P2VP spheres
D	1BR (inner)	<0.1, down with concentration or keep constant	Low	Very strong intersolute affinity and weak solute-substrate affinity	NaHLA
E	1BR (outside)	1	High	Very strong solute-substrate affinity during solidification	Coffee, black tea
F	Central round hill 0 (or +1BR(outside))	(1)	Very low	Associated nonspherical solutes, (<10) structured solutes	Bentonite, tungstic acid, PGK gelatin, ARS, PBA spheres

AIBA-P2VP, CAIBA-P2VP, PEGMA-P2VP and mPEGMA-P2VP, cationic gel spheres of poly(2-vinylpyridine); arrow-root starch; *BR,* broad ring; *lpNIPAm,* linear-type poly(*N*-isopropylacrylamide); *NaHLA,* sodium salt of hyaluronic acid; *PGK,* palygorskite; *PBA,* poly(n-butyl acrylate); *pNIPAm,* thermo-sensitive gel spheres of poly(*N*-isopropylacrylamide).

The broad rings are seen in the outside area of the dried film in Fig. 4.44 (e.g., *D*). The flamelike spoke line is one kind of spoke line. One of the most convenient parameters showing the strength of the intersolute or solute-substrate affinities is d_f/d_i, where d_f is the final size of the broad ring in the dried film and d_i is the size of the initial liquid drop in diameter. The macroscopic drying in solutions and suspensions are grouped into six typical cases, as summarized in Table 4.8.

First, as shown in Fig. 4.44, *A,* one broad ring forms at the outside edge. When solute concentration is high, d_f/d_i is unity. However, the size of the broad ring decreases with decreasing solute concentration (i.e., d_f/d_i decreases as concentration decreases in the range of low concentrations). Most spherical particles such as colloidal polystyrene and silica spheres are grouped in this category. Whenever interparticle and particle-substrate affinities are weak, one size-changeable broad ring appears. Second, as shown in Fig. 4.44, *B,* two broad rings form at the outside edge and inner area. Furthermore, the size of the inner ring increases as initial solute concentration increases. This type of ring occurs when the associates form during the convectional process, but the solute–substrate interactions are weak. Formation of associates results in the appearance of the inner broad ring. pNIPAm, linear-type poly(*N*-isopropylacrylamide) (lpNIPAm), PEGMA-P2VP, and μPEGMA-P2VP are grouped in this category. Third, as shown in Fig. 4.44, *C,* two broad rings form at the outside edge and inner area as in Fig. 4.44, *B.*

However, the size of the inner rings decreases with increasing initial solute concentration. Small and cationic gel spheres and CAIBA-P2VP and AIBA-P2VP spheres are grouped here. For pNIPAm(70-5) spheres at high sphere concentrations, large associates form in part, and central round hills like those in Fig. 4.44, *F,* form in addition to the group *C* patterns. Fourth, as shown in Fig. 4.44, *D,* one small, broad ring appears in the central area of the dried film ($d_f/d_i < 0.1$) when very strong intersolute affinity exists, and the solute-substrate affinity is low. Sodium salt of hyaluronic acid is one of the typical examples. Fifth, as is shown in Fig. 4.44, *E,* one broad ring appears at the outside edge, irrespective of the initial solute concentration. This type of broad ring forms whenever a strong solute-substrate affinity exists, especially during the solidification process. The drying patterns of coffee and black tea are grouped in this category. Sixth, as shown in Fig. 4.44, *F,* a central round hill appears in addition to or instead of the broad ring. The patterns appear when large associates such as gelation, rubber structure, or aggregates of platelike particles occur. Diffusion of solutes is strongly restricted during drying. Suspensions of bentonite particles and solutions of gelatin and arrowroot starch are the typical systems in group *F.*

Sharpness of the thickness profiles of the broad ring is one of the convenient parameters showing the diffusivity of the solutes, intersolutes, and solute-substrate affinities during the course of the convectional process. Fig. 4.45 shows the thickness profiles of the dried film of pNIPAm(70-5) at 25° C.[226] A sharp, round hill appears in the central area in addition to the broad ring at the outside edge. These hills have not been observed for the typical colloidal spheres, except for strong interparticle affinity systems such as poly(methyl methacrylate) spheres. The rotational and translational

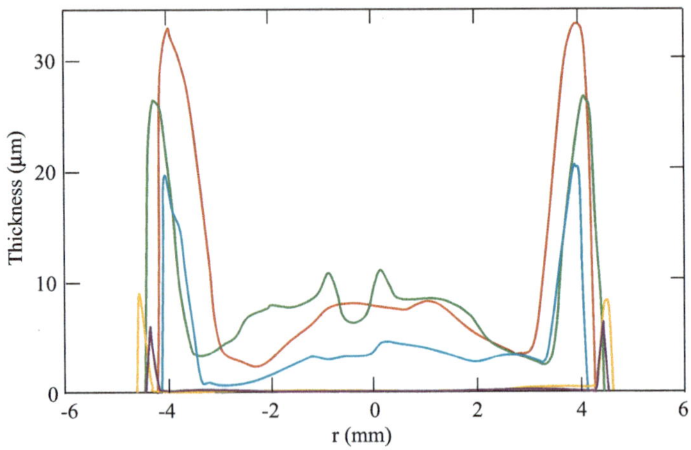

Figure 4.45
Thickness profiles of the dried film of pNIPAm(70-5) on a cover glass at 25° C. In water, 0.05 ml. *Red curve:* 0.02 g/ml; *green curve:* 0.01; *blue curve:* 0.0075; *orange curve:* 0.001; *purple curve:* 0.0005. *Edited from Okubo T, Itoh E, Tsuchida A, Kokufuta E:* Colloid Polym Sci *285:339–349, 2006.*

movement will be highly restricted for pNIPAm(70-5) gel spheres because they are hairy, very soft, and affinitive to the substrate. Similar round hills appeared only for the nonspherical colloidal particles such as bentonites[231] and gelated solutions of gelatin and arrow-root starch.[232] Thus pNIPAm(70-5) gel spheres are grouped into *C* and *F*.

Sharpness of the thickness profiles (*S*) of the macroscopic broad rings in the dried film is evaluated by the ratio of film size in diameter (*D*) against the full width at half maximum (*FWHM*) given in Equation 25.

$$S = D/FWHM \qquad 25$$

When the peak profile is normal distribution, *FWHM* is given by Equation 26, where *s* is the corresponding standard deviation.[233]

$$FWHM = 2.355\ s \qquad 26$$

Another kind of sharpness parameter, *S′*, is also evaluated from the ratio of the dried film size against the width of the broad ring (*WBR*) at the bottom layer in the thickness profiles shown in Fig. 4.45.

$$S' = (D/WBR) \times 2 \qquad 27$$

Here, *FWHM* is assumed to be half of *WBR*; in other words, the peak profiles are approximated to be isosceles triangle–shaped. It has been proved from comparison of the peak profiles of the broad rings and their thickness at the bottom layer for the many drying systems that *S* and *S′* give similar values within experimental errors to each other (see Chapter 5). *S* and *S′* are convenient parameters showing the degree of diffusion of the colloidal particles, including gel spheres, during the course of drying on a substrate. Fig. 4.46 shows the typical examples of *S* values of colloidal and gel particles as a function of their volume fraction in the deionized aqueous suspension. *S* values decreased as concentration of solutes increased in most cases, as is clear in Fig. 4.46. *S* values changed significantly depending on the kind of the particles, as is shown in the figure. *S* values in Fig. 4.42 are divided into four groups, A-C, D, E, and F, as is shown in the figure. Here, the each groups A through F are the same as those cited in Table 4.8. It should be mentioned here that the differences in the *S* values among the groups A, B, and C look similar in the figure. However, there exist rather large differences among the groups, when the *S* values of each group are discussed more in detail later in this chapter.

Fig. 4.47 shows the typical examples of the macroscopic drying patterns of aqueous suspensions of pNIPAm(200-5) gels at 20° C (*A-D*) and 45° C (*E-H*) as a function of gel concentration ranging from 0,039 wt% to 2.5 wt%.[227] Several important findings are clear from the figures. First, two kinds of broad rings appeared: the transparent broad rings at the outside edge and whitish or bluish ones at the inner area, especially at low temperature. Many parts of the drying patterns were transparent and the pictures were not

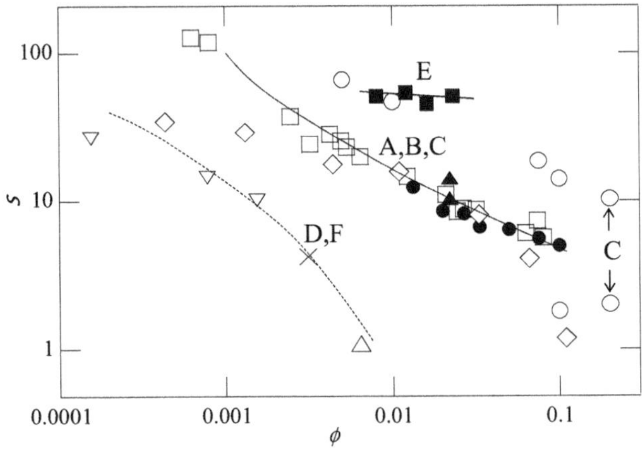

Figure 4.46
Sharpness of the broad ring (*S*) as a function of volume fraction of gel colloidal particles. 0.1 ml, 25° C; *circles:* pNIPAm(70-5); *squares:* PMMA; *solid circles:* CS300; *solid triangles:* D1W52, DIB28; *solid squares:* coffee, black tea with cream; *crosses:* palygorskite; *triangles:* tungstic acid; *reversed triangles,* bentonite, diamonds PBA.

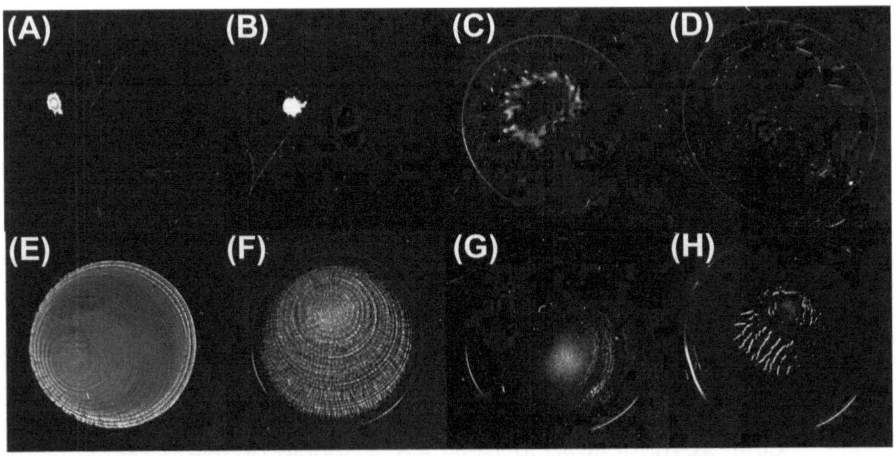

Figure 4.47
Drying patterns of pNIPAm(200-5) gel spheres on a cover glass at 20° C (*A-D*) and 45° C (*E-H*). Humidity = 35 %, 0.1 ml; **A, E,** 0.03888 wt%; **B, F,** 0.194 wt%; **C, G,** 0.78 wt%; **D, H,** 2.52 wt%. *From Okubo T, Suzuki D, Yamagata T, Katsuno A, Mizutani M, Kimura H, Tsuchida A:* Colloid Polym Sci *289:807–816, 2011.*

clear. Coexistence of the two broad rings supports the finding that the single pNIPAm gels and agglomerated gel particles formed at the outside edge and the inner regions during the course of drying. The broad rings at the outside edge were insensitive to the gel concentrations. This supports the finding that the single (nonagglomerated) pNIPAm gel spheres are hygroscopic and the drying frontier started and stayed at the outside edge of the liquid during the whole course of drying by the rather strong affinitive interaction of the hygroscopic gels with the glass substrate. It is highly plausible that the large pNIPAm gels are much hygroscopic than the small ones. The broad ring patterns of small pNIPAm(70-5) gels were composed of the two broad rings, as described previously (see Figs. 4.44 and 4.45).[226] The inner rings of pNIPAm(200-5), on the other hand, decreased as the gel concentration decreased. pNIPAm(200-5) gel spheres are grouped into type B, especially at 20° C. The agglomerated particles of pNIPAm(200-5) gels are highly plausible to be formed during the drying processes of the crystalline suspension by the cooperated forces between the hydrophobic intersphere and also hydrophilic sphere–substrate interactions.

The flamelike spoke lines were observed especially at low temperatures and at high gel concentrations, although the patterns were distorted substantially. This supports the finding that the movement of the gel spheres is disturbed by the weak affinitive interactions between the extended gel spheres in the convectional flow processes at low suspension temperatures. As described previously, the beautiful patterns of the flamelike spoke lines were observed at the high temperatures for pNIPAm(70-5) gel spheres.[226] It should be mentioned here that the dried films of pNIPAm(200-5) gel spheres at 45° C were whitish-blue in most cases. In the dried state, the effective size of gel spheres is close to size at 45° C (110 nm in diameter). These sizes are too small to observe the iridescent colors by Bragg diffraction. However, formation of the crystal-like 2D distribution is highly likely to occur on the substrate. The crystal structures were maintained during the course of drying processes in most cases. Fig. 4.48 compares the

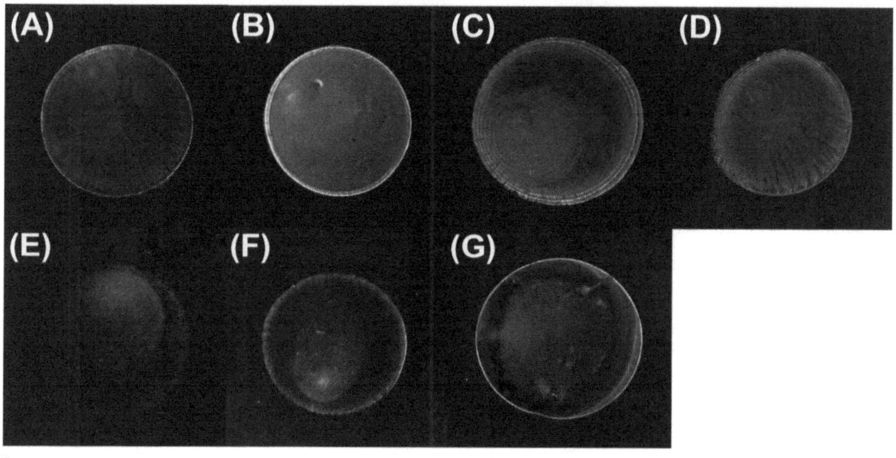

Figure 4.48
Drying patterns of **A,** pNIPAm(200-0.5) (0.084 wt%); **B,** pNIPAm(200-2) (0.039 wt%); **C,** pNIPAm(200-5) (0.039 wt%); **D,** pNIPAm(200-10) (0.047 wt%); **E,** pNIPAm(400-5) (0.078 wt%); **F,** pNIPAm(600-5) (0.078 wt%); and **G,** pNIPAm(1500-5) (0.041 wt%) at 45° C.

macroscopic drying patterns of pNIPAm(200-0.5),[228] pNIPAm(200-2),[229] pNIPAm(200-5),[227] pNIPAm(200-10),[229] pNIPAm(400-5),[228] pNIPAm(600-5),[228] and pNIPAm(1500-5)[228] at high temperature. The flamelike spoke lines are observed for most of the samples (see especially Fig. 4.48, *A, D-F*). The faint iridescent colors were observed for all the samples, especially for large gels, which support the finding that each single state of gel spheres keep the crystal structure even at the dried film.

Fig. 4.49 shows the ratios of the broad ring size of the dried film, d_f, against the size of the initial liquid, d_i. The open and closed symbols indicate the ratios at the outside edge and those at the inner area, respectively. Coexistence of the broad rings at the outside edge and a round hill at the inner area for pNIPAm(200-5)[227] is similar to the observation of pNIPAm(70-5).[226] pNIPAm gel spheres are hygroscopic and the drying frontier begins and stays at the outside edge of the initial liquid during the whole course of drying because of the strong affinitive interaction of the hygroscopic gels with the

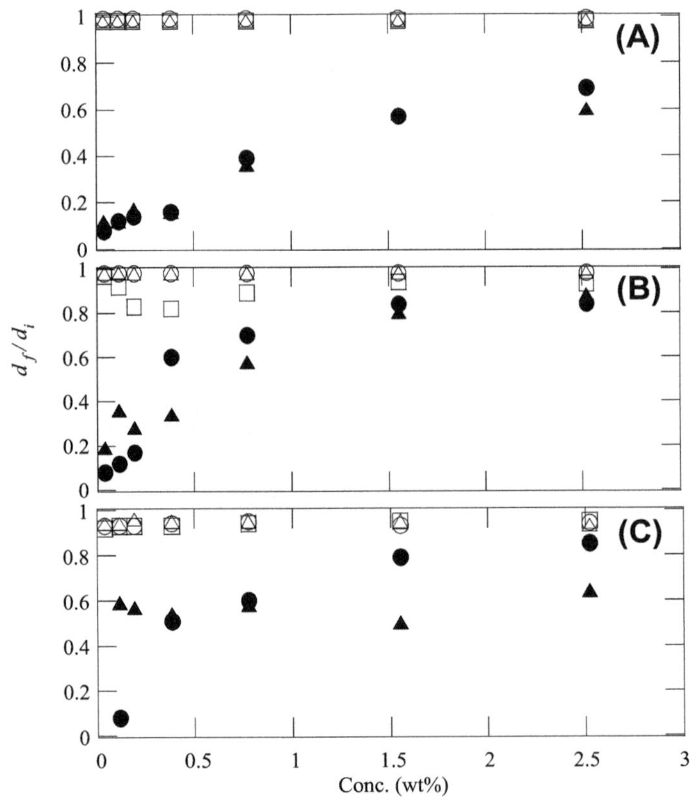

Figure 4.49
d_f/d_i values of pNIPAm(200-5) gel spheres on a cover glass (*A*), a watch glass (*B*), and a Petri glass dish (*C*) at 20° C (*open circles:* outer broad ring; *filled circles:* inner broad ring), 33° C (*open triangles:* outer; *filled triangle:* inner), and 45° C (*open square:* outer; *filled square:* inner). *From Okubo T, Suzuki D, Yamagata T, Katsuno A, Mizutani M, Kimura H, Tsuchida A:* Colloid Polym Sci *289:807–816, 2011.*

glass substrate. The d_f/d_i values at the inner area increased as the gel concentration increased. This tendency has been observed often for most colloidal suspensions and polymer solution. The agglomeration of pNIPAm(200-5) gels is clarified from the microscopic patterns, as described later in this chapter. The broad ring inner area of the films corresponds to the formation of the agglomerates during the course of drying. It should be noted here that the coexistence of the two broad rings and an increased tendency of d_f/d_i with increasing sphere concentrations were observed for most pNIPAm gel spheres examined.[226-230] Thus pNIPAm gel spheres show type B macroscopic patterns. Formation of the two kinds of broad rings has been observed in a few systems so far among the nongel systems—i.e., polymer complex suspensions of poly(acrylic acid) and poly(ethylene glycol),[234] binary and ternary mixtures of monodispersed colloidal silica spheres of different sizes,[235,236] and colloidal crystals of poly(methyl methacrylate) spheres.[237]

Sharpness parameters S' of a series of pNIPAm are shown in Figs. 4.50 and 4.51 as a function of sphere concentration in weight percent. The influence of the degree of cross-linking is shown in Fig. 4.50. S' increased and then turned to decrease as the degree of cross-linking increased in the order of Equation 25 irrespective of suspension temperature.

$$\text{pNIPAm(200-0.5)} \leq \text{pNIPAm(200-2)} < \text{pNIPAm(200-5)} > \text{pNIPAm(200-10)} \qquad 25$$

Spherical pNIPAm gels with a low degree of cross-linking are considered to be more affinitive to the cover glass substrate. The first increase in S' with a

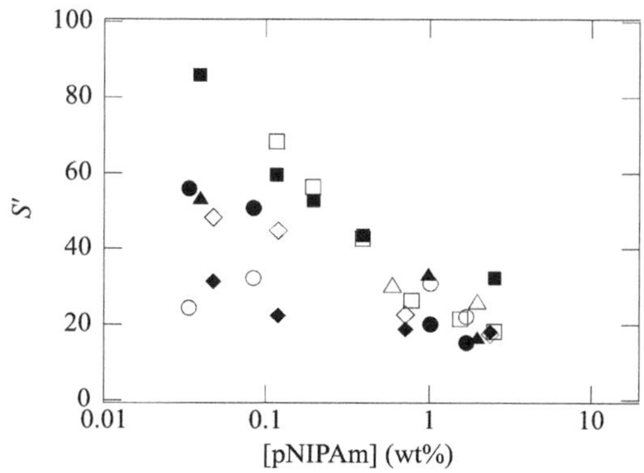

Figure 4.50
Sharpness parameter, S' of pNIPAm as a function of sphere concentration in weight percent. 0.1 ml, *open symbols:* at 20° C; *solid symbols:* at 45° C; *circles:* pNIPAm(200-0.5); *triangles:* pNIPAm(200-2); *squares:* pNIPAm(200-5); *diamonds:* pNIPAm(200-10).

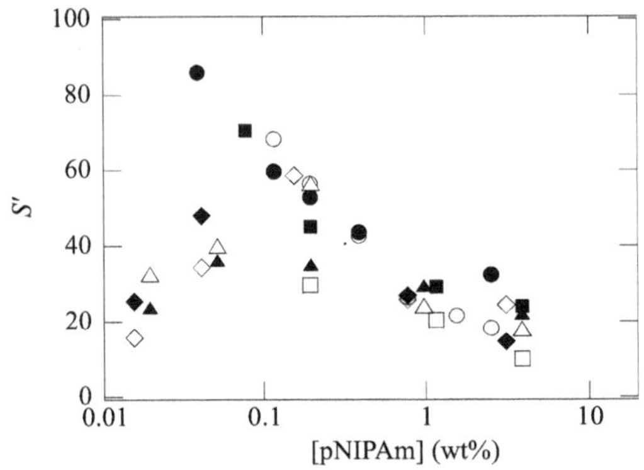

Figure 4.51
Sharpness parameter, S' of pNIPAm as a function of sphere concentration in weight percent. 0.1 ml, *open symbols:* at 20° C; *solid symbols:* at 45° C; *circles:* pNIPAm(200-5); *triangles:* pNIPAm(400-5); *squares:* pNIPAm(600-5); *diamonds:* pNIPAm(1500-5).

degree of cross-linking is mainly caused by a decrease in the gel sphere-substrate affinity, when gel sizes are similar. In relation to turning to decrease in S' from 5 to 10 mol% of degree of cross-linking, formation of agglomerates is most plausible by the rather high intersphere affinity for pNIPAm(200-10).

Influence of pNIPAm gel size upon S′ values is shown in Fig. 4.51. Interestingly, S' decreased as gel size increased, irrespective of suspension temperature, as in Equation 26.

$$\text{pNIPAm(200-5)} > \text{pNIPAm(400-5)} \geq \text{pNIPAm(600-5)} \approx \text{pNIPAm(1500-5)} \quad (26)$$

Equation 26 is consistent with a decrease in the diffusivity of pNIPAm spheres with increasing sphere size.

Interestingly, formation of the monodispersed agglomerated particles (10 to 50 μm in most cases) and their ordered arrays were observed only for pNIPAm and poly(*N*-2-vinylpyridine) P2VP gel crystals. These microscopic ordered structures were (1) ordered spoke lines, (2) ordered rings, (3) net structure, and (4) lattice-like ordered structures of the agglomerated particles. Figs. 4.52 and 4.53 compile the images of these four typical kinds of microscopic patterns for pNIPAm and P2VP gel spheres. The net and lattice structures formed more favorably at high degrees of cross-linking, at high concentrations of gels, at high temperatures, and for large gel spheres. Clear-cut ordered spoke lines and the ordered rings were observed only for small hydrogel spheres (e.g., see Fig. 4.52, *A-C* and *I*). It is plausible that these four kinds of microstructures form successively throughout the

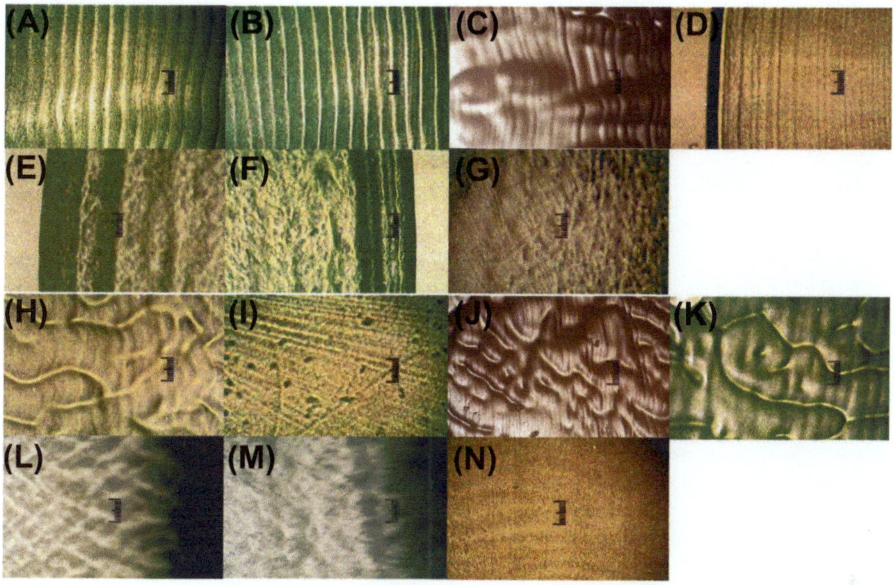

Figure 4.52
Microscopic drying patterns of pNIPAm gel spheres showing ordered ring (*A-G*) and ordered spoke line (*H-N*) structures at 45° C. **A,** pNIPAm(200-0.5), 0.084 wt%, watch glass; **B,** pNIPAm(200-2), 0.039, glass dish; **C,** pNIPAm(200-5), 1.55, watch; **D,** pNIPAm(200-10), 2.37, cover glass; **E,** pNIPAm(400-5), 3.89, cover; **F,** pNIPAm(600-5), 3.88, cover; **G,** pNIPAm(1500-5), 3.10, watch; **H,** pNIPAm(200-0.5), 1.01, cover; **I,** pNIPAm(200-2), 0.59, dish; **J,** pNIPAm(200-5), 1.55, watch; **K,** pNIPAm(200-10), 0.71, watch; **L,** pNIPAm(400-5), 0.97, dish; **M,** pNIPAm(600-5), 3.88, dish; **N,** pNIPAm(1500-5), 0.78, dish.

convectional and sedimentary processes, as has been discussed previously.[227-230,238-240] The ordered rings and ordered spoke lines form for the first stage of convection and sedimentation. Next, the net structure form by the coupling of the ordered rings and the ordered spoke lines. The final and completed stage is the formation of the lattice pattern of the agglomerates. Depending on the experimental conditions, such as gel sphere size, gel concentration, degree of cross-linking of the gels, and the substrates used, for example, one or several patterns are fixed in the dried film. Here, the cooperative interaction between the agglomerated gel spheres and substrate is also important during the course of drying. It should be noted here that the microscopic patterns on a glass dish and a watch glass were clearer compared with those on a cover glass. The main causes for the appearance of the ordered microstructures are the important role of the convectional flow processes of the gels and the stable formation of the electrical double layers around the agglomerated pNIPAm and P2VP gel spheres.

Figs. 4.54 and 4.55 demonstrate the microscopic drying patterns taken by a scanning probe microscope (SPM), except Fig. 4.55, *F*, which was taken by

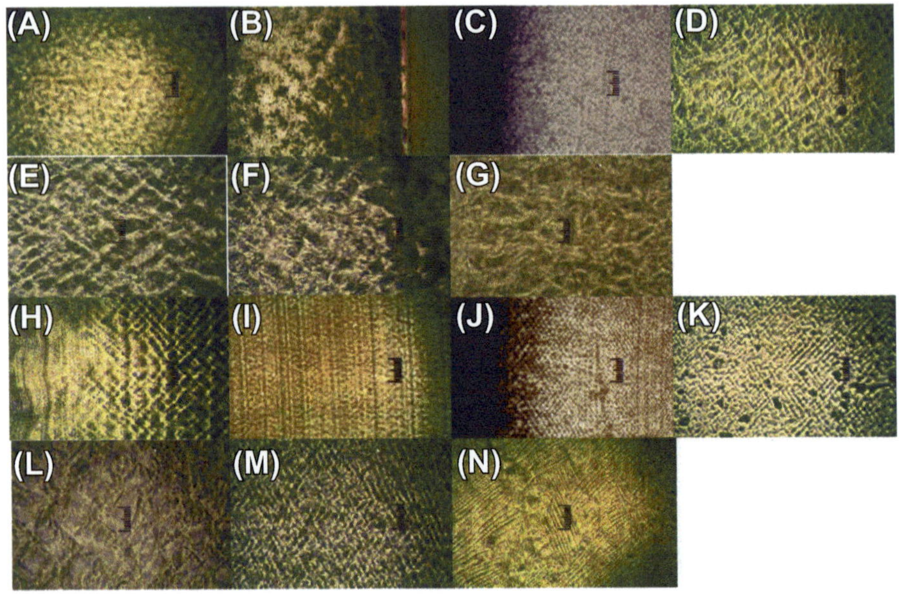

Figure 4.53
Microscopic drying patterns of pNIPAm gel spheres showing "net" (*A-G*) and lattice (*H-N*) structures at 45° C. **A,** pNIPAm(200-0.5), 1.68 wt%, glass dish; **B,** pNIPAm(200-2), [NaCl] = 0.008 M, 0.099 wt%, watch glass; **C,** pNIPAm(200-5), 1.55, glass dish; **D,** pNIPAm(200-10), 2.37, dish; **E,** pNIPAm(400-5), 3.89, dish; **F,** pNIPAm(600-5), 3.88, dish; **G,** pNIPAm(1500-5), 3.10, cover glass; **H,** pNIPAm(200-0.5), 1.01, watch; **I,** pNIPAm(200-2), 0.59, dish; **J,** pNIPAm(200-5), 0.78, dish; **K,** pNIPAm(200-10), 0.71, dish; **L,** pNIPAm(400-5), 3.89, watch; **M,** pNIPAm(600-5), 3.88, dish; **N,** pNIPAm(1500-5), 3.10, dish.

an optical microscope. Fig. 4.48, *A* and *B,* are observed for pNIPAm(200-10) at 0.0474 wt% and dried at 45° C.[229] The lengths of the full scales in the figures are 1 μm and 100 nm, respectively. Fig. 4.48, *A,* shows the array of the agglomerated particles (0.2 to 0.5 μm in diameter). Fig. 4.48, *B,* shows the appearance of single gel particles ranging 100 to 200 nm in diameter in the dried state. The microscopic examination of pNIPAm(200-10) dried at 20° C are shown in Fig. 4.48, *C* and *D,* where the lengths of the full scales are also 1 μm and 100 nm, respectively. The arrays of the agglomerated particles and the single gel particles are observed, respectively, although the pictures dried at 20° C (*C* and *D*) were not so clear compared with those at 45° C (*A* and *B*). The similar SPM observations were obtained for pNIPAm(200-2) samples, although the figures were less clear compared with those of pNIPAm(200-10). The reason is ascribed to the fact that pNIPAm gel spheres with a lower degree of cross-linking are softer than the spheres with a higher degree of cross-linking in suspension and also in the dried state.

Fig. 4.55 shows the typical examples of the microscopic pictures of pNIPAm(400-5) (*A, B*), pNIPAm(600-5) (*C, D*), and pNIPAm(1500-5) (*E, F*); Fig. 4.55, *A* through *E,* were observed by SPM, and Fig. 4.55, *F,* was observed

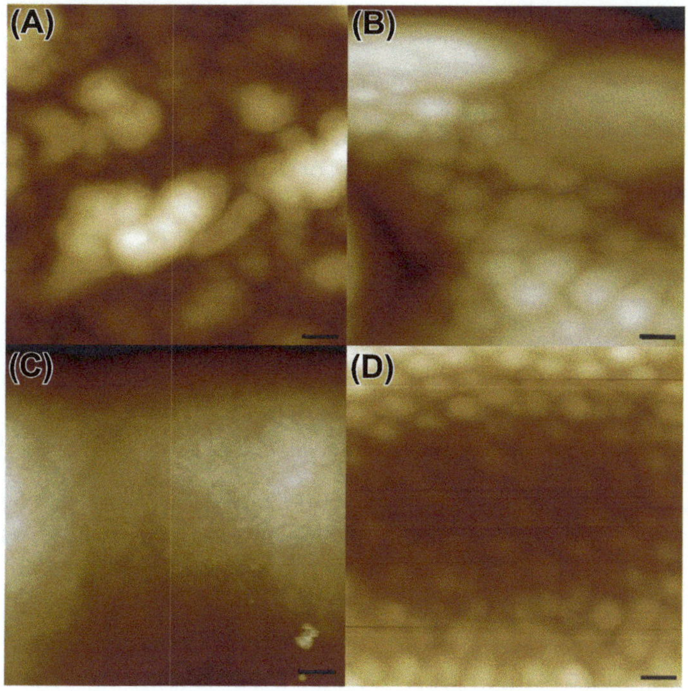

Figure 4.54 Microscopic drying patterns observed by SPM of pNIPAm(200-10) spheres on a cover glass at 45° C (*A, B*) and at 20° C (*C, D*), 0.047 wt% (*A, B*), 0.71 wt% (*C, D*); full scale is 1.15 μm (*A, C*) or 115 nm (*B, D*). *From Okubo T,Suzuki D, Tsuchida A:* Colloid Polym Sci *290:867–877, 2012.*

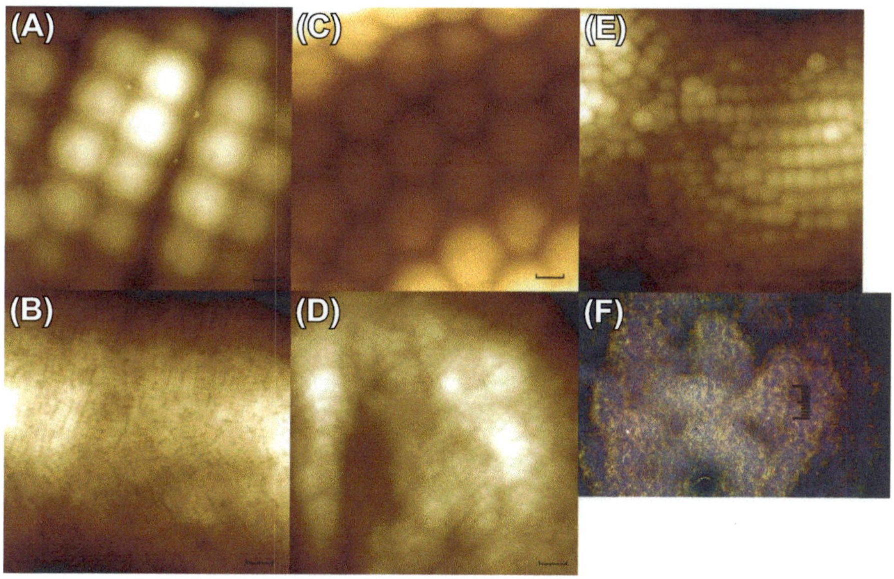

Figure 4.55 Microscopic drying patterns observed by SPM of pNIPAm(400-5) spheres (*A, B*); pNIPAm(600-5) (*C, D*); and pNIPAm(1500-5) (*E, F*); cover glass; **A, B,** 0.97 wt%; **C, D,** 1.16 wt%; **E, F,** 0.78 wt%; full scale is 100 μm (*A, C*), 1 μm (*B, D, E*) or 10 μm (*F*). *From Okubo T, Suzuki D, Tsuchida A:* Colloid Polym Sci *290:1901–1911, 2012.*

with a conventional microscope on a cover glass.[230] Fig. 4.55 *A, C,* and *E* show the crystal-like ordering of the single gel spheres, whereas Fig. 4.55 *B, D,* and *F* show the distribution of the agglomerates composed of the corresponding single gel spheres. Agglomeration of the single gel spheres took place for all sizes of pNIPAm gel spheres ranging from 200 nm to 1500 nm and also different degrees of cross-linking (0.5 to 10 mol%). The important role of the cooperative convectional diffusion flow of the single gel spheres keeping the weak interaction with the substrate must be essential for the formation of the monodispersed agglomerates in the convectional processes during drying.

Thus it is clarified that monodispersed agglomerates formed only for gel systems among many solutions and suspensions examined hitherto. As is shown in Fig. 4.56 schematically, the ordering of the agglomerates takes place in the following order: (1) ordered ring, and (2) ordered spoke lines form; by the coupling of the ordered patterns (1) and (2), (3) net structure and finally the (4) lattice structures form. These four kinds of ordering modes of the agglomerates seem to proceed favorably at high temperature and sphere concentration and for larger gel spheres in general.

Linear-type homopolymer, lpNIPAm, is well known to show lower critical solution temperature (LCST) in an aqueous solution at approximately 33° C. Furthermore, a clear transition of conformation from coil to globule has been clarified above the critical temperature. Colloidal particles of the globule aggregates form reversibly from lpNIPAm above LCST. Below LCST, colloidal suspension transforms to the transparent solution of coil-state lpNIPAm. To understand the similarity and difference between pNIPAm gel spheres and the aggregates of lpNIPAm, the drying patterns were studied for lpNIPAm at 22° C and 50° C.

Fig. 4.57 shows the histogram of the size distribution of the globule aggregates of lpNIPAm measured at 47.5° C from DLS measurements.[241] The histogram at 152 nm was sharp. ζ-Potential was −22.7 mV from ELS measurements. The negatively charged lpNIPAm aggregates originate from the initiators used in the polymer synthesis processes and are further due to the spontaneous charging effects at the surfaces of the globule aggregates in water, as discussed in Chapter 2.

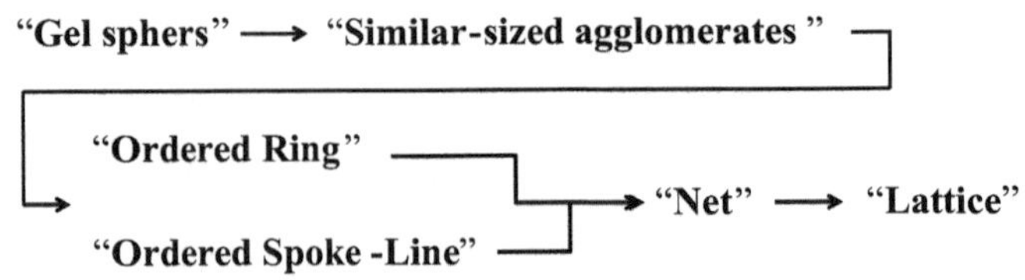

Figure 4.56
Schematic representation of ordering of agglomerates during drying gel crystals.

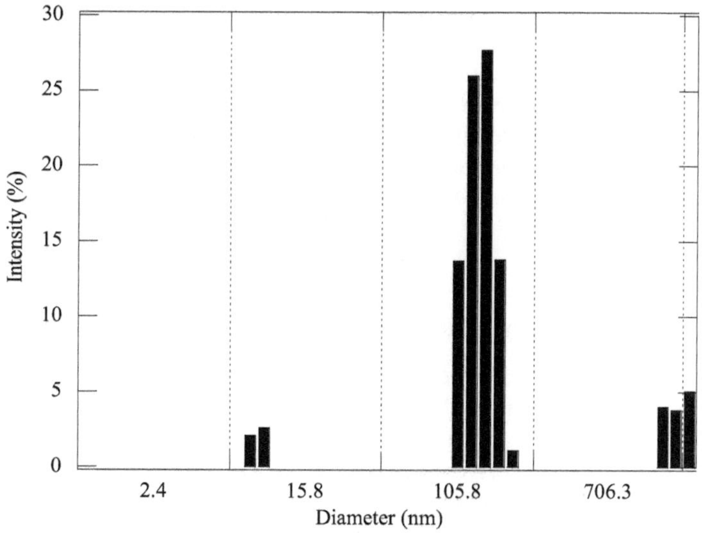

Figure 4.57 Size distribution histogram of aggregates of IpNIPAm at 47.5° C. *From Okubo T, Tsuchida A:* Colloid Polym Sci *292:133–141, 2014.*

Fig. 4.58 shows typical examples of the macroscopic drying patterns of IpNIPAm solution on a watch glass at 22° C (*A-D*) and 50° C (*E-H*).[241] At low temperature, two broad rings appeared as polymer concentration increased, which is typical and has been observed often. This observation supports the excluded volume effects of large associates originated from the coil-state polymers in solution. Interestingly, on a watch glass, the size of the larger broad ring decreased with increasing polymer concentration. This tendency is ascribed to the fact that interpolymer interaction between the coil state of

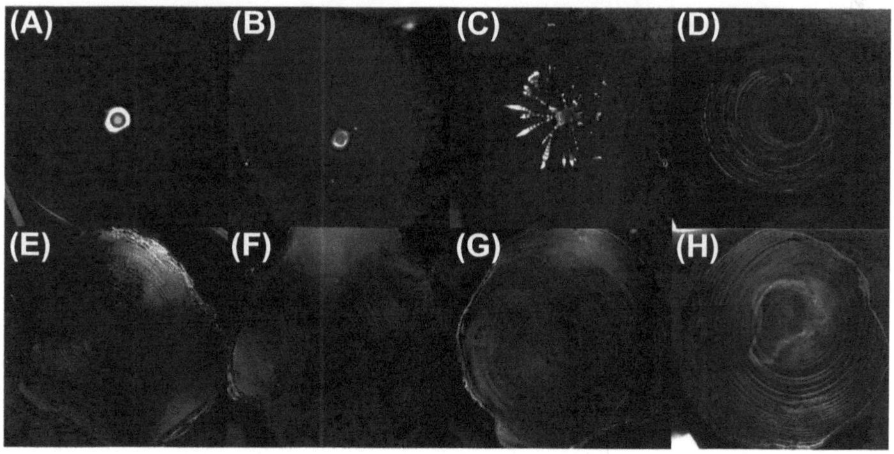

Figure 4.58 Drying patterns of IpNIPAm on a watch glass (0.7 ml, 33 mm × 33 mm) at 22° C (*A-D*) and 50° C (*E-H*). **A, E,** [IpNIPAm] = 0.05 monoM; **B, F,** 0.02 monoM; **C, G,** 0.005 monoM; **D, H,** 0.0005 monoM. *Edited from Okubo T, Tsuchida A:* Colloid Polym Sci *292:133–141, 2014.*

lpNIPAm polymers is strong and the intercoil association takes place more easily during the course of drying on the substrate. Thus most polymers are in an associated state and move toward the central area. The interpolymer interaction should be more or less weakened at the low initial polymer concentrations, because the neighbored interpolymer distance increases with decreasing polymer concentration. At high temperature, drying patterns are whitish and wide compared with those at 22° C. Globule agglomerates of lpNIPAm become significantly compacted compared with the coil-state polymer chain in solution state at low temperature. It is plausible that the interagglomerates' affinitive interaction may still remain. Their drying patterns are consistent with the coil-globule aggregation transition.

Fig. 4.59 shows typical examples of the microscopic drying patterns of lpNIPAm from the central to the outside edge on a cover glass (*A-D*), a watch glass (*E-H*), and a Petri glass dish (*I-L*) at 50° C.[241] Several macroscopically ordered structures are observed. First, the ordered rings are recognized partially (see *C, G,* and *I*). The ordered ring patterns appeared much more clearly for the thermosensitive gel spheres of pNIPAm.[227-230] These rings support the finding that the agglomerates–substrate interaction is strong during drying. It is plausible that the agglomerated particles of the globule aggregates are already formed in the beginning stage of the drying processes and their ring-array forms first. These ordered rings form mainly by the convectional flow of the agglomerated particles during the course of drying. Second, the ordered spoke lines are observed (e.g., see *C, D, F,* and *J*). The inter–spoke line distances were long in the order of several ten micrometers. Third, the net structures appeared on all the substrates. The size of the net structures may

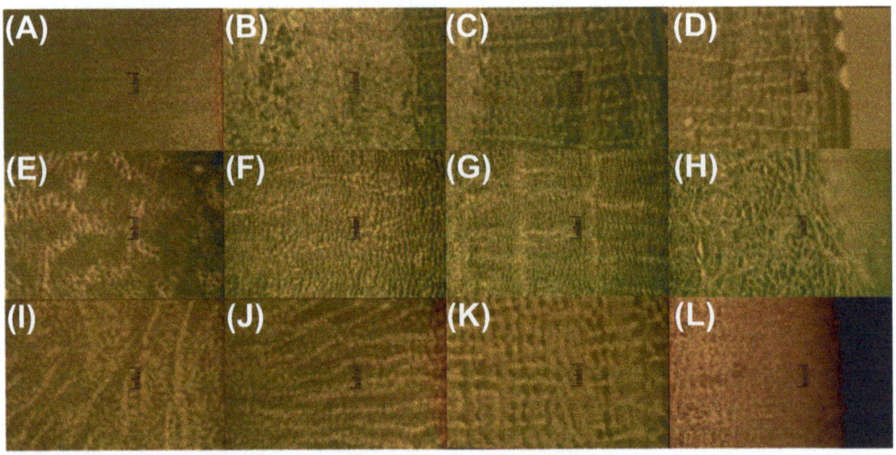

Figure 4.59
Microscopic drying patterns of lpNIPAm on a cover glass (*A-D*), a watch glass (*E-H*), and a glass dish (*I-L*) at 50° C. [lpNIPAm] = 0.05 monoM, **A, E, I** to **D, H, L** are from the center to the right edge; the length of the full scale inserted in each picture is 100 μm. *From Okubo T, Tsuchida A:* Colloid Polym Sci *292:133–141, 2014.*

correspond to the size of the agglomerates. Cooperative convectional flow of the large agglomerates resulted in the large scale of the patterns. It should be noted here that most net structures formed along the spoke lines or rings, as shown in Fig. 4.59, *D, G,* and *K,* for example. The net structures seem a bit vague compared with those of pNIPAm gel spheres. Fourth, the lattice structures are not observed for lpNIPAm. This result is consistent with the incomplete convection flow movement of lpNIPAm agglomerates compared with corresponding pNIPAm spherical gels. Thus patterns of agglomerated particles were observed to be similarly oriented for lpNIPAm and pNIPAm gel spheres. This supports the finding that the orientation of the agglomerated spheres is common between the two systems. The main cause is that the chemical components of the homopolymers and gel spheres are almost the same. The role of the electrical double layers around the agglomerates and the interaction with the substrates are also important for the ordering. lpNIPAm is a type B solute, as are pNIPAm gel spheres.

Fig. 4.60 shows typical examples of very large dendritic patterns (from 50 to 600 μm in size) formed in the presence of NaCl at 22° C and 50° C. These microscopic drying patterns of gel spheres, including aggregates of lpNIPAm, are characteristic to the gel systems exclusively and different from those of typical linear-type polymers and also from typical colloidal spheres.

Drying patterns of pNIPAm spheres were reported to show the iridescent colors on a polystyrene plate.[242] Horigome and Suzuki prepared 2D gel crystals of pNIPAm by drying on a polystyrene plate. Hydrodynamic diameter was 1190 and 430 nm at 25° C and 40° C, respectively, and sphere

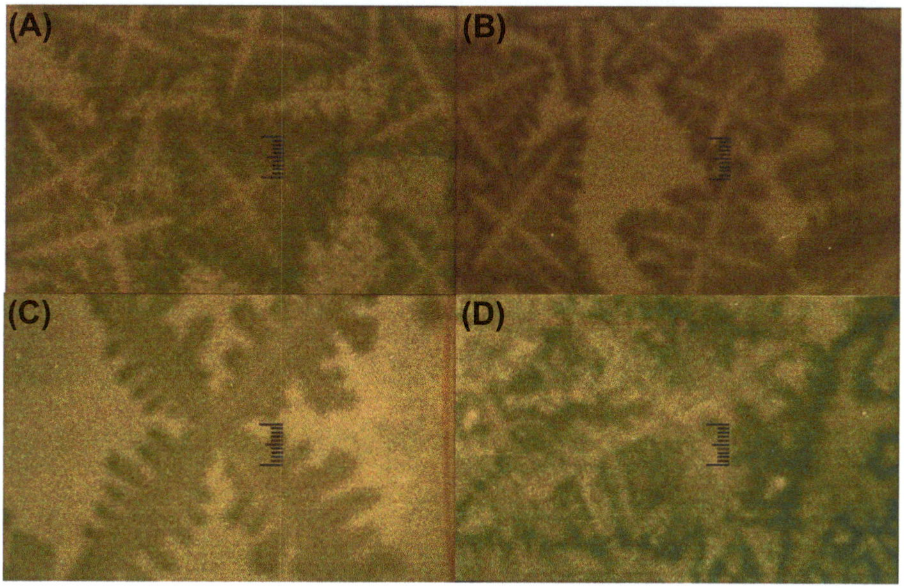

Figure 4.60 Microscopic drying patterns of lpNIPAm + NaCl. **A,** cover glass, [NaCl] = 0.00167 M, 22° C; **B,** cover glass, 0.00167, 22° C; **C,** watch glass, 0.005 M, 22° C; **D,** glass dish, 0.00833 M, 50° C. [lpNIPAm] = 0.00833 monoM. *Edited from Okubo T, Tsuchida A:* Colloid Polym Sci *292:133–141, 2014.*

Figure 4.61
Drying pattern of pNIPAm gel spheres (diameters are 1190 nm and 430 nm at 25° C and 40° C, respectively) on a polystyrene plate at 25° C. 0.0009 wt%. *Courtesy Professor D. Suzuki.*

concentrations were in the range of 0.0006 wt% to 0.002 wt%. A typical example is shown in Fig. 4.61. The picture is courtesy Professor D. Suzuki. A broad ring seems to exist at the outside edge, but it is not recognized clearly.

We now turn to the macroscopic drying patterns of cationic P2VP gel spheres, which are mainly composed of P2VP. Their characteristics are listed in Table 4.3. The hydrodynamic size of these gel spheres was clearly larger in acidic compared with alkaline (basic) conditions.[122,216] The samples are grouped into the gel spheres. Furthermore, fractions of water in P2VP spheres in the suspension state were estimated to be in the range of 0.16 to 0.75 (see f_W values in Table 4.3). Changes in the dissipative patterns during the course of drying for CAIBA-P2VP(0.5), AIBA-P2VP(0.5), PEGMA-P2VP(0.5) and μPEGMA-P2VP(0.5) are shown in Figs. 4.60 to 4.63, respectively. These suspensions of cationic gel spheres were dried on a watch glass at 22° C or 25° C. Clearly, the patterns vary greatly depending on sphere size. For CAIBA-P2VP(0.5), spokelike and clusterlike convectional patterns appeared from 3 hours, 30 minutes (Fig. 4.62, *B*) to 5 hours, 40 minutes (Fig. 4.62, *C*). Furthermore, the sedimentary pattern including clusters and bundles is observed in Fig. 4.62, *D*. The bluish colors of the suspension state are ascribed to the Bragg diffraction of white light by the lattice structures in suspension. A small, whitish broad ring exists at the outside edge, although the film is slightly transparent and is not recognized clearly. The width of the main ring at the outside edge is observable in part around the letter *E* in the figure.

Fig. 4.63 shows typical examples of change in the macroscopic patterns during the course of drying of AIBA-P2VP(0.5) spheres. Whole regions of the suspension state displayed beautiful iridescent colors during drying. Several patterns induced by the convectional flow are recognized from change in the

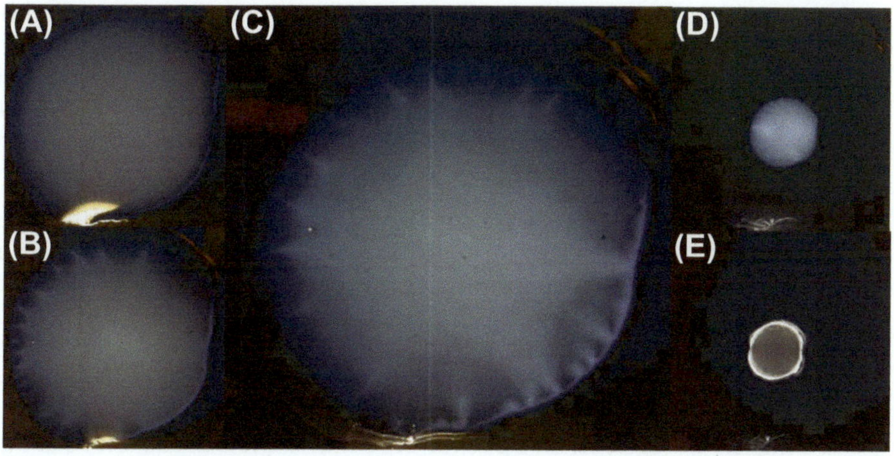

Figure 4.62
Dissipative patterns of CAIBA-P2VP(0.5) on a watch glass at 22° C. 0.7 ml, 0.337 wt%, **A,** 30 minutes after setting; **B,** 3 hours, 30 minutes; **C,** 5 hours, 40 minutes; **D,** 11 hours, 20 minutes; **E,** 16 hours, 20 minutes.
Edited from Fujii S, Yamashita Y, Nakamura Y, Okubo T: Colloid Polym Sci *292:2621–2631, 2014.*

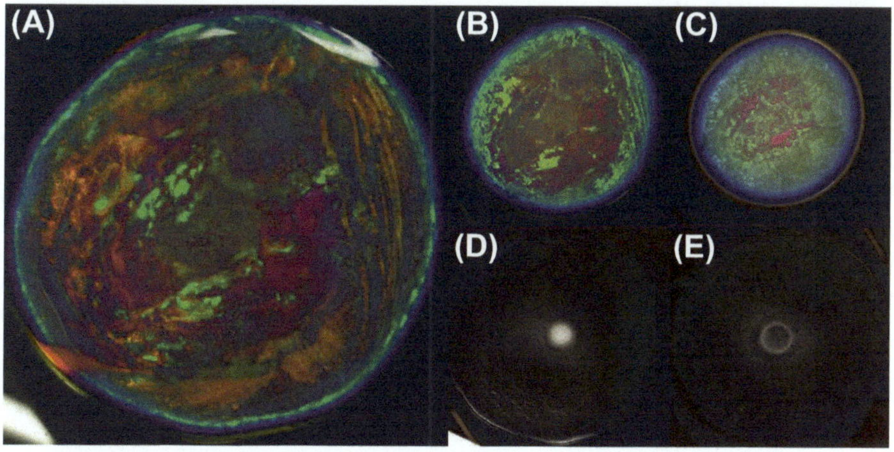

Figure 4.63
Patterns of AIBA-P2VP(0.5) (9.65 wt%) during drying on a watch glass at 22° C 0.7 ml, **A,** 20 minutes after setting; **B,** 2 hours, 20 minutes; **C,** 3 hours, 50 minutes; **D,** 5 hours, 30 minutes; **E,** 16 hours, 10 minutes.
Okubo T, Fujii S, Nakamura Y: Colloid Polym Sci *291:2805–2813, 2013.*

iridescent colors even in the initial stage of convection (see Fig. 4.63, *A*). Colors changed from reddish to bluish with time, which demonstrates that gel concentrations increase with time. The dried areas were whitish. This was due to the enhanced intensity of the multiple scattering from the dried areas compared with that from the Bragg diffraction in the suspension state. The other reason for the weakened iridescent colors in the film state was the small size of gel spheres (≈170-180 in diameter) compared with the light wavelengths. Formation of the two broad ring patterns is clearly observed from change in the colors. When initial concentrations of the gels are low (i.e., below the critical concentration of crystallization), the suspensions showed the iridescent colors with time and then became white at the final stage of drying.

For PEGMA-P2VP(0.5) spheres in Fig. 4.64, whitish convectional spoke-line patterns appeared again (see Fig. 4.64, *C* and *D*). The bluish and transparent broad rings and the multiple broad rings coexisted at the outside edge and the central area in the drying patterns (see Fig. 4.64, *E*). Formation of the multiple-broad rings is characteristic compared with other P2VP gel spheres. It should be mentioned here that the iridescent colors of the dried film shown in Fig. 4.64, *E*, were observed using a fluorescent lamp tube as the incident light. However, when a pocket-type flash light is used instead as light source, much stronger iridescent colors are observed. Iridescent colors change greatly depending on the kind of light source and angles of incident and observation light.

The patterns of large μPEGMA-P2VP(0.5) spheres were weakly iridescent but whitish throughout the drying processes because of the strong multiple-scattering effects (Fig. 4.65). The broad rings at the outside edge of the initial suspension appeared first during the course of drying. However, the main broad ring shifted toward the inner area to the central area. Iridescent colors were observed over the whole film area. Furthermore, spoke line–like cracks formed, especially on a cover glass and a watch glass. These cracks formed by the strain induced during the solidification process on the base of spoke lines formed during convectional flow of the spheres.

Fig. 4.66 displays typical convectional patterns from CAIBA-P2VP to μPEGMA-P2VP spheres. The spoke lines to clusterlike convectional patterns

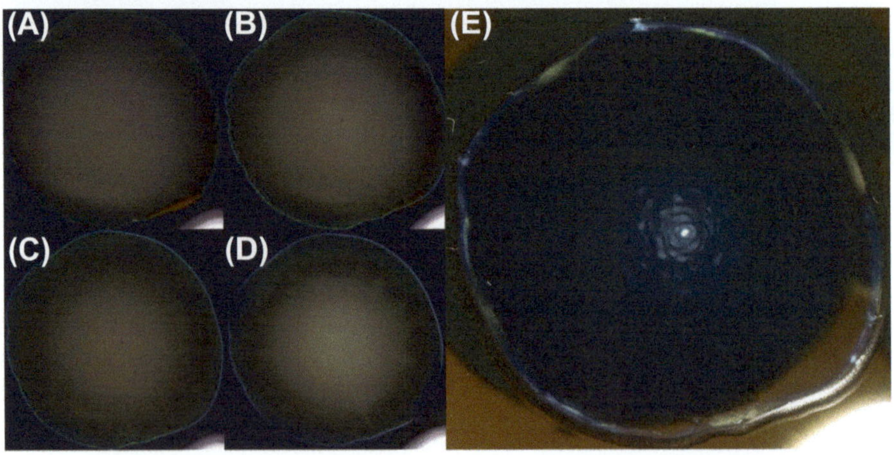

Figure 4.64 Dissipative drying patterns of PEGMA-P2VP(0.5) on a watch glass at 25° C. 0.7 ml, 0.310 wt%, **A,** 40 minutes after setting; **B,** 2 hours, 40 minutes; **C,** 4 hours, 15 minutes; **D,** 6 hours; 50 minutes; **E,** 16 hours.

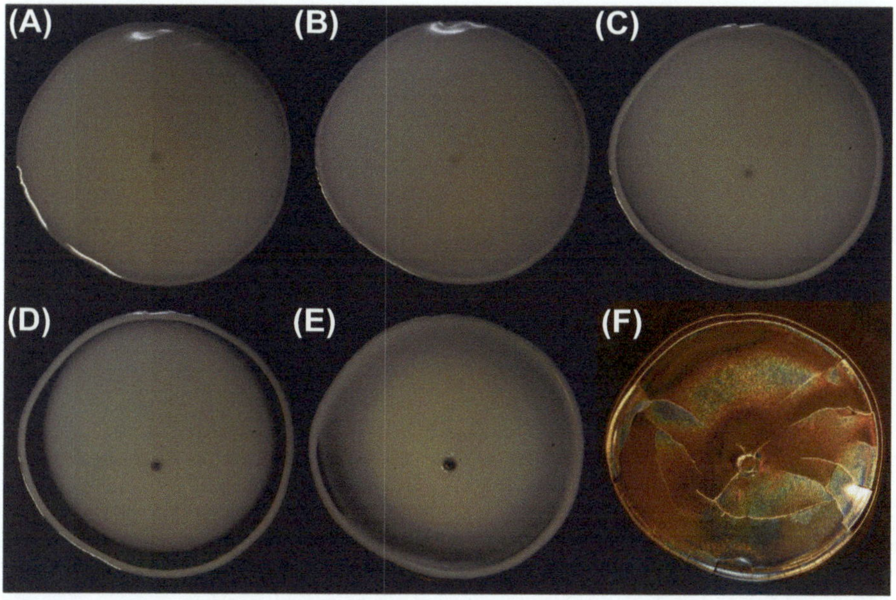

Figure 4.65
Dissipative drying patterns of μPEGMA-P2VP(0.5) on a watch glass at 25° C. 0.7 ml, code 5636, [μPEGMA-P2VP(0.5)] = 5.86 wt%, **A,** 1 hour, 20 minutes after setting; **B,** 2 hours, 45 minutes; **C,** 4 hours, 25 minutes; **D,** 6 hours, 45 minutes; **E,** 11 hours, 10 minutes; **F,** 14 hours, 5 minutes. *Taken in part from Fujii S, Kobayashi K, Nakamura Y, Okubo T:* Colloid Polym Sci, *in press. http://dx.doi.org/10.1007/s00396-015-3575-y.*

appeared for CAIBA-P2VP, AIBA-P2VP, and PEGMA-P2VP suspensions. Furthermore, the convectional patterns of these gel spheres were similar to the patterns of coffee with cream.[243] In drying experiments, 0.7 ml of cationic gel sphere suspension was set on a watch glass. Growth of the short spoke lines appeared from the outside edge toward the central area between 3 and 10 hours. Furthermore, clusters formed by coupling the several spoke lines. At the final stage of convectional flow processes, further growth toward bundles took place from the clusters. Bundles are also one of the typical sedimentary patterns. For μPEGMA-P2VP spheres, clear-cut convectional patterns were not recognized, as seen in Fig. 4.66 *J, K,* and *L.*

Fig. 4.67 shows the typical sedimentary patterns observed for the suspensions of CAIBA-P2VP(0.1), CAIBA-P2VP(1), and AIBA-P2VP(0.1) spheres. For comparison, a part of the sedimentary patterns of pNIPAm(200-2) and pNIPAm(600-5) gel spheres are also presented in Fig. 4.67, *E* and *F.* The starlike bundles of sedimentary patterns are observed. The drying spoke-line patterns are grown from the sedimentary ones. Interestingly, for Fig. 4.67, *A* through *D,* the bundle patterns are still dynamic, although the time scale of fluctuation is long, several minutes, for example. The convectional patterns

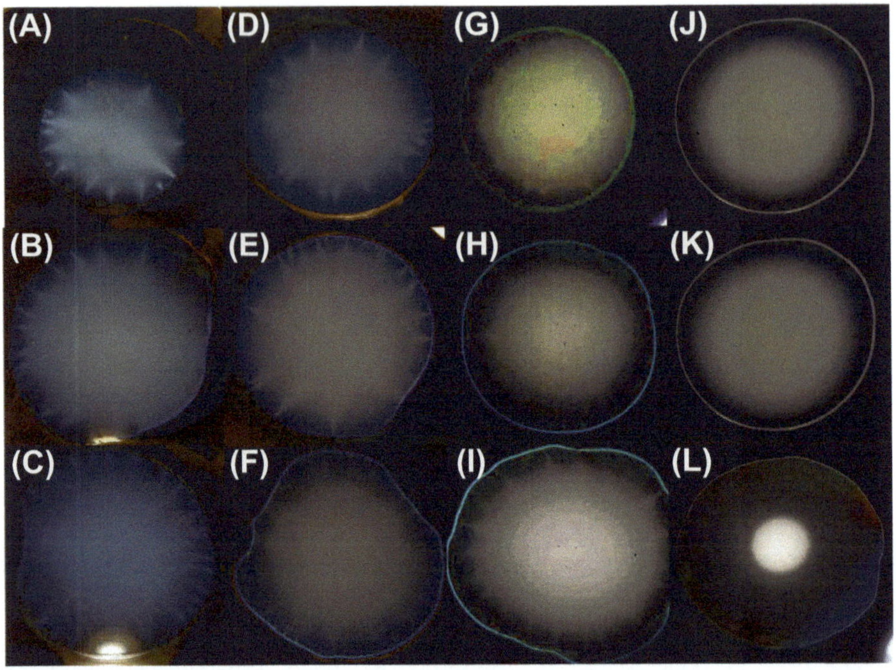

Figure 4.66
Convectional patterns of CAIBA-P2VP (*A-C*); AIBA-P2VP (*D-F*); PEGMA-P2VP (*G-I*); and μPEGMA-P2VP (*J-L*) on a watch glass at 22° C. 0.7 ml. **A,** CAIBA-P2VP(0.1), 0.22 wt%, 11 hours after setting; **B,** CAIBA-P2VP(0.5), 0.34, 3 hours, 30 minutes; **C,** CAIBA-P2VP(1), 0.25, 4 hours, 45 minutes; **D,** AIBA-P2VP(0.1), 0.35 wt%, 7 hours, 55 minutes; **E,** AIBA-P2VP(0.5), 0.39 wt%, 5 hours, 30 minutes; **F,** AIBA-P2VP(1), 0.42 wt%, 10 hours, 50 minutes; **G,** PEGMA-P2VP(0.1), 2.1 wt%, 6 hours, 30 minutes; **H,** PEGMA-P2VP(0.5), 0.51, 6 hours, 30 minutes; **I,** PEGMA-P2VP(1), 0.51, 4 hours, 50 minutes; **J,** μPEGMA-P2VP(0.1), 0.015 wt%, 6 hours, 30 minutes; **K,** μPEGMA-P2VP(0.5), 2.9, 4 hours, 25 minutes; **L** μPEGMA-P2VP(0.5), 0.064, 8 hours, 15 minutes.

of pNIPAm gel suspensions were not observed so often compared with those of the cationic gel spheres. Fig. 4.67, *E,* shows the inner ring/outer cluster–type sedimentary pattern of pNIPAm(200-2), which was observed clearly at the liquid surface of coffee with cream. Dense clusters are clearly observed in Fig. 4.67, *F,* for pNIPAm(600-5).

Fig. 4.68 shows the typical drying patterns of CAIBA-P2VP, AIBA-P2VP, and PEGMA-P2VP gel spheres as a function of initial sphere concentration. Two kinds of broad rings—the broad rings at the outside edge of the initial suspension and at the inner area—appeared. The main broad rings were located at the outside edges irrespective of the substrates among cover glass, watch glass, and glass dish and gel concentration. Coexistence of two broad rings supports the finding that the single state of gel spheres and their agglomerates are segregated with each other at the outside edge and at the inner area during the course of drying. It is interesting to note that the size of the

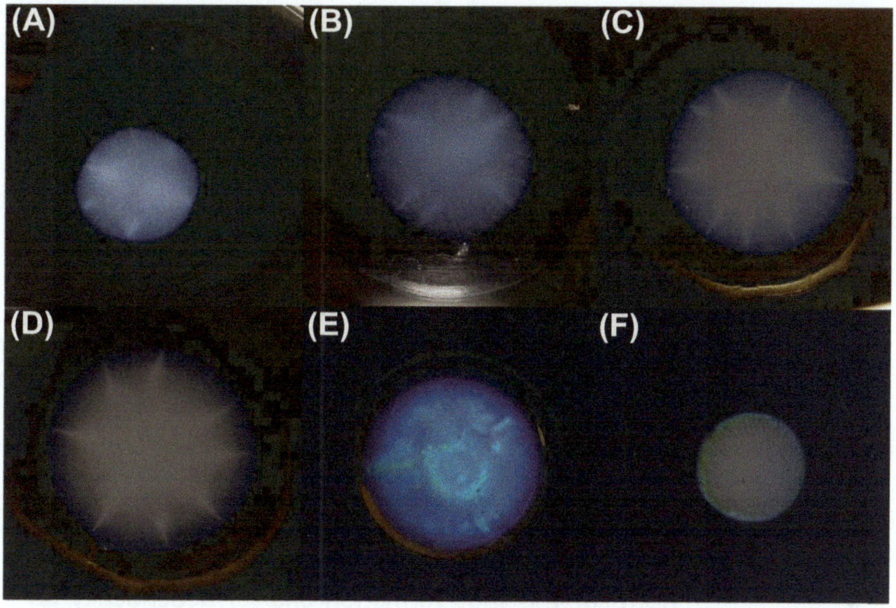

Figure 4.67 Sedimentary patterns of **A,** CAIBA-P2VP(0.1), 0.217 wt%, 13 hours, 40 minutes after setting, on a watch glass; **B,** CAIBAP2VP(1), 0.250, 12 hours, 45 minutes; **C,** AIBA-P2VP(0.1), 0.424, 14 hours, **D,** AIBA-P2VP(1), 0.345, 12 hours, 50 minutes; **E,** pNIPAm(200-2), 1.97, 2 hours, 30 minutes, cover glass, **F,** pNIPAm(600-5), 1.16 wt%, 4 hours, 5 minutes).

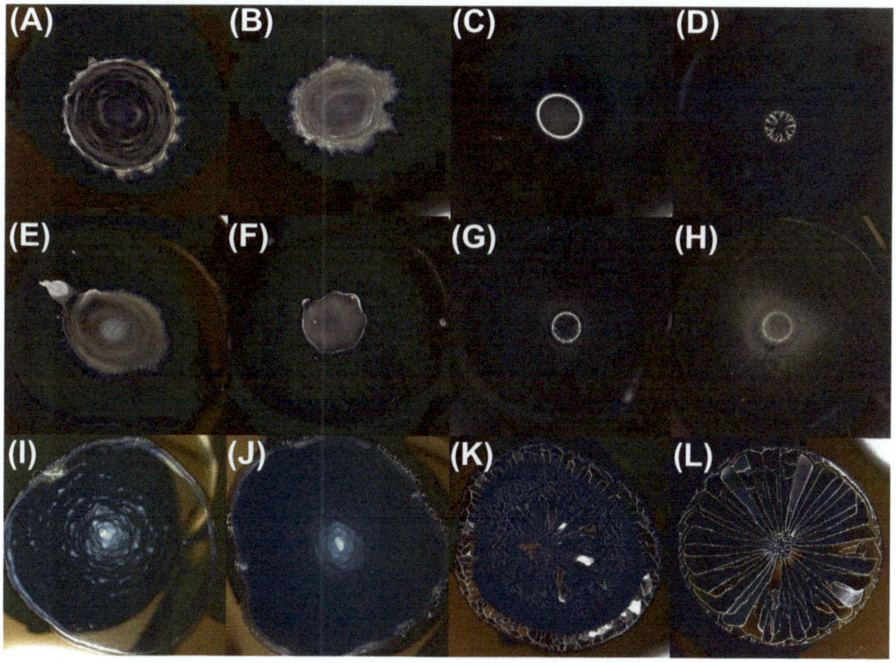

Figure 4.68 Drying patterns of CAIBA-P2VP(0.5) (*A-D*), AIBA-P2VP(0.5) (*E-H*), and PEGMA-P2VP(1) (*I-L*), on a watch glass at 22° C. **A,** 0.034 wt%; **B,** 0.101; **C,** 0.84; **D,** 2.53; **E,** 0.19; **F,** 0.58; **G,** 1.93; **H,** 9.65; **I,** 0.21; **J,** 0.51; **K,** 2.1; **L,** 5.1.

main patterns of the inner rings decreases as the initial concentration increased for CAIBA-P2VP and AIBA-P2VP, whereas the size for PEGMA-P2VP and μPEGMA-P2VP increased oppositely. These shifts of the inner-ring patterns are discussed later in this chapter using d_f/d_r-parameters. The spoke lines appeared both on a cover glass and a watch glass at the high initial gel concentrations (e.g., see Fig. 4.68, *G, H,* and *L*). The convectional movement of the agglomerates is disturbed by the weak affinitive interactions between the intergel spheres and further between the gel sphere and substrate surface during the convectional flow processes.

Fig. 4.69 shows the typical examples of the macroscopic drying patterns of μPEGMA-P2VP(0.5) on a cover glass.[125] Initial gel concentrations ranges from 0.059 wt% (Fig. 4.69, *A*) to 5.9 wt% (Fig. 4.69, *D*). Two kinds of broad rings—the rings close to the outside edge of the initial suspension and the ones at the inner area—appeared. Over the whole film area, strong iridescent colors were observed, especially when the initial concentrations were high. The iridescent hexagonal patterns formed at high sphere concentrations (see Fig. 4.69, *C* and *D*). The hexagonal patterns originated from arrays of the (111) planes of *fcc* crystals along the glass wall. Tomita et al. clarified the mechanisms of the spot formation and its dependency on the sphere size and concentration.[244] Similar patterns were observed for the drying films of colloidal crystals of silica spheres (e.g., see Chapter 3, Fig. 3.20).[245] Coexistence of the two broad rings supports again the fact that the single gel spheres and their agglomerated particles are segregated to each other toward the outside and inside during the course of drying. It is interesting to note that the iridescent broad ring areas at the outside regions were

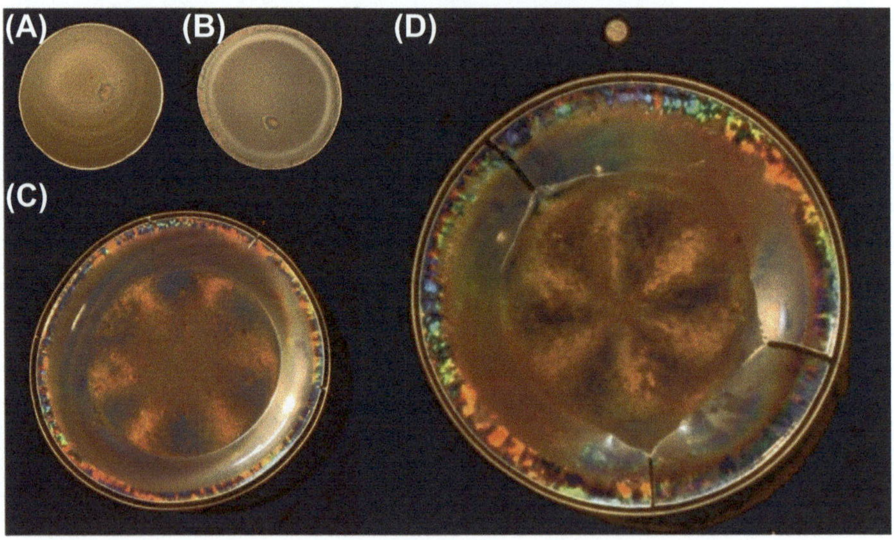

Figure 4.69 Drying patterns of μPEGMA-P2VP(0.5) on a watch glass at 25° C. 0.1 ml, **A,** 0.059 wt%; **B,** 0.39 wt%; **C,** 2.93 wt%; **D,** 5.86 wt%. *Taken in part from Fujii S, Kobayashi K, Nakamura Y, Okubo T:* Colloid Polym Sci, *in press. http://dx.doi.org/10.1007/s00396-015-3575-y.*

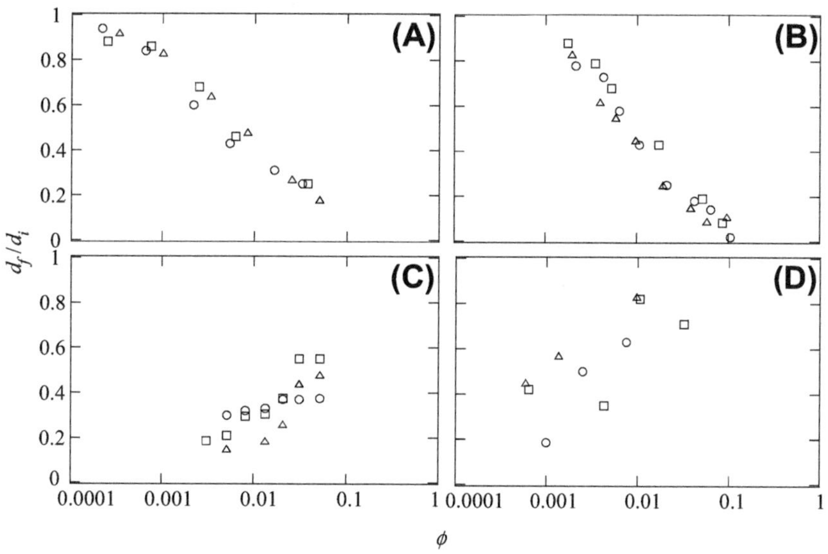

Figure 4.70
d_f/d_i of the inner rings of CAIBA-P2VP (*A*), AIBA-P2VP (*B*), PEGMA-P2VP (*C*), and μPEGMA-P2VP (*D*) as a function of sphere concentration on a cover glass at 25° C. *Circles:* P2VP(0.1), *triangles:* P2VP(0.5), *squares:* P2VP(1). *Taken in part from Fujii S, Kobayashi K, Nakamura Y, Okubo T:* Colloid Polym Sci, *in press. http://dx.doi.org/10.1007/s00396-015-3575-y.*

dispatched from the substrate especially for the gel spheres with a high degree of cross-linking and at high sphere concentrations. The similar macroscopic drying patterns also were observed on a cover glass and a glass dish.

Fig. 4.70 shows the ratio of the final size of dried film (d_f) against the initial size of liquid drop (d_i) for the inner broad rings for the deionized suspensions of CAIBA-P2VP (Fig. 4.70, *A*),[240] AIBA-P2VP (Fig. 4.70, *B*),[239] PEGMA-P2VP (Fig. 4.70, *C*),[238] and μPEGMA-P2VP (Fig. 4.70, *D*) on a cover glass. Main broad rings of these samples exist at the outside edges, $d_f/d_i = 1$. The plots of the data of the main rings were omitted in the figures. Interestingly, the d_f/d_i values of the inner broad rings decreased for CAIBA-P2VP and AIBA-P2VP as sphere concentration increased, whereas those of PEGMA-P2VP and μPEGMA-P2VP increased. Decrease of d_f/d_i with concentration is not familiar. Strong electrostatic affinitive interactions of these small cationic spheres with the anionic substrate is plausible. Small gel spheres contain a lot of water molecules and are soft. So it is imagined that agglomeration should take place favorably near the substrate. The electrostatic affinity between small cationic gel spheres and anionic substrate is also effective. As the size of the gel spheres increases, however, the diffusive movement and the convectional flow begin to occur freely, because the sphere-substrate

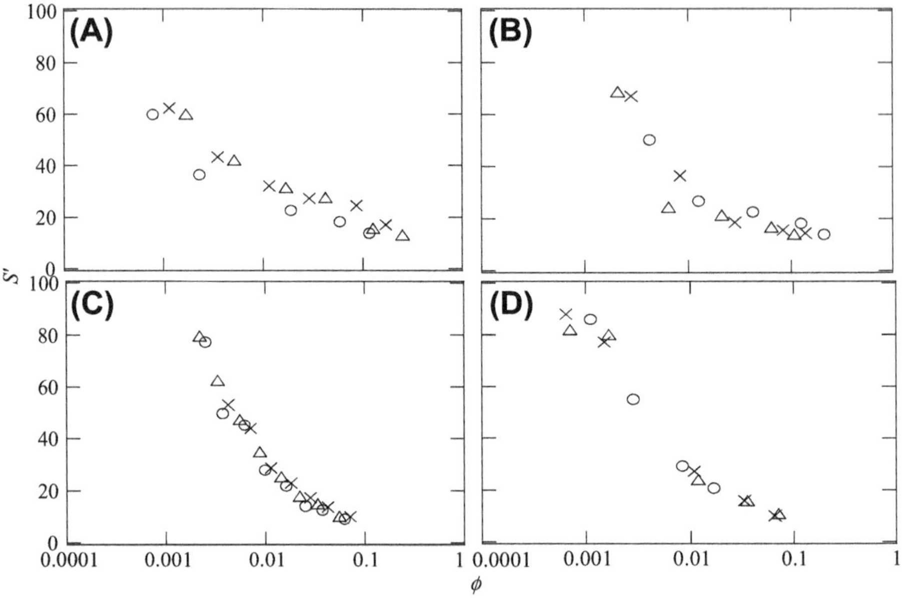

Figure 4.71 Sharpness of the broad ring (S') as a function of volume fraction of CAIBA-P2VP (*A*), AIBA-P2VP (*B*), PEGMA-P2VP (*C*), and μPEGMA-P2VP (*D*) on a cover glass at 22° C (*A, B*) and 25° C (*C, D*). *Taken in part from Fujii S, Kobayashi K, Nakamura Y, Okubo T:* Colloid Polym Sci, *in press. http://dx.doi.org/10.1007/s00396-015-3575-y.*

affinity forces are weakened. The effects of the degree of cross-linking of gel spheres on the d_f/d_i were not observed clearly in the figures. CAIBA-P2VP and AIBA-P2VP are type-C solutes, and PEGMA-P2VP and μPEGMA-P2VP are type B gel solutes, respectively.

Fig. 4.71 shows the S' values of gel crystals of P2VP as a function of sphere concentration in volume fraction. S' values were not sensitive to the degree of cross-linking except CAIBA-P2VP, where the values increased slightly as the degree of cross-linking increased. The S' values of P2VP gel spheres at $\phi =$ 0.001 are roughly 60, 80, 110, and 80, for CAIBA-P2VP, AIBA-P2VP, PEGMA-P2VP, and μPEGMA-P2VP respectively. Clearly, S' increased first with increasing sphere size and then decreased again. These tendencies of the S' values support the finding that CAIBA-P2VP gel spheres (110 nm in mean diameter) are not so diffusive during drying owing to the strong gel-substrate affinitive interactions from both electrostatic (between cationic spheres and anionic glass wall) and hydrophilic attraction forces. With increasing gel size to AIBA-P2VP (170 nm) and then to PEGMA-P2VP (390 nm), the affinitive interactions are weakened because the spheres become more compact and diffusive as the ρ values increase sharply as sphere size increases. For μPEGMA-P2VP (1150 nm), lowering of the diffusion by the size effects should predominate over the effects of the weakening of affinitive interactions mentioned previously. In conclusion, the S' (or S) value is one of the convenient parameters indicating the diffusive properties of gel spheres during the course of drying.

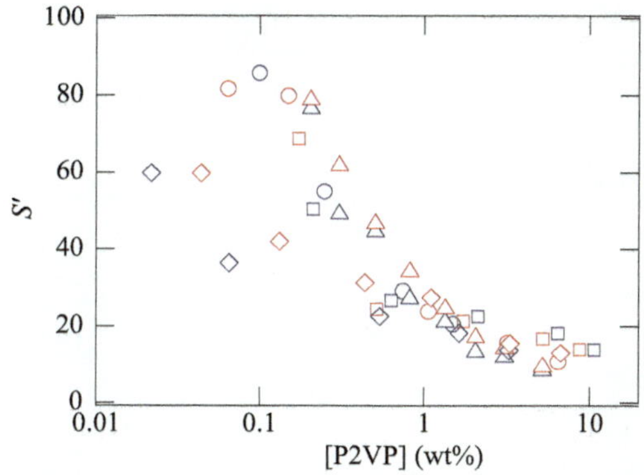

Figure 4.72
S′ values of μPEGMA-P2VP(0.1) (*blue circles*), μPEGMA-P2VP(1) (*red circles*), PEGMA-P2VP(0.1) (*blue triangles*), PEGMA-P2VP(1) (*red triangles*), AIBA-P2VP(0.1) (*blue squares*), AIBA-P2VP(1) (*red squares*), CAIBA-P2VP(0.1) (*blue diamonds*) and CAIBA-P2VP(1) (*red diamonds*).

Fig. 4.72 shows the S′ values of P2VP gels at the degree of cross-linking at 0.1 and 1 mol% as a function of sphere concentration in weight percent. To compare the S′ values with a series of P2VP spheres, the plots against concentration in weight percent are meaningful, because weight percent of spheres is insensitive to the degree of cross-linking. Several findings are clear. First, S′ values of AIBA-P2VP, PEGMA-P2VP, and μPEGMA-P2VP are similar. However, S′ values of CAIBA-P2VP are substantially low compared with other P2VP spheres. This finding is the same as that in Fig. 4.71. The strong gel sphere–substrate affinity is highly plausible again. Second, the influence of the degree of cross-linking was not so significant except CAIBA-P2VP, where the S′ values are high at high temperature. An increase in diffusivity with increasing temperature is the main cause for the observation.

Typical microscopic drying patterns of cationic P2VP gel spheres showing the ordered arrays of the agglomerates are displayed in Fig. 4.73. These four kinds of drying patterns are quite similar to those of pNIPAm gel spheres. Typical ordered rings appeared often for small spheres, CAIBA-P2VP especially at low temperatures and low sphere concentrations, a tendency similar to pNIPAm gel spheres (see Fig. 4.73, *A*). Ordered spoke lines also formed at the similar conditions as ordered rings (see Fig. 4.73, *E*) The net and lattice structures formed often at high sphere concentrations. The clear-cut pictures showing formation of the agglomerates were observed especially in large P2VP spheres. Fig. 4.73, *O* supports the fact that the single crystals composed of the agglomerates are surrounded by the grain boundaries. It is highly plausible that during the dissipative drying processes, single crystal lattice structures form and further orientation along the convectional flow is followed. It is surprising

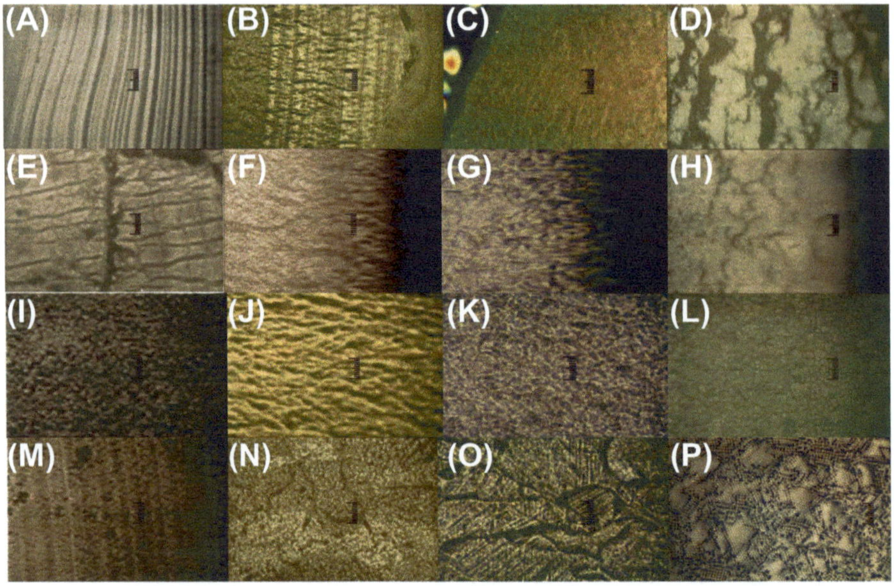

Figure 4.73
Typical microscopic drying patterns of **A,** CAIBA-P2VP(1), 0.25 wt%, 22° C; **B,** AIBA-P2VP(1), watch glass, 3.45 wt%, 22° C; **C,** PEGMA-P2VP(0.1), cover glass, 5.1; **D,** μPEGMA-P2VP(1), watch glass, 0.064; **E,** CAIBA-P2VP(0.5), 5.06, cover, 22° C; **F,** AIBA-P2VP(0.1), dish, 4.24; **G,** PEGMA-P2VP(0.1), dish, 5.1, 25° C; **H,** μPEGMA-P2VP(1), dish, 0.064; **I,** CAIBA-P2VP(1), 1.88, dish, 22° C; **J,** AIBA-P2VP(1), watch glass, 3.45, 22° C; **K,** PEGMA-P2VP(0.1) + NaCl, dish, 3.1, 0.002 M NaCl; **L,** μPEGMA-P2VP(0.5), watch glass, 5.86; **M,** CAIBA-P2VP(0.5), 2.53, dish, 22° C; **N,** AIBA-P2VP(0.1), dish, 4.24; **O,** PEGMA-P2VP(0.5) + NaCl, dish, 3.1, NaCl 0 M, 25° C; **P,** μPEGMA-P2VP(0.5), cover, 2.93, 25° C, full scale. **A-O,** 100 μm, **P,** 10 μm. Ordered broad ring, ordered spoke line. net, and lattice structures are shown mainly in **A-D, E-H, I-L,** and **M-P,** respectively.

to note that the highly monodispersed agglomerates form during drying and they further form partially oriented single crystals.

Fig. 4.74 demonstrates the crystal arrays of μPEGMA-P2VP(0.1) (*A-C*), μPEGMA-P2VP(0.5) (*D-F*), and μPEGMA-P2VP(1) (*G-I*) spheres from the central area to the right edge on a cover glass. Full scale is 10 μm in microscopic observation. Agglomerates, which correspond to the single crystals in this case, are distributed in a ringlike pattern, although roughly and in the two separated planes. These morphologic characteristics of crystals are due to the strain induced during the course of drying. Table 4.9 compiles the drying patterns that appeared for the gel crystal suspensions.

Orientation of the agglomerated spheres is common for both the anionic (pNIPAm) and cationic gel spheres (P2VP). At a glance, agglomeration of the single gel spheres appears similar to the aggregation of the gel spheres.

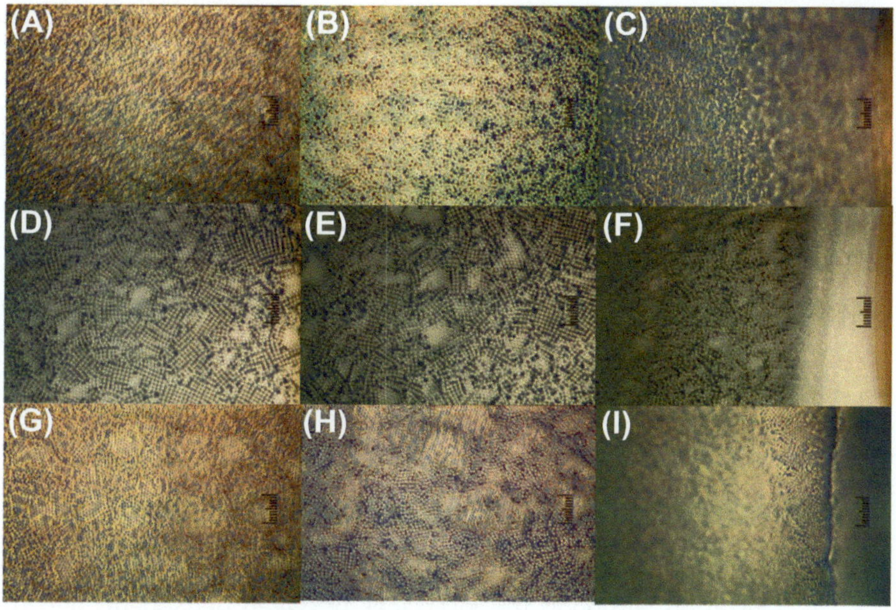

Figure 4.74
Microscopic drying patterns of μPEGMA-P2VP(0.1) (1.49 wt%, *A-C*), μPEGMA-P2VP(0.5) (2.93 wt%, *D-F*) and μPEGMA-P2VP(1) (6.37 wt%, *G-I*) on a cover glass at 25° C, 0.1 ml, **A, D, G** and **C, F, I** from the center to the right edge, full scale is 10 μm. *Taken in part from Fujii S, Kobayashi K, Nakamura Y, Okubo T:* Colloid Polym Sci, *in press. http://dx.doi.org/10.1007/s00396-015-3575-y.*

Table 4.9 Typical Drying Patterns of Gel Spheres

Macroscopic Patterns		
BR	1BR (outside or inner)	pNIPAm, P2VP
	2BR (outside and inner)	pNIPAm, P2VP
Multiple (ordered BR)	pNIPAm, P2VP	
Spoke lines	Normal	pNIPAm, P2VP, lpNIPAm
	Flamelike	
Clusters	Including flamelike	pNIPAm(70-5)
Hexagonal bundles		
Central round hill		
Microscopic Patterns		
Lattice (single spheres)		pNIPAm, P2VP
Ordered spoke lines (agglomerates)		pNIPAm, P2VP
Ordered broad rings (agglomerates)		pNIPAm, P2VP
Netlike (agglomerates)		pNIPAm, P2VP
Latticelike (agglomerates)		pNIPAm, P2VP
Dendritic (with NaCl)		lpNIPAm
Flowerlike		PEGMA-P2VP(0.5)
Blocklike		

BR, Broad rings; *lpNIPAm,* linear-type poly(*N*-isopropylacrylamide); *NaCl,* sodium chloride; *P2VP,* cationic gel spheres of poly(2-vinylpyridine); *pNIPAm,* thermo-sensitive gel spheres of poly(*N*-isopropylacrylamide).

However, according to my view, agglomeration and aggregation are different: The former takes place by the weak and cooperative affinitive interactions among the monomer gel components, resulting in the monodispersed associates, whereas the latter forms by the strong intersphere affinitive forces, resulting in the polydispersed associates. We should pay attention to the importance of the ordered convectional flow of the single agglomerated particles and also weak interaction between the single spheres and the substrate on the formation of the monodispersed agglomerates. We should cite here the important work on drying patterns. Multicolored micropatterning of thin films of dry gels were prepared by Klajn et al.[246] Imprinted photonic gel crystals for lithography have been studied intensively.[247,248]

4.13 Conclusion and Future Perspectives

Several characteristic properties for gel crystals exist: Types 1, 2, and 3. Type 1 crystals are similar to the typical colloidal crystals in the origin of the crystallization, although the interfaces between gel spheres and the medium phase are more vague and fuzzy than those between hard colloidal spheres and medium phase. The critical concentration of melting of Type 1 gel crystals is sensitive to the ionic concentration of the gel suspension, as is clear in the sharp change of the critical concentration of melting with and without the resins. Furthermore, morphologic characteristics of the single crystals of gel spheres surprisingly resemble those of typical colloidal spheres such as polystyrene and colloidal silica spheres. Turbidity of the liquid-state suspensions of the gel spheres was high to an extent similar to that of the typical colloidal spheres having a similar hydrodynamic size. Turbidity of the gel supports the finding that a certain interface exists separating two phases: gel spheres and the medium molecules having different refractive indices, respectively. This work clearly demonstrates that the crystallization of the gel spheres, especially in the deionized aqueous system, takes place by the excluded volume effect of the gels themselves, and also by the contribution of the extended electrical double layers formed around each gel sphere.

I have believed hitherto that the gel spheres of pNIPAm do not form the interfaces with water phase because of the significant diffusion of water and ionic species through the gels. However, we should note here that Ballauff et al. found that the nearest-neighbored intergel distances in the crystal structures of core-shell spheres, having pNIPAm as a shell at rather high sphere concentration without resins, were slightly larger than the hydrodynamic size of the gels.[109-111] They attributed this difference to the existence of the local charges on the gel spheres, which originated from the charges of initiators used during polymerization of the core-shell spheres. I believe that their explanation is correct. Additionally, their experiments are also consistent with the importance of the electrical double layer effect on the crystallization, although the extent of the contribution of the double layers is small from their experimental condition without deionization of the suspension. It should be noted most of my experiments have been made in the deionized condition, where the contribution of the electrical double layers are substantially high. The important contribution of the electrical double layers on crystallization is clearly demonstrated by the experiments of crystallization at 45° C, where the excluded diameter of the gel spheres (d) is very small compared with the intersphere distance (l), $d << l$. Even in this condition, crystallization took place.

Crystallization phenomena of the mixtures of pNIPAm gels and colloidal silica spheres were studied in the deionized state by my group.[107,108] The crystallization was discussed by the coupling effects of the extended electrical double layers around the colloidal silica spheres and the excluded-volume effects of the pNIPAm gels. It should be noted here that the pNIPAm gel spheres have a lot of water molecules in the inner region of the sphere, especially at low temperature, as described previously. Then the gel spheres in the suspension state are assumed to be liquid phase, which is distinguished from the bulk water phase. It is plausible that the gel spheres and water molecules interacted strongly with each other by the solvation and polar-polar interaction between gels and water, and behaves like a kind of liquid different from the bulk water. In other words, the gel suspensions are one of the micro-emulsion systems, where the electrical double layers extend from the liquid-phase spheres.

The excluded diameter of the gel sphere has been assumed to be equal to the hydrodynamic diameter estimated from the DLS measurements. If the excluded diameter was sensitive to the ionic concentration of the suspensions especially at low temperatures, 25° C, for example, the possibility of the crystallization being caused by the extension of the gel spheres themselves still remains even in the deionized state. In other words, the possibility of crystallization taking place only by the excluded-volume effect still remains, especially at low temperatures. However, at high temperatures, at 45° C as described, gel size is small compared with the intergel distances by the substantial shrinking of the gels, and the contribution of the excluded volume effect becomes quite small. Thus the crystallization at high temperatures should not take place without the contribution of the electrical double layers. Accurate evaluation of the size of gel spheres in the deionized and low concentrations, where gel spheres do not interact with each other, is important for further advances in the study of Type 1 gel crystallization.

Type 2 gel crystals are ideal for coupling the mechanical and optical properties in a gel system (i.e., preparation of the optical devices responsible to the several types of external fields such as temperature, electric field, pressure, centrifugal field and salt addition). However, gel crystals are more or less restricted from free thermal movement of the crystal structures, as described in Section 4.9. Balancing between flexibility and responsibility is important for further advances in relevant studies.

Type 3 gel crystals are exciting systems for making various devices responsible for external fields in the future. Microscopic crystal structures ranging from submicron to micrometers are often formed spontaneously on the substrate simply drying gel suspensions, as described in Section 4.12 in detail. Application-directed study of Type 3 gel crystals is anticipated.

References

1. Flory PJ. *Principles of polymer chemistry*. Ithaca, NY: Cornell University Press; 1953.
2. Tanaka T, Fillmore DJ. *J Chem Phys* 1979;**70**:1214–8.
3. Tanaka T, Fillmore D, Sun ST, Nishio I, Swislow G, Shah A. *Phys Rev Lett* 1980;**45**:1636–9.
4. Hirokawa Y, Tanaka T. *J Chem Phys* 1984;**81**:6379–80.
5. Suzuki A, Tanaka T. *Nature* 1990;**346**:345–7.
6. Otake K, Inomata H, Konno M, Saito S. *Macromolecules* 1990;**23**:283–9.
7. Tokita M, Tanaka T. *J Chem Phys* 1991;**95**:4613–9.
8. Tokuhiro T, Amiya T, Mamada A, Tanaka T. *Macromolecules* 1991;**24**:2936–43.
9. Park TG, Hoffman AS. *Macromolecules* 1993;**26**:5045–8.
10. Shibayama M, Norisuye T, Nomura S. *Macromolecules* 1996;**29**:8746–50.
11. Suzuki A, Yoshikawa M. *J Chem Phys* 1999;**111**:360–7.
12. Takahashi K, Takigawa T, Masuda T. *J Chem Phys* 2004;**120**:2972–9.
13. Kojima H, Tanaka F. *Macromolecules* 2010;**43**:5103–13.
14. Erman B, Flory PJ. *Macromolecules* 1986;**19**:2342–53.
15. Suzuki Y, Nozaki K, Yamamoto T. *J Chem Phys* 1992;**97**:3808–12.
16. Zhang J, Pelton R, Deng Y. *Langmuir* 1995;**11**:2301–6.
17. Nakamura T, Hattori M, Kawasaki H, Miyamoto K, Tokita M, Komai T. *Phys Rev E* 1996;**54**:1663–8.
18. Lele AK, Hirue MM, Badiger MV, Mashelkar RA. *Macromolecules* 1997;**30**:157–9.
19. Ikkai F, Shibayama M. *Phys Rev E* 1997;**56**:R51–4.
20. Tanaka N, Matsukawa S, Kurosu H, Ando I. *Polymer* 1998;**39**:4703–6.
21. Suetoh Y, Shibayama M. *Polymer* 2000;**41**:505–10.
22. Varga I, Gilanyi T, Meszaros R, Fillipsei G, Zrinyi M. *J Phys Chem B* 2001;**105**:9071–6.
23. Stieger M, Richtering W. *Macromolecules* 2003;**36**:8811–8.
24. Suzuki A, Yoshikawa M. *J Chem Phys* 2006;**125**:174901 (6p).
25. Okubo T. *Colloid Surf A* 1996;**109**:77–88.
26. Alexander S, Chaikin PM, Grant P, Morales GM, Pincus P, Hone D. *J Chem Phys* 1984;**80**:5776–81.
27. Okubo T. *Prog Polym Sci* 1993;**18**:481–517.
28. Palberg T, Kottal J, Bitzer F, Simon R, Wurtham M, Leiderer P. *J Colloid Interf Sci* 1995;**169**:85–9.
29. Garbow N, Evers M, Palberg T, Okubo T. *J Phys Condens Matter* 2004;**16**:3835–42.
30. Vanderhoff W, van de Hul HJ, Tausk RJM. Overbeek JThG. In: Goldfinger G, editor. *Clean surfaces: their preparation and characterization for interfacial studies*. New York: Dekker; 1970. p. 15–44.
31. Hiltner PA, Papir YS, Krieger IM. *J Phys Chem* 1971;**75**:1881–6.

32. Kose A, Ozaki M, Takano K, Kobayashi Y, Hachisu S. *J Colloid Interf Sci* 1973;**44**:330–8.
33. Williams R, Crandall RS, Wojtowicz PJ. *Phys Rev Lett* 1976;**37**:348–51.
34. Mitaku S, Ohtsuki T, Enari K, Kishimoto A, Okano K. *Jap J Appl Phys* 1978;**17**:305–13.
35. Lindsay HM, Chaikin PM. *J Chem Phys* 1982;**76**:3774–81.
36. Pieranski P. *Contemp Phys* 1983;**24**:25–73.
37. Ottewill RH. *Ber Bunseng Phys Chem* 1985;**89**:517–25.
38. Aastuen DJW, Clark NA, Cotter LK, Ackerson BJ. *Phys Rev Lett* 1986;**57**:1733–6.
39. Pusey PN, van Megen W. *Nature* 1986;**320**:340–2.
40. Okubo T. *Acc Chem Res* 1988;**21**:281–6.
41. Sood AK. *Solid State Phys* 1991;**45**:1–73.
42. Okubo T. *J Chem Soc Faraday Trans* 1. 1988;**84**:1163–9.
43. Lowen H, Palberg T, Simon R. *Phys Rev Lett* 1993;**70**:1557–60.
44. Okubo T. In: Schmitz KS, editor. *Macro-ion characterization. From dilute solutions to complex fluids ACS Symposium Series 548*. Washington, DC: American Chemical Society; 1994. p. 364–80.
45. Okubo T, Tsuchida A. *Forma* 2002;**17**:141–53.
46. Okubo T. In: Hubbard A, editor. *Encyclopedia of surface and colloid science*. New York: Marcell Dekker; 2002. p. 1300–9.
47. Okubo T. In: Kinoshita S, Yoshioka S, editors. *Structural colors in biological systems*. Osaka, Japan: Osaka University Press; 2005. p. 267–86.
48. Okubo T. *Polym J* 2008;**40**:882–90.
49. Russel WB. *Phase Transition* 1990;**21**:127–37.
50. Dhont JKG, Smits C, Lekkerkerker HNW. *J Colloid Interf Sci* 1992;**152**:386–401.
51. Verhaeghe NAM, van Blaaderen A. *Langmuir* 1994;**10**:1427–38.
52. Butler S, Harrowell P. *Phys Rev E* 1995;**52**:6424–30.
53. Okubo T. *Naturwissenschaften* 1992;**79**:317–20.
54. Okubo T. *Colloid Polym Sci* 1993;**271**:190–6.
55. Okubo T. *Langmuir* 1994;**10**:1695–702.
56. Okubo T, Tsuchida A. *Colloid Polym Sci* 2002;**280**:438–45.
57. Pelton RH, Chibante P. *Colloids Surf* 1986;**20**:247–56.
58. Pelton R. *Adv Colloid Interf Sci* 2000;**85**:1–33.
59. McPhee W, Tam KC, Pelton R. *J Colloid Interf Sci* 1993;**156**:24–30.
60. Ito S, Ogawa K, Suzuki H, Wang B, Yoshida R, Kokufuta E. *Langmuir* 1999;**15**:4289–94.
61. Jones CD, Lyon LA. *Macromolecules* 2000;**33**:8301–6.
62. Vo CD, Kuckling D, Adler HJP, Schohoff M. *Colloid Polym Sci* 2002;**280**:400–9.
63. Hoare T, McLean D. *J Phys Chem B* 2006;**110**:20327–36.
64. Hoare T, McLean D. *Macromol Theory Simulation* 2006;**15**:619–32.
65. Gao J, Wu C. *Macromolecules* 1997;**30**:6873–6.
66. Arleth L, Xia X, Hjelm RP, Wu J, Zhibinc HU. *J Polym Sci B Polym Phys* 2005;**43**:849–60.
67. Mason TG, Lin MY. *Phys Rev E* 2005;**71**:040801.
68. Ru G, Wang N, Huang S, Feng J. *Macromolecules* 2009;**42**:2074–8.
69. Rodriguez BE, Wolfe MS, Fryd M. *Macromolecules* 1994;**27**:6642–7.
70. Davis KA, Ackerson BJ, Yu H, Ford WT, Xia JL, Dubin PL. *Langmuir* 1994;**10**:2145–50.
71. Ohshima H. *J Colloid Interf Sci* 1994;**163**:474–83.
72. Kiminta DMO, Luckham PF, Lenon S. *Polymer* 1995;**36**:4827–31.
73. Kawaguchi S, Yekta A, Winnik MA. *J Colloid Interf Sci* 1995;**176**:362–9.
74. Kokufuta E, Wang B, Yoshida R, Khokhlov AR, Hirata M. *Macromolecules* 1998;**31**:6878–84.
75. Flint NJ, Gardebrecht S, Swanson L. *J Fluoresc* 1998;**8**:343–53.
76. Senff H, Richtering W. *J Chem Phys* 1999;**111**:1705–11.
77. Crowther HM, Saunders BR, Mears SJ, Cosgrove T, Vincent B, King SM, Yu GE. *Colloids Surf A* 1999;**152**:327–33.

78. Senff H, Richtering W. *Colloid Polym Sci* 2000;**278**:830–40.
79. Guillermo A, Addad JPC, Bazile JP, Duracher D, Elaissari A, Pichot C. *J Polym Sci B Polym Phys* 2000;**38**:889–98.
80. Daly E, Saunders BR. *Phys Chem Chem Phys* 2000;**2**:3187–93.
81. Gan DJ, Lyon LA. *J Am Chem Soc* 2001;**123**:8203–9.
82. Kratz K, Hellweg T, Eimer W. *Polymer* 2001;**42**:6631–9.
83. Woodward NC, Chowdhry BZ, Snowden MJ, Leharne SA, Griffiths PC, Winnington AL. *Langmuir* 2003;**19**:3202–11.
84. Gan DJ, Lyon LA. *Anal Chim Acta* 2003;**496**:53–63.
85. Stieger M, Richtering W. *Macromolecules* 2003;**36**:8811–8.
86. Hoare T, Pelton R. *Langmuir* 2004;**20**:2123–33.
87. Stieger M, Lindner P, Richtering W. *J Phys-Condens Matter* 2004;**16**:s3861–72.
88. Saunders BR. *Langmuir* 2004;**20**:3925–32.
89. Stieger M, Richtering W, Pedersen JS, Lindner P. *J Chem Phys* 2004;**120**:6197–206.
90. Seidel J, Pinkrah VT, Mitchel JC, Chowdhry BZ, Snowden MJ. *Thermochim Acta* 2004;**414**:47–52.
91. Hoare T, Pelton R. *Macromolecules* 2004;**37**:2544–50.
92. Christensen ML, Keiding K. *Colloids Surf A* 2005;**252**:61–9.
93. Hoare T, Pelton R. *Polymer* 2005;**46**:1139–50.
94. Mason TG, Lin MY. *Phys Rev E* 2005;**71**:040801.
95. Hoare T, Pelton R. *J Colloid Interf Sci* 2006;**303**:109–16.
96. Matsumura Y, Iwai K. *J Colloid Interf Sci* 2006;**296**:102–9.
97. Hoare T, Pelton R. *Langmuir* 2006;**22**:7342–50.
98. Hoare T, Pelton R. *J Phys Chem B* 2007;**111**:1334–42.
99. Ikkai F, Suzuki T, Karino T, Shibayama M. *Macromolecules* 2007;**40**:1140–6.
100. Hoare T, Pelton R. *J Phys Chem B* 2007;**111**:11895–906.
101. Hoare T, Pelton R. *Cur Opinion Colloid Interf Sci* 2008;**13**:413–28.
102. Xia Y, Gates B, Yin Y, Lu Y. *Adv Mater* 2000;**12**:693–713.
103. Hellweg T, Dewhurst CD, Bruckner E, Kratz K, Eimer W. *Colloid Polym Sci* 2000;**278**:972–8.
104. Debord JD, Lyon LA. *J Phys Chem B* 2000;**104**:6330–1.
105. Xia Y. *Adv Mater* 2001;**13**:369.
106. Gao J, Hu Z. *Langmuir* 2002;**18**:1360–7.
107. Okubo T, Hase H, Kimura H, Kokufuta E. *Langmuir* 2002;**18**:6783–8.
108. Okubo T, Mizutani T, Okamoto J, Kimura K, Tsuchida A, Tauer K, Khrenov V, Kawaguchi H, Tsuji S. *Colloid Polym Sci* 2006;**285**:351–8.
109. Crassous JJ, Ballauff M, Drechsler M, Schmidt J, Talmon Y. *Langmuir* 2006;**22**:2403–6.
110. Crassous JJ, Wittemann A, Siebenburger M, Schrinner M, Drechsler M, Ballauff M. *Colloid Polym Sci* 2008;**286**:805–12.
111. Crassous JJ, Rochette CN, Wittemann A, Schrinner M, Ballauff M, Drechsler M. *Langmuir* 2009;**25**:7862–71.
112. Okubo T, Suzuki D, Yamagata T, Katsuno A, Sakurai M, Kimura H, Tsuchida A. *Colloid Polym Sci* 2011;**289**:291–9.
113. Okubo T, Suzuki D, Yamagata T, Horigome K, Shibata K, Tsuchida A. *Colloid Polym Sci* 2011;**289**:1273–81.
114. Suzuki D, Horigome K, Yamagata T, Shibata K, Tsuchida A, Okubo T. *Colloid Polym Sci* 2011;**289**:1799–808.
115. Suzuki D, Yamagata T, Horigome K, Shibata K, Tsuchida A, Okubo T. *Colloid Polym Sci* 2012;**290**:107–17.
116. Okubo T, Suzuki D, Shibata K, Tsuchida A. *Colloid Polym Sci* 2012;**290**:1403–12.
117. Okubo T, Suzuki D, Tsuchida A. *Colloid Polym Sci* 2013;**291**:661–8.
118. Okubo T, Suzuki D. *Colloid Polym Sci* 2013;**291**:1721–7.

119. Okubo T, Fujii S, Aono K, Nakamura Y, Tsuchida A. *Colloid Polym Sci* 2013;**291**:1201–10.
120. Okubo T, Fujii S, Aono K, Nakamura Y. *Colloid Polym Sci* 2013;**291**:2569–77.
121. Fujii S, Yamashita Y, Nakamura Y, Tsuchida A, Okubo T: 292:1627–1637, 2014.
122. Fujii S, Dupin D, Araki T, Armes SP, Ade H. *Langmuir* 2009;**25**:2588–92.
123. Okubo T. *J Chem Soc Faraday Trans* 1. 1986;**82**:3163–73.
124. Eckert T, Richtering W. *J Chem Phys* 2008;**129**:1–6. 124902.
125. Fujii S, Kobayashi K, Nakamura Y, Okubo T. *Colloid Polym Sci*, in press. http://dx.doi.org/10.1007/s00396-015-3575-y.
126. Okubo T. *Colloid Polym Sci* 2006;**285**:245–50.
127. Okubo T. *Langmuir* 1994;**10**:3529–35.
128. Okubo T. *J Colloid Interf Sci* 1995;**171**:55–62.
129. Okubo T, Fujita H. *Colloid Polym Sci* 1996;**274**:368–74.
130. Okubo T, Fujita H, Kiriyama K, Yamaoka H. *Colloid Polym Sci* 1996;**274**:73–80.
131. Okubo T, Yoshimi H, Shimizu T, Ottewill RH. *Colloid Polym Sci* 2000;**278**:469–74.
132. Okubo T, Kimura H, Hase H, Lovell PA, Errington N, Thongnoi S. *Colloid Polym Sci* 2005;**283**:393–401.
133. Gottwald D, Likos CN, Kahl G, Lowen H. *Phys Rev Lett* 2004;**92**:1–4. 068301.
134. Okubo T, Kiriyama K, Nemoto N, Hashimoto H. *Colloid Polym Sci* 1996;**274**:93–104.
135. Okubo T, Kiriyama K. *Ber Bunsenges Phys Chem* 1996;**100**:849–56.
136. Okubo T, Kiriyama K. *J Mol Liquids* 1997;**72**:347–64.
137. Okubo T, Okada S, Tsuchida A. *J Colloid Interf Sci* 1997;**189**:337–47.
138. Pusey PN, van Megen W, Bartlett P, Ackerson BJ, Rarity JG, Underwood SM. *Phys Rev Lett* 1989;**63**:2753–6.
139. von Berlepsch H, Strey R. *Colloid Polym Sci* 1994;**272**:577–83.
140. Hartl W, Versmold H. *Z Phys Chem Neue Folge* 1987;**153**:1.
141. Okubo T. *J Colloid Interf Sci* 1988;**125**:380–5.
142. Baker JA, Henderson D. *J Chem Phys* 1967;**47**:2856–61.
143. Wadachi M, Toda M. *J Phys Soc Jpn* 1972;**32**:1147.
144. Hachisu S, Kobayashi Y, Kose A. *J Colloid Interf Sci* 1973;**42**:342–8.
145. Brenner SL. *J Phys Chem* 1976;**80**:1473–7.
146. Weissman JM, Sunkara HB, Tse AS, Asher SA. *Science* 1996;**274**:959–60.
147. Reese CE, Mikhonin AV, Kamenjicki M, Tikhonov A, Asher SA. *J Am Chem Soc* 2004;**126**:1493–6.
148. Okubo T. *J Chem Soc Faraday Trans* 1. 1988;**84**:1171–80.
149. Okubo T. *J Chem Soc Faraday Trans* 1. 1988;**84**:1949–59.
150. Okubo T. *J Chem Soc Faraday Trans* 1. 1988;**84**:4161–8.
151. Wilson HW. *Phil Mag* 1900;**50**:238–50.
152. Frenkel J. *Phys Z Sowjetunion* 1932;**1**:498–500.
153. Okubo T, Tsuchida A, Okada S, Fujitsuna K, Ishikawa M, Morita T, Tada T. *Colloid Surf A* 1999;**160**:311–20.
154. Tsuchida A, Takyo E, Taguchi K, Okubo T. *Colloid Polym Sci* 2004;**282**:1105–10.
155. Haw MD, Poon WCK, Pusey PN. *Phys Rev E* 1998;**57**:6859–64.
156. Crandall RS, Williams R. *Science* 1977;**198**:293–5.
157. Mitaku S, Ohtsuki T, Okano K. *Jpn J Appl Phys* 1980;**19**:439–48.
158. Mitaku S, Ohtsuki T, Kishimoto A, Okano K. *Biophys Chem* 1980;**71**:411–6.
159. Tauer K, Khrenov V. *Macromol Symp* 2002;**179**:27–52.
160. Tsuji S, Kawaguchi H. *Langmuir* 2004;**20**:2449–55.
161. Asher SA, Holtz J, Liu L, Wu Z. *J Am Chem Soc* 1994;**116**:4997–8.
162. Yoshinaga K, Fujiwara K, Tanaka Y, Nakanishi M, Takesue M. *Chem Lett* 2003;**32**:1082–3.
163. Yoshinaga K, Satoh S, Mouri E, Nakai A. *Colloid Polym Sci* 2006;**285**:694–8.
164. Asher SA, Kimble KW, Walker JP. *Chem Materials* 2008;**20**:7501–9.
165. Muscatello MMW, Asher SA. *Adv Funct Mater* 2008;**18**:1186–93.

166. Tian E, Wang J, Zheng Y, Song Y, Jiang L, Zhu D. *J Mater Chem* 2008;**18**:1116–22.
167. Arunbabu D, Sannigrahi A, Jana T. *Soft Matt* 2011;**7**:2592–9.
168. Lin FY, Yu LP. *Anal Methods* 2012;**4**:2838–45.
169. Wang JH, Hu Y, Deng R, Liang R, Li W, Liu S, Zhu J. *Langmuir* 2013;**29**:8825–34.
170. Nakayama D, Takeoka Y, Watanabe M, Kataoka K. *Angew Chem Int Ed* 2003;**42**: 4197–200.
171. Takeoka Y, Watanabe M. *Langmuir* 2002;**18**:5977–80.
172. Takeoka Y, Watanabe M. *Adv Mater* 2003;**15**:199–201.
173. Takeoka Y, Watanabe M. *Langmuir* 2003;**19**:9104–6.
174. Matsubara K, Watanabe M, Takeoka Y. *Angew Chem Intern Ed* 2007;**46**:1688–92.
175. Ueno K, Matsubara K, Watanabe M, Takeoka Y. *Adv Mater* 2007;**19**:2807–912.
176. Ueno K, Sakamoto J, Takeoka Y, Watanabe M. *J Mater Sci* 2009;**19**:4778–83.
177. Saito H, Takeoka Y, Watanabe M. *Chem Commun* 2003:2126–7.
178. Ueno K, Inaba A, Sano Y, Kondoh M, Watanabe M. *Chem Commun* 2009:3603–5.
179. Wang J, Hu Y, Deng R, Liang R, Li W, Liu S, Zhu J. *Langmuir* 2013;**29**:8825–34.
180. Suzuki D, Mc Grath JG, Kawaguchi H, Lyon LA. *J Phys Chem C* 2007;**111**:5667–72.
181. Ikeda T, Nakano M, Yu Y, Tsutsumi O, Kanazawa A. *Adv Materials* 2003;**15**:201.
182. Zhao B, Moore JS. *Langmuir* 2001;**17**:4758–63.
183. Zhang XZ, Yang YY, Chung TS, Ma KX. *Langmuir* 2001;**17**:6094–9.
184. Lee YJ, Braun PV. *Adv Mater* 2003;**15**:563–6.
185. Sorrell CD, Lyon LA. *J Phys Chem B* 2007;**111**:4060–6.
186. Shibata K, Kimura H, Tsuchida A, Okubo T, Sato S, Yoshinaga K. *Colloid Polym Sci* 2006;**284**:372–9.
187. Stoimenova M, Dimitrov V, Okubo T. *J Colloid Interf Sci* 1996;**184**:106–11.
188. Tsuchida A, Kuzawa M, Okubo T. *Colloids Surf* 2002;**209**:235–42.
189. Okubo T, Tsuchida A, Okada S, Kobata S. *J Colloid Interf Sci* 1998;**199**:83–91.
190. Okubo T. *J Chem Soc Faraday Trans* 1988;**84**:3377–88.
191. Tsuchida A, Taniguchi T, Tanahashi T, Okubo T. *Langmuir* 1999;**15**:4203–7.
192. Okubo T, Tsuchida A, Tanahashi T, Iwata A. *J Colloid Interf Sci* 1998;**207**:130–6.
193. Tanaka T, Nishio I, Sun ST, Ueno-Nishio S. *Science* 1982;**218**:467–9.
194. Suzuki D, Taniguchi H, Yoshida R. *J Am Chem Soc* 2009;**131**:12058–9.
195. Kanai T, Lee D, Shum HC, Shah RK, Weitz DA. *Adv Mater* 2010;**22**:4998–5002.
196. Honda M, Seki T, Takeoka Y. *Adv Mater* 2009;**21**:1801–4.
197. Suzuki D, McGrath JG, Kawaguchi H, Lyon LA. 2007; **111**:5667–72.
198. Zhou J, Cai T, Tang S, Marquez M, Hu Z. *Langmuir* 2006;**22**:863–6.
199. Debord JD, Lyon LA. *J Phys Chem B* 2000;**104**:6327–31.
200. Sahiner N, Turhan T, Lyon LA. *Energy* 2014;**66**:256–63.
201. Sakai T, Takeoka Y, Seki T, Yoshida R. *Langmuir* 2007;**23**:8651–4.
202. Ngai T, Auweter H, Behrens SH. *Macromolecules* 2006;**39**:8171–7.
203. Ramsden W. *Proc Royal Soc* 1903;**72**:156–64.
204. Pickering SU. *J Chem Soc* 1907;**91**:2001–21.
205. Binks BP. *Curr Opin Colloid Interf Sci* 2002;**7**:21–41.
206. Aveyard R, Binks BP, Clint JH. *Adv Colloid Interf Sci* 2003;**100-102**:503–46.
207. Aveyard R, Binks BP, Fletcher PDI, Peck TG, Rutherford CE. *Adv Colloid Interf Sci* 1996;**64**:67–142.
208. Pugh RJ. *Adv Colloid Interf Sci* 1996;**64**:67–142.
209. Du Z, Bilbao-Montoya M, Binks BP, Dickinson E, Ettelaie R, Murray BS. *Langmuir* 2003;**19**:3106–8.
210. Dickinson E, Ettelaie R, Kostakis T, Murray BS. *Langmuir* 2004;**20**:8517–25.
211. Alargova RG, Warhadpande DS, Paunov VN, Velev OD. *Langmuir* 2004;**20**:10371–4.
212. Binks BP, Horozov TS. *Angew Chem Int Ed* 2005;**44**:3722–5.
213. Fujii S, Ryan AJ, Armes SP. *J Am Chem Soc* 2006;**128**:7882–6.

214. Fujii S, Iddon PD, Ryan AJ, Armes SP. *Langmuir* 2006;**22**:7512–20.
215. Fujii S, Kameyama S, Armes SP, Dupin D, Suzaki M, Nakamura Y. *Soft Matt* 2010;**6**: 635–40.
216. Akamatsu K, Shimada M, Tsuruoka T, Nawafune H, Fujii S, Nakamura Y. *Langmuir* 2010;**26**:1254–9.
217. Geisel K, Richtering W, Isa L. *Soft Matter* 2014;**10**:7968–76.
218. Hofmann CH, Grobelny S, Erlkamp M, Winter R, Richtering W. *Polymer* 2014;**55**: 2000–7.
219. Li Z, Richtering W, Ngai T. *Soft Matter* 2014;**10**:6182–91.
220. Wu Y, Wiese S, Balaceanu A, Richtering W, Pich A. *Langmuir* 2014;**30**:7660–9.
221. Zhang L, Spears MW, Lyon LA. *Langmuir* 2014;**30**:7628–34.
222. Spears MW, Herman ES, Gaulding JC, Lyon LA. *Langmuir* 2014;**30**:6314–23.
223. Clarke KC, Lyon LA. *Langmuir* 2013;**29**:12852–7.
224. Peng Y, Wang Z, Han Y. *J Phys Conf Ser* 2011;**319**:012010.
225. Nozawa I, Suzuki Y, Sato S, Sugibayashi K, Morimoto Y. *J Biomed Mater Res* 1991;**25**: 243–54.
226. Okubo T, Itoh E, Tsuchida A, Kokufuta E. *Colloid Polym Sci* 2006;**285**:339–49.
227. Okubo T, Suzuki D, Yamagata T, Katsuno A, Mizutani M, Kimura H, Tsuchida A. *Colloid Polym Sci* 2011;**289**:807–16.
228. Okubo T, Suzuki D, Tsuchida A. *Colloid Polym Sci* 2012;**290**:411–21.
229. Okubo T, Suzuki D, Tsuchida A. *Colloid Polym Sci* 2012;**290**:867–77.
230. Okubo T, Suzuki D, Tsuchida A. *Colloid Polym Sci* 2012;**290**:1901–11.
231. Yamaguchi T, Kimura K, Tsuchida A, Okubo T, Matsumoto M. *Colloid Polym Sci* 2005;**283**:1123–30.
232. Okubo T, Tsuchida A. *Colloid Polym Sci* 2014;**292**:3187–93.
233. Ida T. New measures of sharpness for symmetric powder diffraction peak profiles. *J Appl Crystallography* 2008;**41**:393–401.
234. Okubo T, Okamoto J, Tsuchida A. *Colloid Polym Sci* 2008;**286**:385–94.
235. Okubo T, Okamoto J, Tsuchida A. *Colloid Polym Sci* 2008;**286**:941–9.
236. Okubo T, Okamoto J, Tsuchida A. *Colloid Polym Sci* 2010;**288**:189–97.
237. Okubo T. *Colloid Polym Sci* 2008;**286**:1527–31.
238. Okubo T, Fujii S, Nakamura Y. *Colloid Polym Sci* 2013;**291**:2805–13.
239. Okubo T, Fujii S, Aono K, Nakamura Y. *Colloid Polym Sci* 2013;**291**:1019–30.
240. Fujii S, Yamashita Y, Nakamura Y, Okubo T. *Colloid Polym Sci* 2014;**292**:2621–31.
241. Okubo T, Tsuchida A. *Colloid Polym Sci* 2014;**292**:133–41.
242. Horigome K, Suzuki D. *Langmuir* 2012;**28**:12962–70.
243. Okubo T, Okamoto J, Tsuchida A. *Colloid Polym Sci* 2009;**287**:351–65.
244. Tomita M, Takano K. *van de Ven: J Colloid Interf Sci* 1983;**92**:367–82.
245. Okubo T. *Ber Bunseng Phys Chem* 1987;**91**:516–9.
246. Klajn R, Fialkowski M, Bensemann IT, Bitner A, Campbell CJ, Bishop K, Smoukov S, Grzybowski BA. *Nature Mater* 2004;**3**:729–35.
247. Kang JH, Moon JH, Lee SK, Park SG, Jang SG, Yang S, Yang SM. *Adv Mater* 2008;**20**:3061–5.
248. Whitcombe MJ. *Nature Chem* 2011;**3**:657–8.

5

Drying Patterns of Dispersions and Solutions

5.1 Cooperative Processes from Wetting to Solidification

Most structural patterns in nature form via self-organization accompanied by the dissipation of free energy and in the nonequilibrium state. Drying dissipative structures are so often observed in nature, because all the dissipative structures form in the condition of dissipation of free energy and in the open atmosphere. Fig. 5.1 shows a typical example of sedimentation to drying patterns of beer bubbles, where the ringlike sedimentation patterns of beer bubbles (lowest ring) change to the drying patterns (highest) appearing inside the wall of a beer mug. One new ring of bubbles appears toward the lower layers whenever we drink. Then characteristic different patterns appear from the lowest to upper layers as a function of time elapsed. The distance between the neighboring ringlike patterns is determined by how much beer was drunk in one sip. The following processes occur: (1) the fine bubbles disappear and a few large spherical bubbles grow; (2) the bubbles change their shape from spherical to rod shaped during the slow flow of bubbles; (3) the rods become slender; (4) the rods split to several spherical liquid spots; and finally (5) dried spots arrayed downward keep a similar distance between neighboring spots during drying. Another typical example of ordering in nature is the freeze-thaw cycles of ground water at the edge of a lake in Norway, which produces convection cells that are traced out by stones on the lake bed.[1] The ordering of the assembled stones is considered to be one of a convectional and sedimentation pattern. It should be noted that these dissipative patterns in nature are not so simple to understand in most cases, because so many factors must be taken into account to understand the pattern formation as a function of time.

To understand the mechanisms of the dissipative self-organization of simple model systems, instead of the much more complex nature itself, studies on the wetting, evaporating, convectional, sedimentary, and solidified dissipative patterns are important during the course of drying suspensions and solutions. However, these five kinds of structures are correlated

Colloidal Organization. http://dx.doi.org/10.1016/B978-0-12-802163-7.00005-2

Figure 5.1
Sedimentary and drying patterns of beer foam at the inside wall of a beer mug observed as a function of time elapsed. The new broad rings formed downward from the initial liquid height in each drinking.

cooperatively and overlapped to each other as is simply shown schematically in Fig. 5.2.[2-4] For example, the size of the drying film is determined during the course of wetting to sedimentary processes, where these processes are correlated intimately and cooperatively. Studies on drying patterns are fundamental and will produce important knowledge in diverse science fields, including the physical chemistry, polymer chemistry, biology, astronomy, and industrial fields, for example.

Wetting of the suspension and solution takes place first at the air–liquid and liquid–substrate interfaces during the course of dryness of liquid. Evaporation takes place almost simultaneously.[5,6] Lowering of the temperature of a liquid close to the interface takes place differently depending on the location, which induces the convectional flow of solvent by the Marangoni effect and gravitational convection. The degree of water temperature lowering at the surface area becomes more than 1 degree even at room temperature. Inhibition of the evaporation of water from the liquid surface is extremely difficult without thorough shielding of the liquid in a container from the atmosphere, and further at the regulated temperature to keep an equilibrium evaporation and condensation of the solvent molecules. It should be noted

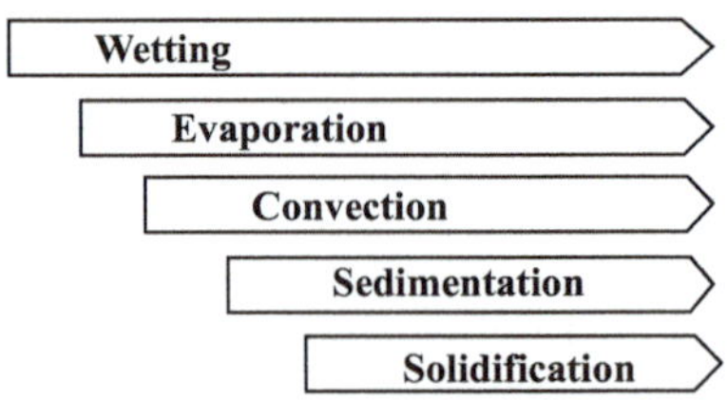

Figure 5.2
Overlapping and cooperative interaction among wetting, evaporating, convectional, sedimentary, and solidified processes during the course of drying dispersions and solutions.

that the accurate measurements of the temperature distribution in the normal and horizontal directions around the interface area is not so easy. For example, the probe for detecting local temperature itself disturbs the local convectional flow and the thermal distribution significantly.

Typical convectional patterns are the Benard cell[7,8] (i.e., the hexagonal circulating pattern) and the Terada cell,[9-11] (i.e., the spoke lines spreading across the whole liquid surface accompanied with the huge number of cell convections in the normal direction of the spoke lines). Recently, the processes of convectional patterns have been studied experimentally.[12-16] Theoretical studies of the convectional patterns have been reported mainly using Navier-Stokes equations.[17-21] However, these are not always successful when the theories are compared with the actual observation.

Sedimentary dissipative patterns during the course of drying suspensions have been studied in detail on a cover glass, a watch glass, a glass dish, and other surfaces, for the first time, in our laboratory.[14,22-29] Several types of the sedimentary patterns, the broad ring, for example, formed already in suspension state. In most cases, the sedimentary particles are suspended in a vessel by the electrical double layers formed around the particles, and moved always by the balancing of the force fields between the convectional flow and the gravitational sedimentation during the course of thermal diffusion. Dynamic clusters and bundlelike" sedimentary patterns form cooperatively from the spokelike convectional structures, which have been observed for coffee[15] and black tea,[16] for example.

The final solidified drying patterns have been studied for many kinds of solutes: colloidal particles, linear-type synthetic and bio-polyelectrolytes, water-soluble neutral polymers, ionic and nonionic surfactants, gels, colloidal polymer-complex particles, and dyes by many researchers hitherto.[2-4,30-36] The macroscopic broad ring patterns of the hill accumulated with the solutes formed most frequently. Macroscopic spoke-line patterns (i.e., spokelike cracks or fine hills including flickering spokelike ones) were also observed for many solutes. Beautiful fractal patterns such as branchlike, arclike, blocklike, starlike, crosslike, stringlike and others were often observed in the microscopic scale, especially in the central and thin area of the dried film.

We should note several important findings on the drying processes. First, the primitive vague sedimentary patterns form before solidification already in the concentrated suspensions or solutions before dryness, and they grow further toward fine structures during the course of solidification.[28] Second, information about the suspensions and solutions such as shape and size, conformation of solutes, atmospheric humidity and temperature, and other factors are transferred into the drying patterns. Some information is transferred to the drying patterns accompanied with the selection and amplifying

Table 5.1 Characteristics of Typical Colloidal Particles

Spheres	d (nm) ± s (nm)	s/d	Charge Density (strong) (μC/cm²)
Polystyrene Spheres			
D1W52	88 ± 6.2	0.071	1.14
D1B28	212 ± 3	0.014	1.29
Colloidal Silica Spheres			
CS22p	29 ± 6	0.21	0.48
CS45	56.3 ± 7.6	0.13	0.30
CS82	103 ± 13	0.13	0.38
CS91	110 ± 4.5	0.041	0.48
CS161	183 ± 18.6	0.101	0.47
CS300	305 ± 9.1	0.030	—
CS301	311 ± 22	0.071	0.40
CS501	507 ± 51	0.10	0.59
CS550	560 ± 17.6	0.031	—
CS1001	1090 ± 45	0.041	—
CS1000A	1205 ± 14	0.012	—
Poly(methyl methacrylate) Spheres			
GW1	100 ± 7	0.07	
GW6	200 ± 10	0.05	
GW8	300 ± 12	0.04	
PM1000	1000 ± 20	0.02	
Polymer Gel Spheres			
CAIBA-P2VP(0.1)	113 nm (25° C)		Cross-linked poly(2-vinylpyridine)
CAIBA-P2VP(0.5)	112 nm		
CAIBA-P2VP(1)	107 nm		
AIBA-P2VP(0.1)	170 nm,		Cross-linked poly(2-vinylpyridine)
AIBA-P2VP(0.5)	170 nm		
AIBA-P2VP(1)	180 nm		
pNIPAm(70-5)	72 nm (20° C), 28 nm (50° C)		Poly(*N*-isopropylacrylamide)
Anisotropic-Shaped Particles			
Kaimei	32.8 ± 9.2		Chinese black ink
Fueki	33.0 ± 6.4		Chinese black ink
Green tea	≈5-20 (μm)		Platelike particle
Coffee with cream	212 ± 1		−52 mV (ζ-potential)
Coffee without cream	455 ± 1		−24 mV
Miso paste	≈10 (μm)–1 (mm)		
Black tea	320		
2p7f (bentonite)	324		
TA (tungstic acid)	TA3 (2*c* = 3 μm, 2*b* = 1.3 μm), TA4 (4, 2.1), TA5 (5, 2.3), TA6 (8.9, 2.6), TA7 (7.1, 3.1)		
CD1 (Diamond)	1.5 ± 0.32 (μm)		
SHB*	688 (long axes), 158 (short)		−12.3 mV (ζ-potential)
Lycopodium spore	30.6 ± 5.6 (μm)		
Adhesive Colloidal Spheres			
PBA†	320 ± 30		−36.1 mV (ζ-potential)

**Spindle-shaped hematite particles coated with polymer brush.*
†Poly(n-butyl acrylate)

effects by the static and dynamic interaction of solutes with other solutes or substrate surfaces. Nowadays, molecular and colloidal information are available in part from the drying patterns themselves, as described later in this chapter. Third, dissipative crystallization of synthetic and biological polyelectrolytes has been clarified.[37-43] Accumulation, uneven distribution, symmetric distribution, ordering of the single crystals, coupling of the single crystals with the dissipative broad ring patterns, and others have been found. As is well known, the three major groups of biopolymers are polypeptides, polysaccharides, and polynucleotides. The chemical structures of these main chains are each different. It is interesting to discuss how these structures influence the drying patterns.

5.2 Convectional Patterns

Tables 5.1 and 5.2 compile the colloidal particles and water-soluble solutes appearing in this chapter. In Table 5.1, *d* and *s* denote the diameter and the dispersion of the diameter of particles, respectively. Gelatinous solutions of COL31 and arrow-root starch (ARS) in Table 5.2 are typical examples of the

Table 5.2 Abbreviations of Typical Water-Soluble Solutes

Abbreviation	Solutes
PEG	poly(ethylene glycol) PEG0.2K, PEG1K, PEG2KI, PEG4K, PEG8K, PEG20K, PEG500K, PEG2000K, PEG3500K
PVA	polyvinyl alcohol
PVP	poly(*N*-vinyl pyrrolidone), PVP25K, PVP30K, PVP90K
NaPMA	sodium salt of poly(methacrylic acid)
HPAA	HPAA25K + PEG20K, HPAA25K + PVP25K poly(acrylic acid) + polyethylene glycol (polyvinyl pyrrolidone)
PAL	poly(allylamine hydrochloride)
RPVP	poly(4-vinyl-*N*-alkyl-pyridinium halide), alkyl: ethyl, butyl, benzyl, hexadecyl-co-benzyl, C2PVP, C4PVP, BzPVP, C16BzPVP
NaP*l*GA	sodium poly(α,*L*-glutamate)
NaP*d*GA	sodium poly(α,*D*-glutamate)
P*l*L.HBr	poly(*L*-lysine hydrobromide)
P*d*L.HBr	poly(*D*-lysine hydrobromide)
HPC	hydroxypropyl cellulose, HPC1, HPC2, HPC3, HPC4
NaHLA	sodium salt of hyaluronic acid
NaDSS	sodium salt of dextran sulfate, NaDSS1, NaDSS2, NaDSS3
NaCMC	sodium salt of carboxymethyl cellulose, NaCMC1, NaCMC2, NaCMC3
αCD	alpha-Cyclodextrin
βCD	beta-Cyclodextrin
γCD	gamma-Cyclodextrin
NaDNA	sodium salt of deoxyribonucleic acid
KPolyA	potassium salt of poly(riboadenylic acid)
KPolyU	potassium salt of poly(ribouridylic acid)
KPolyG	potassium salt of poly(riboguanylic acid)
KPolyC	potassium salt of poly(ribocytidylic acid)
COL31	gelatin (block, Non-spherical)

(Continued)

Table 5.2 Abbreviations of Typical Water-Soluble Solutes—cont'd

Abbreviation	Solutes
ARS	arrow-root starch (block, nonspherical)
10TMAB	*n*-Decyltrimethylammonium bromide
12TMAB	*n*-Dodecyltrimethylammonium bromide
12TMAC	*n*-dodecyltrimethylammonium chloride
14TMAB	*n*-Tetradecyltrimethylammonium bromide
16TMAB	*n*-Hexadecyltrimethylammonium bromide
S6S	*n*-hexyl sulfate
S8S	*n*-octyl sulfate
S10S	*n*-decyl sulfate
S12S	*n*-dodecyl sulfate
S16S	*n*-hexadecyl sulfate
S18S	*n*-octadecyl sulfate
POEmCn	polyoxyethylene-*n*-cetylether (m = 2, 5, 10, 20, n = 16)
POEmCn	polyoxyethylene-*n*-stearylether (m = 5, 10, 20, n = 18)
Dyes	Rhodamine 6G, uranine, 7-hydroxy coumarin, 7-amino-4-(trifluoro methyl) coumarin
Simple salt	KCl, NaCl, $CaCl_2$, $LaCl_3$

netted bulk gels formed at low temperatures, although they are in a solution state above their critical temperatures. The most famous convectional pattern is the hexagonal circulating Benard cell, which has been observed when liquids contain platelike colloidal particles as monitors and are heated homogeneously in a plain pan (Fig. 5.3).[7,8] Distorted Benard cells were observed in miso soup, coffee, and black tea. Another typical convectional pattern is spoke lines, which were observed in the whole area at the liquid surface (Fig. 5.4, A) and also appeared in various substrates, often accompanied by a huge number of small cell convections (Fig. 5.4, *B*). The spoke-line patterns with cell convections were observed formerly for the membranes of Chinese black ink on the surface of water by Terada et al.[9-12,44] We call this

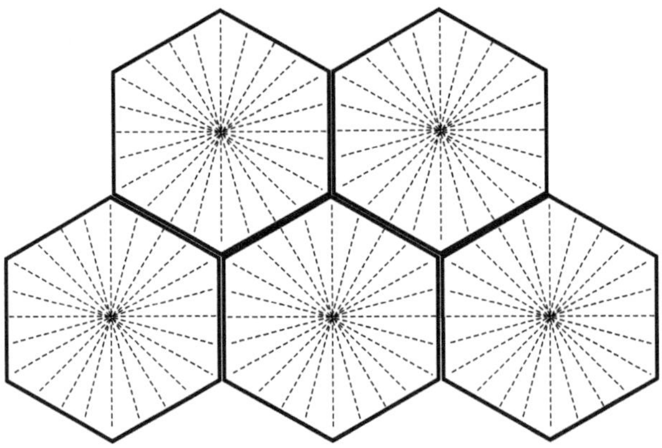

Figure 5.3 Schematic representation of Benard cell. Liquid streams upward at the center then to the hexagonal outside edge in the upper layers near the air–liquid interface.

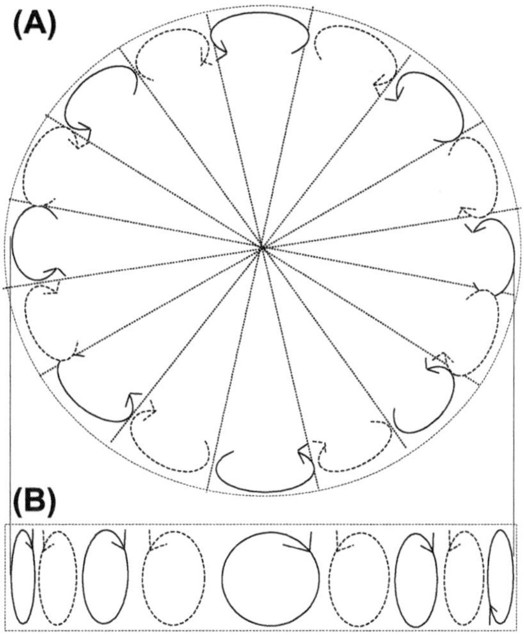

Figure 5.4 Schematic representation of a Terada cell. **A** and **B**, the plane and side views of Terada cell of convection; **B**, cell convections.

the pattern a *Terada cell.* The convectional patterns, especially Terada cells, were observed directly in the initial course of drying of the Chinese black ink in a glass dish,[12] 100% ethanol suspensions of colloidal silica spheres,[13] a cup of miso soup,[14] coffee,[15] black tea,[16] and colloidal crystals of poly (methyl methacrylate) (PMMA) spheres.[45,46] For the 100% ethanol suspensions of colloidal silica spheres, Terada cell–type convectional flow was observed clearly with the naked eye, and the convectional patterns changed dynamically toward the sedimentary patterns with time. Deegan et al.[47,48] and others have reported the traces of spoke-like patterns in the suspensions of polystyrene spheres (1 μm in diameter).

Fig. 5.5 shows typical examples of Terada cell patterns during convectional processes of Chinese black ink observed in a glass petri dish and its schematic presentation (upper and lower pictures of *A*, respectively). Fig. 5.5, *B*, shows the drying patterns formed eventually and the schematic picture showing spoke lines and multiple rings. Macroscopic multiple-ring structures were also observed in the dried film beside the spoke lines. The linear relationship between the number of spoke lines appeared and the number of the cell convections at the outside edge was observed.[12] For Chinese black ink, the other two kinds of convectional patterns, grainlike and smokelike patterns appeared besides spoke lines (Terada cell). The former is shown in Fig. 5.6. The drying patterns corresponding to the latter convectional patterns were

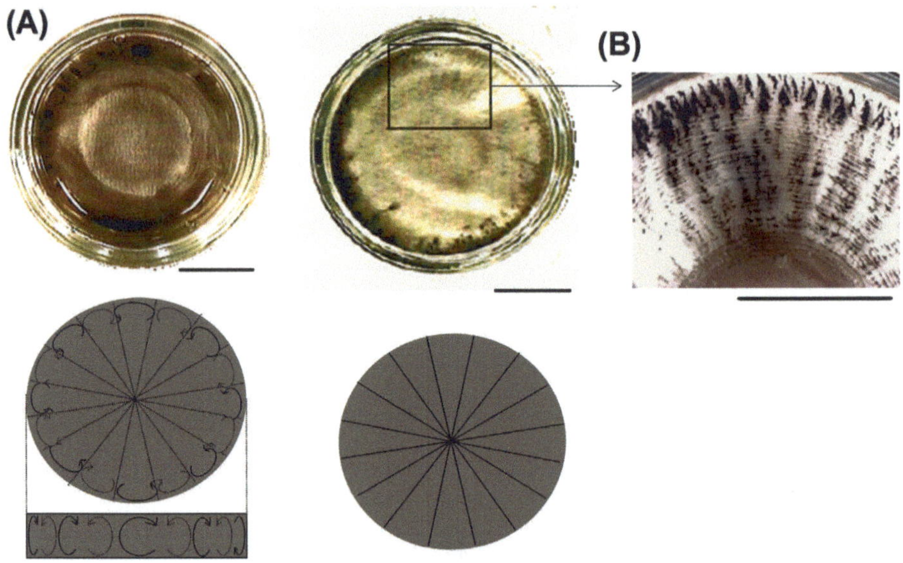

Figure 5.5 Patterns formed in the drying process of Chinese ink in a petri glass dish at 80°. $v = 2$ ml, $w = 1.0 \times 10^{-5}$ g/ml, **A,** 21 minutes after setting; **B,** dried film. Length of the bar is 10 mm. *From Okubo T, Kimura H, Kimura T, Hayakawa F, Shibata T, Kimura K:* Colloid Polym Sci *283:1-9, 2005.*

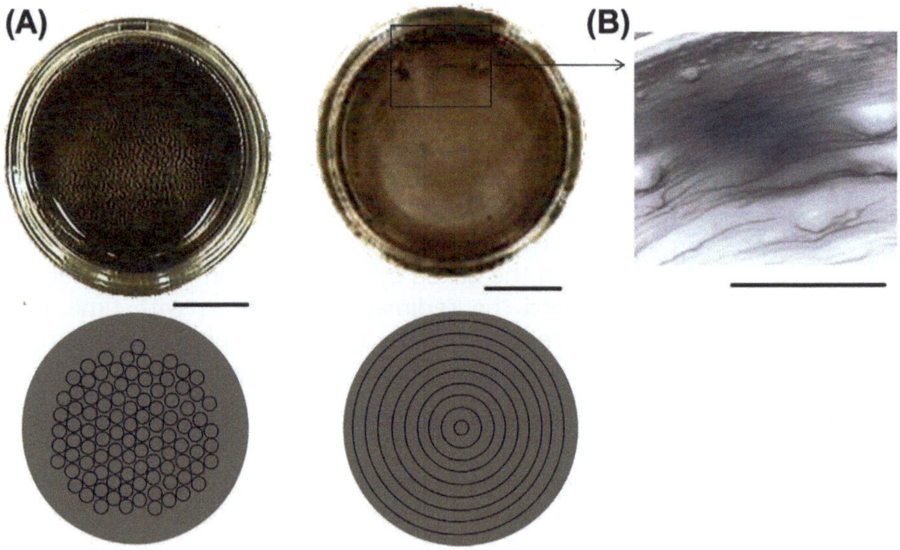

Figure 5.6 Patterns formed in the drying process of Chinese ink in a petri glass dish at 80°. $v = 2$ ml, $w = 5.0 \times 10^{-5}$ g/ml, **A,** 15 minutes after setting, **B,** dried film. Length of the bar is 10 mm. *From Okubo T, Kimura H, Kimura T, Hayakawa F, Shibata T, Kimura K:* Colloid Polym Sci *283:1-9, 2005.*

not observed clearly on the black dried film with the naked eye. These three types of convectional patterns appeared by the delicate balancing of the hydrodynamic and optical contributions, such as the difference in the flow rates between the particles and water, viscosity and Reynolds number of the dispersion, amount of suspension, shape and size of the particles, interparticle distance (particle concentration), atmospheric temperature and humidity, and other factors.

We should note here that the spoke lines in the convectional patterns are just the base of the spoke lines forming in the drying film. When the film is hard enough to keep film from breaking against the shearing forces caused during the solidification process, the number of drying spoke lines decreases. The number of drying spoke lines changed significantly depending on sphere size. In general, the number decreased sharply with the increasing size of colloidal spheres, as is described in Section 5.4.2.[45,46] Furthermore, the arrayed grain patterns in the convectional processes changed into multiple circles with similar intervals, as is shown in Fig. 5.6. In conclusion, drying patterns form as a final event in a series of patterns during cooperative processes: wetting, evaporation, convection, sedimentation, and solidification.

Fig. 5.7 shows a typical example of the convectional patterns of miso soup in a bowl. Several important findings are clear. First, the distorted Benard cells grew up within 30 seconds after setting. Colloidal particles of miso soup are highly polydisperse in their size and shape. And flow birefringence, which affects the multiple scattering, is strong. This high birefringence effect was observed with the naked eye. Benard cells have been seldom observed, appearing only in Chinese black ink in a glass dish[12] and coffee (or black tea) with cream in our laboratory,[15] although the patterns are highly distorted from the normal hexagonal patterns. On the other hand, Terada cells (i.e., the spoke-line patterns on the whole liquid surface) were often observed for Chinese black ink in a glass dish,[12] colloidal crystals of PMMA spheres on a cover glass,[49,50] and coffee with cream,[15] for example. However, the Terada cells were not observed for miso soup. Second, in the beginning stage of convection from 30 seconds to 1 minute, the holes formed and grew large at the cross points of the neighboring three Benard cells. The miso particles rise up from the central area of each Benard cell and flowed down at the outside edges, especially at the holes. Furthermore, the holes moved slowly toward the outside first within approximately 1 minute after setting. Third, the enlarged holes came together at 1 to 6 minutes. Reversal of the flow direction of the holes between 1 and 2 minutes after setting is clear.

When the hot suspensions were set in the substrates, rapid circulations take place by the gravitational buoyancy effect. Then the temperature gradient forms where temperatures of the upper and lower layers of the suspensions are high and low, respectively. When these gradients form stably, the reversal of the global flow of suspensions takes place by the Marangoni effect from the top toward the bottom, especially at the central area of the

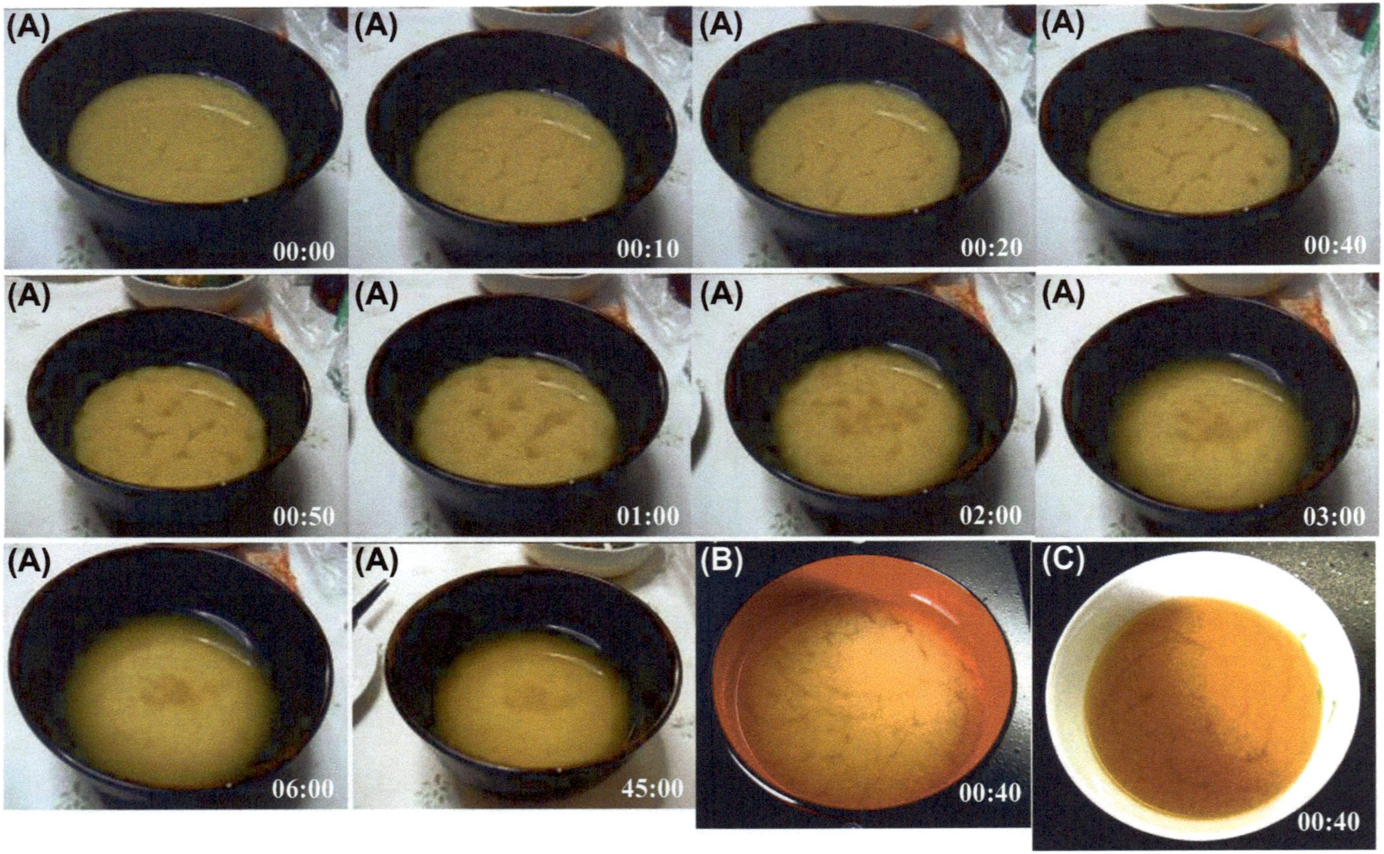

Figure 5.7 **A,** Convectional patterns of miso soup I in a bowl from approximately 60° C to 30° C. Exp. 1, time (min:sec), Miso I: 0.06 g/ml, 180 ml. **B,** Convectional patterns of type II miso soup in a bowl from approximately 60° C to 30° C, Experiment 3, time (min:sec), Miso II: 0.030 g/ml, 180 ml. **C,** Convectional patterns of type II Miso soup in a tea cup from approximately 40° C to 20° C; Exp. 6, time (min:sec), Miso II: 0.038 g/ml, 100 ml. *From Okubo T:* Colloid Polym Sci *287:167-178, 2009.*

liquid. The reversal of the global convection flow has been observed as described previously. Clearly, growth of the distorted hexagonal Benard cells was always accompanied with the outward total flow at the liquid surfaces. On the other hand, the reversed inward flow is accompanied with the formation of a large hole at the central area and, further, a broad, ringlike sedimentation pattern forms mainly by the outward global flow of colloidal particles in the lower layers of liquid. The broad ring–like sedimentation patterns form by the balancing between the outward (or upward) flow of convection in the lower layers of liquid and the inward (or downward) flow of the sedimentation of the colloidal particles on a cover glass or a watch glass. Fig. 5.7, *B* and *C*, are other examples of the distorted Benard cells as they appeared in miso soup. For *C*, slow clockwise rotation of the total liquid remained 40 minutes after setting, and the Benard cells appear to be substantially distorted.

Next, let's discuss the formation of spoke lines and their transformation to clusters and further to the sedimentary bundles patterns. Systems of coffee with cream were used often to study the convectional flow. The mean sizes of coffee and coffee with cream colloidal particles were 460 nm and 210 nm, respectively, and the former was twice as large as the latter. This strongly supports the finding that the coffee particles are apt to aggregate in a suspension state. However, each coffee particle is coated with the cream layers (mainly composed of oils and fats) and forms small, stable dispersions. Zeta potentials of coffee without and with cream were negative and −24 mV and −52 mV, respectively. These values of the zeta potential also support that the coffee particles coated with cream are more stable than the bare coffee particles in a suspension state, because the colloidal particles have large absolute values of zeta potential are more stably dispersed in general. Actually, significant sedimentation of coffee without and with cream in a test tube did not take place for about 1 and 3 days at 8° C, respectively.

When a large amount of coffee with cream suspensions were set on a watch glass, (1) the circulating line patterns were observed within 30 minutes (see Fig. 5.8 and Fig. 5.9). (2) Between 4 and 30 minutes, the global flow of suspension at the liquid surface was in the direction from the center to the outside edge. (3) The global flow was reversed from the outside edge toward the center after 30 minutes, and (4) growth of the spoke lines toward the central area and the formation of clusters and bundles were observed (see Figs. 5.8 and 5.9). (5) After approximately 7 hours, the large bundles further grew into approximately three blocks of bundles (see the time 12:25 and 15:21 of Figs. 5.8 and 5.9). Interestingly, the big bundle is already a dynamically stable sedimentation pattern and further the drying patterns are originated from the bundlelike sedimentation patterns.

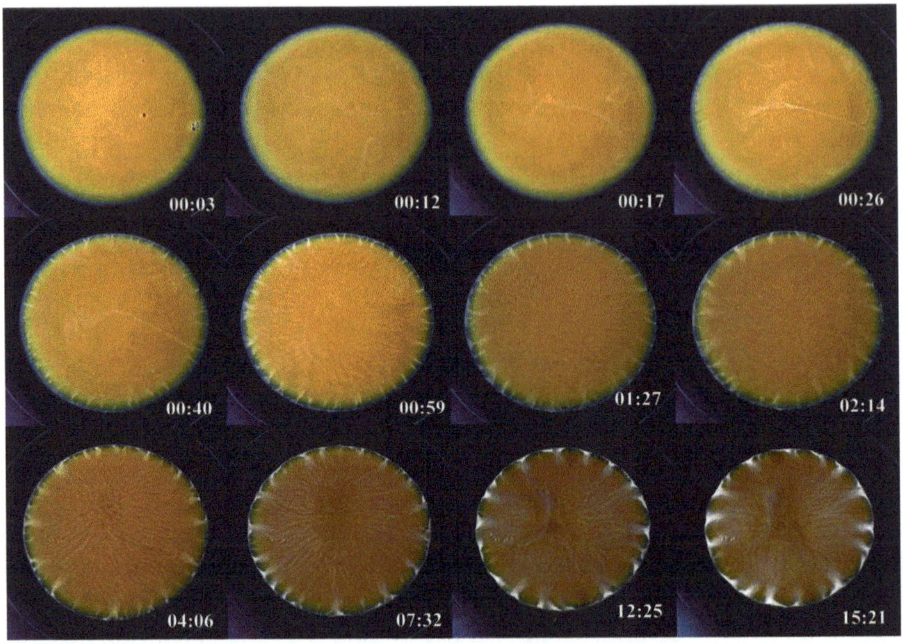

Figure 5.8 Time dependency of the convectional patterns of coffee with cream in a large watch glass. Liquid temperature changes from approximately 70° C to 20° C. *From Okubo T, Okamoto J, Tsuchida A:* Colloid Polym Sci *287:351-365, 2009.*

Fig. 5.10 shows the convectional, sedimentary, and drying patterns of coffee with cream on a cover glass at 25° C. The clear convectional patterns appeared. Twenty minutes after setting (see Fig. 5.10, *A*), the vague, blocklike or broad ring–like patterns appeared from the central area to the outside region of the air–liquid interfaces. Furthermore, a large number of short spoke lines were observed at the outside edge of the liquid. The cooperative convectional patterns of the spoke lines formed clearly. After 65 minutes (see Fig. 5.10, *B*), cooperative clustering of several spoke lines began, the number of spoke lines decreased, and the spoke lines became thick and clear. Furthermore, bundles of the clusters were recognized in the inner area of the liquid surface. The bundles from the clustered spoke lines grew with time and approximately 10 main bundles were formed after 24 hours, 40 minutes (see Fig. 5.10, *E*). These bundles further grew and seven main bundles were observed after 3 hours, 35 minutes (see Fig. 5.10, *G*). Similar convectional patterns showing formation of the clusters and bundles were observed also for other suspensions of coffee with cream (e.g., the stock suspensions and diluted suspensions, in which one part of the stock suspensions were diluted with nine parts of pure water).

The vigorous changes in the convectional flow of colloidal silica spheres in ethanol also supported the formation of the clusters and bundles, as shown in Fig. 5.11, A through I.[13] It should be mentioned again that the bundles are

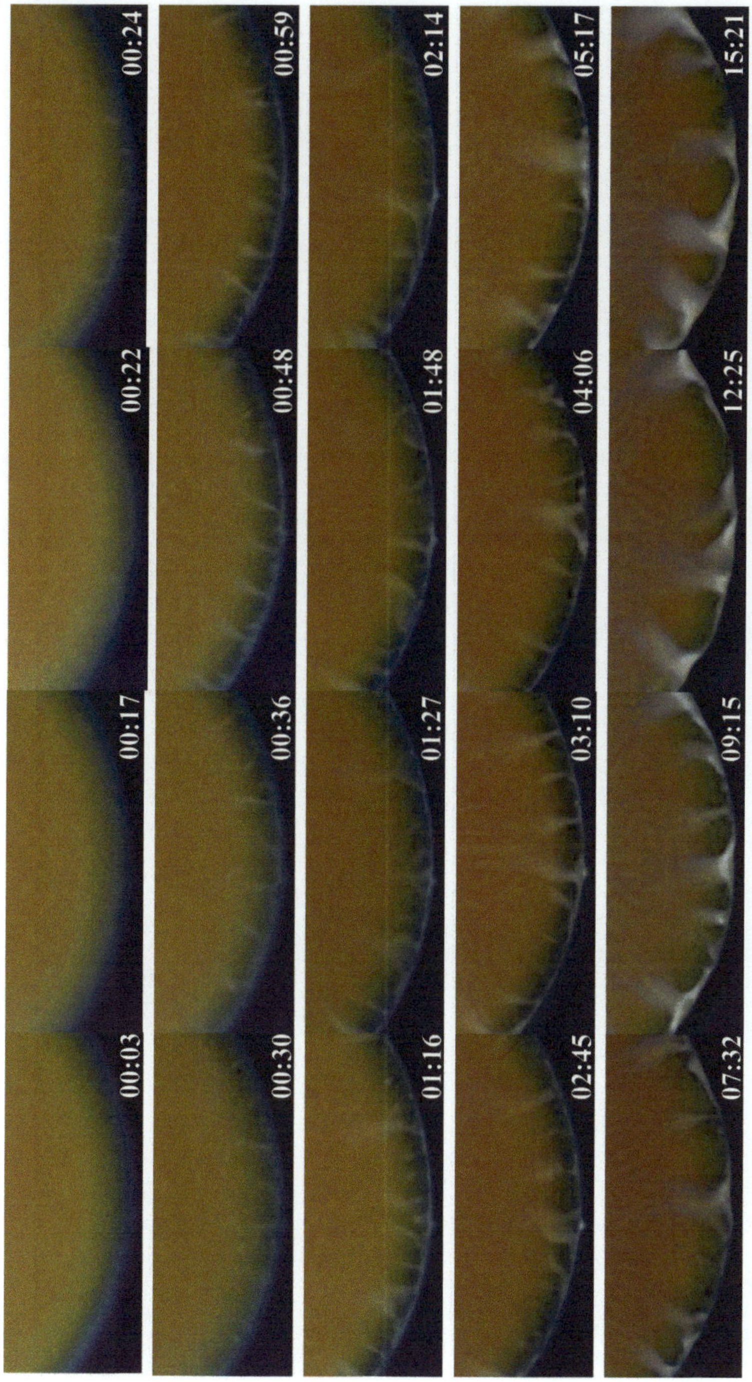

Figure 5.9 Time dependencies of the convectional patterns of coffee with cream in a large watch glass. Extended area of the lower outside region. Liquid temperature changes from approximately 70° C to 20° C. Experiment 5. *From Okubo T, Okamoto J, Tsuchida A:* Colloid Polym Sci *287:351-365, 2009.*

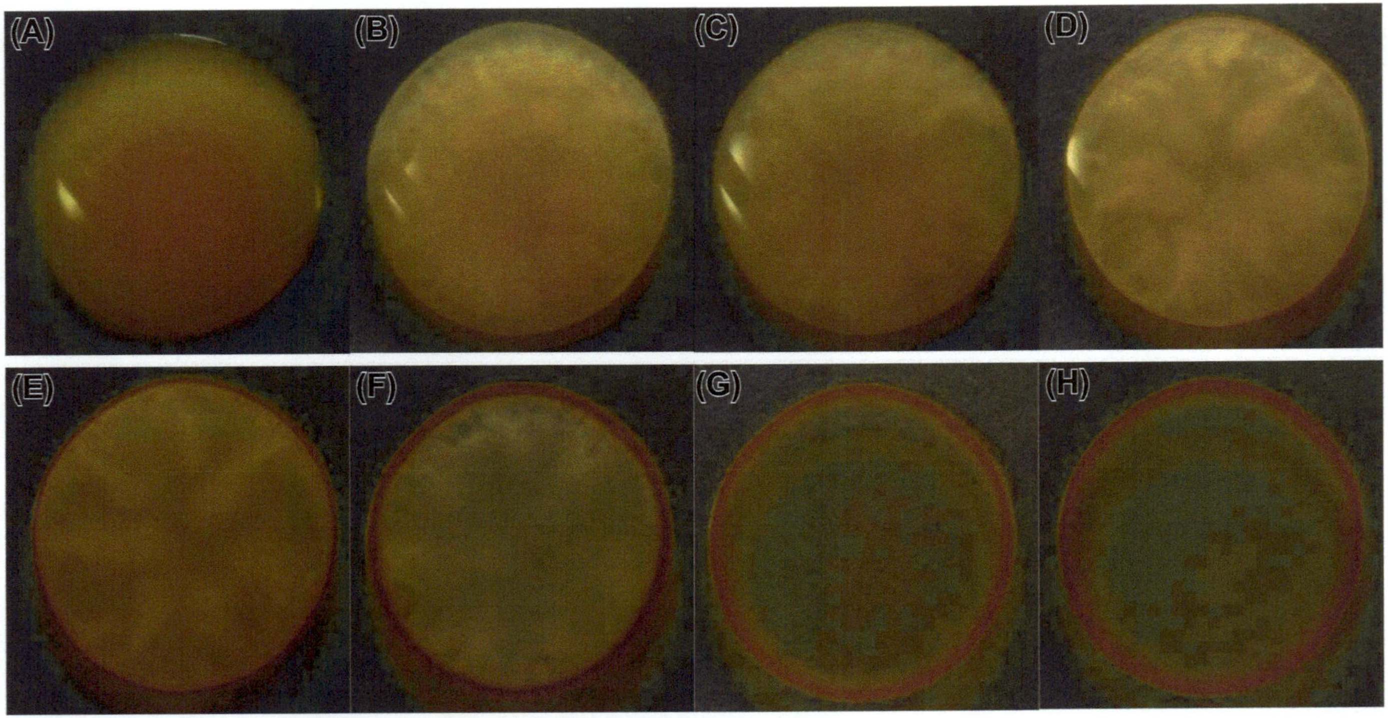

Figure 5.10 Convectional, sedimentation and drying patterns of coffee with cream on a cover glass at 25° C, 0.70 wt%, 0.1 ml, **A,** 20 minutes after setting; **B,** 1 hour, 5 minutes; **C,** 1 hour, 35 minute; **D,** 2 hour, 10 minutes; **E,** 2 hour, 40 minutes; **F,** 3 hour, 15 minutes; **G,** 3 hours, 35 minutes; **H,** 6 hours, 30 minutes. *From Okubo T, Okamoto J, Tsuchida A:* Colloid Polym Sci *287:351-365, 2009.*

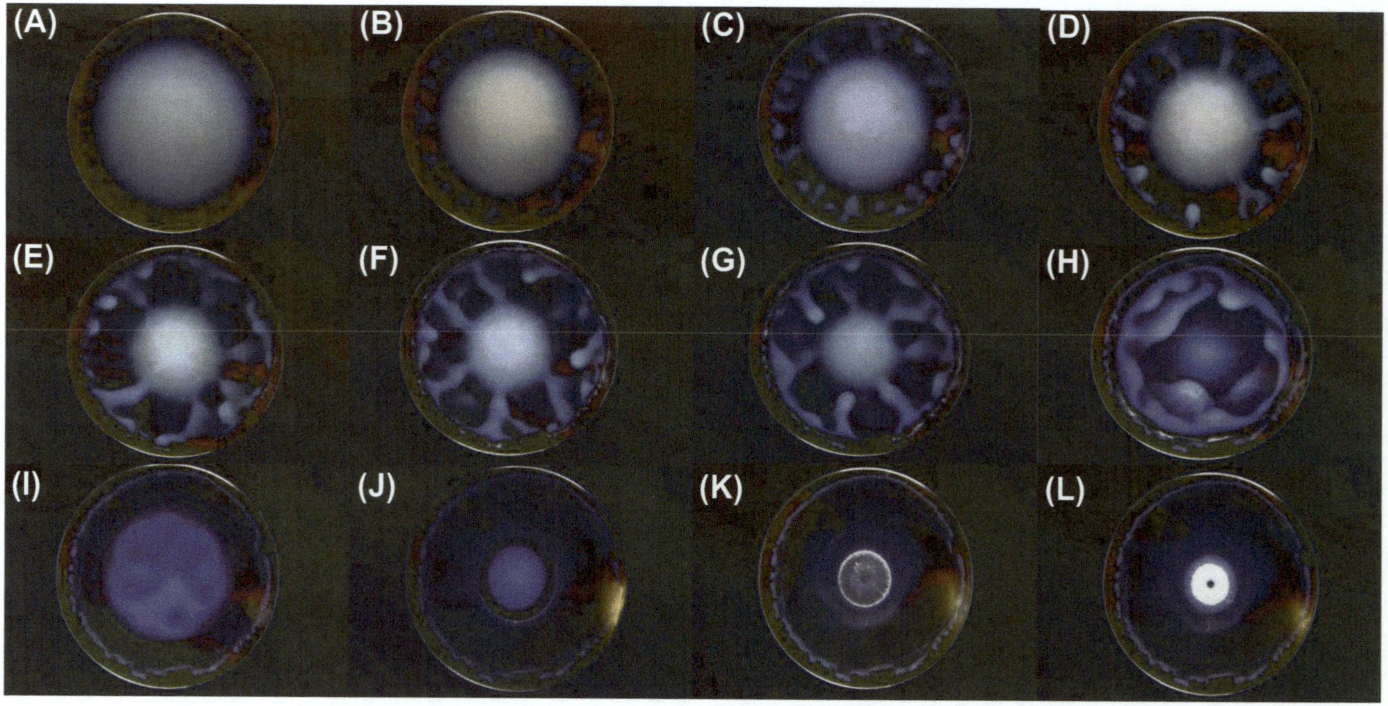

Figure 5.11 Sedimentation and drying dissipative patterns of 100% ethanol suspensions of CS91 spheres in a watch glass at 27° C. $\phi = 0.0064$, 4 ml, **A,** after 7 minutes; **B,** 25 minutes; **C,** 40 minutes; **D,** 50 minutes; **E,** 60 minutes; **F,** 62 minutes; **G,** 65 minutes; **H,** 70 minutes; **I,** 75 minutes; **J,** 90 minutes; **K,** 122 minutes; **L,** 125 minutes.

sedimentation patterns already in the final stage of convection. The clusters and the bundles originate from the spoke lines and always change their forms dynamically and cooperatively. Let's consider the cooperativity in the convectional flow using Fig. 5.11. Vigorous convectional flow gave the beautiful dynamic change in the convectional patterns. In the beginning stage, within 50 minutes, the spoke-line patterns formed outside from the mother suspension in the central area. The suspension drops in the outer region were sometimes bridged with the mother suspension in the central area and disappeared when most of the drop suspensions flowed down to the mother suspension. Then the suspension flow took place from the mother suspension pool toward the outside edge simultaneously, resulting again in the appearance of the spoke-line suspension drops. These dynamic changes in the patterns and the bluish Bragg diffraction colors were beautiful. Between 60 and 75 minutes, the mother suspension area decreased with time and the initial patterns changed drastically to the different kind of patterns, like the section of lotus root (a kind of bundles). After 90 minutes, the suspension remained in the central area only, and the broad rings formed in the final course of drying, which are similar to the patterns so often observed previously in aqueous suspensions.

Fig. 5.12 shows the typical examples of the convectional patterns of PMMA spheres. These patterns were not observed clearly with the naked eye during experiments because the patterns were too small to be recognized. However, the convectional patterns appeared in the extended pictures. Several important findings are obvious in these pictures. First, the lines appeared to develop toward the central area accompanied by an increase in the number of spoke lines. The cell convections, which have been observed for Chinese black ink, were not observed. However, it is highly plausible that each spoke line is composed of the valleys and hills with the upward and downward flow of the circular cell convections (see Fig. 5.4, *B*). Second, the number of spoke lines was small at the initial stage of the growth of the spoke lines. However, the number increased with time and then decreased, passing the maximum number near the final stage of convectional flow. Fig. 5.13 shows the counted results of spoke-line numbers as a function of time for the suspensions of GW1, GW6, and GW8 on a cover glass.[49] The number clearly increased as sphere concentration increased. Third, the round spots of the spoke lines appeared at the central area, though some of the spots existed without accompanied spoke lines and some of the spoke lines did not have spots. We believe that all the round spots are the hollows and not hills. Downward convectional flow at the hollows is highly plausible, although the direct observation on the flow direction was difficult in this observation. As is shown in Table 5.3, several kinds of convectional patterns appeared. It should be noted here that most of the convectional patterns appeared for the anisotropic-shaped colloidal particles, because the particle orientation along

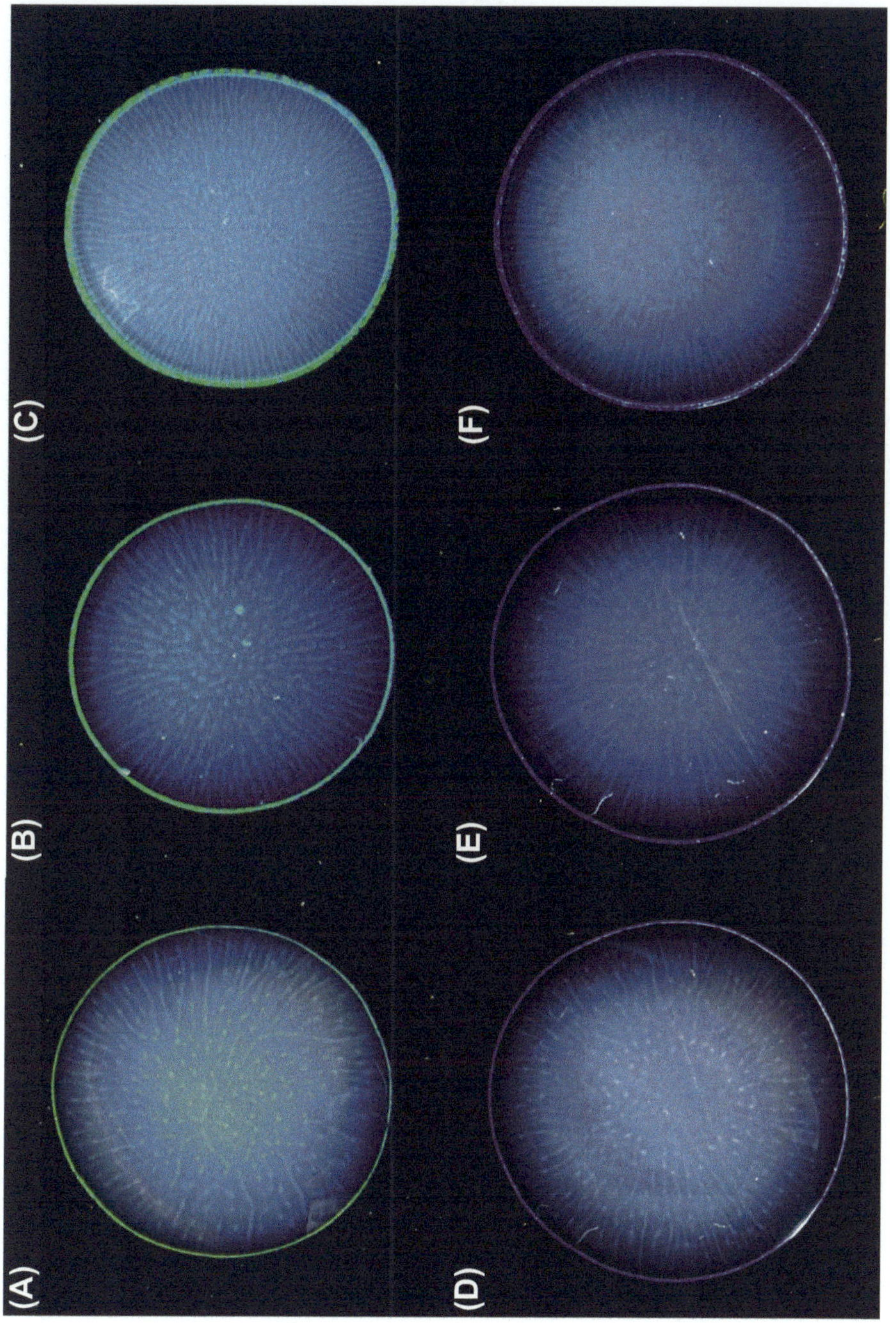

Figure 5.12 Convectional patterns of GW6 (*A-C*) and GW8 (*D-F*) spheres on a cover glass at 24° C, **A,** $\phi = 0.00021$, 3 hours, 15 minutes after setting; **B,** 0.00063, 4hours; **C,** 0.0021, 4 hours; **D,** 0.00025, 3 hours, 20 minutes; **E,** 0.00074, 3 hours, 20 minutes; **F,** 0.00074, 4 hours, 15 minutes.

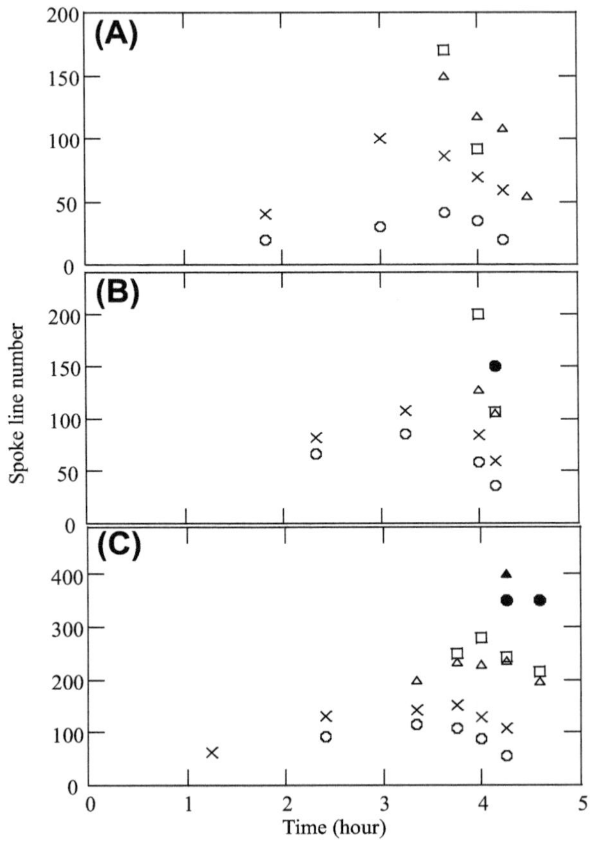

Figure 5.13 Plots of spoke line number of GW1 (*A*), GW6 (*B*), and GW8 (*C*) on a cover glass against time elapsed. **A,** *open circle* $\phi =$ 0.000318, *cross* 0.00095, *triangle* 0.00318, *square* 0.0064; **B,** *open circle* $\phi =$ 0.00021, *cross* 0.00063, *triangle* 0.0021, *square* 0.0042, *solid circle* 0.0105; **C,** *open circle* $\phi =$ 0.000245, *cross* 0.00074, *triangle* 0.00245, *square* 0.0049; *solid circle* 0.0123; *solid triangle* 0.0245. *From Okubo T, Okamoto J, Tsuchida A:* Colloid Polym Sci *286:1123-1133, 2008.*

Table 5.3 Typical Convectional Patterns Observed

Macroscopic Patterns	
Spoke lines (Terada cell)	Coffee, black tea, PMMA
Hexagonal cell (Benard cell), Normal	Metal flakes in silicon oil
Hexagonal cell (Benard cell), Distorted	Miso soup
Smokelike cells	Chinese black ink
Grainlike	Chinese black ink
Microscopic Patterns	
Cell convection (Terada cell)	Chinese black ink

PMMA, Poly(methyl methacrylate).

the spoke lines by the flow birefringence was the main cause for the appearance of the flow lines observable with the naked eye. Highly monodispersed spheres of PMMA are considered to show no convectional flow patterns. It is plausible that the temporal and reversible aggregates form by the weakly hydrophobic intersphere attraction forces during the convectional processes.

In conclusion, the convectional processes are analyzed as six steps, as is shown in Fig. 5.14. Here, time scales given in the figure are based on the experiments of coffee with cream. (1) At the initial stage, the irregular circulating lines appeared at random in their direction. (2) Global flow of convection at the surface layers at the initial stage was from the center toward the outside edge. (3) Furthermore, the distorted Benard cells formed at the liquid surface. (4) Meanwhile, at the middle stage of convections, the global flow of convection was reversed in direction from the outside edge to the central area at the liquid surface, and the outward direction remained for a long time until solidification took place and also induced the broad, ringlike sedimentation structure. (5) At the same time, a few short spoke lines appeared at the outside edge. Then the number of spoke lines increased and grew toward the center. (6) Growth of the spoke lines and formation of the clusters and bundles took place from the middle to the final stages. It should be noted that one or two patterns among six steps were not recognized, depending on the systems. This is because the shape, size, and polydispersity values of the particles influence the quality of the optical observation substantially by

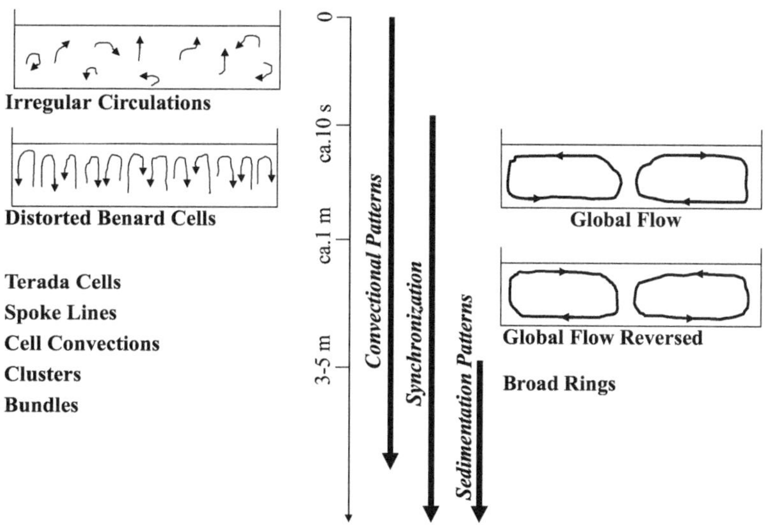

Figure 5.14 Schematic presentation of formation processes of convectional and sedimentary patterns with time. *From Okubo T, Okamoto J, Tsuchida A:* Colloid Polym Sci *287:351-365, 2009.*

the change in the magnitudes of strength of the multiple scattering, for example.

Studies on the convectional flow have been reported hitherto.[17-21,47,48] Calchile et al.[51,52] reported on the droplets of completely wet liquids deposited on a thoroughly smooth, wet surface for which no contact line anchoring occurs. The convectional flow of solvent and solutes is most important among the wetting, evaporation, convection, sedimentation, and solidification processes during the course of drying. It should be noted further that information on the size, shape, conformation, and flexibility of particles and polymers is transformed from the initial convectional process toward the succeeding sedimentation and then solidification processes cooperatively and accompanied with the amplification and selection processes. The important role of Marangoni convection in the metal colloidal crystallization was proposed by Pileni et al.[53] Experimental and theoretical studies on the convectional flow of colloidal particles have been reported, although these studies are not satisfactory for discussing drying patterns.[20,53-63]

5.3 Sedimentary Patterns

When green tea is prepared by dipping the tea bag dipped into hot water in a cup, the broad ring always formed within 10 minutes when green tea in a cup was left to stand on a disk. Fig. 5.15, *A* and *B*, shows one of the typical examples of the broad ring patterns of green tea.[29] The broad ring, which is the hill accumulated with green tea colloidal particles, formed in the inclined inner wall of the cup, and not in the deepest central area. The broad-ring structures were clarified to form by the delicate balancing of the upward (or outward) convectional flow of the colloidal particles near an inner wall and downward gravitational sedimentation (or inward convectional flow).

The broad ring–like sedimentary patterns were observed in the final stage of the convectional flow of miso particles, as is shown in Fig. 5.15, *C* (also see Fig. 5.7, *A*, 45 hours after setting.[14] Sedimentary patterns were observed also in the diluted suspension, as is clearly shown in Fig. 5.15, *D*.

Several other different sedimentary patterns were observed in the final stages of convectional patterns represented by the bundles, which were often grown from the spoke lines and then from the clusters by the cooperative interactions among the spoke lines. Fig. 5.16, *A* through *D*, shows the typical examples of the sedimentary patterns of bundles observed for coffee with cream on a cover glass (*A*, *B*) and a watch glass (*C*, *D*).[15] Fig. 5.16, *E* and *F*, show the broad ring–like sedimentary patterns accompanied with the clusters outside the broad ring (i.e., inner broad ring and outer clusters), which were observed in a coffee cup and in an evaporating dish, respectively.

Figure 5.15
Typical sedimentary patterns of green tea (*A, B*) and miso soup (*C, D*). ***A** and **B** from Okubo T:* Colloid Polym Sci *285:331-337, 2006;* ***C** and **D** from Okubo T:* Colloid Polym Sci *287:167-178, 2009.*

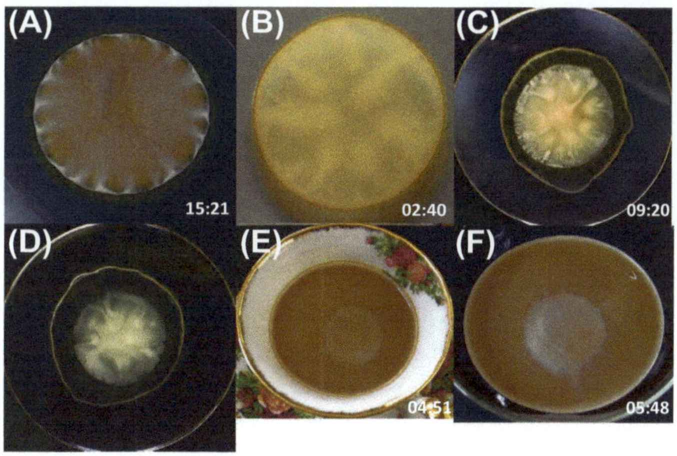

Figure 5.16
Typical sedimentary patterns of coffee with cream on a cover glass (*B*), a watch glass (*A, C, D*), a coffee cup (*E*), and evaporating dish (*F*). **A,** Large watch glass, 2.3 wt%, 40 ml, 70° C to 20° C; **B,** cover glass, 0.70 wt%, 25° C; **C,** 0.70 wt%, 3/10, 25° C; **D,** 10 hours, 10 minutes, 1 ml; **E,** 2.3 wt%, 130 ml, 70° C to 25° C, 4 hour, 51 minutes; **F,** evaporating dish. *From Okubo T, Okamoto J, Tsuchida A:* Colloid Polym Sci *287:351-365, 2009.*

Starlike, clusterlike, and bundlelike sedimentary patterns were observed for PMMA spheres on a cover glass and a watch glass, as is shown in Fig. 5.17.[49,50] Continuity from the sedimentary to the drying patterns are clear in Fig. 5.17, *E* and *F*. Approximately 10 iridescent colored bundles formed in the suspension phase in Fig. 5.17, *E*, are also kept in the drying patterns of Fig. 5.17, *F*, at the same positions. This observation supports the existence of the transferring effects of the suspension's information during the course from the sedimentary to solidification processes.

The starlike bundles of sedimentation patterns were observed often for cationic gel spheres of poly(2-vinylpyridine (P2VP) (i.e., CAIBA-P2VP[64] and AIBA-P2VP[65]) on a cover glass and a watch glass (see also Fig. 4.67, *A-D,* in Chapter 4). Characteristics of CAIBA-P2VP and AIBA-P2VP is given in Table 4.3 of Chapter 4. Most sedimentary patterns were starlike bundles. The number of main bundles was between six and eight. In general, the number of bundles in the convectional patterns does not change so much during the sedimentary processes. However, the number of spoke lines during the convectional patterns increased first and then decreased, accompanied by

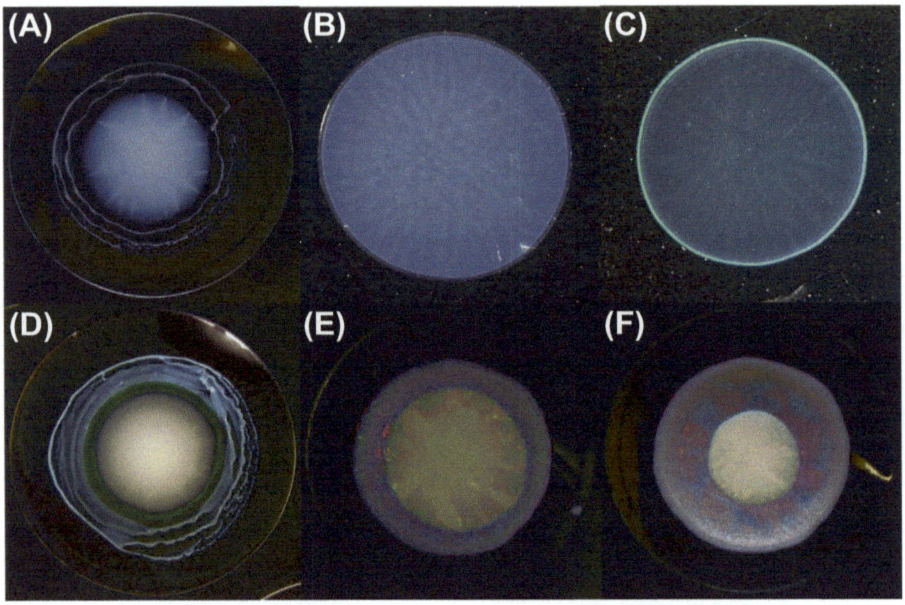

Figure 5.17
Sedimentary patterns of PMMA spheres at 25° C. **A,** GW1 26 hours after setting, $\phi = 0.00095$, watch glass; **B,** GW1, 0.00095, cover glass, **C,** GW6, 0.00021, 4 hours, 10 minutes, cover glass; **D,** GW6, 0.0021, 30 hours, 50 minutes, watch glass; **E,** GW8, 0.074, 17 hours, 15 minutes, watch glass; **F,** GW8, 0.074, 20 hours, 25 minutes, watch glass. *From Okubo T, Okamoto J, Tsuchida A:* Colloid Polym Sci *286:1123-1133, 2008; Okubo T:* Colloid Polym Sci *286:1307-1315, 2008.*

the formation of sedimentary patterns. The sedimentary patterns of poly(*N*-isopropylacrylamide) (pNIPAm) gel spheres[66] were not observed clearly compared with the clear-cut patterns of P2VP gel spheres. Fig. 4.67, *E* and *F*, show the examples of inner broad ring and outer clusters and the intermediate patterns of bundles and clusters of pNIPAm gel spheres, respectively.

In ethanol suspensions of colloidal silica spheres, sedimentary patterns that look like a section of lotus root were observed between 60 and 75 minutes after the suspension was set.[13] Three examples of the sedimentary patterns are shown in Fig. 5.18, *A* through *C*, for CS91 spheres. The sedimentary patterns were still dynamic and the patterns changed slowly with time. In Fig. 5.18, *D* through *F*, typical examples of the sedimentary patterns showing the intermediate states of spoke lines and clusters are given for aqueous suspensions of poly(*n*-butyl acrylate) (PBA) spheres.[67]

Let's discuss the dynamic nature of the sedimentary patterns. Fig. 5.19 shows the sedimentary patterns of the deionized suspensions of CS300 spheres in a glass dish.[24] We controlled the room temperature and humidity by turning the air conditioner on and off. The experiments started 45 hours after the suspension was set, when the stable sedimentation state was achieved without air conditioning. After a Fig. 5.19, *A*, was taken (at 27° C and 67% humidity), the air conditioner was turned on. After 15,

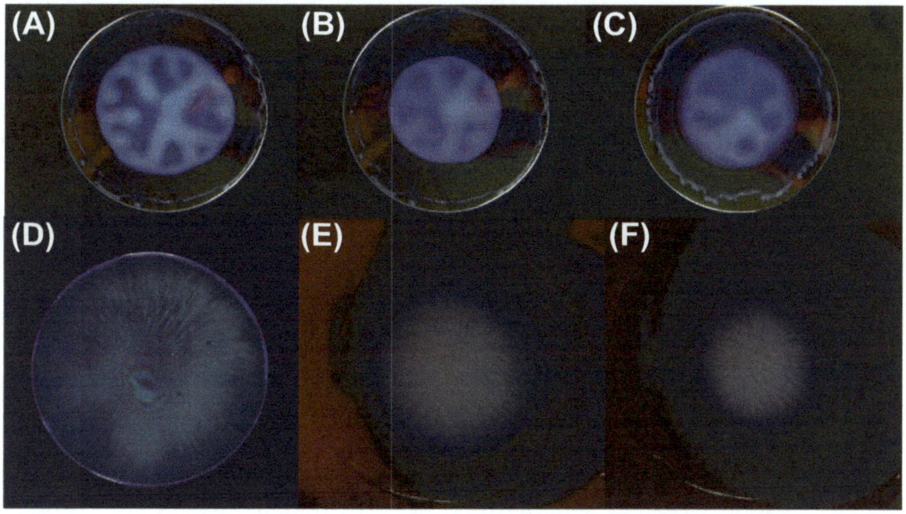

Figure 5.18
Sedimentary patterns of **A,** CS91 (in 100% ethanol, $\phi = 0.00427$, on a watch glass, 4 ml, 70 minutes after setting, at 25° C); **B,** CS91 ($\phi = 0.00427$, 75 minutes); **C,** CS91 ($\phi = 0.0064$, 4 ml, 75 minutes); **D,** PBA (0.025 wt%, 3 hours, 50 minutes, on a cover glass); **E,** PBA (0.048 wt%, on a watch glass, 12 hours, 30 minutes); **F,** PBA (13 hours, 20 minutes).

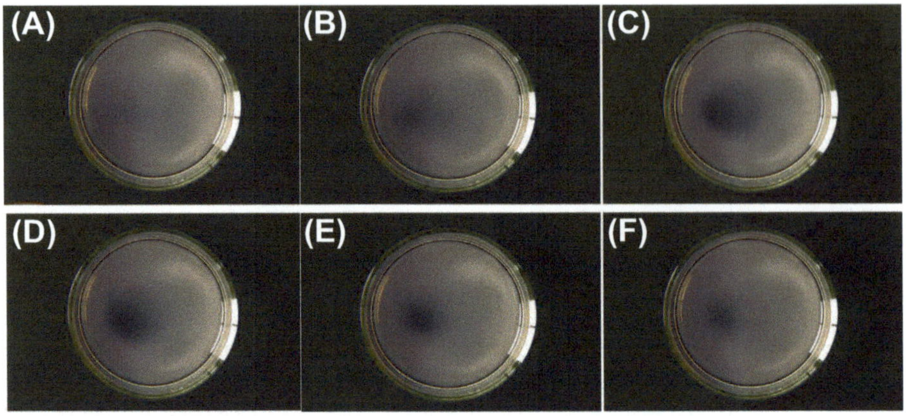

Figure 5.19
Change in the sedimentation patterns of CS300 silica suspensions in a glass dish 45 hours after setting at 27° C. $\phi = 0.00038$, 10 mL, **A,** 0 min, 27.0° C, 67%, ON; **B,** 15 minutes, 26.6° C, 65 %; **C,** 35 minutes, 25.5° C, 61%; **D,** 45 minutes, 25,2° C, 60%, OFF; **E,** 67 minutes, 25.2° C, 63%; **F,** 70 minutes, 25.1° C, 66%. *From Okubo T, Okamoto J, Tsuchida A:* Colloid Polym Sci *285:967-975, 2007.*

35, and 45 minutes, Figs. 5.19, *B, C,* and *D* were taken. At these times, the values of the room temperature and humidity decreased to 26.6° C and 65% humidity, 25.5° C and 61% humidity, and 25.2° C and 60% humidity. Then the sedimentary patterns became sharp with time. After taking Fig. 5.19, *D,* the air conditioner was turned off. The patterns became slightly vague after 67 minutes (*E*) and 70 minutes (*F*), respectively, as the temperature and humidity of the room decreased only slightly and increased, respectively. The thermal movement of spheres and water molecules are responsive to the change in temperature and humidity at the air–suspension interfaces. The thermal movement of the spheres and water molecules is the main cause for the broad-ring sedimentary patterns. It should be mentioned here that the slopes, not the absolute values, of the temperature and humidity against time are important for the sedimentary pattern formation.

The relationship between the sedimentary and solidification patterns is discussed in more detail. Fig. 5.20, *A* through *D,* shows a bentonite suspension on a cover glass after 46 minutes, 30 seconds; 48 minutes; 48 minutes, 45 seconds; and 49 minutes, 50 seconds.[28] The drying frontier moves from right to left with time in the pictures. The border between the liquid and solid regions is the frontier zone of drying. Soon after setting the suspensions, vague wrinkled sedimentary patterns appear, which are composed of the accumulated particles the had already formed in the suspension phase, although they were not so fine and clear. As time elapsed, the drying frontier moved to the left side and the vague patterns formed in the liquid phase became clearer and finer during the course of dryness. These observations

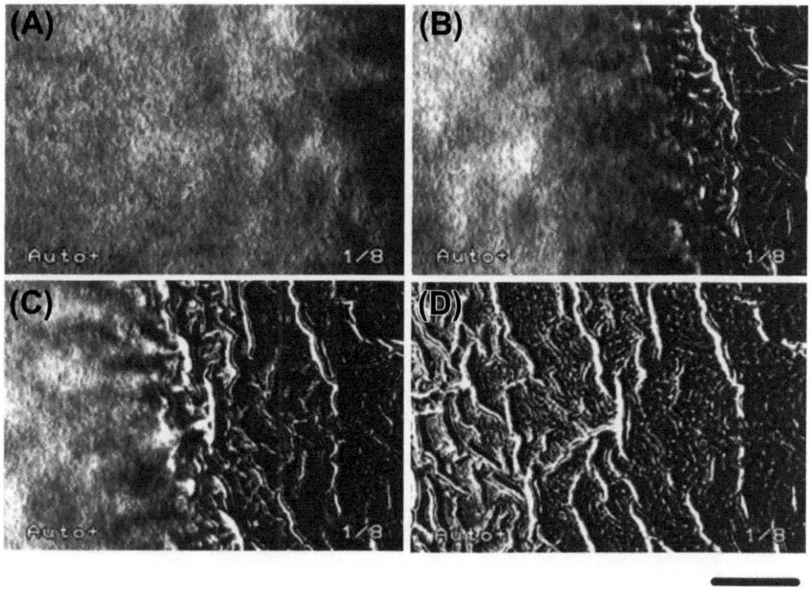

Figure 5.20 Drying patterns of bentonite particles on a cover glass at 25° C. w = 4.6 mg/ml, 0.05 ml, **A,** 46 minutes, 30 seconds after dropping the suspension; **B,** 48 minutes; **C,** 48 minutes, 45 seconds; **D,** 49 minutes, 50 seconds, length of the bar is 10 mm. *From Yamaguchi T, Kimura K, Tsuchida A, Okubo T, Matsumoto M:* Colloid Polym Sci *283:1123-1130, 2005.*

support the finding that that the drying patterns form more clearly and finely based on the vague sedimentary patterns. Interparticle attraction forces induced by the capillary forces is one of the main causes for the sharp drying pattern formation during the phase transition.

Fig. 5.21 shows the growth processes of the spoke-line cracks with time during the course of drying suspensions of CS82 colloidal silica spheres on a rinsed cover glass.[46] Here, the sedimentary patterns are not observable clearly. At t = 13 seconds, only the region of the outside edges were dried up, but most of the area was the blue suspension, indicating the colloidal crystal state. At t = 18 minutes, 15 seconds, the crack formation started to grow. The parallel crack lines developed first toward the left-hand side, and then turned to grow further toward the central region. The growth of the cracks was very fast and stopped around the middle point between the outside edge and the frontier of the suspension area. It is highly plausible that the black area between the growth stopping points and the suspension frontier is wetted with water, although the area appears to the naked eye to be dry. The black area where the cracks appeared is, therefore, really dried. When the growth of the preceding crack line stopped, the growth of the new crack took place at the left-hand side neighbor cooperatively. The growth again stopped at the middle point. Then another crack continued to grow at the left-hand side neighbor, and so on. After all the cracks were completed around the middle points, they grew toward the central area slowly and simultaneously with similar rates as the withdrawal of the

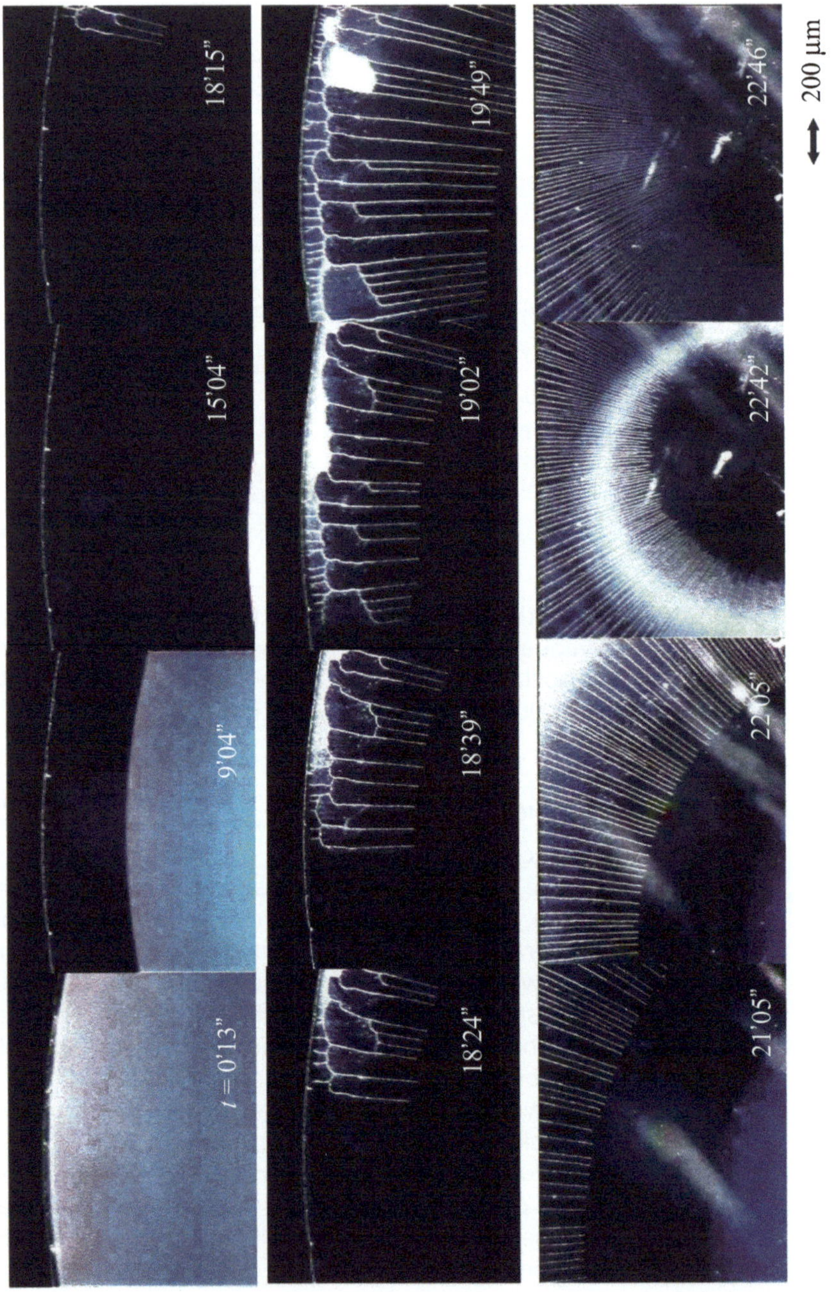

Figure 5.21 Processes of the drying patterns of colloidal crystals of CS82 spheres on a rinsed cover glass at 25° C. 10 ml, $\phi = 0.033$.
From Okubo T, Nozawa M, Tsuchida A: Colloid Polym Sci *285:827-832, 2007.*

Table 5.4 Typical Sedimentary Patterns Observed

Macroscopic Patterns	
BR	Green tea, miso soup, silica spheres
Clusters	Coffee, black tea, PMMA spheres, PBA spheres
Bundles	Coffee, black tea, PMMA spheres
Starlike bundles	P2VP gel spheres
BR inner and clusters outer	Coffee, black tea
Cross-section of lotus root	Colloidal silica spheres in 100% ethanol
Microscopic Patterns	
Vague wrinkled patterns	Bentonite

BR, Broad ring; *P2VP*, poly(2-vinylpyridine); *PBA*, poly(*n*-butyl acrylate); *PMMA*, poly(methyl methacrylate).

suspension frontier. The fast and slow growth rates of the crack formation were 6.2 mm/s and 9.8 μm/s, respectively. The former rate was approximately 630-fold faster than the latter. It should be noted that the cooperative fine-crack formation keeping intervals similar to its neighbors support the finding that the fine spoke-line drying patterns formed based on the vague sedimentary patterns, which are also formed based on the convectional patterns. This finding clarifies the fact that the drying patterns form cooperatively on the basis of the sedimentation patterns, which also originate from the convectional patterns in the suspension state, of course.

The interesting sedimentation and drying patterns also formed during the course of dryness of binary and ternary colloidal crystal mixtures among the colloidal silica spheres of 183 nm, 305 nm, and 1.21 μm in diameter in an aqueous suspensions on a watch glass, a glass dish, and a cover glass.[26,27] Colorful macroscopic broad-ring drying patterns appeared on the basis of the sedimentation structures. In a watch glass, for example, macroscopic sedimentation and then drying patterns were composed of the outer and inner layers of small and large spheres, respectively, for the binary mixtures by the segregation effects.[26] Typical sedimentary patterns observed during the course of drying are compiled in Table 5.4. Continuity of the drying patterns from the convectional, sedimentary, and solidification patterns is supported from the observation described above.

5.4 Drying Patterns of Colloidal Hard Spheres

5.4.1 Monodispersed Polystyrene Spheres

In most cases, 0.1 ml of the aqueous sample suspension was dropped carefully and gently on a microscopic cover glass (30 mm × 30 mm), thickness number 1 (≈0.12-0.17 mm) set in a plastic dish. The cover glass

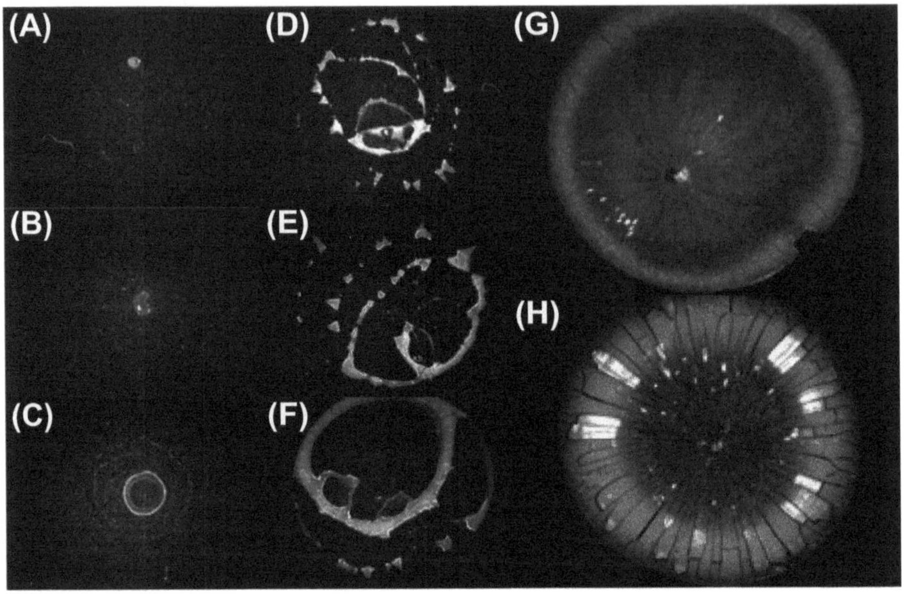

Figure 5.22 Drying patterns of monodispersed polystyrene spheres (D1W52) in the deionized aqueous suspension on a cover glass at 25° C. 0.1 ml, **A,** $\phi = 8.9 \times 10^{-10}$; **B,** 8.9×10^{-8}; **C,** 8.9×10^{-6}; **D,** 8.9×10^{-4}; **E,** 0.0044; **F,** 0.0089; **G,** 0.010; **H,** 0.022.

was used without further rinse in many cases. The extrapolated value of the contact angle for pure water was $31 \pm 0.2°$ from the drop profile of a small amount of water (0.2, 0.4, 0.6 and 0.8 ml) on the cover glass. Some of the experiments were made on a rinsed cover glass with use of the mixed acid sulfuric acid solution of potassium dichromate. The extrapolated contact angle of the rinsed cover glass for pure water was $11 \pm 0.2°$. In most experiments, 0.7 ml was set on a watch glass dish (50 mm in diameter) and a petri glass dish (27 mm in inner diameter and 15 mm in height). Different sizes of cover glasses, watch glasses, and glass dishes were also used for the different samples. Drying patterns were observed macroscopically and microscopically using a close-up camera, a digital HD microscope, a video charge-coupled device camera, and a metallurgical microscope.

Fig. 5.22 shows the typical examples of the drying patterns of monodispersed polystyrene spheres.[68] The pictures were taken by a video recorder and were not so clear. Above 0.01 in volume fraction (ϕ) clear-cut broad rings formed at the outside edge. Spoke lines corresponding to the cracks formed at the broad ring areas. At sphere concentrations lower than $\phi = 0.0089$, the broad rings at the outside edge became small. And the broad rings shifted toward the inner area, although the spots arrayed in the rings or fine ring patterns remained outside the broad ring. The main cause for the remaining accumulation area outside the broad ring in Fig. 5.22, *D* through *F*, was the high hydrophobic attraction forces between the hydrophobic polystyrene spheres and a part of the stained substrate glass surface, where lipophilic impurities might be adsorbed locally. The excluded volume effect is supported, as is shown in Fig. 5.22. These broad rings formed clearly based on the broad ring–like sedimentation patterns prior to solidification. A main cause for the broad ring formation in the drying pattern is the convectional flow of colloidal spheres and water in the different rates (i.e., the motion of colloidal spheres from the central area toward the outside edge in the lower layer of the liquid, which was observed with the naked eye) for Chinese black ink in a glass dish and a drop of PMMA suspension on a cover glass and a watch glass, as discussed in Section 5.2. The convectional flow is enhanced by the evaporation of water at the liquid surface, resulting in the lowering of the temperature at the upper region of the liquid. When the spheres reach the edge of the drying frontier, a part of the spheres will turn upward and go back to the central area. However, the movement of most spheres may stop at the frontier region, because of the capillary effects proposed by Nagayama et al.[69,70] and the disappearance of water molecules. These processes must be followed by the broad ring–like accumulation of the spheres at the outside edge.

The position where the broad ring–like accumulation starts is not always at the outside edge of the initial liquid, as proposed by Diegan et al.[47,48] (i.e., pinning effect at the contact line). The size of the broad ring is determined mainly by the excluded-volume effects of the colloidal spheres, instead, which is roughly given by the sphere number multiplied by the effective volume of colloidal spheres, $(4/3)\pi(d_{eff}/2)^3$, where d_{eff} is the sum of real diameter of the sphere plus twice the Debye screening length (l_D) ($d_{eff} = d + 2 \times l_D$). When the total excluded volume in suspension state is larger than the volume of the initial suspension, the broad ring forms at the outside edge. However, the broad ring size decreases when the total excluded volume is lower than the initial liquid volume. At low sphere concentration, for example, the broad ring size should decrease until the broad ring area reaches the total excluded volume. However, the change in the broad ring size took place only when the sphere substrate interaction is negligibly small. When the affinity of spheres with the substrate is strong, the broad ring formation takes place always at the outside edge of the initial suspension. The patterns in Fig. 5.22 support the finding that affinity between polystyrene spheres and glass substrate is weak.

Fig. 5.23 shows the ratio of the broad ring size of the dried film in diameter d_f against the size of the initial liquid d_i for polystyrene spheres, D1W52 and D1B28. At the sphere concentration above $\phi \approx 0.01$, the d_f / d_i values kept unity, whereas the ratio decreased significantly as sphere concentration decreased below $\phi \approx 0.01$.

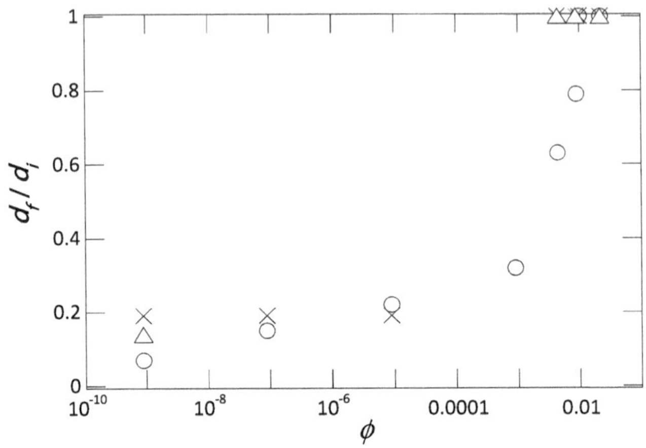

Figure 5.23 Values of d_f/d_i of D1W52 (*circles,* 0.1 ml and *crosses,* 0.01 ml) and D1B28 spheres (*triangles,* 0.1 ml) as a function of sphere concentration.

Fig. 5.24 shows the thickness profiles of the dried film of polystyrene spheres as a function of the distance from the center of the dried film on an arbitrary diameter line. Sharp broad-ring patterns were observed at 0.0221 in volume fraction for D1W52 and D1B28 spheres. For D1W52 (88 nm in diameter), a low but hill-like accumulation area was observed in addition to the sharp broad ring at the outside edge. However, the hill disappeared for D1B28, in which sphere size is 212 nm in diameter. It is highly plausible that the small D1W52 spheres are apt to aggregate slightly or that some hydrophobic affinitive interaction exists between D1W52 spheres and the glass wall of the substrate.

Sharpness of the peak profiles (S) was evaluated by the ratio of film size in diameter (d_f) against the full width at half maximum ($FWHM$) in the thickness

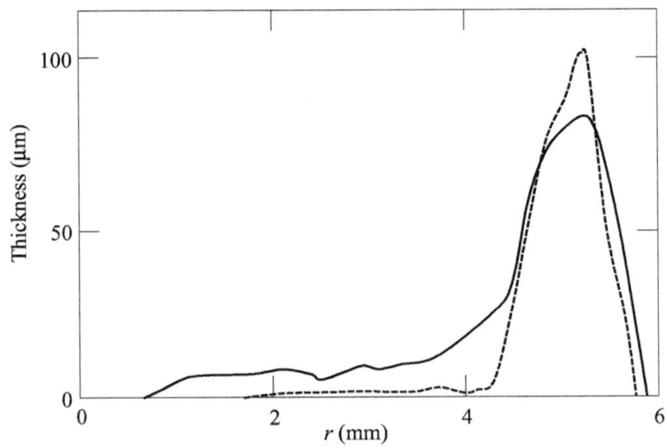

Figure 5.24 Thickness profiles of dried films as a function of the distance from the center at 25° C. In deionized water, *solid curve:* D1W52; *dotted curve:* D1B28. ϕ = 0.0221, 0.1 ml.

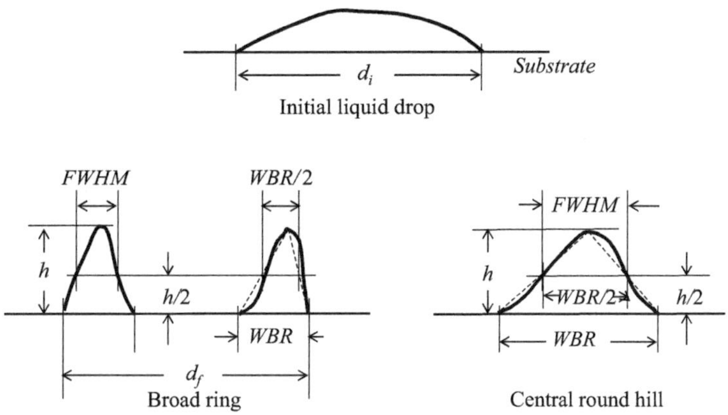

Figure 5.25 Schematic picture showing the method of evaluation of the sharpness parameters S and S'.

profile shown in Fig. 5.24, for example. Methods of evaluation of the sharpness parameters S and S' are shown schematically in Fig. 5.25.

$$S = d_f / FWHM \quad (1)$$

When the peak profile is normal distribution, *FWHM* is given by Equation 2, where s is the corresponding standard deviation.[71]

$$FWHM = 2.355\, s \quad (2)$$

Another kind of sharpness parameter, S', was also evaluated from the ratio of the dried film size against the width of the broad ring (*WBR*) at the bottom layer in the thickness profiles shown in Fig. 5.25, for example. S' is evaluated conveniently from the macroscopic patterns directly without measurements of the thickness profiles.

$$S' = (d_f / WBR) \times 2 \quad (3)$$

Here, *WBR* is assumed to be twice *FWHM* (i.e., the peak profiles are approximated to be shaped like an isosceles triangle). Fig. 5.26 shows the S or S' values of polystyrene spheres (D1W52 and D1B28) as a function of sphere concentration. The agreement between S and S' values look excellent. Furthermore, S or S' of D1B28 (open or solid triangles) are clearly larger by approximately 50% than those of D1W52 spheres, even the experimental errors in S (or S') are taken into account. This result supports the finding that the small spheres are more affinitive to the substrate or are apt to aggregate.

Fig. 5.27 shows the typical example of the reflection spectra during the course of drying at the center of the initial liquid drop. The peak wavelength

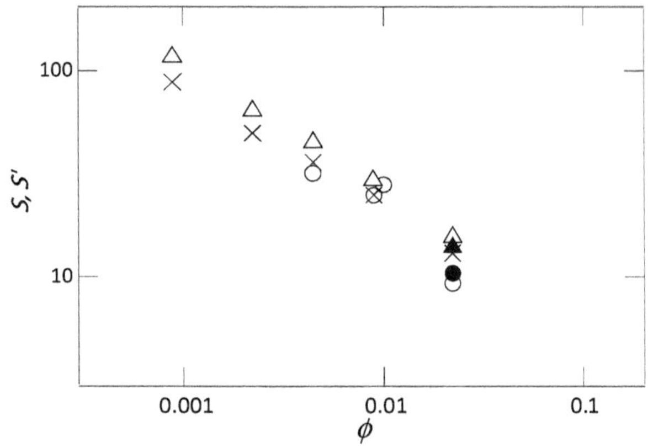

Figure 5.26
Sharpness parameters of polystyrene spheres at 25° C. Circle *S'* of D1W52, 0.1 ml, *Crosses S'* of D1W52, 0.01 ml; *triangle S'* of D1B28, 0.1 ml; *solid circle S* of D1W52, 0.1 ml; *solid triangle S* of D1B28, 0.1 ml.

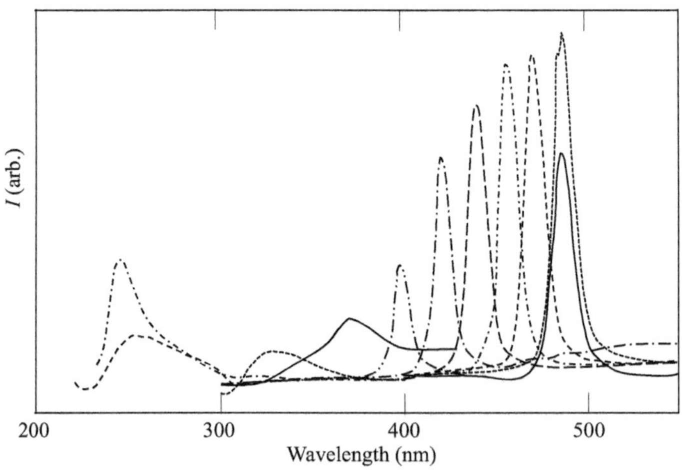

Figure 5.27
Reflection spectra during the drying processes of D1W52 spheres on a cover glass at 25° C. $\phi = 0.0332$, 0.05 ml, $r = 0$ mm. *Solid curve:* $t = 0$ minutes; *dotted:* 20 minutes; *broken:* 40 minutes; *dot-broken:* 60 minutes; *long broken:* 80 minutes; *dot-long broken:* 100 minutes; *dot-dot-long broken:* 120 minutes; *solid:* 140 minutes; *dotted:* 160 minutes; *broken:* 180 minutes; *dot-broken:* 12 days. *From Okubo T, Kimura K, Kimura H:* Colloid Polym Sci *280:1001-1008, 2002.*

(λ_p) decreased with time, especially drastically at the final stage of dryness at 160 minutes to 180 minutes. The λ_p values at $r = 2.7$ mm and 3.5 mm shifted slightly to the shorter wavelengths during the course of dryness. The intersphere distance (l) from the reflection spectroscopy at $r = 0$ mm decreased from 225 nm at the initial time to 98.8 nm 12 days after setting (D1W52 spheres, $\phi = 0.0332$). This value of the dried film is larger by approximately 10% than the diameter of the colloidal spheres as reported by the manufacturer. It is plausible that the neighbored spheres are close but not in contact with each other, keeping a crystal-like structure. The reason why a slight difference occurs is not definite yet. However, several potential causes are suggested. First, the film at $t = 12$ days still contains water in the cavities of the film. Second, intersphere repulsion forces via electrical double layers in the suspension were still effective just before solidification. Third, expansion of the sphere size by the hairy chains of polystyrene at the sphere surface is plausible. Fourth, the size determined with an electron microscope experimentally is smaller than the real size in suspension state. Fifth, large errors must be considered for the assumption of several parameters in reflection spectroscopy, especially the dielectric constant of polystyrene spheres in suspension state. It should be mentioned here that for D1B28 polystyrene spheres, agreement between the diameter and the intersphere distance determined reflection spectroscopy was excellent, 212 and 213 nm, respectively.

Fig. 5.28 shows the nearest-neighbored intersphere distance (*l*) observed from the reflection spectroscopy shown in Fig. 5.27. At the initial process of drying, drying rates are slightly slow compared with those at outer region. Furthermore, at the final stage, the intersphere distance closed sharply to the diameter of the suspended spheres (shown by an *arrow*).

When close-up observation was made based on Fig. 5.29, *A*, for polystyrene spheres D1W52, the microscopic patterns were observed from *B* through *D*. Most extended drying patterns of D1W52 spheres are shown in Fig. 5.30. Earth worm–like fractal patterns are observed. It is clear that the stratified structures form from macro- to microscales.

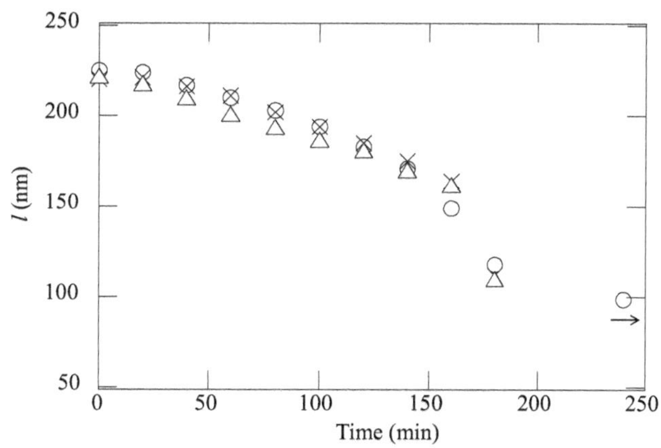

Figure 5.28 *l* versus *t* plots during the drying processes of D1W52 spheres on a cover glass at 25° C. $\phi = 0.0332$, 0.05 ml, $r = 0$ mm.

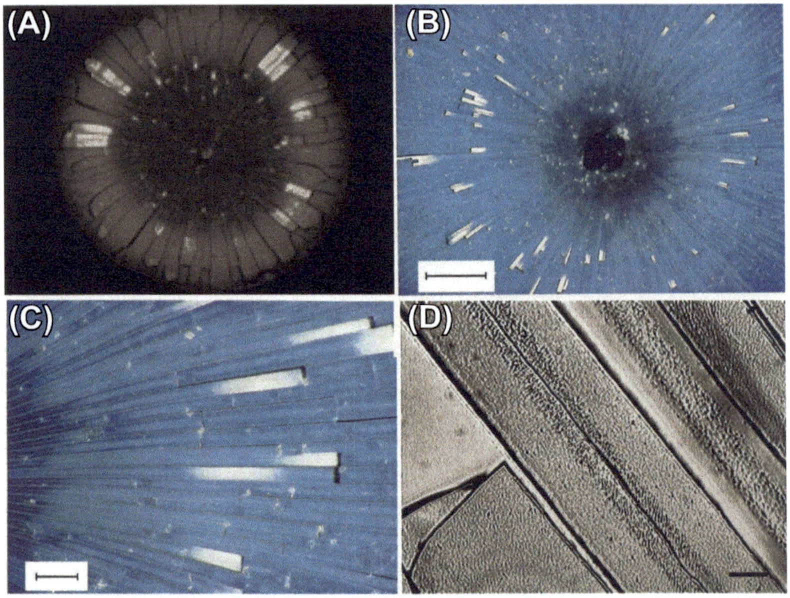

Figure 5.29 Drying patterns from macroscopic (*A*) to microscopic patterns (*B-D*) of D1W52 spheres on a cover glass at 25° C. **A,** $\phi = 0.022$, 0.1 ml; **B,** 0.011, 0.2 ml, length of the bar: 1.0 mm; **C,** 0.011, 0.2 ml, 200 μm; **D,** 0.011, 0.1 ml, 25 μm.

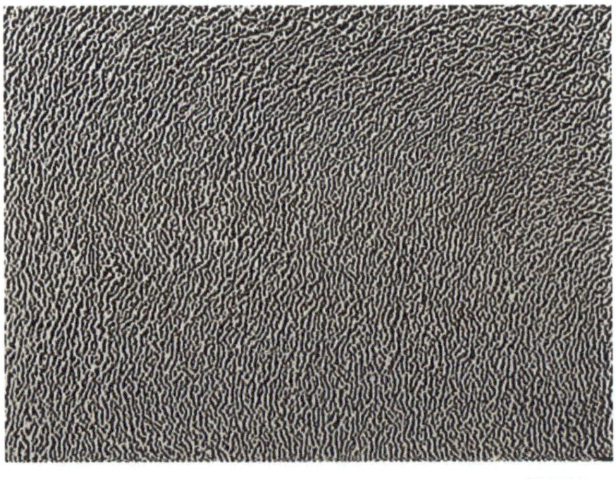

Figure 5.30 Most extended microscopic drying patterns of D1W52 spheres at 25° C. $\phi = 0.011$, 0.1 ml length of the bar: 25 μm.

5.4.2 Colloidal Silica Spheres

Fig. 5.31 shows the drying patterns of monodispersed colloidal silica spheres ranging from 29 nm to 1.1 μm in diameter. Sphere concentrations are the same, 0.0333 in volume fraction. An unrinsed cover glass at room

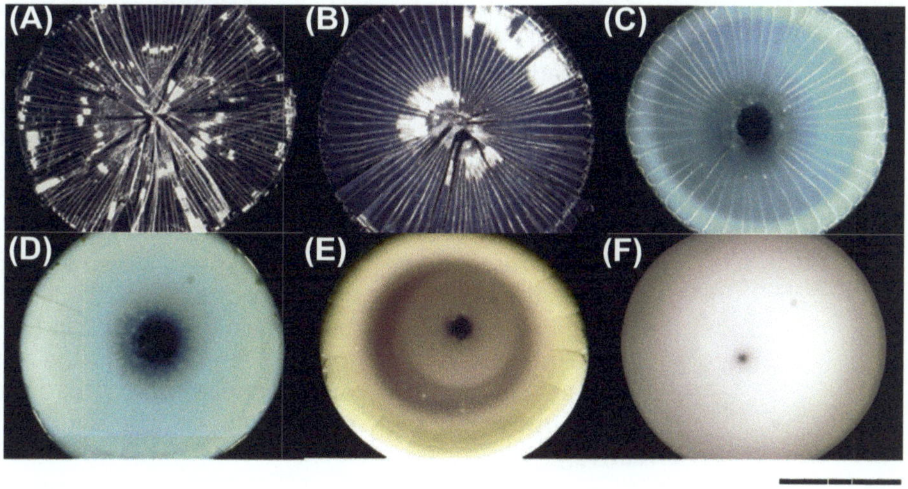

Figure 5.31 Drying patterns of colloidal silica spheres at 25° C. $\phi = 0.0333$, 0.1 ml; **A,** CS22p, **B,** CS45; **C,** CS82; **D,** CS161; **E,** CS301; **F,** CS1001; length of the bar is 5.0 mm. *From Okubo T, Yamada T, Kimura K, Tsuchida A:* Colloid Polym Sci *283:1007-1015, 2005.*

temperature of 25° C was used to dry 0.1 ml of the sample suspensions.[45] Two kinds of macroscopic patterns, broad ring and spoke line, are observed with the naked eye. These two macroscopic patterns are typical and have been observed for most colloidal dispersions and most solutes in a solution state except the nondiffusible particles or solutes forming gel structures or large aggregates. There were quite few spheres in the central region, whereas the broad-ring region was occupied with many spheres. Clearly, the cracks formed during the process of shrinking of the wetted films.[46] The cracks curved at the outside edge and then developed straight toward the center of the film. It is impressive that the new cracks developed successively, keeping the similar angle with the adjacent elder cracks.[46] Clearly, the film is transparent and so many cracks are observed for small spheres. When sphere size increases, the film becomes whitish and the number of the cracks decreases sharply. Increase in sphere size will result in the increase in the elastic modulus of the dried film and then the number of cracks will decrease. It should be noted that so many potential cracks form already in the suspension state by cell convection and the spoke-line convectional flow of spheres and solvent molecules. As was described Section 5.2, a main cause for the broad ring formation is the convectional flow of solvent and colloidal spheres from the central area toward the outside edge in the lower layers of the liquid drop, which was observed directly from the movement of the temporary aggregates of the particles of Chinese black ink,[12] PMMA spheres,[49,50] and others. The dried films in Fig. 5.31 emit weak colors from bluish to yellow and faint pink as sphere size increased, which supports the finding that crystal structures are maintained through the drying processes, as was demonstrated by reflection spectroscopy of polystyrene spheres.

Fig. 5.32 shows the drying patterns of CS300 spheres on a cover glass (*A-D*), a watch glass (*E-H*), and a glass dish (*I-L*).[24,45] The dried films emit iridescent greenish, reddish, and pinkish colors by white light. The colors were strong in a petri glass dish. Furthermore, the broad rings formed and their size (d_f value) decreased as sphere concentration decreased. Colloidal silica spheres are the same as the substrates in the basic substances, glass. Thus it is highly plausible that affinitive interaction exists between silica spheres and the substrate comparing with the interaction between hydrophobic polystyrene spheres and the hydrophilic glass substrate. However, the sharp decrease in the broad-ring size glass-glass affinity will be not so strong, and the broad-ring size is determined mainly by the excluded-volume effects, as discussed for polystyrene spheres. Interestingly, the netlike (or wavelike) patterns appeared outside of the inner shifted broad ring (see Fig. 5.32, *I*). However, most of the spheres moved toward the inner broad ring effectively.

Fig. 5.33 demonstrates the strong iridescent colors for the dried films of CS300 silica spheres in a glass dish.[24] When the incident light angles are changed, the drastic change in the iridescent colors was observed clearly (compare Fig. 5.33, *A*, to Fig. 5.33, *B*).

Table 5.5 compiles the peak wavelengths (λ_p) of the reflection spectra of the dried films of CS300 without and with sodium chloride (NaCl) and the nearest-neighbored intersphere distance (l) estimated from the spectroscopy. The l values were estimated from λ_p using Equation 4.[72,73]

$$l = 0.6124\lambda_p/n \qquad 4$$

$$n = [\text{refractive index of silica}] \times 0.74 + [\text{refractive index of air}] \times 0.26 \qquad 5$$

Refractive indices of silica and air are assumed to be 1.5 and 1, respectively. The reflection spectra of the dried film without salt and with salt lower than 0.0005 M of NaCl at the initial suspension state were sharp. This supports the finding that the films are composed with the crystal arrays of CS300 spheres even in the presence of a tiny amount of salt. The broad reflection peaks of the films were observed from the suspensions containing NaCl higher than 0.003 M to even 0.03 M. These salt effects show that the crystal structures remained but distorted slightly by a large amount of salt. The l values observed were close to the diameter of CS300 spheres, 305 nm $\pm$ 9.1 nm, which supports the neighbored spheres almost contacting each other. The spheres in the dried film have been reported to distribute in the close-packed (or face-centered cubic [*fcc*]) structure from the transmitted-light intensity measurements.[72] Values for λ_p and l shifted slightly toward the shorter wavelengths by the addition of salt, which is explained by the drying processes of the initial suspensions with the distorted structures of the *fcc*

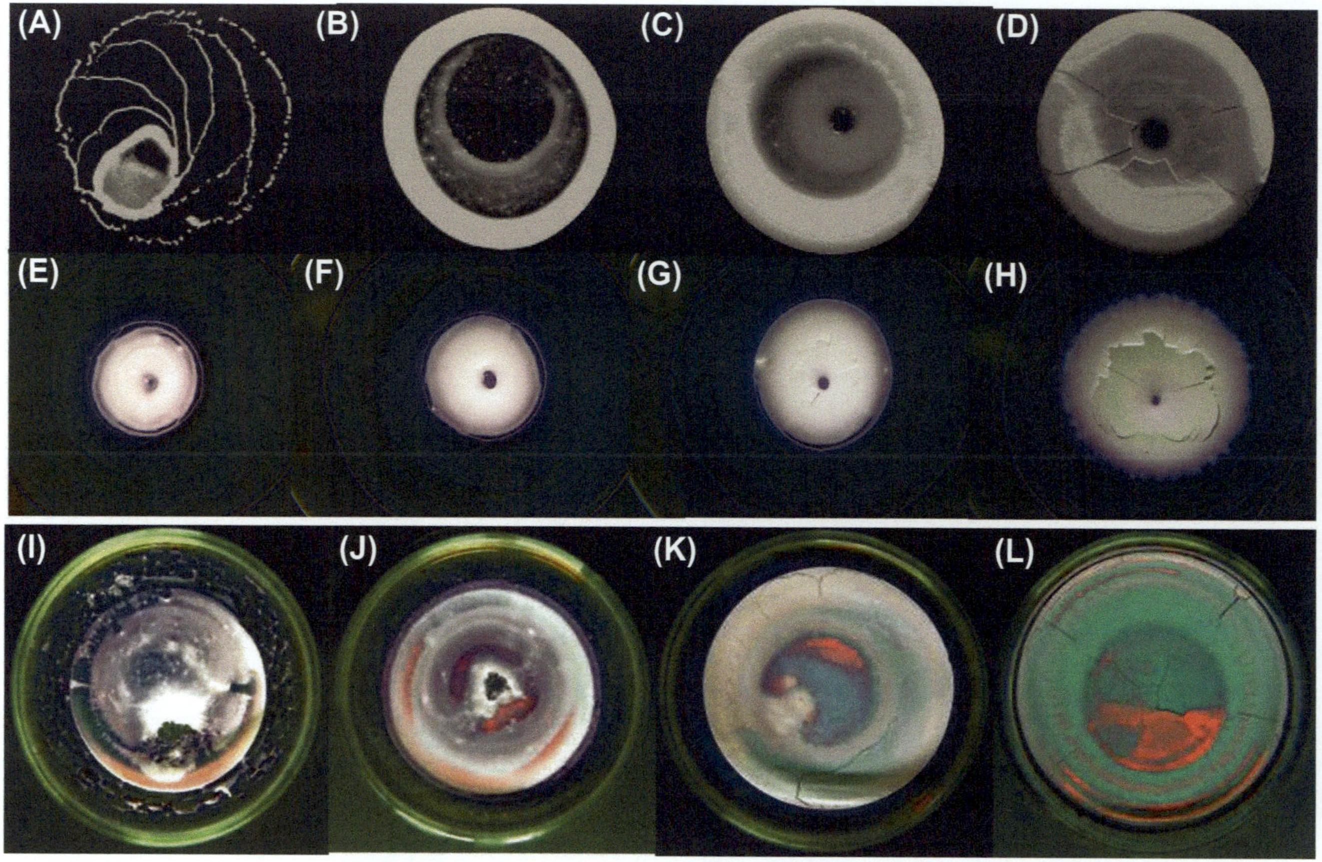

Figure 5.32 Drying patterns of CS300 colloidal silica spheres on a cover glass (0.1 ml, *A-D*), a watch glass (4 ml, *E-H*), and a glass dish (10 ml, *I-L*) at 25 (*A-D*) and 27° C (*E-L*). **A,** $\phi = 0.000133$; **B,** 0.0133; **C,** 0.0333; **D,** 0.1; **E, I,** 0.00038; **F, J,** 0.00076; **G,** 0.00228; **H,** 0.0076; **K,** 0.00455; **L,** 0.0152.

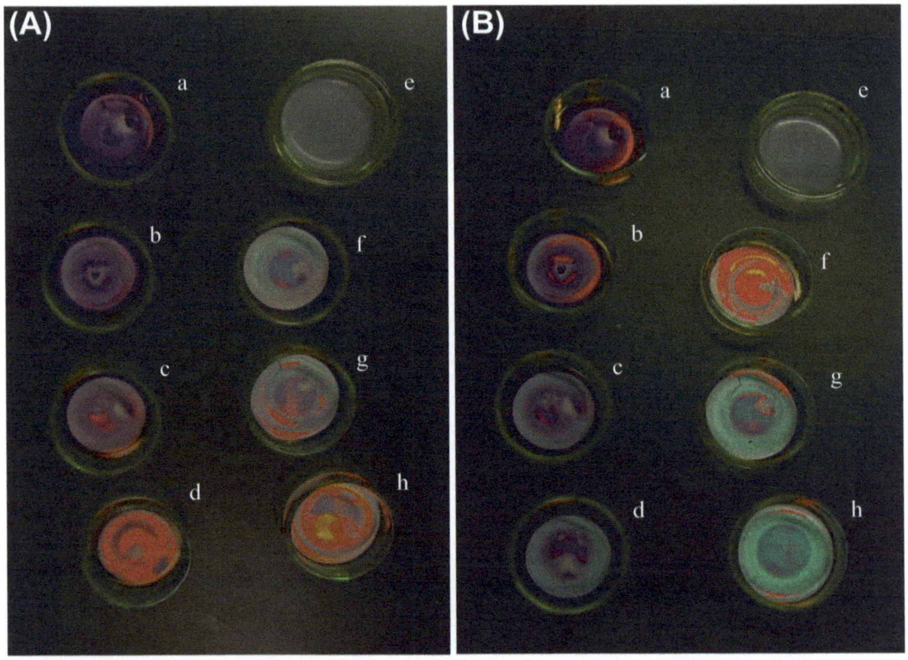

Figure 5.33 Drying patterns of CS300 silica spheres in a glass dish at 27° C. After 8.4 days, dry, 10 ml, **A,** ϕ = 0.00038; **B,** 0.00076; **C,** 0.00152; **D,** 0.00228; **E,** 0.00228; with cap; **F,** 0.00455, **G,** 0.0076; **H,** 0.0152, incident light angles differ between A and B.

toward body-centered cubic structures and further melted liquid structures with the salt higher than 0.03 M. In Table 5.5, the λ_p and l values decreased slightly as the initial sphere concentrations increased. This also supports the idea that the spheres contact each other more tightly as the sphere

Table 5.5 Reflection Peak Wavelengths (λ_p) and the Nearest-Neighbored Intersphere Distances (l) of the Dried Crystal Film of CS300 Spheres from Reflection Spectroscopy

ϕ of Initial Suspension	[NaCl] (M)	λ_p (nm)	l (nm)
0.00038	0	660	300
0.00076	0	652	292
0.00152	0	649	290
0.00228	0	647	289
0.00455	0	633	283
0.0076	0	629	281
0.0152	0	624	279
0.00038	0	661	301
0.00038	0.0001	700	313
0.00038	0.0003	658	294
0.00038	0.0005	647	289
0.00038	0.003	660 (broad)	300
0.00038	0.01	640 (broad)	290
0.00038	0.03	630 (broad)	280

NaCl, Sodium chloride.

concentration increases by the thinning of the electrical double layers formed around the spheres in suspension state.

Fig. 5.34 shows d_f/d_i values of CS45, CS82, and CS301 as a function of sphere concentration on a cover glass.[45] At the higher concentrations than approximately 0.01 in volume fraction, the broad rings formed at the outside edge of a series of colloidal silica spheres having different sphere sizes. These results support the finding that size of the broad ring is determined by the excluded-volume effects for colloidal silica spheres as observed for polystyrene spheres.

Fig. 5.35 shows the thickness profiles of the dried films of CS22p, CS45, CS82, CS161, CS301, and CS1001.[45] The sphere concentrations were fixed at $\phi = 0.1$ (*solid curve*), 0.067 (*dotted*), 0.0333 (*broken*), and 0.0133 (*dot-broken*), respectively. The profiles became broad as sphere concentration increased in general. With increasing sphere concentration, furthermore, the thickness peak shifted from the outside edge of the dried film to the intermediary positions between the outside edge and the central area of the dried film.

The *S* values of colloidal silica spheres (*circles*) and polystyrene spheres (*crosses*), both group A systems, are compared in Fig. 5.36, *A*. Important findings are clear. First, the *S* values of polystyrene spheres were larger than those of colloidal silica spheres, when a comparison was made at the same sphere size. Second, *S* values increase with sphere size and are saturated for CS300. A decrease of *S* in the small colloidal silica spheres was already observed for polystyrene spheres. It is highly plausible that affinitive interaction of the small spheres with substrate is strengthened and the

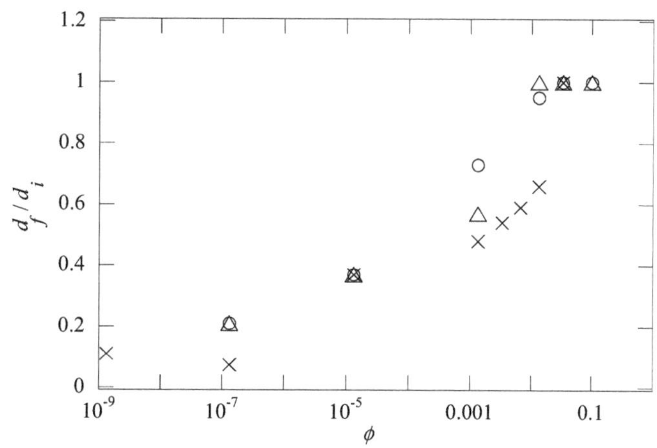

Figure 5.34
Values of d_f/d_i of CS45 (*circles*, 0.1 ml), CS82 (*crosses*, 0.2 ml), and CS301 (*triangles*, 0.1 ml) on a cover glass at 25° C.

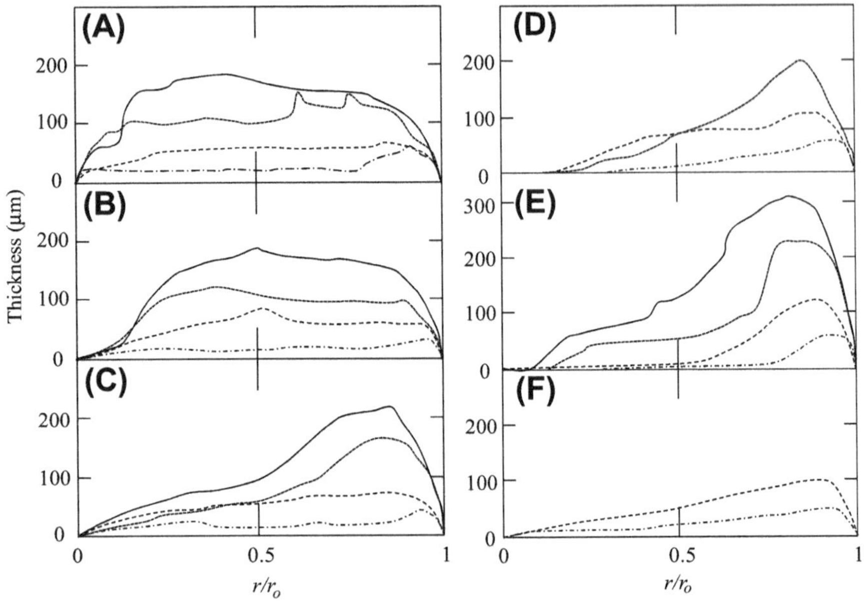

Figure 5.35 Thickness profiles of colloidal silica spheres at 25° C. 0.1 ml, **A,** CS22p; **B,** CS45; **C,** CS82; **D,** CS161; **E,** CS301; **F,** CS1001; *solid line:* $\phi = 0.1$; *dotted line:* 0.0667; *broken line:* 0.0333; *dotted broken line:* 0.0133.

convectional diffusion of the small spheres is retarded. Decreasing tendency of the diffusion for large spheres is normal, because the translational coefficients, D_t, is given by $k_BT/3\pi\eta d$ (see Equation 1 in Chapter 2) and should decrease as sphere size increases. By coupling these two factors, the maximum S value will appear for CS301 spheres.

Fig. 5.36, *B*, shows temperature dependency of the sharpness parameter, S, of the broad rings for CS82 spheres (Ito M, Tsuchida A, Okubo T, unpublished data). The S values increased sharply as suspension temperature increased, but began to decrease at the high temperature of 75.3° C. An increase in S with temperature is understandable, because the convectional diffusion constant of spheres increases linearly with an increase in the absolute temperature. Enhancement of the outward convectional flow of spheres will result in the effective accumulation of colloidal spheres at the outside edge. It should be mentioned here that the dried film size, d_f, decreased sharply from 5.4 mm to 4 mm as suspension temperature was raised from 23.4° C to 75.3° C. Furthermore, the inner area from the broad ring became a thin layer within 1 μm in thickness. A decrease in the S value at the highest temperature is due to a decrease in the film size or disturbance of the convection of the spheres from a regular laminar flow to an irregular flow by the violent evaporation of solvent molecules. The S values of D1W52 (*cross*) and D1B28 (*triangle*) are again larger than that of CS82 (*circles*) (see Fig. 5.36, *A*).

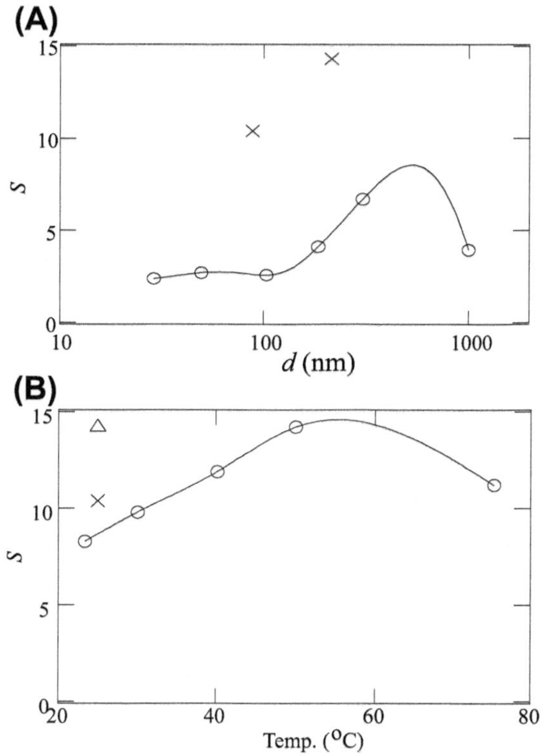

Figure 5.36
A, Sharpness of the broad ring as a function of size of colloidal spheres. 0.1 ml, 25° C, *circles* CS spheres, $\phi = 0.0333$, *triangles* PS spheres, $\phi = 0.0221$. **B,** Sharpness parameters of the broad ring of polystyrene and silica spheres as a function of temperature of suspension. *Circles:* CS82, $\phi = 0.02$, 0.01 ml, *cross:* D1W52, 0.0221, 0.1 ml, *triangle:* D1B28, 0.0221, 0.1 ml.

Now, let's see the microscopic patterns of colloidal silica spheres. Fig. 5.37 demonstrates one of the typical color pictures of microscopic drying patterns of CS300 from the left edge (*A*) to the right (*I*) in a Petri glass dish at 27° C.[24] Beautiful, colorful microstructures were observed. These iridescent colors are ascribed to the Bragg diffraction of light by the crystal array of the colloidal spheres in the dried film. The orange area surrounded with brownish bands in Fig. 5.37, *A* and *I*, are the broad ring areas. The black areas of *A* and *I* are regions outside the dried film.

Fig. 5.38 shows typical microscopic patterns of the dried films of CS550 spheres, where the initial sphere concentrations range from 0.00041 to 0.0083 in volume fraction.[25] Ten ml of the sample suspensions were set in a glass dish. Beautiful, colorful microstructures were observed. The microstructures corresponding to the macroscopic patterns of spoke lines (*A, C,* and *F*) and multiple rings (*B, D,* and *E*) are clear. It can be deduced from the *B* that the latticelike rectangular fine structures are formed by the coupling of the fine spoke lines and fine rings.

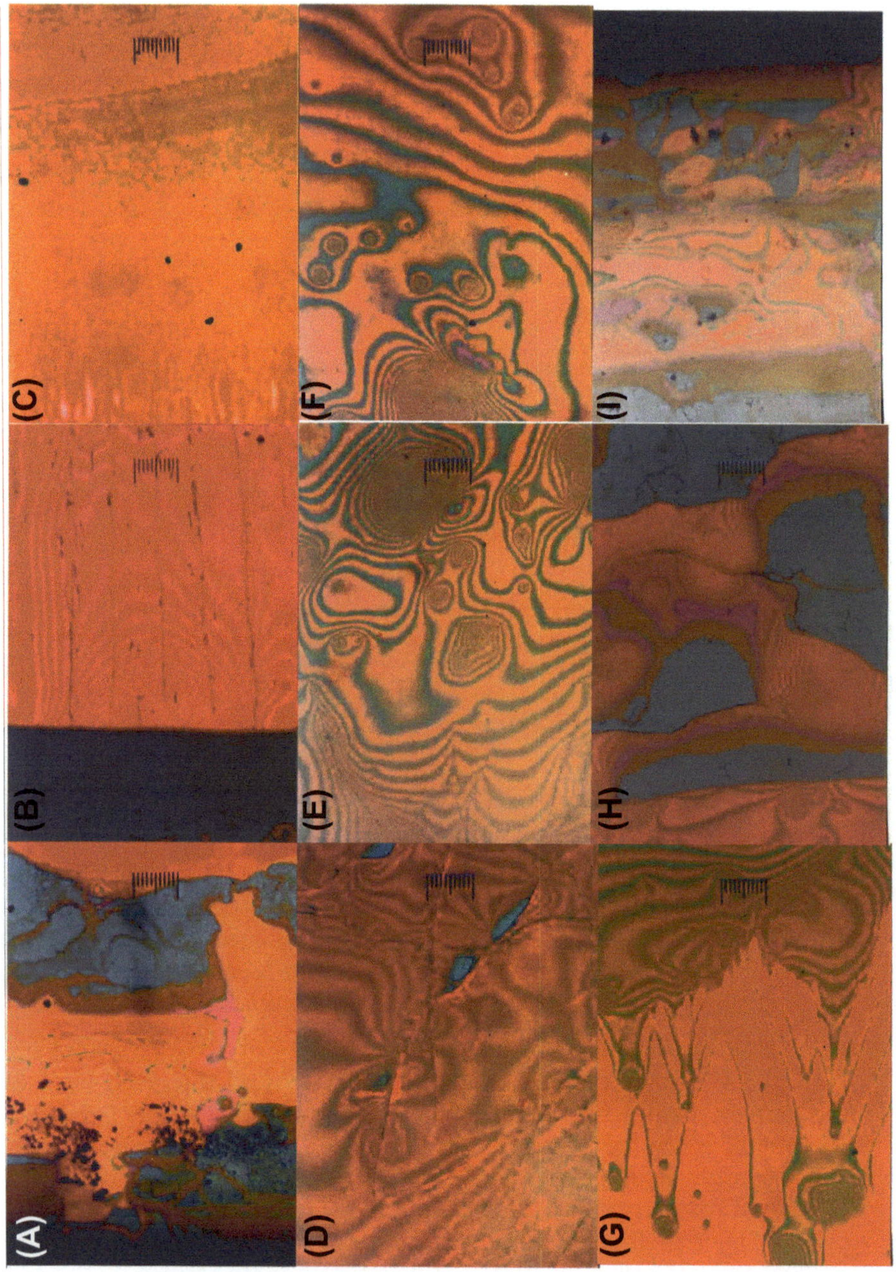

Figure 5.37 Microscopic drying dissipative patterns of CS300 silica spheres in a glass dish at 27° C. $\phi = 0.00038$, 10 ml, dry, from left edge (*A*) to right (*I*), full scale = 0.2 mm. *From Okubo T, Okamoto J, Tsuchida A:* Colloid Polym Sci *285:967-975, 2007.*

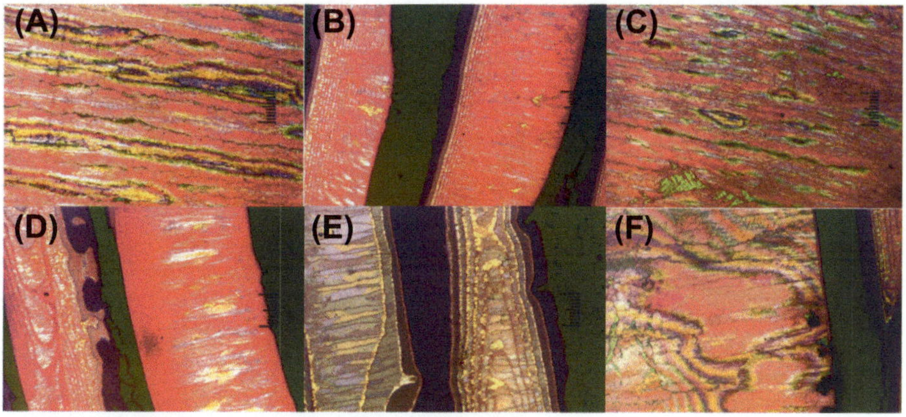

Figure 5.38 Microscopic drying dissipative patterns of CS550 silica spheres in a glass dish at 27° C. **A,** $\phi = 0.00041$; **B,** 0.00165; **C,** 0.0025; **D,** 0.0041; **E,** 0.0058; **F,** 0.0083, 10 ml, after 122.5 hours, dry. *From Okubo T:* Colloid Polym Sci *285:1495-1503, 2007.*

Fig. 5.39 shows the microscopic pictures of the drying patterns of CS1000A from the central area to the right side in a very small glass dish (0.4 ml of the sample was set) and a small watch glass (1.5 ml was set).[22] Full scale is 200 μm. Similar fine, white stringlike ring structures were observed irrespective of the cell type and size. These patterns clearly show the traces of the movement of the drying frontier zones toward the center. Formation of the small accumulates of the colloidal spheres looks like white snow spots in Fig. 5.39, *A* and *B*.

5.4.2.1 Influence of Substrates

Fig. 5.40 shows the dissipative patterns during the drying CS301 suspensions on a poly(tetrafluoroethylene-co-perfluoroalkyl vinyl ether) (PFA) plate. After 133 minutes, 17 seconds, the strongly iridescent and doughnut-like drying pattern was obtained (courtesy Professor A. Tsuchida). On the lipophilic substrate, the suspension drop has a large contact angle with the substrate, and the liquid drop of the initial stage is spherical, as is shown in the picture after 1 minutes, 2 seconds. Approximately 2 hours later, the central area of the liquid drop sank slightly, and the hole completed after 2 hours, 2 minutes, 31 seconds. After 133 minutes, 17 seconds, the drying processes completed. It should be mentioned that the iridescent colors were observed during the whole course of drying. The thickness profiles of the dried film on a PFA plate and that of the same sample on a cover glass are compared in Fig. 5.41. When the different scales of the height axes are taken into account, the contact angle of the dried film on a PFA plate is clearly larger than those on a cover glass. The sharpness parameters, *S* value of CS301 spheres on PFA (2.5) was smaller than that on glass substrate (3.5).

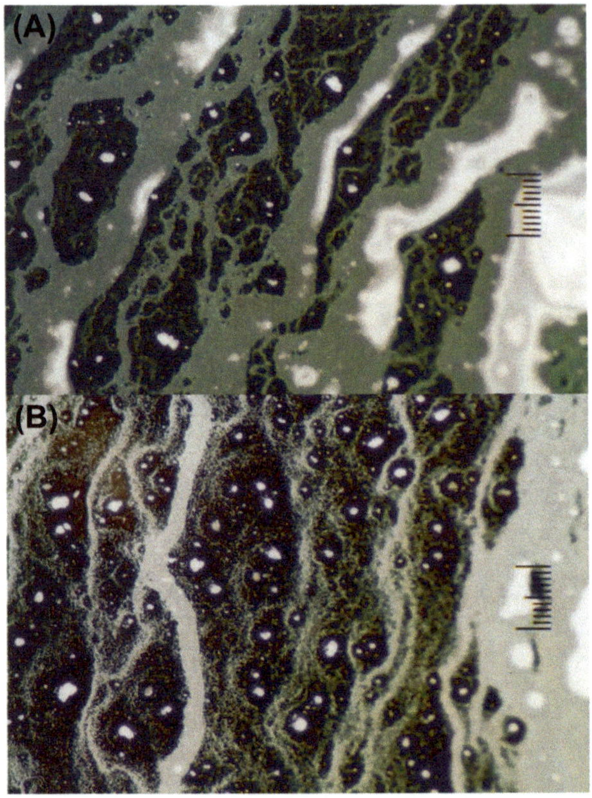

Figure 5.39 Drying dissipative patterns of CS1000A silica spheres in a watch glass at 24° C. $\phi = 0.00129$, **A**, very small watch glass, 0.4 ml; **B**, small watch glass, 1.5 ml, full scale = 0.2 mm. *From Okubo T:* Colloid Polym Sci *284:1191-1196, 2006.*

5.4.2.2 Influence of Salt Addition

Fig. 5.42 shows the microscopic structures of CS300 spheres in the presence of 0.003 M of NaCl.[24] The pictures were taken from the left outside edge to the right. The black and bluish patterns are composed of the silica spheres and the whitish patterns are the salt. These beautiful branchlike (dendritic) patterns demonstrate that the synchronous cooperative interactions existing among the spheres, the salt molecules, and the substrate play an important role for the pattern formation. Complexation of the silica spheres and the salt is clear especially in Fig. 5.42, *D-G.*

Fig. 5.43 shows the microscopic observation of the complexation between CS550 and NaCl in a glass dish.[25] Concentration of NaCl coexisted in suspension ranged from 0.0001 M to 0.03 M. The pinkish, bluish, or greenish leaflike patterns support the complexation taking place by the synchronous cooperative interactions among the spheres, salt, and the glass wall of the substrate during drying processes. A large cubic crystal of NaCl is observed at

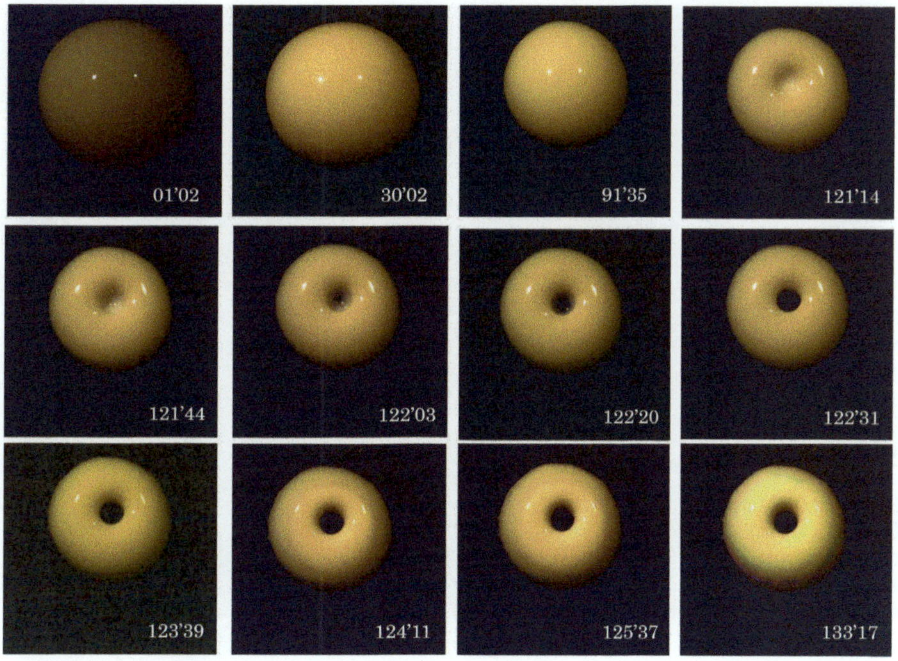

Figure 5.40
Close-up observation of drying processes of CS301 spheres on PFA plate at 25° C. 0.1 ml, $\phi = 0.03$, time (min:sec). *Courtesy Professor A. Tsuchida.*

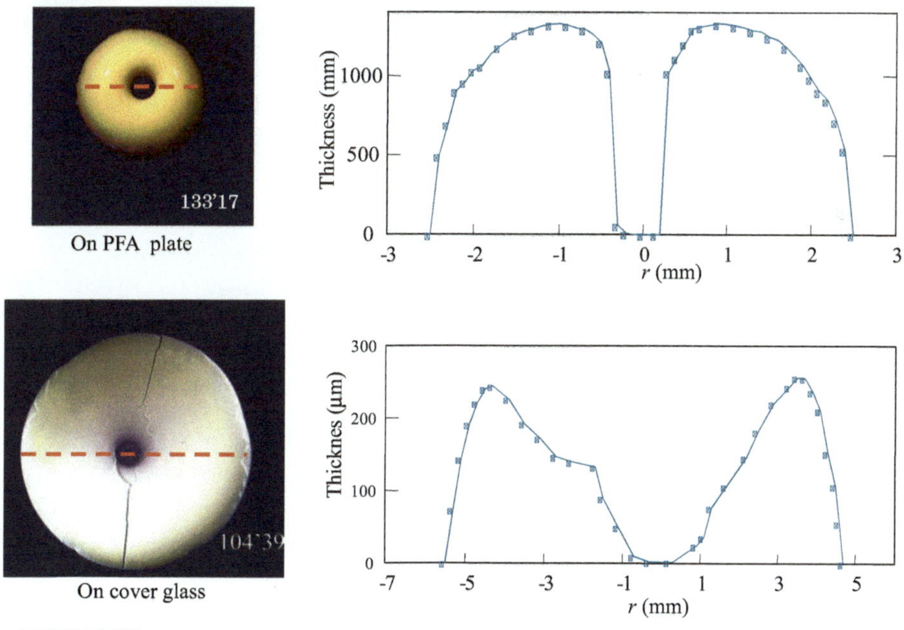

Figure 5.41
Comparison of thickness profiles in the drying patterns of CS301 on a PFA plate and a cover glass at 25° C. Length of the bar is 5 mm. *Courtesy Professor A. Tsuchida.*

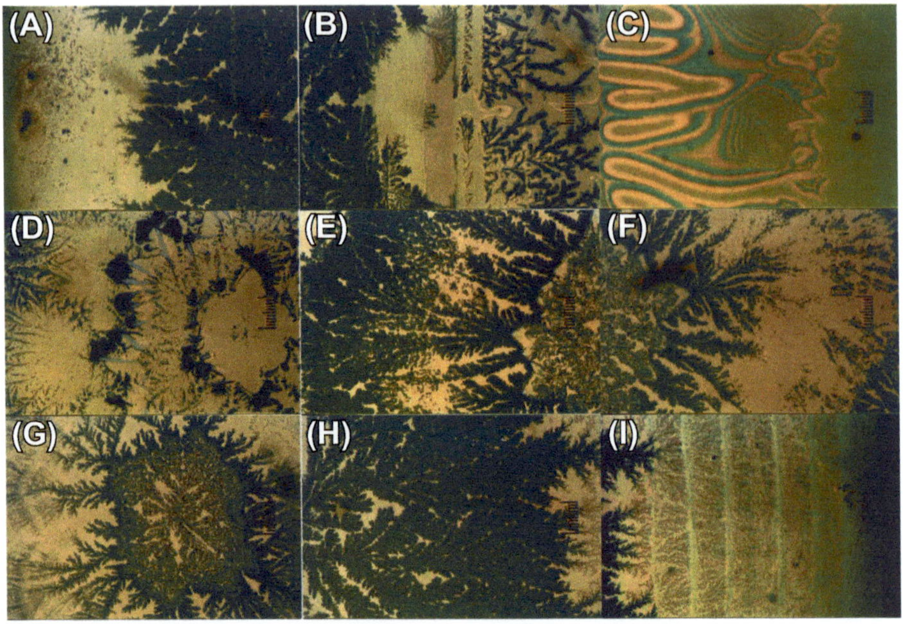

Figure 5.42 Microscopic drying dissipative patterns of CS300 silica spheres in a glass dish at 27° C. ϕ= 0.00038, [NaCl] = 0.003 M, 10 ml, dry, from left edge (*A*) to right (*I*).

the central area of the dried film (see *F*). The excess amount of NaCl for the complexation was crystallized independently with the help of the segregation effect with the complexes.

Microscopic drying patterns of CS1000A suspensions containing different amounts of NaCl ranging from 0.0003 M to 0.003 M are shown in

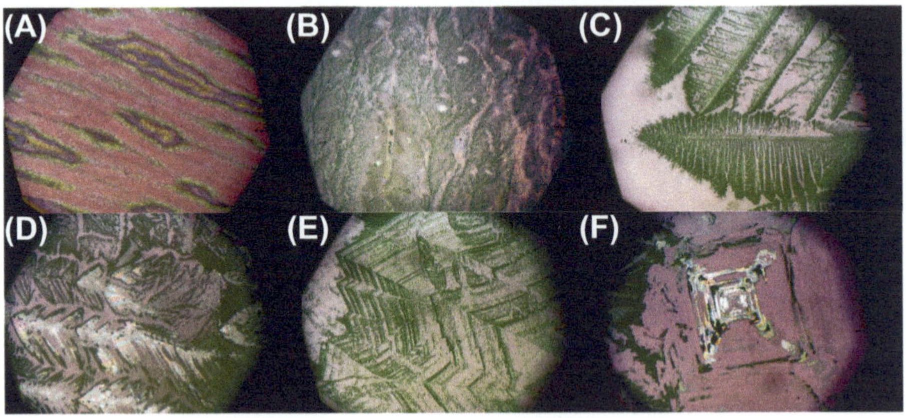

Figure 5.43 Microscopic drying dissipative patterns of CS550 silica spheres in a glass dish at 27° C. ϕ = 0.000827, 10 ml, after 165 hours, dry; **A,** [NaCl] = 0.0001 M; **B,** 0.001 M; **C,** 0.003 M; **D,** 0.01 M; **E, F,** 0.03 M. *From Okubo T:* Colloid Polym Sci *285:1495-1503, 2007.*

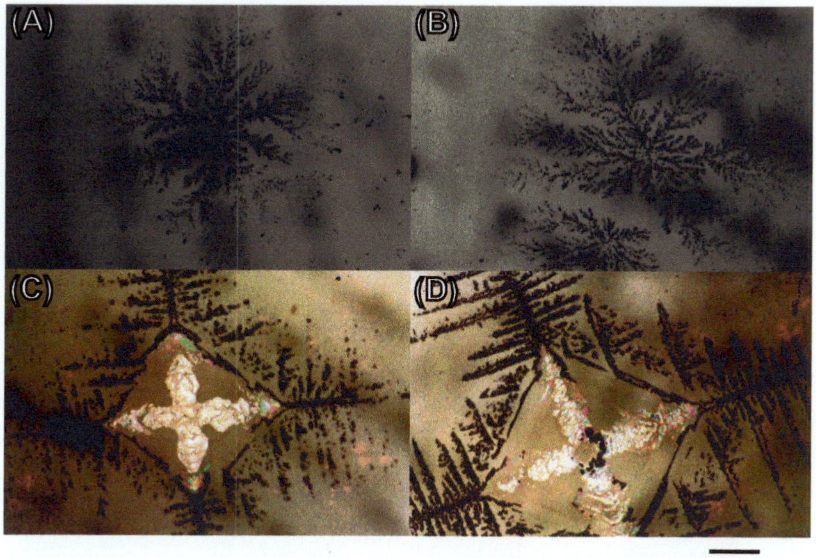

Figure 5.44 Microscopic drying dissipative patterns of CS1000A silica spheres in a glass dish at 24° C. ϕ = 0.00129, 10 ml; **A,** [NaCl] = 0.0003 M; **B,** 0.0006 M; **C,** 0.001 M; **D,** 0.003 M; length of the bar is 0.2 mm. *From Okubo T:* Colloid Polym Sci *284:1395-1401, 2006.*

Fig. 5.44.[23] The corresponding macroscopic patterns showed starlike spots. Black dendritic patterns in the picture are composed of the silica spheres and are not observed for the deionized CS1000A suspensions. When a small amount of NaCl coexists, the dendritic patterns of the complexes of the silica spheres and the salt were observed, as shown in Fig. 5.44, *A* and *B*. Furthermore, when a large amount of NaCl is added to the suspension, the colorful crosslike patterns of NaCl were surrounded symmetrically with the dendritic patterns of silica spheres. Figs. 5.42 to 5.44 support the finding that the synchronous cooperative complexations take place for the mixtures of colloidal silica spheres (300 nm to 1.2 μm in diameter) and NaCl in the dissipative processes of dryness on a substrate, especially in a glass dish.

5.4.2.3 Binary and Ternary Mixtures

Drying patterns of two- and three-component mixtures of monodispersed colloidal silica spheres have been studied.[26,27] Fig. 5.45 shows the macroscopic drying patterns of mixtures of CS1000A + CS161 (*A, E, I*), CS1000A + CS300 (*B, F, J*), CS161 + CS300 (*C, G, K*), and CS1000A + CS300 + CS161 (*D, H, I*) in a watch glass (*A-D*), a glass dish (*E-H*), and a cover glass (*I-L*) at 24° C. The concentration of each sphere was kept at 0.0069 in volume fraction.[26,27] The main dried areas of CS1000A, CS300, and CS161 were whitish, pinkish, and bluish, respectively. The dried patterns of the binary mixtures were composed of the outer broad layers of small spheres and

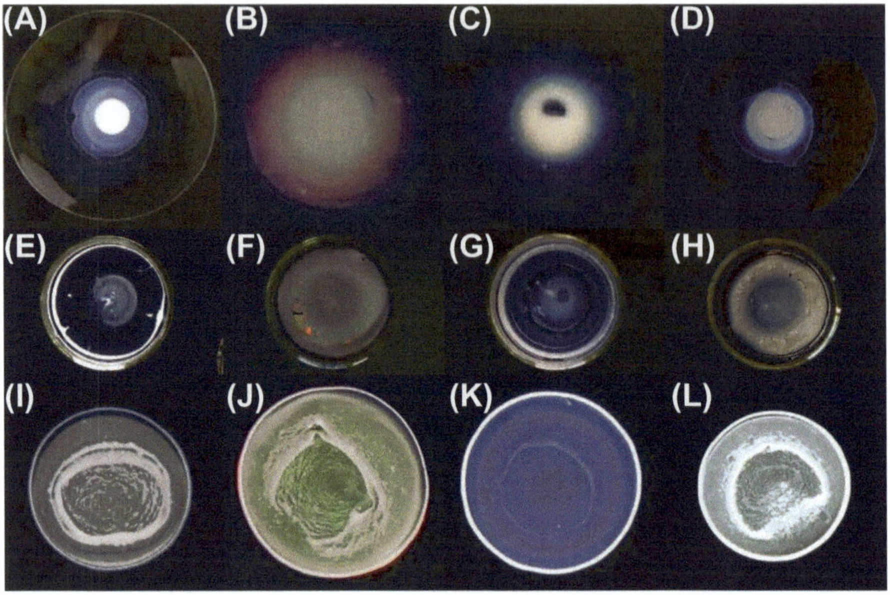

Figure 5.45
Drying patterns of mixed suspensions of CS1000A(1) + CS161(3) (*A, E, I*), CS1000A(1) + CS300(2) (*B, F, J*), CS161(3) + CS300(2) (*C, G, K*), and CS1000A(1) + CS300(2) + CS161(3) (*D, H, L*) in a watch glass (*A-D*), a glass dish (*E-H*), and a cover glass (*I-L*) at 24° C. $\phi_{CS1000A} = \phi_{CS300} = \phi_{CS161} = 0.0069$ (*A-C, E-L*), 0.0128 (*D*). *From Okubo T, Okamoto J, Tsuchida A:* Colloid Polym Sci *286:941-949, 2008.*

the inner layer of large spheres. It is interesting to note that the width ratio (r) of each layer in the binary mixtures changed roughly in proportion to the mixing ratio (x) of the small and large spheres, as is shown in Fig. 5.46. The experimental errors were large. Thus analysis of the mixing ratio of each sphere in the mixtures is possible semiquantitatively from the r values on a watch glass.

Fig. 5.47 shows typical examples of the microscopic pictures of the binary mixtures on a cover glass for CS1000A + CS161 (*A-E*), CS1000A + CS300 (*F-J*), and CS300 + CS161 (*K-O*) mixtures at the outside areas of the dried films. The broad rings at the outermost edges are colored corresponding to the smaller spheres in the mixtures: blue (CS161), red (CS300), or white (CS1000A). Furthermore, the WBRs (L) increased in proportion as the mixing ratio (x) of the smaller spheres increased. Especially for mixtures of CS1000A decreased sharply with mixing (see Fig. 5.47, *F-J*). Fig. 5.48 shows the plots of the L values estimated from the microscopic pictures against x. The proportionality between L and x is excellent. Thus it is clear that the mixing ratio in the initial mixtures is analyzable from the width of the outermost broad rings of the dried film.

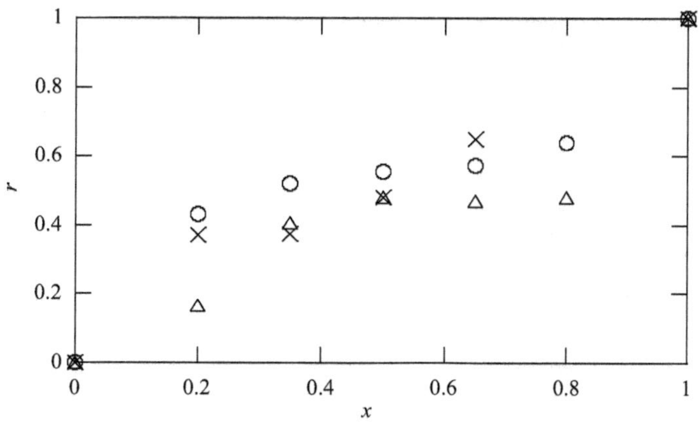

Figure 5.46
Width ratio (*r*) in the segregated areas in a watch glass as a function of mixing ratio in the two-component sphere mixtures. *Circles:* CS1000 + CS161; *crosses:* CS1000 + CS300; *triangles:* CS161 in CS300 + CS161. *From Okubo T, Okamoto J, Tsuchida A:* Colloid Polym Sci *286:385-394, 2008.*

Fig. 5.45, *D, H,* and *L*, shows the examples of the drying patterns of the ternary mixtures of CS1000A, CS300, and CS161 at the condition of the concentrations of each spheres being equal to each other on a watch glass, a glass dish, and a cover glass. In Fig. 5.45, *D*, the three different colored layers (white, slightly pinkish, and bluish) appeared from the center to the outside regions. The patterns of the ternary mixtures are composed of the inner layer of large spheres, middle layer of middle size of spheres, and the outer layer of small spheres. This size-segregation effect for the ternary mixtures was supported also from the reflection spectroscopy and atomic force microscopy observation.[27] It should be noted that the rather sharp size-segregation effect was observed clearly in a watch glass compared with those on the other substrates. This is ascribed to the effective balancing between upward (or outward) convectional flow and the downward (or inward) sedimentation of spheres. Small spheres move upward fast by the convectional flow and also move downward slowly by the sedimentation compared with large spheres. These segregation effects also play an important role for transferring information about suspensions and atmosphere to drying patterns through selection and amplification. Furthermore, these experiments support the fact that the size distribution of the colloidal mixtures is analyzed in principle, from the simple experiments of the drying pattern observation.

5.4.2.4 Influence of Organic Solvents

Fig. 5.49, *A* through *E*, show the drying patterns in pure ethanol (*A*), pure water (*E*), and their mixtures (*B-D*).[74] Clearly, these pictures show the continuous change of the patterns from the major broad rings in the inner region for 100% ethanol toward the broad rings in the outside edge for water. Furthermore,

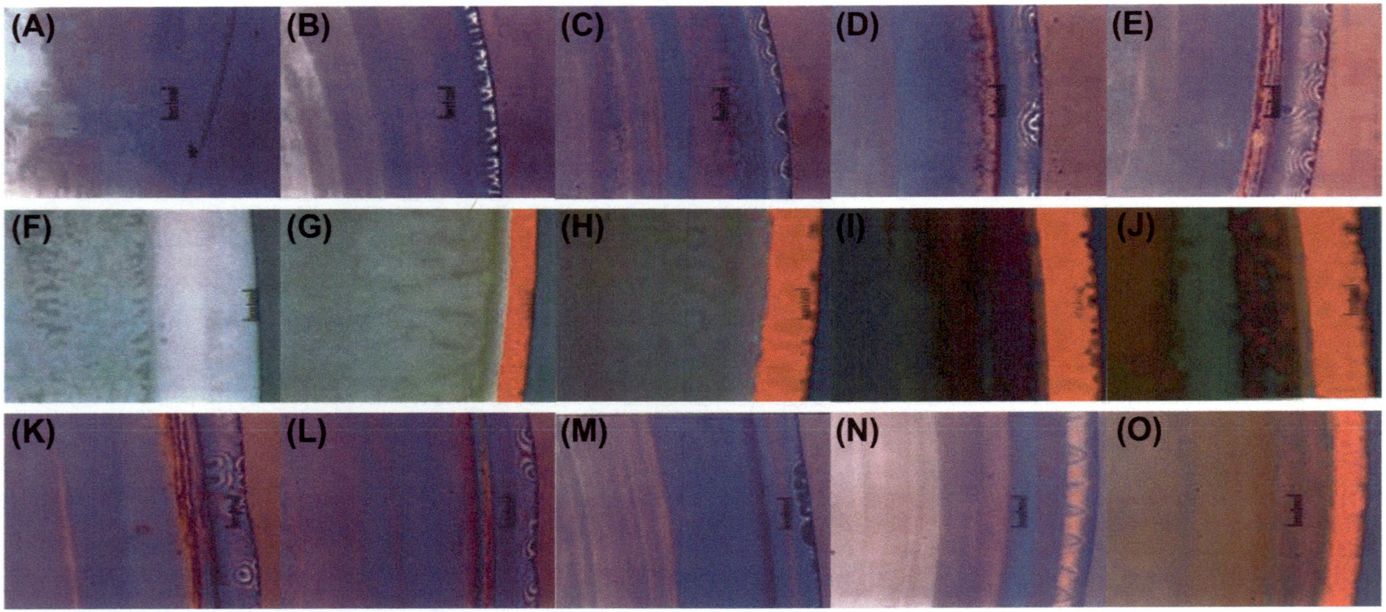

Figure 5.47 Microscopic drying dissipative patterns of mixed suspensions of CS1000A(1) + CS161(3) (*A-D*), CS1000A(1) + CS300(2) (*F-J*) and CS161(3) + CS300(2) (*K-O*) on a cover glass at 25° C. 0.1ml, **A,** $\phi_1 = 0.01375$, $\phi_3 = 0$, **F,** $\phi_1 = 0.01375$, $\phi_2 = 0$, **K,** $\phi_3 = 0.01375$, $\phi_2 = 0$, **B, G, L,** 0.0110, 0.00275; **C, H, M,** 0.00687, 0.00687; **D, I, N,** 0.00275, 0.0110; **E, J, O,** 0, 0.01374, after 6 hours, 10 minutes, dry, at the right edge, full scale is 0.2 mm. *From Okubo T, Okamoto J, Tsuchida A:* Colloid Polym Sci *286:385-394, 2008.*

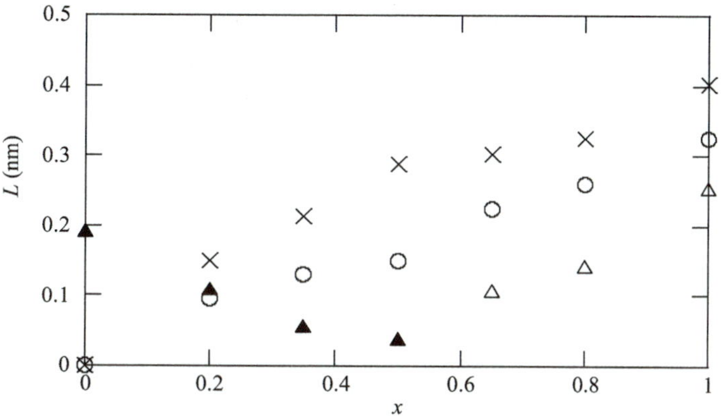

Figure 5.48
Width of the broad rings (L) on a cover glass as a function of mixing ratio (x) in the two-component sphere mixtures. *Circles:* CS161 in CS1000 + CS161; *crosses:* CS300 in CS1000 + CS300; *triangles:* CS161 in CS300 + CS161; *solid triangles:* CS300 in CS300 + CS161. *From Okubo T, Okamoto J, Tsuchida A:* Colloid Polym Sci *286:385-394, 2008.*

film size decreased as the content of water increased. These pattern changes are explained with the increase in surface tension and change in the initial shapes of the liquid drop on a cover glass from the mountain with a wide, thin skirt toward the mountain with the sharp slope in the outside edge as the

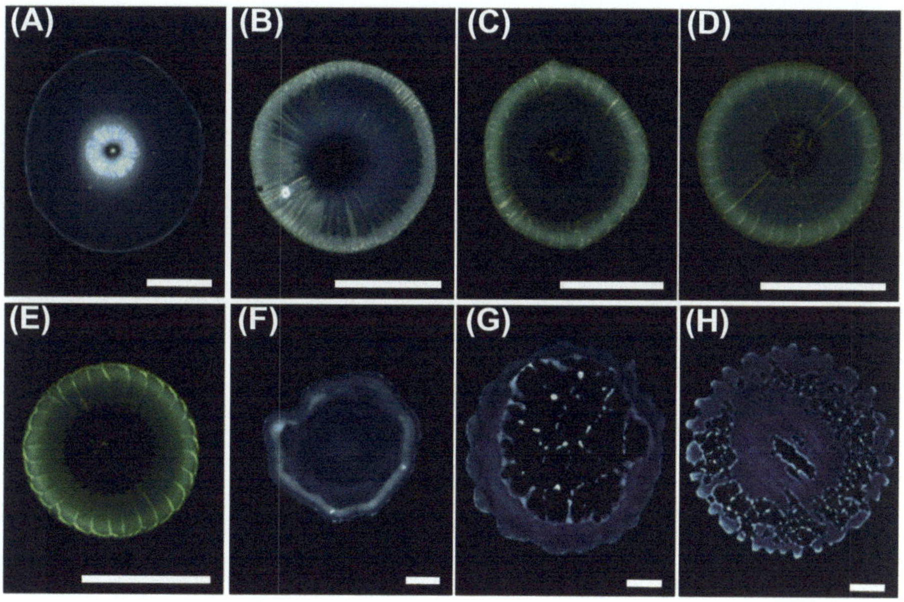

Figure 5.49
Drying patterns of CS91 spheres in aqueous organic solvent mixtures on a cover glass at 25° C. $\phi = 0.02$, $v = 7\ \mu l$, **A,** EtOH; **B,** EtOH:water = 3:1; **C,** EtOH:water = 1:1; **D,** EtOH:-water = 1:3; **E,** water; **F,** EtO-H:ether = 3:1; **G,** EtOH:ether = 1:1; **H,** EtOH:ether = 1:3; length of the bars = 3 mm. *From Okubo T, Nakagawa N, Tsuchida A:* Colloid Polym Sci *285:1247-1255, 2007.*

content of water increases. The spokelike cracks always appeared in the broad ring area irrespective of the content of the two solvents. The number of spoke lines decreased as the water content increased.

Fig. 5.50, *A* through *D*, shows the macroscopic drying patterns in 100% ethyl alcohol and the mixtures with methyl alcohol. The overall patterns are similar irrespective of the mixing ratio. The differences in the values of boiling point and surface tension between ethyl alcohol and methyl alcohol are small. The viscosity of the former is about twice that of the latter. However, this difference seems not to affect the patterns so much. Thickness profiles of the dried film for ethyl alcohol, methyl alcohol, and their mixtures were also similar (compare the *broken* and *dot-dot-broken curves* in Fig. 5.51). Fig. 5.50, *E* through *G*, are the drying patterns of the mixtures of ethyl alcohol and 1-propyl alcohol. The size of the drying patterns (d_f) decreased slightly as content of 1-propyl alcohol increased, which is explained with a slight increase in the surface tension of 1-propyl alcohol (24 dyn/cm) compared with that of ethyl alcohol (22 dyn/cm). Interestingly, the broad rings in the inner area shifted toward the outside edges continuously when the content of 1-propyl alcohol increased, as is also shown in Fig. 5.51 (compare the *dot-broken* and *dot-dot-broken curves*). These changes in the patterns are explained with the fact that the surface tension and viscosity of 1-propyl alcohol are larger than those of ethyl alcohol (i.e., these parameters are favorable for shrinking of the area of a liquid drop and thickening of the initial liquid layer in the outside area).

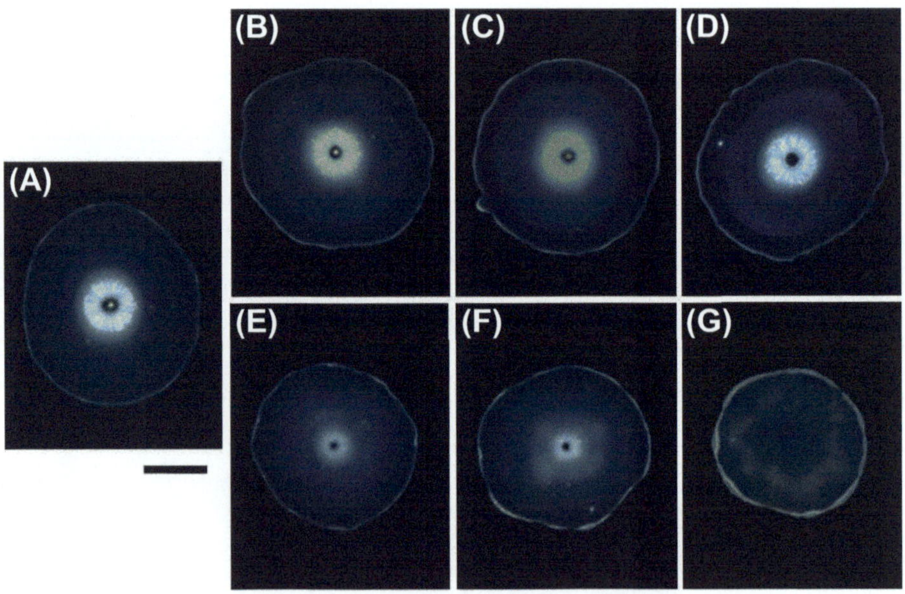

Figure 5.50 Drying patterns of CS91 spheres in aqueous organic solvent mixtures on a cover glass at 25° C; ϕ = 0.02; v = 7 µl; **A,** EtOH; **B,** EtOH:-MeOH = 3:1; **C,** EtOH:MeOH = 1:1; **D,** EtOH:MeOH = 1:3; **E,** EtOH:PrOH = 3:1; **F,** EtOH: PrOH = 1:1; **G,** EtOH:PrOH = 1:3; length of the bars = 3 mm. *From Okubo T, Nakagawa N, Tsuchida A:* Colloid Polym Sci *285:1247-1255, 2007.*

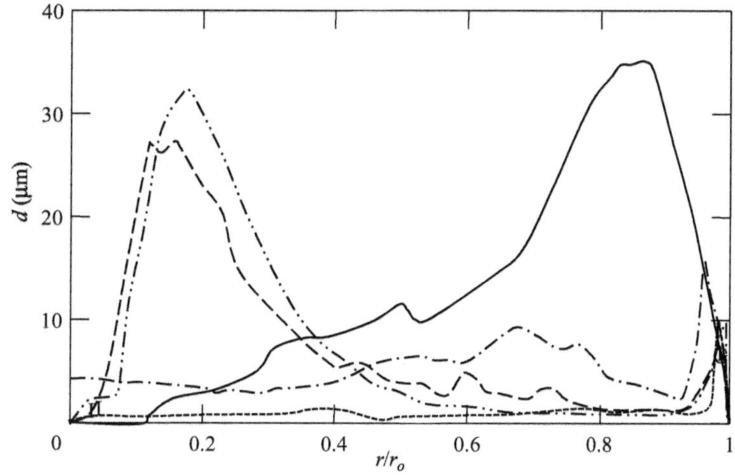

Figure 5.51
Thickness profiles of drying films of CS91 spheres in aqueous organic solvent mixture as a function of normalized distance from the pattern center (r/r_o) at 25° C. $\phi = 0.02$, $v = 7$ μl, *solid curve:* EtOH:water = 1:3; *dotted:* EtOH:ether = 1:3; *broken:* EtOH:MeOH = 1:3; *dot-broken:* EtOH:PrOH = 1:3; *dot-dot-broken:* EtOH.
Modified from Okubo T, Nakagawa N, Tsuchida A: Colloid Polym Sci *285:1247-1255, 2007.*

The macroscopic drying patterns of ethanol and its mixtures with diethyl ether were studied in Fig. 5.49, *F* through *H*. The film size (d_f) increased substantially by the addition of diethyl ether, which is quite consistent with the sharp decrease in the surface tension of the mixtures by the addition of diethyl ether. Broad rings at the outside edges always appeared irrespective of the content of diethyl ether. However, the heights of the rings were quite low in these mixtures, as is shown in the dotted curve of Fig. 5.51. Fig. 5.51 also shows that the dried films formed from ethyl alcohol and diethyl ether mixtures were thin and flat except the broad rings in the outside edges. Thin, flat drying patterns support the idea that the initial shape of the liquid is flat and thin, and furthermore that very fast evaporation takes place during the course of drying, because most the parameters of boiling point, surface tension, and viscosity of diethyl ether are the lowest among five kinds of solvents examined.

Fig. 5.52 shows the sharpness parameter *S* of the broad rings in ethanol with other solvent mixtures, which were evaluated from the thickness profiles shown in Fig. 5.51 and others.[74] *S* values of the outer broad rings at the ethanol mixtures with other solvents (1:3 in volume fraction) are in the order of Equation 6.

$$H_2O \ll \text{methanol} \approx \text{propanol} < \text{diethyl ether} \qquad 6$$

This order among the solvents shows the order of the magnitudes in diffusivity of CS91 spheres on the glass substrate. The order of Equation 6

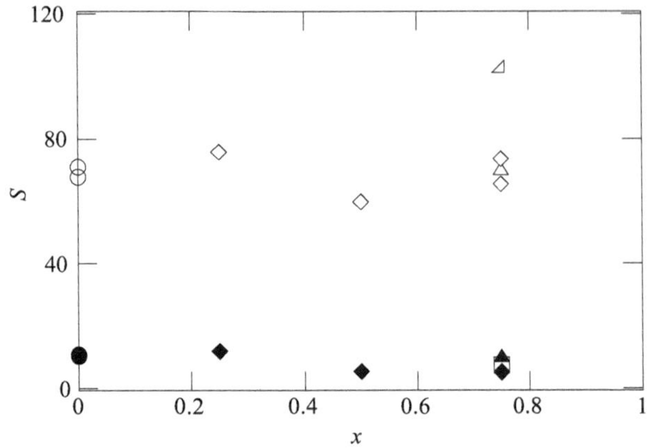

Figure 5.52 Sharpness of the broad ring as a function of mixing ratio (x) of organic solvents on a cover glass at 25° C. *Circles:* 100% ethanol; *triangles:* methanol; *diamond:* propanol; *square:* water; *open symbols:* outer broad ring; *solid symbols:* inner.

corresponds to that of the boiling temperatures of the solvents and inversed order of the viscosities. S values are one of the indicators showing the strength of the affinity between colloidal spheres (solutes) and substrate surfaces.

Fig. 5.53 shows typical examples of the microscopic drying patterns of 100% ethanol suspension of CS91 spheres on a watch glass from the center to the right edge.[13] Coupling patterns of fine spoke lines and ringlike cracks are beautiful (e.g., see *D* and *F*). Fast convectional flow of spheres and solvents appears to be "memorized" in the dried film.

5.4.3 Poly(Methyl Methacrylate) Spheres

Thanks to recent advancement in the polymer synthesis techniques, a wide range of sizes and monodispersed PMMA colloidal spheres are available commercially. Fig. 5.54 shows typical examples of the macroscopic drying patterns of GW8 spheres (300 nm in diameter) on a cover glass.[49] Initial suspension concentrations range from $\phi = 0.00074$ to 0.074. On a cover glass, the broad rings, where the widths increase with increasing sphere concentration, are observed clearly. Interestingly, the inner broad rings appeared when sphere concentration is high, as shown in Fig. 5.54, *D* through *F*.

Fig. 5.55 shows the d_f/d_i values of PMMA spheres at the inner area on a watch glass.[50] On a watch glass, d_f/d_i decreases as ϕ decreases. Two broad rings appeared for PMMA spheres, and these systems are grouped into type B in Table 5.6. The decreasing tendency was most substantial for the largest PMMA spheres. Furthermore, at the higher sphere concentrations than certain critical concentration m^*, d_f/d_i values remained the same, and inner broad rings were fused with those at the outside edges. The existence of the associates of the PMMA spheres and their segregation to the inner broad ring areas are highly plausible, especially at high sphere concentrations.

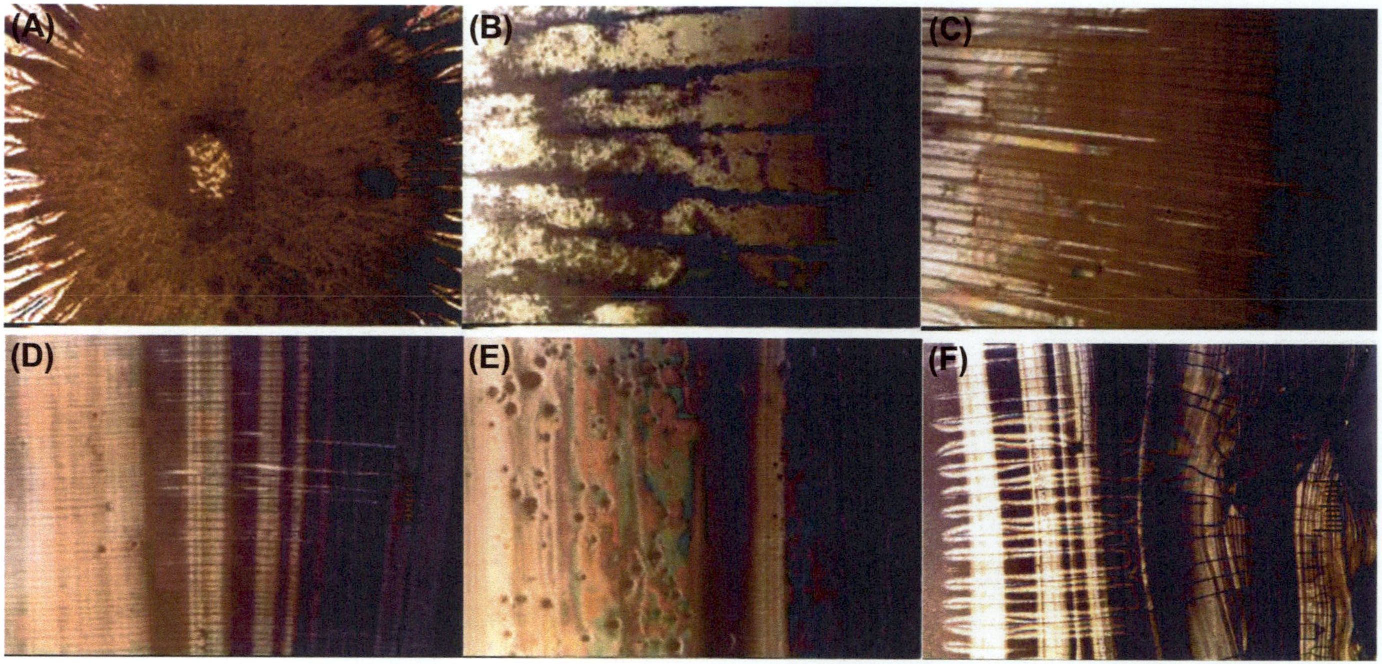

Figure 5.53 Microscopic drying dissipative patterns of 100% ethanol suspension of CS91 spheres in a watch glass at 20° C, ϕ= 0.0064, 4 ml, dry, from center (*A*) to right edge (*F*), full scale = 0.2 mm.

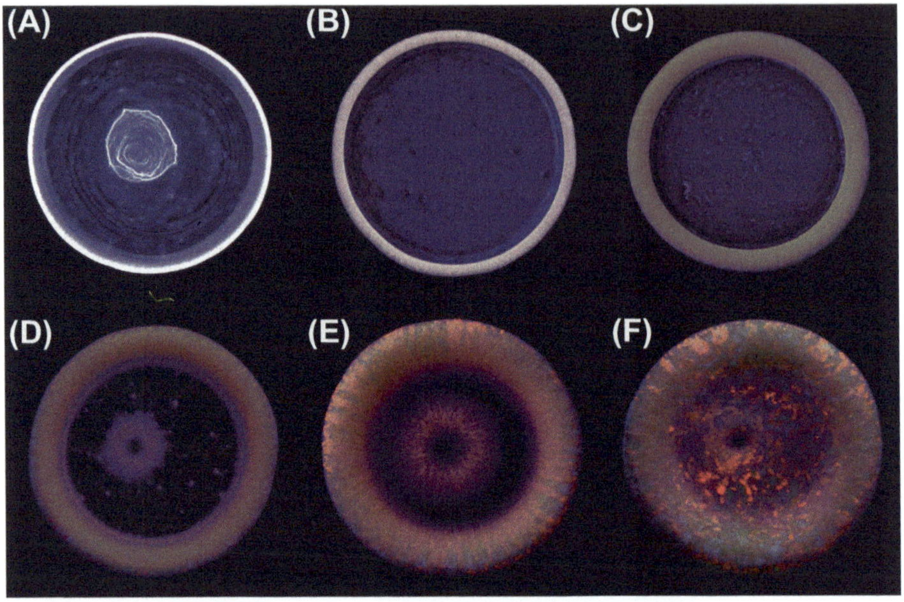

Figure 5.54 Drying patterns of GW8 spheres on a cover glass at 24° C, **A,** $\phi =$ 0.00074; **B,** 0.0049; **C,** 0.0123; **D,** 0.0245; **E,** 0.049; **F,** 0.074. *From Okubo T, Okamoto J, Tsuchida A:* Colloid Polym Sci *286:1123-1133, 2008.*

Fig. 5.56, *A* through *Q*, show the thickness profiles of the dried films of GW1 (*A-D*), GW6 (*E-H*), GW8 (*I-M*), and PM1000 spheres (*N-Q*) at the various initial sphere concentrations.[49] Several findings are clear in the figures. First, the sharp broad rings formed at the low sphere concentrations irrespective of the sphere size of PMMA spheres. Second, the broad ring became broad as

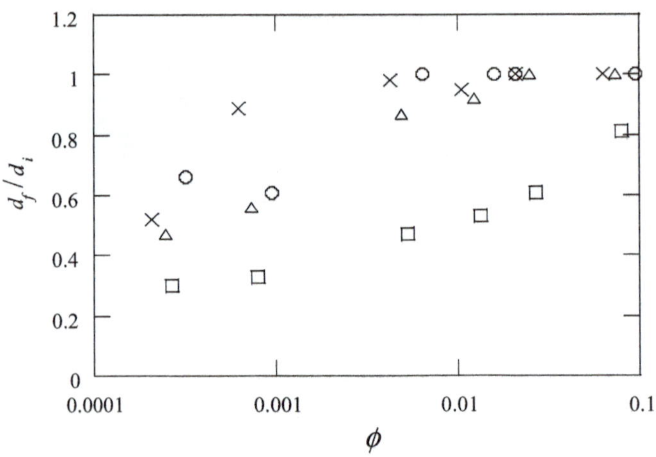

Figure 5.55 Values of d_f/d_i of the inner broad rings of GW1 (*circles*), GW6 (*crosses*), GW8 (*triangles*) and PM1000 (*squares*) as a function of sphere concentration in volume fraction on a watch glass at 25° C. 0.1 ml.

Table 5.6 Six Groups of Macroscopic Patterns Appearing in the Drying Patterns and Their Relations with d_f/d_i and S Values

Group	Broad ring (BR)	d_f/d_i	S	Affinitive Interaction	Typical Solutes
A	1BR	≈0-1, Up with concentration	High	Weak intersolutes, solute- substrate affinity	Polystyrene-, silica-spheres, PEG, NaPMA, PlHBr, rhodamine 6G, uranine
B	2BR (outside) (inner)	1 ≈0-1, up with concentration	High Low	Strong solute-substrate affinity and weak intersolutes affinity	PMMA-spheres, NaPGA, CD1 NaCMC, NaDSS, HPC, CD, pNIPAm, lpNIPAm, (μ) PEGMA-P2VP spheres
C	2BR (outside) (inner)	1 ≈0-1, down with concentration	High Low	Strong solute-substrate affinity and strong intersolutes affinity	pNIPAm(70-5), CAIBA-P2VP, AIBA-P2VP spheres KPolyA, KPolyA+KPolyU
D	1BR	<0.1, down with concentration or keep concentration	Low	Very strong intersolutes affinity and weak solute-substrate affinity	NaHLA, NaDNA
D′	1BR	≈0.25-0.5, up with concentration	—	Strong intersolutes affinity and weak solute-substrate affinity	NaCl, KCl, $CaCl_2$, $LaCl_3$
E	1BR	1	High	Very strong solute-substrate Affinity during solidification	coffee, black tea
F	Central round hill (or +1BR [outside])	0 (1)	Very low (<10)	Associated nonspherical solutes, structured solutes	Bentonite, tungstic acid, PGK, gelatin, ARS, PBA-spheres, POE5C18

AIBA-P2VP and CAIBA-P2VP, cationic gel spheres of poly(2-vinyl pyridine); *ARS*, arrow-root starch; *$CaCl_2$*, calcium chloride; *HPC*, hydroxypropyl cellulose; *KCl*, potassium chloride; *KPolyA*, potassium salt of poly(riboadenylic acid); *KPolyU*, potassium salt of poly(ribouridylic acid); *$LaCl_3$*, lanthanum chloride; *lpNIPAm*, linear-type poly(*N*-isopropylacrylamide); *NaCl*, sodium chloride; *NaCMC*, sodium salts of carboxymethyl cellulose; *NaDSS*, sodium salt of dextran sulfate; *NaPGA*, sodium poly(alpha glutamate); *NaPMA*, sodium polymethacrylate; *P2VP*, poly(2-vinylpyridine); *PBA*, poly(*n*-butyl acrylate); *PEG*, polyethylene glycol; *PEGMA*, polyethylene glycol; *PGK*, palygorskite; *PlHBR*, poly(*L*-lysine hydrobromide); *PMMA*, poly(methyl methacrylate); *pNIPAm*, poly(*N*-isopropylacrylamide).

sphere concentration increased. Third, the broad ring at the inner area developed in addition to the broad ring at the outside edge, with increasing sphere concentration or decreasing sphere size. The third observation supports again the fact that the intersphere association takes place with the help of affinitive interaction of the sphere-substrate affinity for small spheres or high sphere concentrations. The temporal associates in the convectional processes were supported also from the observation of convectional flow patterns as described in Section 5.2.

The concentration dependencies of S values for GW1, GW6, GW8, and PM1000 are plotted in Fig. 5.57. The solid symbols indicate the S values of colloidal silica (CS300) and polystyrene spheres (D1W52 and D1B28), for comparison. For PMMA spheres, S values decreased sharply as sphere concentration increased, and sphere-size dependence on the S values appears to

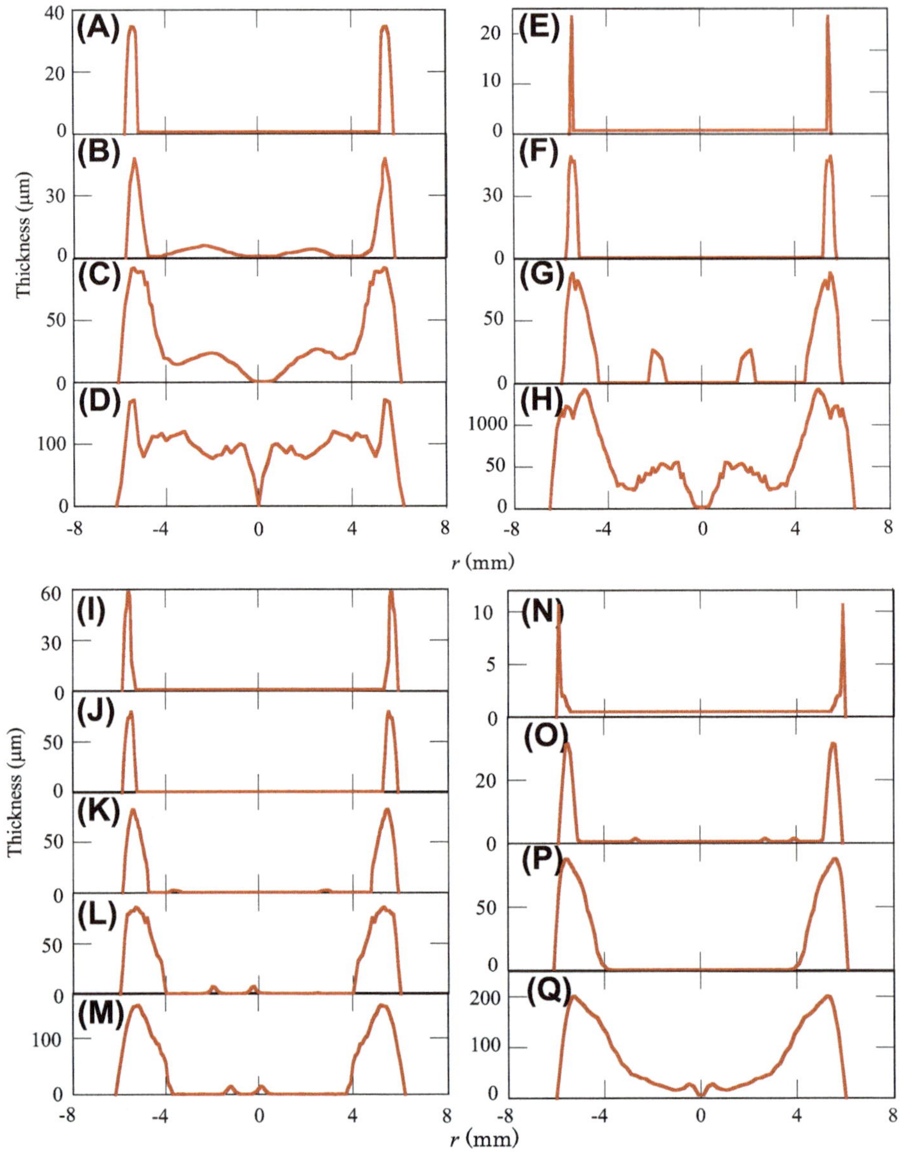

Figure 5.56 Thickness profiles of the dried films of GW1 (*A-D*), GW6 (*E-H*), and GW8 (*I-M*) and PM1000 spheres (*N-Q*) on a cover glass as a function of the distance from the center (*r*). **A,** $\phi = 0.0032$; **B,** 0.0064; **C,** 0.032; **D,** 0.095; **E,** 0.00063; **F,** 0.0042; **G,** 0.021; **H,** 0.063; **I,** 0.0025; **J,** 0.0049; **K,** 0.0123; **L,** 0.025; **M,** 0.074; **N,** 0.00080; **O,** 0.0053; **P,** 0.027; **Q,** 0.080. *From Okubo T, Okamoto J, Tsuchida A:* Colloid Polym Sci *286:1123-1133, 2008.*

be negligible, when the experimental errors are taken into account. Furthermore, the order of Equation 7 appears to hold in *S* values.

$$\text{Silica spheres} \leq \text{PMMA spheres} \leq \text{polystyrene spheres} \qquad 7$$

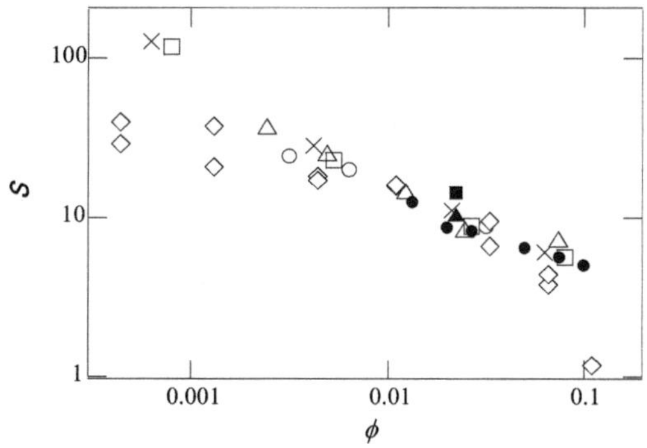

Figure 5.57
Sharpness of the broad ring as a function of volume fraction of colloidal spheres. 0.1 ml, 25° C, *circles* GW1, *crosses* GW6, *triangles* GW8, *squares* PM1000, *solid circles* CS300, *solid triangles* D1W52, *solid squares* D1B28, *diamonds* PBA.

The order of Equation 7 coincides with the order in the strength of the hydrophobicity of colloidal spheres (i.e., a colloidal silica sphere is most hydrophilic and polystyrene spheres are strongly hydrophobic in their nature in a water suspension state).

Typical examples of the microscopic drying patterns of GW6 PMMA spheres at high sphere concentrations, $\phi = 0.042$ are shown in Fig. 5.58 on a cover glass (*A-D*), a watch glass (*E-H*), and a petri glass dish (*I-L*).[49,50,75] The central areas shown in panels *A, E,* and *L* are thin, and the broad ring regions are shown in panels *D, H,* and *I.* Beautiful colors are due to the Bragg diffraction of light by the crystal array of PMMA spheres in the dried film. A large number of spoke lines (cracks) were at the central area, whereas cracks at the broad ring regions were few but originated from the spoke lines and fine rings. In other words, the crack number increased as the thickness of the dried film decreased, which is similar to observations reported previously for colloidal silica and polystyrene spheres.[45,46,71,76]

Quite recently, Tsuchida and Sekita observed the accumulation of the flat layered crystal structures around the outside edge of GW4 spheres on a cover glass, as is shown in Fig. 5.59. These pictures, courtesy of Professor A. Tsuchida, teach us the solidification mechanisms of the colloidal crystals during the course of drying.

5.5 Poly(*N*-Butyl Acrylate) Spheres

Monodispersed colloidal spheres of PBA spheres in deionized aqueous suspension kept their crystal structure during drying, but the beautiful colors disappeared rapidly at the final stage of solidification.[67] Fig. 5.60 shows

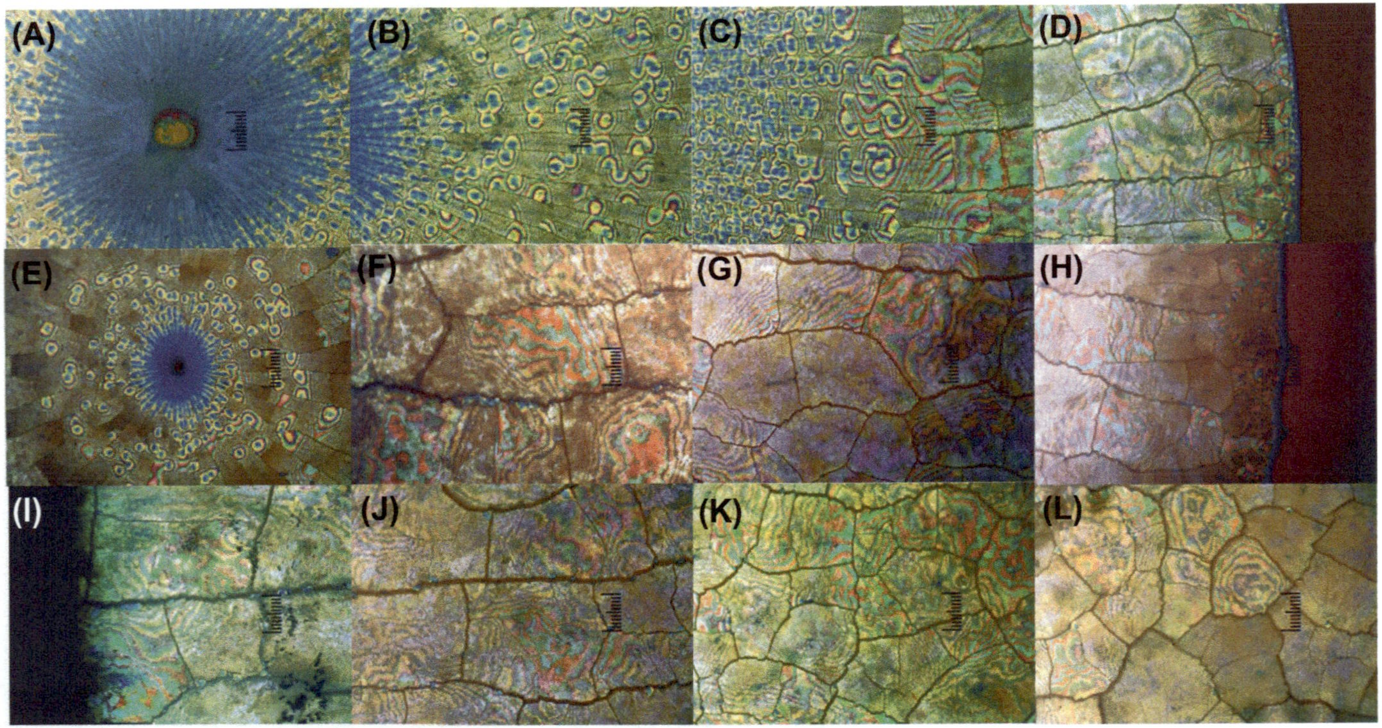

Figure 5.58 Microscopic drying patterns of GW6 PMMA spheres on a cover glass (0.1 ml, *A-D*), a watch glass (4 ml, *E-H*) and a glass dish (5 ml, *I-L*) at 25° C. $\phi = 0.042$, from the center (*A, E*) to right (*D, H*) and the left edge (*I*) to the center (*L*), full scale = 200 μm. *Modified from Okubo T, Okamoto J, Tsuchida A:* Colloid Polym Sci *286:1123-1133, 2008; Okubo T:* Colloid Polym Sci *286:1307-1315, 2008; Okubo T:* Colloid Polym Sci *286:1527-1535, 2008.*

Figure 5.59
SPM images of the drying film of GW4 spheres at the outside edge at 25° C. $\phi = 0.03$, 0.1 ml, length of the bar = 150 nm.

typical examples of the change in the dissipative patterns on a cover glass (*A-D*), a watch glass (*E-H*), and a glass dish (*I-L*) at the sphere concentration of 6.2 wt%. All the suspensions emitted iridescent colors and single crystals. However, at the solidification stage, all the colors and single crystals

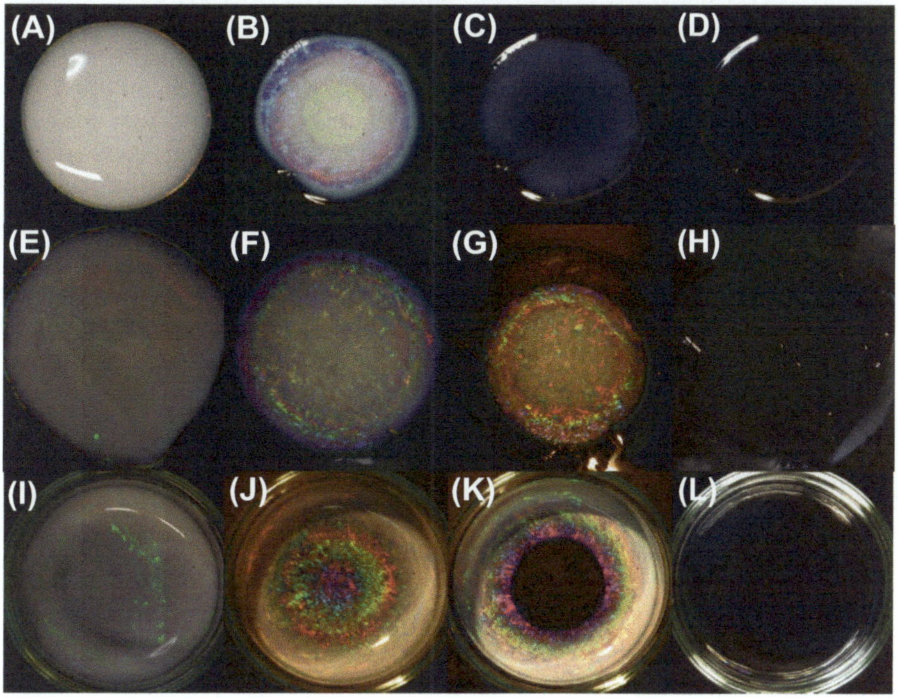

Figure 5.60
Dissipative patterns of PBA with elapsed time on a cover glass (*A-D*), a watch glass (*E-H*), and a glass dish (*I-L*) at 22° C. [PBA] = 6.2 wt%; **A,** 40 minutes after setting; **B,** 5 hours, 30 minutes; **C,** 5 hours, 45 minutes; **D,** 6 hours, 45 minutes; **E,** 40 minutes after setting; **F,** 7 hours, 40 minutes; **G,** 9 hours; **H,** 11 hours, 20 minutes; **I,** 40 minutes after setting; **J,** 13 hours, 35 minutes; **K,** 22 hours; **L,** 32 hours, 45 minutes. *Modified from Fujii S, Nakamura Y, Tsuchida A, Okubo T:* Colloid Polym Sci *292:2303-2310, 2014.*

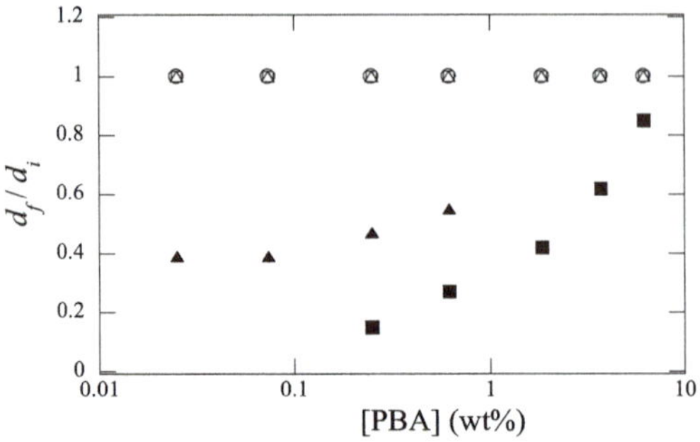

Figure 5.61
Value of d_f/d_i of PBA on a cover glass (*circles*), a watch glass (*triangles*), and a glass dish (*squares*) at 22° C. *Modified from Fujii S, Nakamura Y, Tsuchida A, Okubo T:* Colloid Polym Sci *292:2303-2310, 2014.*

disappeared. Fusion of the crystals and the adhesive solidification of PBA spheres took place.

Fig. 5.61 shows the size ratio, d_f/d_i as a function of PBA concentration. On a cover glass, the broad rings appeared always at the outside edge, when the sphere concentration is low. This supports the strong affinity of PBA spheres with the surfaces of the substrate during the fusion by the strong intersphere affinity interaction. On a watch glass and a glass dish, on the other hand, the inner rings formed in addition to the outer broad rings as shown in the solid triangles and solid squares. The size of the inner rings increased with sphere concentration, which is ascribed to the contribution of the excluded volume effects of each PBA sphere before fusion.

Fig. 5.62 shows the thickness profiles of the dried film of PBA as a function of the distance from the center of the film on an arbitrary line. Interestingly, the sharp broad ring patterns at the low sphere concentration became broad and more round, hill-shaped patterns as initial sphere concentration increased.

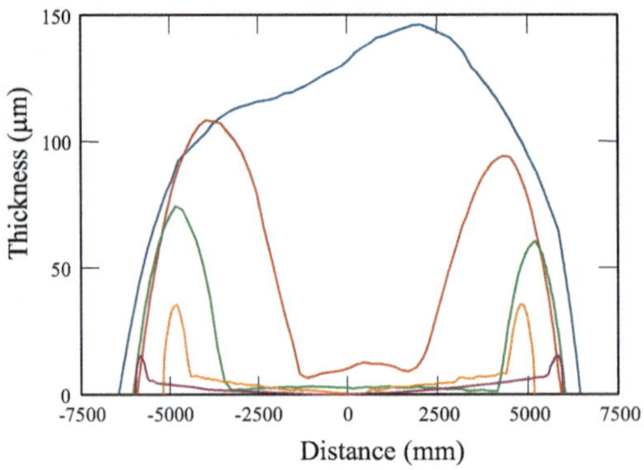

Figure 5.62
Thickness profiles of the dried films of PBA as a function of the distance from the center. At 22° C, *blue curve* $\phi = 0.11$, *red* 0.066, *green* 0.033, *orange* 0.0044, *purple* 0.00044. *From Fujii S, Nakamura Y, Tsuchida A, Okubo T:* Colloid Polym Sci *292:2303-2310, 2014.*

A round, hill-shaped profile is unique and has been observed also for the aggregates of the anisotropic-shaped particles such as plate-shaped bentonite and associative biopolymers such as gelatin of collagen and ARS, as is described later in this chapter. The main cause is the highly restricted diffusion of the solutes, especially during the convectional processes.

Macroscopic and microscopic drying patterns of the monodispersed spherical colloidal particles were discussed previously. Most familiar macroscopic patterns were broad rings and spoke lines. The broad rings are often observed in the inner area in addition to the outside edge of the dried film. Values of d_f/d_i were clarified to be one of the most convenient parameters showing the strength of the intersolute and solute-substrate affinities. From the systematic studies on the d_f/d_i values of suspensions and solutions in my laboratory, the broad ring patterns in the dried film are classified into six typical groups in general, as is shown in Table 5.6. In group A, one broad ring forms at the outside edge. When solute concentration is high, d_f/d_i is unified. However, the size of the broad ring decrease with decreasing solute concentration (i.e., d_f/d_i decreases as concentration decreases in the range of low concentrations). Most spherical particles such as colloidal polystyrene and silica spheres are grouped in this category. Whenever interparticle and particle-substrate affinities are weak, size-changeable 1 broad ring (1BR) appears. In group B, two broad rings form at the outside edge and inner area. Furthermore, the size of the inner ring increases as initial solute concentration increases. This type takes place when the associates form in part but the interassociates affinity is weak during the convectional process. Furthermore, the solute–substrate interactions are strong. Formation of associates results in the appearance of the inner broad ring. Colloidal spheres of PMMA is grouped in this category. Many polysaccharides are also grouped into B, as is discussed in Section 5.8.2. In group C, two broad rings form at the outside edge and inner area like group B. However, the size of the inner rings decrease with increasing initial solute concentration. Small cationic gel spheres, CAIBA-P2VP and AIBA-P2VP are grouped here, as described in Chapter 4. For thermo-sensitive gel spheres of pNIPAm(70-5) at high sphere concentrations, large associates form in part, and group F–like central round hills form in addition to the group C patterns. A part of polynucleotides, potassium salt of poly(riboadenylic acid) (KPolyA) and KPolyA + potassium salt of poly(ribouridylic acid) (KPolyU), for example, are group C, as is described in Section 5.8.3. Group D has single, small broad rings in the central area of the dried film ($d_f/d_i < 0.1$), when very strong intersolute affinity exists, although the solute-substrate affinity is low. Sodium salt of hyaluronic acid (NaHLA) is a typical examples. When d_f/d_i values of the single broad rings of solutes are in the range of approximately 0.2 to 0.5 and increase with solute concentration, these are sorted into group D. The simple electrolytes such as NaCl and calcium chloride ($CaCl_2$) are in this group. In group E, one broad ring appears at the outside edge ($d_f/d_i = 1$), irrespective of the initial solute concentration. This type of broad ring forms whenever there is very strong solute-substrate affinity, especially during the solidification process. The drying patterns of coffee and black tea are grouped in this category. Group F shows a central round hill in addition to or instead of the broad ring at the outside edge. The patterns appear whenever large associates such as a gel and rubber structure or aggregates of platelike particles take place. Diffusion of solutes is strongly restricted during drying. Suspensions of bentonite particles and solutions of gelatin and ARS are the typical systems of group F.

5.6 Nonspherical Particles

Macroscopic and microscopic drying patterns of nonspherical (anisotropic-shaped) particles differ in several points from those of spherical particles in general. Spoke lines of nonspherical particles become broad or disappear. Furthermore, a round hill (type F) pattern forms often on the substrate such as a cover glass. The main cause for the unique pattern formation is the restricted diffusivity of solutes in a suspension state, especially during the convectional flow process. Intersolute network structural formation is a typical example. Another cause is the strengthened interparticle affinitive interactions (e.g., aggregates formation) during sedimentary or solidification processes, especially for platelike particles.

5.6.1 Palygorskite, Tungstic Acid, and Bentonite Particles

Fig. 5.63 shows typical examples of the macroscopic drying patterns of needle-shaped palygorskite (PGK) and ellipsoidal, platelike tungstic acid (TA) particles.[77] Several findings are clear in the figure. First, broad rings formed and their size decreased as particle concentration decreased. Fig. 5.64 shows the d_f/d_i values of PGK and TA6. The values are insensitive to the amount of liquid drops ranging 0.01 ml to 0.1 ml and increased as particle concentration increased. The values reach unity at approximately the concentrations of 0.01 and 10 wt% for PGK and TA6, respectively. The decreasing tendency of d_f/d_i values from unity supports that the PGK-substrate affinity is weak, and the d_f values are determined mainly by the excluded-volume effects of the solutes themselves in suspension state. Influence of the particle size on the d_f/d_i values for TA is shown in Table 5.7. The d_f/d_i values decreased as particle size increased. This tendency is consistent with decreased diffusivity of the particles or the strengthened interparticle affinity as particle size increased.

The second finding in Fig. 5.63 is that the very broad ring patterns appeared for PGK and TA. Appearance of the broad rings is shown clearly in the

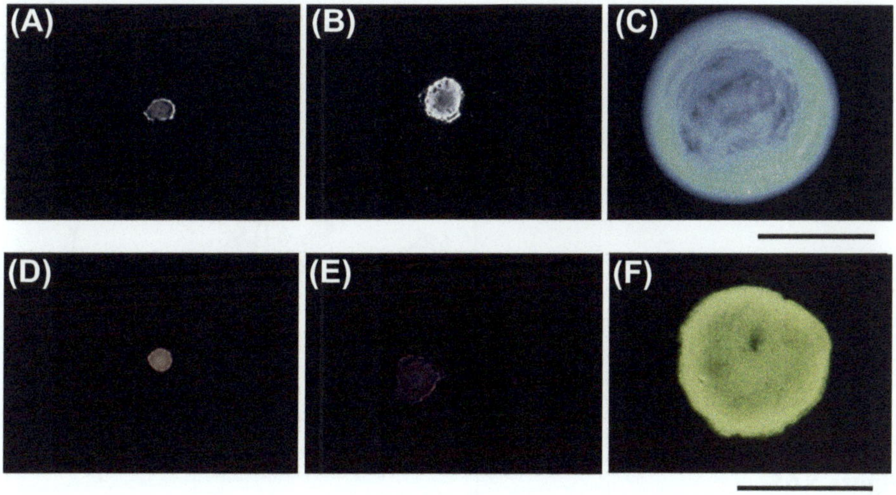

Figure 5.63 Drying patterns of PGK (*A-C*) and TA (*D-F*) suspensions on a cover glass at 25° C. 9.05 ml, **A,** 6.2 × 10^{-9} wt%; **B,** 6.2 × 10^{-5} wt%; **C,** 0.62 wt%; **D,** 3.6 × 10^{-6} wt%; **E,** 3.6 × 10^{-4} wt%; **F,** 0.36 wt%, length of the bar is 5 mm. *From Okubo T, Otake A, Tsuchida A:* Colloid Polym Sci *287:1435-1444, 2009.*

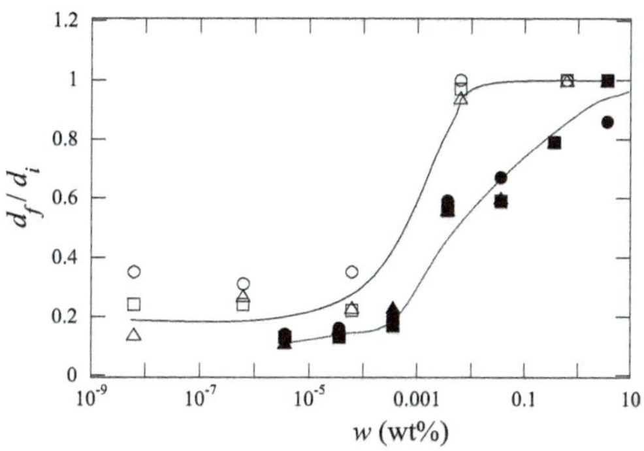

Figure 5.64 Value of d_f/d_i of PGK (*open symbols*) and TA6 (*solid symbols*) on a cover glass at 25° C. *Circles* 0.01 ml, *triangles* 0.05 ml, *squares* 0.1 ml.

Table 5.7 Influence of Size on d_f/d_i Values of TA at 25° C in water, 0.05 ml

TA	2*c* (μm)	2*b* (μm)	d_f/d_i			
			0.000165 (wt%)	0.0165 (wt%)	0.165 (wt%)	1.65 (wt%)
TA3	3.0	1.3	0.22	0.62	0.84	1.0
TA4	4.0	2.1	0.13	0.57	0.78	0.97
TA5	5.0	2.3	0.13	0.55	0.74	0.97
TA6	8.9	2.6	0.11	0.52	0.76	0.96
TA7	7.1	3.1	0.11	0.50	0.83	0.91

TA, Tungstic acid.

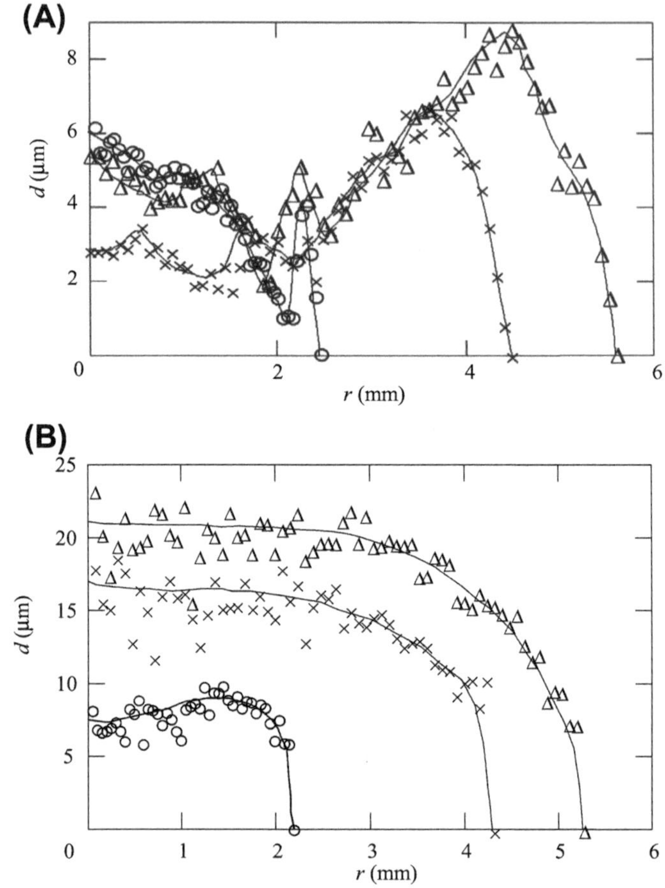

Figure 5.65
Thickness profiles of PGK (*A*) and TA6 (*B*) on a cover glass at 25° C. **A,** 0.62 wt%, circles 0.01 ml, crosses 0.05 ml, triangles 0.1 ml; **B,** 3.6 wt%, circles 0.01 ml, crosses 0.05 ml, triangles 0.1 ml, *Modified from Okubo T, Otake A, Tsuchida A:* Colloid Polym Sci *287:1435-1444, 2009.*

thickness profiles of the dried films of PGK shown in Fig. 5.65. Thickness profiles of the broad ring of colloidal silica spheres of CS82 at the high concentration (14.7 wt%) was similar to that formed at very low particle concentrations, 0.62 wt% for PGK. Peak profiles of CS82 spheres corresponding particle concentrations should be sharp, as discussed in Section 5.4. The broadening of the ring structure is also ascribed to the restricted diffusivity of the anisotropic-shaped particles compared with that of spherical particles and also with the formation of the stacking associates of PGK and TA. The stacking is known to take place especially for platelike particles in aqueous suspension state.

The third finding is the formation of the central round hill in the drying films of PGK and TA. The central round hill patterns have formed for gel solutions such as gelatin of collagen and ARS (described in Section 5.8.4) and further

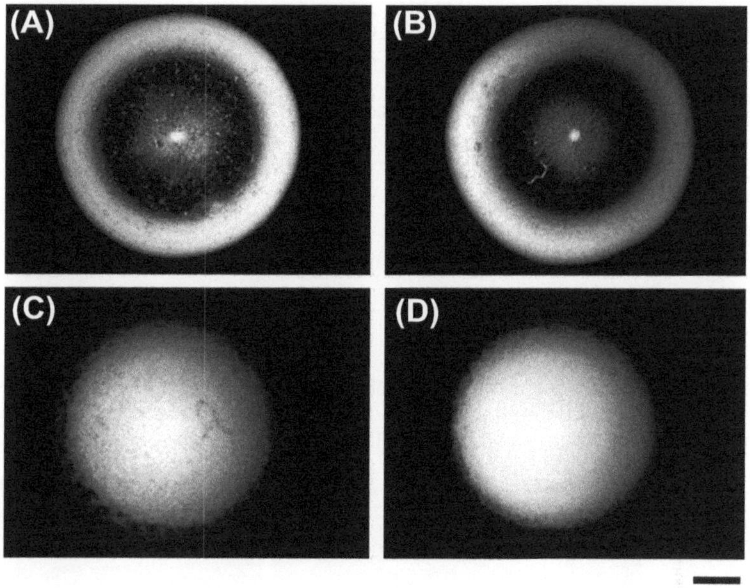

Figure 5.66 Drying patterns of 2p7f bentonite at 25° C. 0.05 ml; 4.6 mg/ml; **A,** [NaCl] = 0 M; **B,** 1×10^{-5} M; **C,** 1×10^{-4} M; **D,** 0.001 M; length of the bar is 2 mm. *From Yamaguchi T, Kimura K, Tsuchida A, Okubo T, Matsumoto M:* Colloid Polym Sci *283:1123-1130, 2005.*

for small cationic gel spheres of pNIPAm(70-5) (discussed in Section 4.12 of Chapter 4). Thus the central round hill patterns form often when the associates of nonspherical solute or highly netted solutes form. Fig. 5.66 shows salt effects on the macroscopic drying patterns of platelike bentonites.[28] The macroscopic drying patterns changed sharply from the broad ring patterns to the central round hill shape when the salt concentration changed from 0M (*A*) to 0.001 M (*D*). It is highly plausible that the stacking association of the platelike bentonite particles takes place, resulting in the formation of the large aggregates.

Thus the macroscopic drying patterns of PGK, TA6, and bentonite are similar, and categorized in group F, as is shown in Table 5.6. *S* values of group F are small and come close to unity at high particle concentration (see Fig. 5.67). Concentration dependencies of *S* values for PGK, TA, and bentonite are surprisingly similar, as is shown in Fig. 5.67. Typical examples of the microscopic drying patterns of PGK and TA are shown in Fig. 5.68. Similar patterns are also observed from the microscopic observation. The surfaces of the broad rings of PGK (*A*) and TA6 (*B*) formed in the outside regions look very rough, which may support the inhomogeneous accumulation of the needlelike and platelike particles.

Furthermore, fourth, regular branchlike (or netlike) patterns demonstrate that these particles are stacked regularly and in the fractal manner, forming the hills oriented circularly. No spoke-line patterns, which have been observed for spherical colloidal particles, are shown clearly in Fig. 5.68.

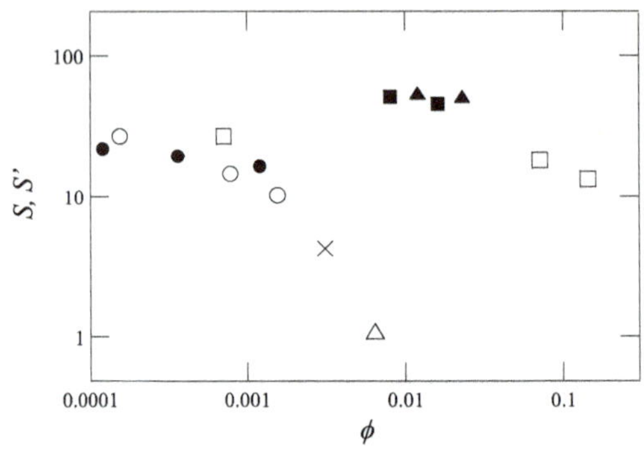

Figure 5.67
Sharpness parameter of nonspherical colloidal particles as a function of volume fraction at 25° C. 0.1 ml, *circles* bentonite, *crosses* palygorskite, *triangles* tungstic acid, *squares* Chinese black ink, *solid circles* SHB, *solid triangles* coffee with cream, *solid squares* black tea with cream.

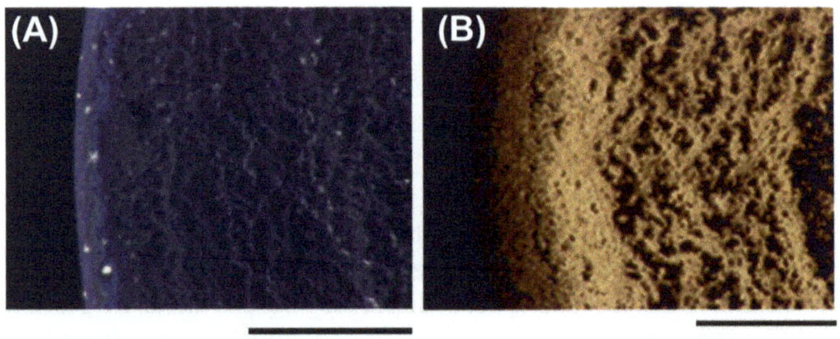

Figure 5.68
Microscopic drying patterns of PGK (*A*) and TA (*B*) on a cover glass at 25° C. 0.05 ml, **A,** 0.062 mg/ml, length of the bar is 0.7 mm; **B,** 0.36 mg/ml; bar length is 0.5 mm. *Modified from Okubo T, Otake A, Tsuchida A:* Colloid Polym Sci *287:1435-1444, 2009.*

5.6.2 Spindle-Shaped Hematite Particles Coated with Polymer Brush

Recently, Sasano et al. prepared the highly monodispersed spindle-shaped hematite particles coated with poly(poly(ethylene oxide) methyl ether methacrylate) (SHB) brush.[78] Ohno et al. studied anisotropic colloidal crystals and drying dissipative structures of SHB.[79,80] Fig. 5.69 shows the typical examples of the drying patterns of SHB particles on a cover glass (*A-D*), a watch glass (*E-H*), and a glass dish (*I-L*).

The outer and inner broad rings appeared in the macroscopic drying patterns irrespective of the substrates. The ratios of the final size of the broad ring d_f against size of the initial liquid d_i are shown in Fig. 5.70. Open and closed symbols show the size of the outer and inner broad rings, respectively. The outer rings kept unity irrespective of the particle concentration of the initial suspensions. This means that a part of the SHB particles remained at the

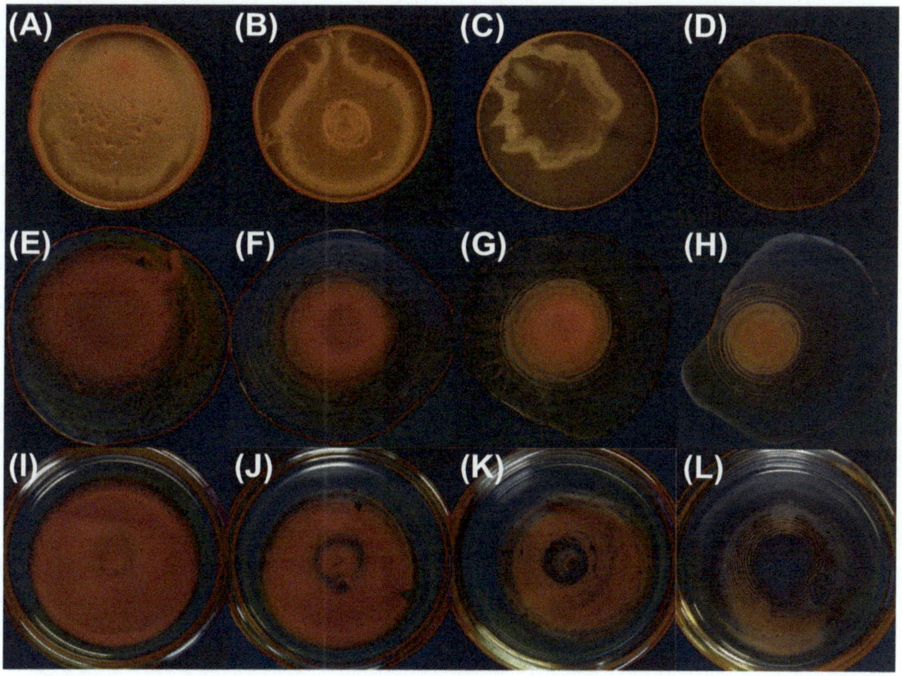

Figure 5.69 Drying patterns of SHB on a cover glass (*A-D*, 0.1 ml), a watch glass (*E-H*, 0.7 ml) and a glass dish (*I-L*, 0.7 ml) at 20° C. **A, E, I,** 0.383 wt%; **B, F, J,** 0.115 wt%; **C, G, K,** 0.0383 wt%; **D, H, L,** 0.0115 wt %. *From Ohno K, Huang Y, Okubo T:* Colloid Polym Sci *292:1143-1151, 2014.*

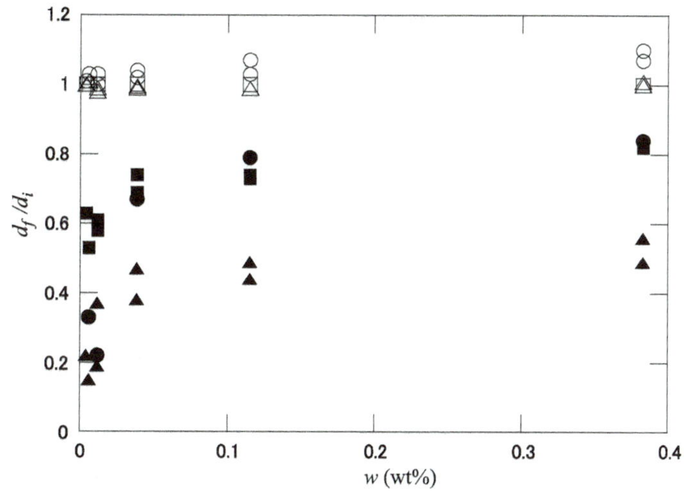

Figure 5.70 Value of d_f/d_i of SHB on a cover glass (0.1 ml, *circles*), a watch glass (0.7 ml, *triangles*) and a glass dish (0.7 ml, *squares*) at 20° C. *Open symbols:* broad ring at the outside edge; *closed symbols:* in the inner area. *From Ohno K, Huang Y, Okubo T:* Colloid Polym Sci *292:1143-1151, 2014.*

outside edge of the initial liquid by the affinitive SHB–substrate interaction. Weak solute-substrate affinity will remain during the course of drying the suspension. Most of the SHB particles formed the inner broad ring area. In general, the size of the broad ring is determined by the sum of the excluded volumes of each SHB including the electrical double layers formed around the particles during dryness. It should be mentioned here that the inner broad rings are broad irrespective of the substrate. The *S* values of the inner rings are estimated roughly and plotted in Fig. 5.67. *S* values of SHB (*solid circles*) were similar to those of PGK (*cross*), TA (*triangle*), and bentonite particles (*circles*).

Fig. 5.71 shows an example of the most magnified microscopic pictures (*A-F*) of SHB particles from the central area (*A*) to the outside edge (*F*) on a cover glass at 0.038 wt%. Fig. 5.71, *G* through *I*, are observed with a scanning electron microscope (SEM) on a cover glass (*G*), a watch glass (*H*), and a glass dish (*I*), respectively, around the inner broad ring at 20° C. The particle-free spots appeared in Fig. 5.71, *A* through *E*, and these areas increased as particle concentration decreased. When Fig. 5.71, *A* and *B*, are examined carefully, most of the long axes of SHB particles appear to be oriented parallel to the substrate plane. In Fig. 5.71, *G* through *I*, clear-cut, long-range ordering of SHB particles was not observed. However, two kinds of local arrays, parallel oriented clusters and flowerlike oriented clusters, of SHB particles were observed. The direction of each cluster appears to be random. The main cause for the random distribution of clusters of SHB particles is that the effective anisotropy of SHB particles (4.4 in the axis ratio) including the electrical double layers comes close to unity ($\approx$1.2), especially in the initial deionized suspension state during the convectional and sedimentary processes.

5.6.3 Chinese Black Ink and Diamond Particles

Chinese black ink is made from soot and glue, and has been used for writing for a long time in China, Japan, and other countries. Terada et al.[9-12,44] clarified that Chinese ink was a typical example of a colloidal suspension. The Terada cell as a typical convectional pattern observed using Chinese black ink was described in Section 5.2. Fig. 5.72 shows typical examples of the macroscopic drying patterns on the unrinsed (*A-D*) and rinsed (*E-H*) cover glasses. The patterns extended and became finer on a rinsed substrate. However, the essential characteristics of the drying pattern don't appear to be changed by rinsing. Macroscopic patterns of the broad rings and spoke lines were recognized clearly for rinsed and unrinsed cover glasses. The spoke lines appeared but became significantly broadened.[12] From the WBRs, S' values were estimated to be 27, 18, and 13 at volume fractions 0.00071, 0.071, and 0.143, respectively. The S' values were larger than the S values of PGK, TA, and bentonite particles. The d_f/d_i value of the broad rings at the outside edges decreased slightly from 1 to 0.79 and from 1 to 0.68 on unrinsed and rinsed cover glasses, respectively, when concentration decreased from 20

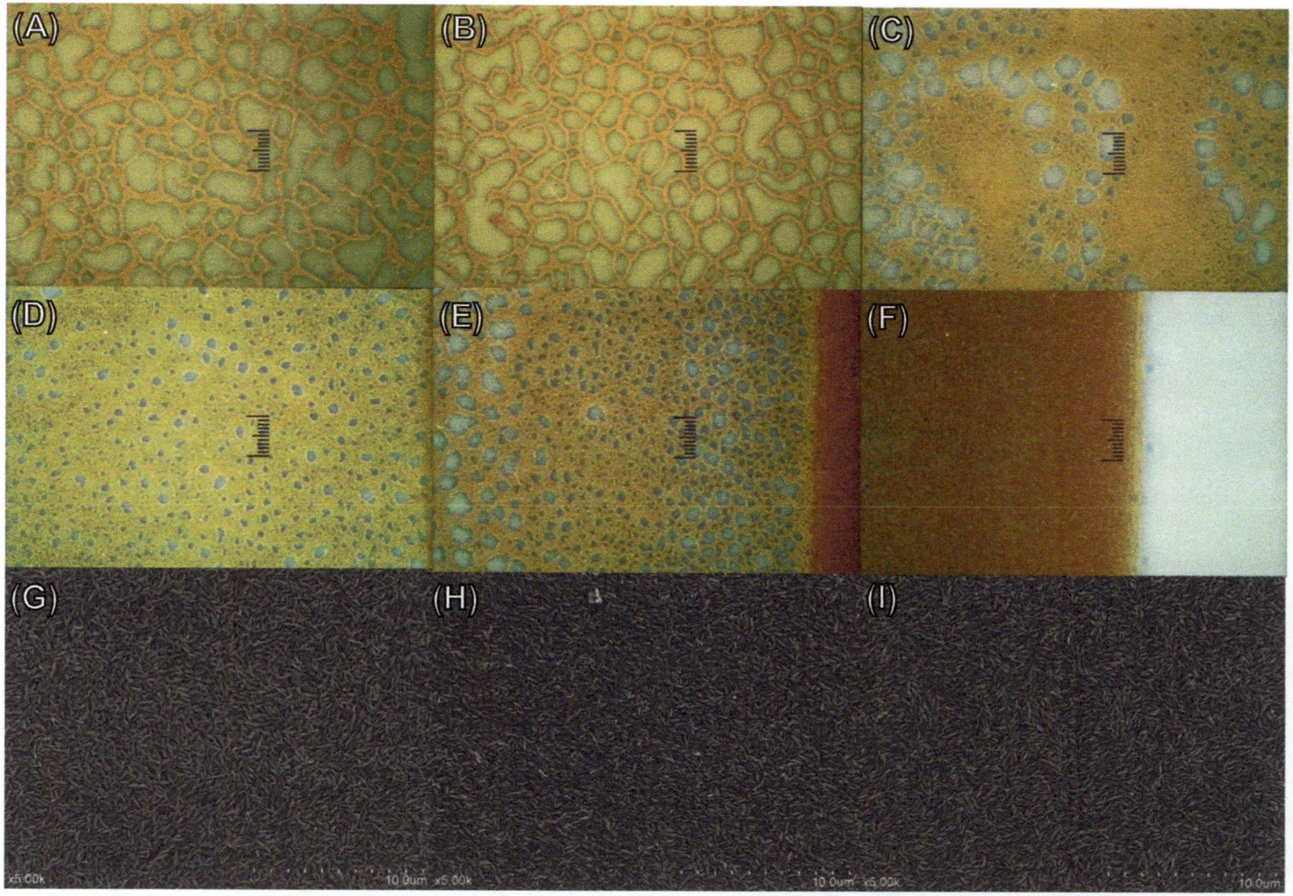

Figure 5.71 Microscopic (**A-F,** 0.0383 wt%) and SEM (**G-I,** 0.383 wt%) drying patterns of SHB at 20° C. **A-F,** On a cover glass, pictures from **A** to **F** are the center to the right edge, full scale = 100 μm;, **G,** on a cover glass, 0.1 ml; **H,** a watch glass, 0.7 ml; **I,** a glass dish, 0.7 ml, 0.383 wt%, full scale = 10 μm (*A-F*). *Modified from Ohno K, Huang Y, Okubo T:* Colloid Polym Sci *292:1143-1151, 2014.*

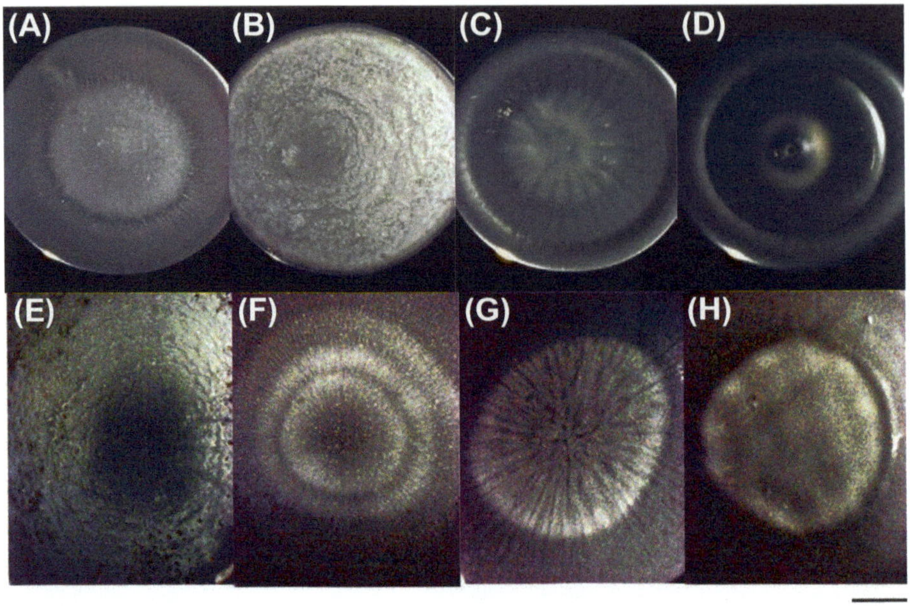

Figure 5.72 Drying patterns of Chinese black ink on an unrinsed (*A-D*) and rinsed (*E*-H) cover glass at 25° C. **A, E,** 0.001 g/ml; **B, F,** 0.01 g/ml; **C, G,** 0.1 g/ml; **D, H,** 0.2 g/ml; full scale is 2.0 mm.

wt% to 0.0001 wt%. The shape of Chinese black ink particles is deduced near spherical and blocklike from comparison of the *S* (or *S'*) values.

The similar-sized colloidal agglomerates of diamonds (CD1, 1.5 μm in diameter) were prepared by the deionization of the aqueous suspensions of preparticles of diamond particles of approximately 4 nm in diameter.[81] The drying patterns of CD1 were observed recently.[82] Two kinds of macroscopic patterns, outer and inner broad rings, and rather broad spoke lines appeared. The d_f/d_i values of CD1 are shown in Fig. 5.73 as a function of particle concentration in wt%. The outer rings kept unity irrespective of the particle concentration of the initial suspensions. This means that a part of CD1 particles remained at the outside edge by the particle-substrate affinitive forces. Most CD1 particles, however, formed the inner broad rings. The size of the inner broad rings increased as particle concentration increased. When the excluded-volume effects are effective, the size of the inner ring is determined by the sum of the excluded volumes of each CD1 particles, including the electrical double layers formed around the particles in suspension state. CD1 is grouped into type B in Table 5.6.

Fig. 5.74 shows typical examples of the microscopic drying patterns of CD1 on a cover glass (*A-D*) and a watch glass (*E-H*). Colorful patterns formed are caused by the reflection from the thin film of the inner area from the broad ring. These patterns are also ascribed to the fact that the films contain a small

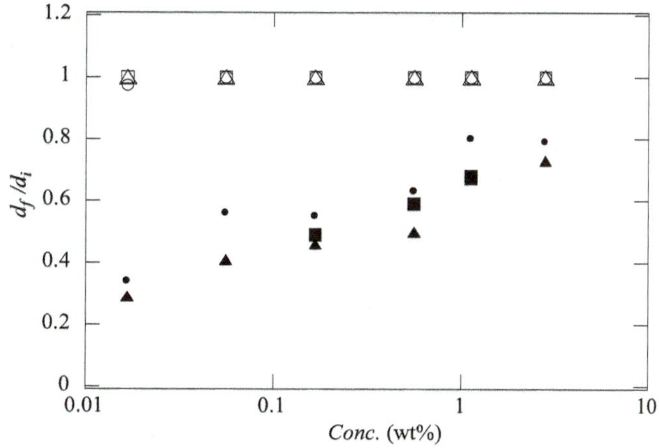

Figure 5.73 Value of d_f/d_i of CD1 on a cover glass (*open circles:* outer rings; *solid circles:* inner rings), a watch glass (*open triangles:* outer; *solid triangles:* inner), and a glass dish (*open squares:* outer; *solid squares:* inner) at 25° C. *From Okubo T:* Colloid Polym Sci *291:1887-1893, 2013.*

amount of water molecules in the dried film, because the drying processes take place at room temperature and the film is slightly hydrophilic in nature.

Fig. 5.74, *I* through *L*, show the formation of the aggregates of CD1 particles by the addition of NaCl on a cover glass (*I, J*), a watch glass (*K*), and a glass dish (*L*). Highly distorted dendritic aggregates of CD1 particles from 100 to 500 μm in size are recognized. The dendritic aggregation took place when the salt concentration was higher than 0.6 mM. CD1-salt complexation is highly plausible.

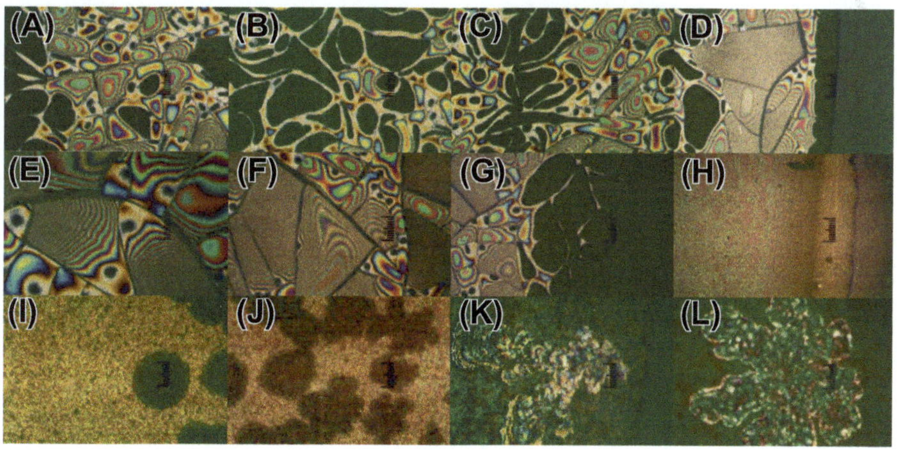

Figure 5.74 Microscopic drying patterns of CD1 on a cover glass (*A-D,* 0.1 ml) and a watch glass (*E-H,* 0.7 ml), and CD1 + NaCl on a cover glass (*I, J,* 0.1 ml), a watch glass (*K,* 0.7 ml), and a glass dish (*L,* 0.7 ml) at 25° C. *Modified from Okubo T:* Colloid Polym Sci *291:1887-1893, 2013.*

5.6.4 Coffee, Black Tea, and Green Tea

Fig. 5.75 demonstrates macroscopic drying patterns of hot (*A, B, E, F*) and iced (*C, D*) coffees with cream (*A, C, E*) and without cream (*B, D, F*) on a cover glass (*A-D*) and a watch glass (*E, F*).[15] On a cover glass, a single broad ring formed at the outside edge. The d_f/d_i values observed for coffee with cream were 0.98 $\pm$ 0.01, 0.98, and 0.99 at $\phi = 0.0023$, 0.007, and 0.023, where density of coffee granules were assumed to be in unity in the calculation of volume fraction.[15] However, on a watch glass, a fine outside ring and very broad inner ring coexisted (see *E* and *F*). Broad ring patterns are insensitive to the substrate used in general. Therefore it is plausible that d_f/d_i values should decrease as solute concentration decreases in the low coffee concentration range even on a cover glass. Observation of the broad ring at lower concentrations than 0.1 wt% is, however, difficult experimentally. In the our experiments, therefore, it is not clear whether the d_f/d_i values keep unity or decrease with further decreasing concentration. If d_f/d_i values keep unity even in the diluted suspensions, there should be very strong solute-substrate affinity, especially at the solidification process. In this case, coffee is grouped in type E. On the other hand, when d_f/d_i values decrease from unity in future experiments, the group will differ from group E and probably be classified in group B. Drying patterns of iced coffee were slightly broader than those of hot coffee. Coffee particles at very low temperatures will be slightly

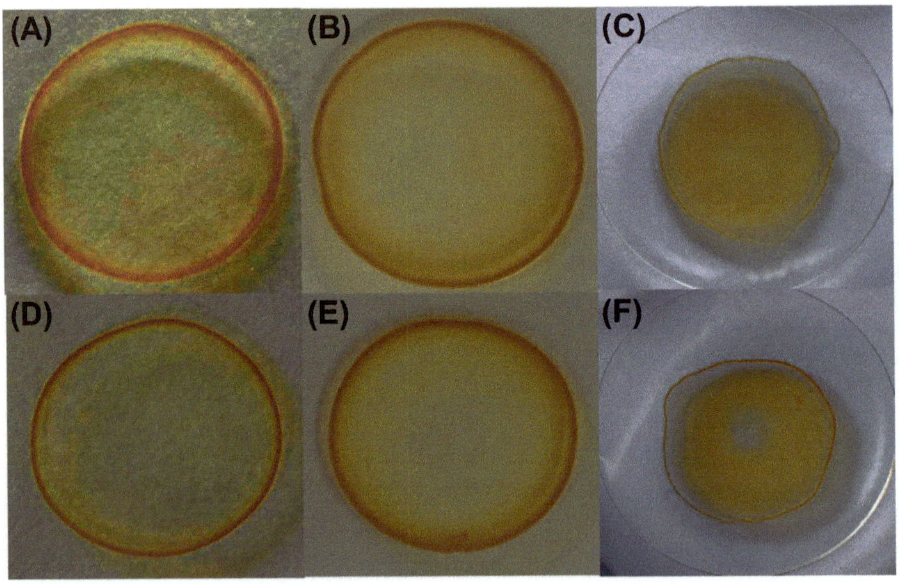

Figure 5.75 Drying patterns of coffee on a cover glass (*A, B,* 0.1 ml), iced coffee on a cover glass (*C, D,* 0.1 ml), and coffee on a watch glass (*E, F,* 1 ml) with cream (*A, C, E*) and without cream (*B, D, F*) at 25° C. **A, E,** 0.70 wt%; **B, F,** 0.29 wt%; **C,** 1.26 wt%; **D,** 0.70 wt %. *Modified from Okubo T, Okamoto J, Tsuchida A:* Colloid Polym Sci *287:351-365, 2009.*

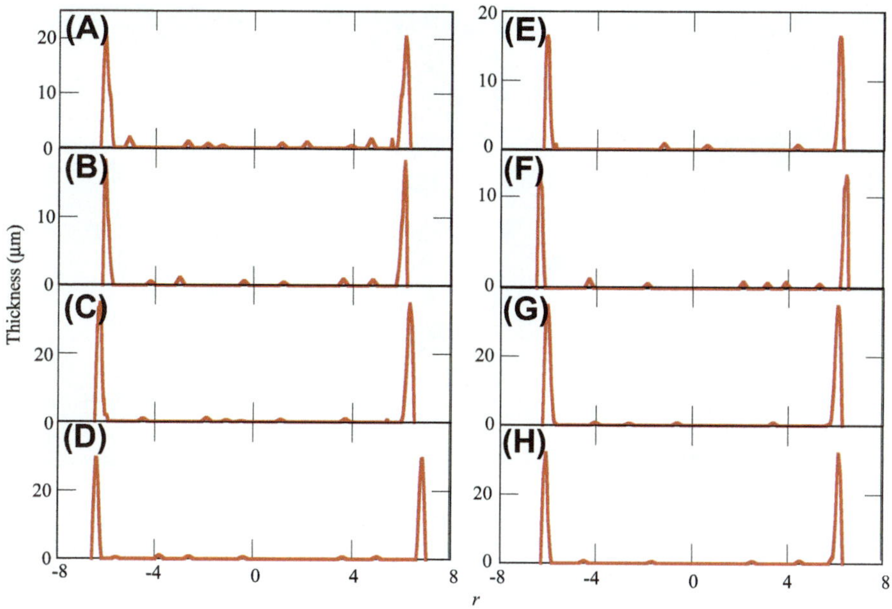

Figure 5.76
Thickness profiles of coffee (*A-D*) and black tea (*E-H*) without cream (*A, B, E, F*) and with cream (*C, D, G, H*) as a function of the distance from the center at 25° C. **A,** 0.96 wt%; **B,** 0.48 wt%; **C,** 2.3 wt%; **D,** 1.2 wt%; **E,** 0.3 wt%; **F,** 0.15 wt%; **G,** 1.6 wt%; **H,** 0.8 wt%. *Modified from Okubo T, Okamoto J, Tsuchida A:* Colloid Polym Sci *287:351-365, 2009; Okubo T, Okamoto J, Tsuchida A:* Colloid Polym Sci *287:645-657, 2009.*

unstable slightly or the diffusivity of the particles during the convectional process will also be low.

Fig. 5.76 shows thickness profiles of the dried films of coffee (*A-D*) and black tea (*E-H*) without cream as a function of the distance from the center at 25° C.[15,16] The profiles with cream were quite similar to those without cream. Two high peaks at the outside edges in Fig. 5.76 correspond to the broad ring. We should note here that the low broad rings coexisted at the inner area of the film even on a cover glass. The large, nonspherical particles of coffee may distribute at the inner area by segregation effects. Similarity in the thickness profiles of coffee and black tea is clear in the figures. The sharpness parameter of the broad ring, *S*′ values of coffee and black tea, are shown in Fig. 5.67. The *S* values were high, which supports strong solute-substrate affinity. It should be noted here that the high *S* values do not mean high diffusivity of the solutes for coffee and black tea systems.

Fig. 5.77 shows the microscopic patterns of coffee with cream (*A-F*) and black tea with cream (*G-L*), respectively. Accumulation of the colloidal particles of coffee and black tea at the inner area of the film was also observed clearly by

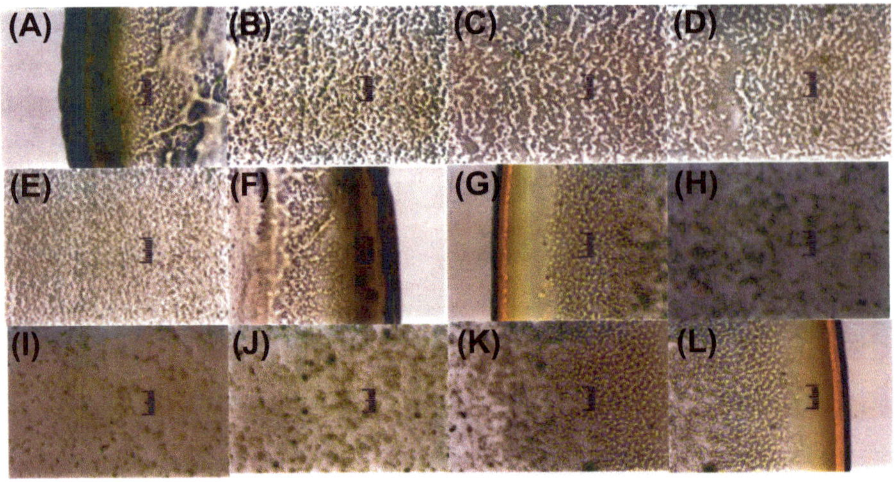

Figure 5.77
Microscopic drying patterns of coffee with cream (*A-F*) and black tea without cream (*G-L*) at 25° C. **A-F** and **G-L** are from the left edge to the right, full scale = 100 μm. *Modified from Okubo T, Okamoto J, Tsuchida A:* Colloid Polym Sci *287:351-365, 2009; Okubo T, Okamoto J, Tsuchida A:* Colloid Polym Sci *287:645-657, 2009.*

the microscopic observation of the dried film. The black and brownish areas in Fig. 5.77, *A, F, G,* and *L*, are the broad rings at the outside edge of the dried film, where thick accumulation of coffee or black tea particles is clear. The patterns of the dried film with cream were much finer in their textures than those of the film without cream. Coating of cream around coffee particles is supported.

Fig. 5.78 shows typical drying patterns of green tea in a watch glass without salt (*A, E*) and with salt (*B-D*).[29] Fig. 5.78, *F*, is the microscopic

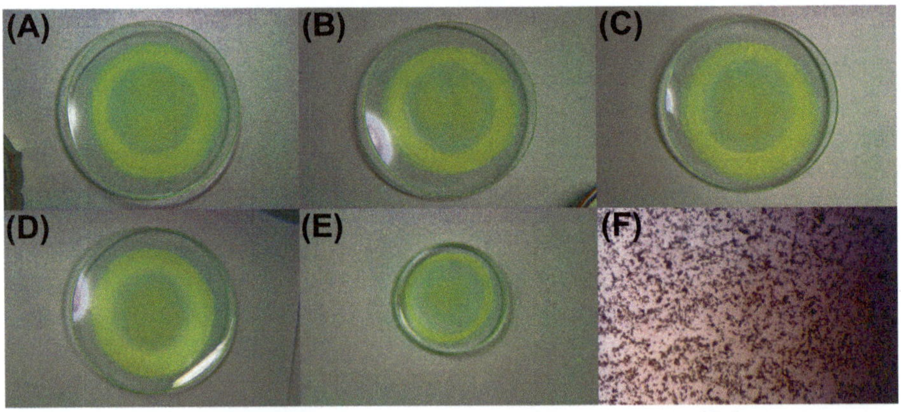

Figure 5.78
Drying patterns of green tea in a watch glass at 24° C. **A,** Macroscopic pattern, 4 ml, [NaCl] = 0 M; **B,** 0.001 M; **C,** 0.006 M; **D,** 0.01 M; **E,** 0 M, in a small watch glass, 0.4 ml; **F,** microscopic pattern, full scale = 200 μm. *Modified from Okubo T:* Colloid Polym Sci *285:331-337, 2006.*

pattern close to the broad ring area. A main broad ring appeared at the outside edge. Broadness of the ring patterns is correlated intimately to the size distribution of the green tea particles. In the figure, another inner ring is also recognized clearly. Existence of the outer and inner broad rings in the pictures will indicate that two main peaks exist in size distribution. The drying patterns were insensitive to the salt addition, but the accumulation of the particles in the central area looks clearer in the presence of the salt. Microscopic drying pattern in Fig. 5.78, *F*, shows the loose linear arrays of tea particles.

5.7 Synthetic Polymers

5.7.1 Water-Soluble Neutral Polymers

Fig. 5.79 shows a series of macroscopic drying patterns of polyethylene glycol (PEG) of various molecular weights from 4000 to 3.5×10^6 and at polymer concentrations ranging from 0.003 monoM to 0.1 monoM.[38] For PEG4K and PEG8K, the liquid-state areas coexisted with the solid pattern areas at 0.003 monoM in the dried film. From PEG20K to PEG3500K, solid drying patterns were observed at any concentration. The broad ring patterns were always observed at the outside edge. The single broad ring patterns formed, and d_f/d_i values increased sharply from 0.4 to unity as the molecular weight of PEG increased from 1000 to 5×10^5. The ratios kept unity for PEG of higher molecular weights (Fig. 5.80). The ratio also increased as polymer concentration increased. The excluded-volume effects play an important role for the broad ring size. Clearly, the macroscopic drying patterns of PEG are one of the typical solutes of group A.

Table 5.8 shows the sharpness parameters S and S' of the broad rings of PEG as a function of the molecular weight. An increase in the parameters with increasing molecular weight is clear. It should be mentioned here that S' values were larger than the corresponding S values.

Fig. 5.81 shows the microscopic drying patterns of polyethylene glycol (PEG500K, PEG2000K, and PEG3500K) observed by the optical microscopy.[38] Clearly, the coupled structures of spherulites and lamellae, with the latter originated from the ringlike drying pattern, are observed. Morphologic characteristics of PEG crystals was influenced greatly during the course of dryness, although the lattice structure should not be changed during the dissipative processes.

Fig. 5.82 shows the typical pictures of the polarizing microscopy of PEG2000K on a cover glass (*A*, *B*), a watch glass (*C*), and a glass dish (*D*) at 24° C.[38] Maltese cross patterns, especially at the central area (*A*) support the existence of spherulite crystal structure. However, the patterns are distorted

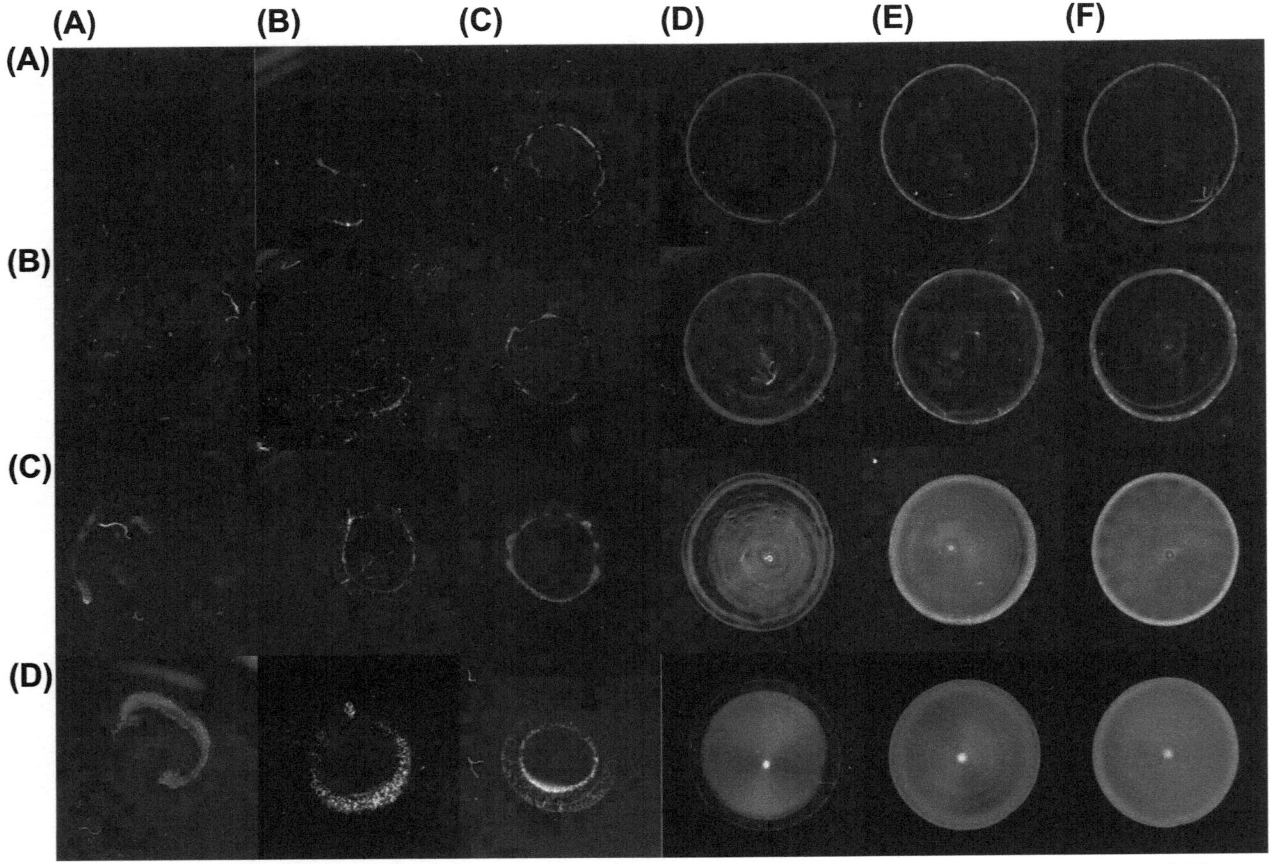

Figure 5.79 Drying patterns of PEG solution on a cover glass at 24° C. 0.1 ml; **A,** 0.003 monoM; **B,** 0.01 monoM; **C,** 0.03 monoM; **D,** 0.1 monoM; **A,** PEG4K; **B,** PEG8K; **C,** PEG20K; **D,** PEG500K; **E,** PEG2000K; **F,** PEG3500K. *From Okubo T, Okamoto J, Takahashi S, Tsuchida A:* Colloid Polym Sci *287:933-942, 2009.*

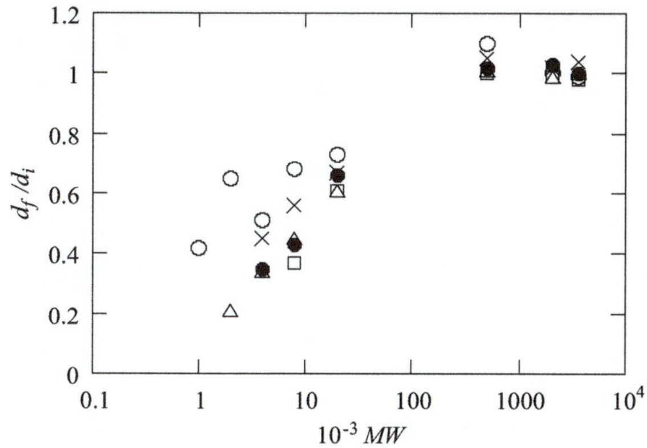

Figure 5.80
Value of d_f/d_i of PEG at 0.1 monoM (*open circles*), 0.03 monoM (*crosses*), 0.01 monoM (*open triangles*) and 0.003 monoM (*open squares*) on a cover glass and at 0.01 monoM (*solid circles*) in a watch glass as a function of molecular weight. *From Okubo T, Okamoto J, Takahashi S, Tsuchida A:* Colloid Polym Sci *287:933-942, 2009.*

Table 5.8 *S* and *S'* of PEG Polymers at 24° C, 0.1 ml

PEG	Concentration (monoM)	*S*	*S'*
PEG8K	0.03	10	15
PEG20K	0.03	14	16
PEG500K	0.03	16	19
PEG2000K	0.03	15	19
PEG3500K	0.03	19	26
PEG20K	0.1	7.7	9.0
PEG500K	0.1	11	18
PEG2000K	0.1	16	23
PEG3500K	0.1	16	20

PEG, Polyethylene glycol.

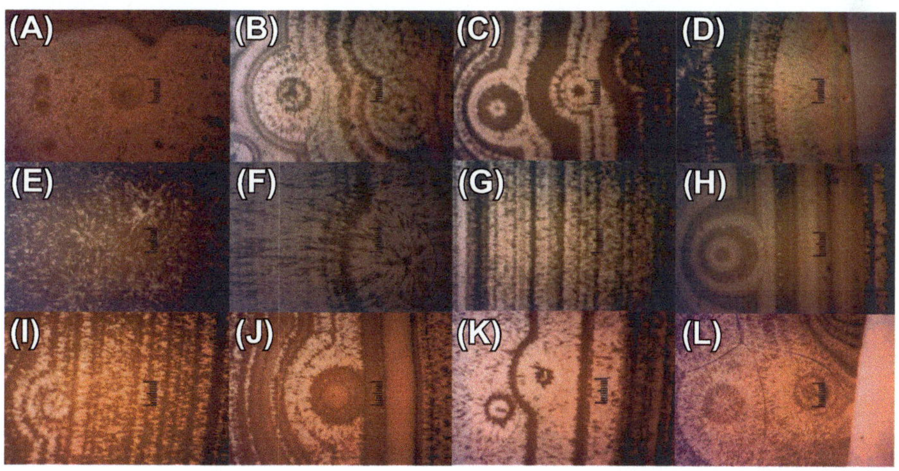

Figure 5.81
Microscopic drying patterns of PEG in a watch glass at 24° C. **A-D,** PEG500K; **E-H,** PEG2000K; **I-L,** PEGMA3500K, from center to right edge, 4 ml, 45 hours, 25 minutes after setting, full scale = 100 μm. *From Okubo T, Okamoto J, Takahashi S, Tsuchida A:* Colloid Polym Sci *287:933-942, 2009.*

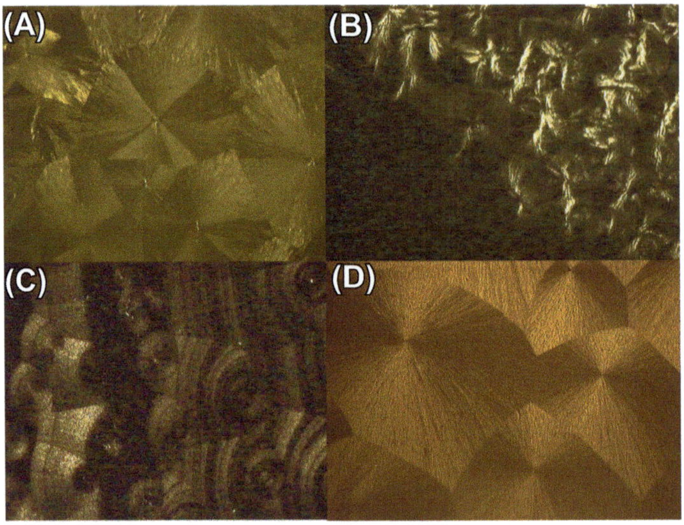

Figure 5.82
Polarizing microscopic drying patterns of PEG on a cover glass (*A, B*), a watch glass (*C*), and a glass dish (*D*) at 24° C. **A,** PEG200K, 0.1 monoM, central area, frame size 1.0 × 1.4 mm; **B,** PEG2000K, 0.1 monoM, outside area, 0.26 × 0.34 mm; **C,** PEG2000K, 0.1 monoM, frame size 2.6 × 3.4 mm; **D,** PEG3500K, 0.01 monoM, 1.0 × 1.4 mm. *From Okubo T, Okamoto J, Takahashi S, Tsuchida A:* Colloid Polym Sci *287:933-942, 2009.*

to some extent, because of the collision or overlapping of the spherulites. In Fig. 5.82, *C*, clear-cut coupling of the spherulites and the lamellae in the broad ring area is apparent. In Fig. 5.82, *D*, the spherulites are distorted clearly by the downward gravitational forces, and anisotropic spherulites formed. Independent experiments on the drying patterns of PEG were also studied previously.[83]

Drying patterns of polyvinyl alcohol (PVA) and poly(*N*-vinyl pyrrolidone) (PVP) were studied (Okubo T, unpublished results). The macroscopic drying patterns of PVP30K and PVP90K were liquidlike or transparent solid–like irrespective of the substrates used (cover glass, watch glass, and glass dish). Thickness profiles of the dried film were typical single broad ring types at the outside edge. The d_f decreased as polymer concentration decreased. The microscopic drying patterns of PVP25K were observed on a cover glass. *S* and *S'* values of the broad rings of PVP25K were 14.2 and 16, respectively, at 0.1 monoM, 0.1 ml, and at 24° C.[36] The coupling effect is one of the typical examples of dissipative crystallization of polymers. It should be mentioned here that monodispersed aggregates of polymer complexes between polyacrylic acid and PEG- or PVP-formed colloidal crystals.[36] Their macroscopic and microscopic drying patterns were similar to those of the monodispersed colloidal spheres. Choi studied drying patterns of PEG polymers with salt on a Teflon plate.[84]

5.7.2 Polyelectrolytes

Characteristic coupling dissipative effects of spherulites (or hedrites) were also observed for one of the anionic polyelectrolytes, sodium polymethacrylate (NaPMA).[40] The ringlike arrays of the large spherulites along the broad ring area, a very large single spherulite, and the bundle of the spherulites at the central area formed on a cover glass, a watch glass, and a glass dish, respectively, as is shown in Fig. 5.83. On a cover glass (*A*), the spherulites were packed densely and arrayed along the broad ring. Furthermore, most of the spherulites were cut sharply by the outside broad ring lines. This supports the finding of strong coupling of the broad ring formation and the spherulite crystallization. On a watch glass (*B*), the fine structure of a single spherulite crystal at the center and the circular arrayed spherulites at the outside edge formed, respectively. Furthermore, the valleys of the spherulites, especially in the inner area of the broad ring, were deep and sharp. On a glass dish, a broad ring appeared at the outside area of the circular pattern in the central region. All the broad rings on a substrate were of the single type, and their sizes d_f or d_f/d_i values increased as polymer concentration increased, just like PEG. Macroscopic drying patterns support NaPMA being categorized as group A.

Microscopic drying patterns of NaPMA were observed by the usual optical microscopy and polarizing microscopy.[40] Fig. 5.84 shows typical polarizing microscopic patterns of NaPMA on a cover glass (*A, B*), a watch glass (*C, D*), and a glass dish (*E, F*). Surprisingly, large and fine spherulites and hedrite crystals appeared. The crystalline structures changed from hedrites to spherulites as polymer concentration increased. The coupled structures of

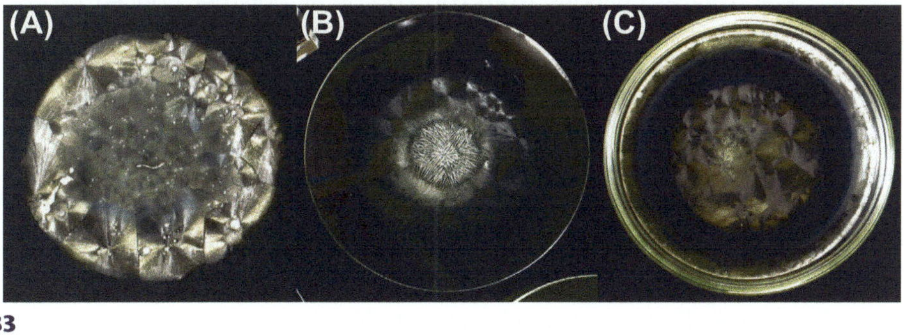

Figure 5.83
Drying patterns of NaPMA at 25° C. 0.02 monoM **A,** on a cover glass; **B,** on a watch glass; **C** on a glass dish.
Modified from Okubo T, Hagiwara A, Kitano H, Okamoto J, Takahashi S, Tsuchida A: Colloid Polym Sci *287:1155-1165, 2009.*

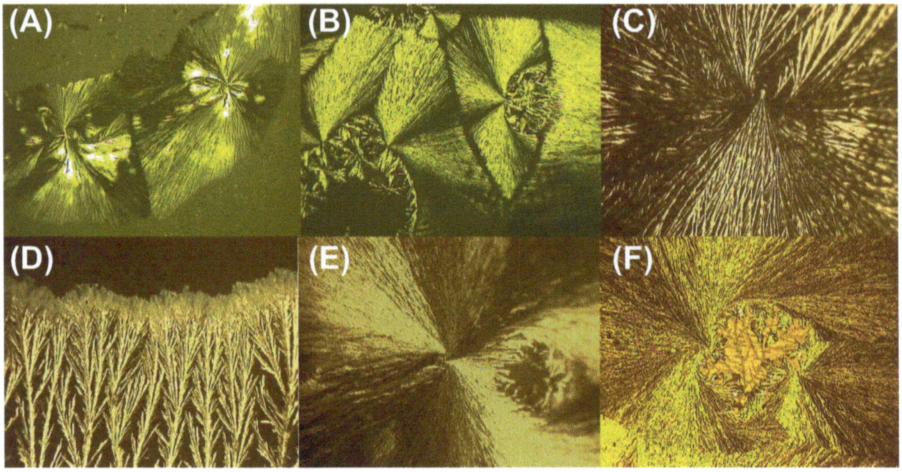

Figure 5.84
Polarizing microscopic patterns of NaPMA on a cover glass (*A, B*), a watch glass (*E, F*). **A,** 0.0006 monoM; **B,** 0.02 monoM; **C, D,** 0.00333 monoM; **E, F,** 0.02 monoM. Frame sizes are 1.0 mm × 1.4 mm (*A*) and 2.6 mm × 3.4 mm (*B-F*). *From Okubo T, Hagiwara A, Kitano H, Okamoto J, Takahashi S, Tsuchida A:* Colloid Polym Sci *287:1155-1165, 2009.*

spherulites and the broad rings appeared at the outside edge, although the patterns were not so clear compared with those of PEG.

Drying patterns of poly(allylamine hydrochloride) (PAL), one of the cationic polyelectrolytes, were studied on a cover glass.[37] The fractal patterns with 1.2 to 1.6 in dimension are shown in Fig. 5.85. Surprisingly, the shape of the polymer crystals changed from crosslike to curved patterns as the initial polymer concentration increased. Furthermore, crystal size increased with increasing concentration, which is the reverse tendency for crystal size change in a solution state. It should be recalled that conformation of polyelectrolytes changes from rod to coil as polymer concentration increases. Thus it is plausible that the shape information of the polymer at the initial solution state was transferred to the drying patterns. In the diluted solution state, rodlike macroions are adsorbed weakly by the electrostatic attraction forces keeping the conformation of the polymer on the substrate surface. As the drying process proceeds, the solution is concentrated. But the rodlike macroion cannot change its conformation to coil-like so easily, because its conformation is fixed weekly by the macroion-substrate affinity. The main cause for the memory effects of the conformation of the polymer at the initial solution is considered to be the cooperative affinity between the macrocations and the substrate during the course of drying, especially in the convectional and sedimentary processes. Several researchers discussed the drying fractal patterns with the use of fractal theory.[85-88]

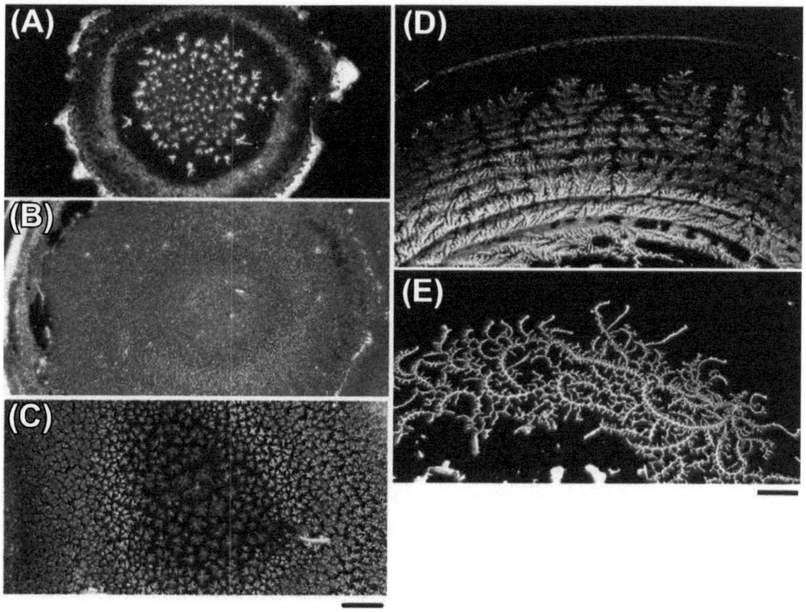

Figure 5.85
Microscopic patterns of PAL on a cover glass at 25° C. 0.1 ml, **A,** 1×10^{-7} monoM; **B,** 1×10^{-6} monoM; **C,** 1×10^{-5} monoM; **D,** 1×10^{-4} monoM; **E,** 0.001 monoM; length of the bar is 200 μm. *From Okubo T, Kanayama S, Ogawa H, Hibino M, Kimura K:* Colloid Polym Sci *282:230-235, 2004.*

Drying patterns of a series of poly(*N*-alkylpyridinium halide) (RPVP) were observed.[89] The patterns were influenced substantially with the hydrophobicities of the polyelectrolytes. Drying processes of highly concentrated polymer solutions have been studied, in which the formation of a round hill was recognized in the central area.[90-92] The thickness profiles of RPVP salts are shown in Fig. 5.86. The broad rings formed at the outside

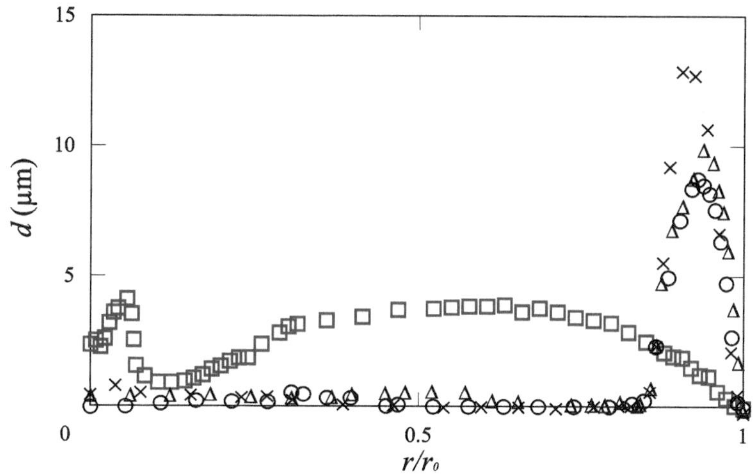

Figure 5.86
Thickness profiles of the dried film of C2PVP (*circles*), C4PVP (*crosses*), BzPVP (*triangles*), and C16BzPVP (*squares*) on a cover glass at 25° C. 0.1 ml, $r_o = 7.2$ mm (C2PVP), 5.2 mm (C4PVP), 5.5 mm (BzPVP), and 6.0 mm (C16BzPVP), 0.01 monoM. *From Okubo T, Ogawa H, Tsuchida A:* Colloid Polym Sci *288:245-256, 2010.*

edge in the dried film. It is well known that the hydrophobicity increases in the following order:

$$\text{C2PVP} < \text{C4PVP} < \text{BzPVP} < \text{C16BzPVP} \qquad 8$$

The sharpness parameters S (S') evaluated were 21 (25), 24 (25), 20 (26), and 2.9 (5.2) for C2PVP, C4PVP, BzPVP, and C16BzPVP, respectively (0.1 ml, 0.01 monoM at 25° C). Most strongly, the hydrophobic polymer of C16BzPVP gave the lowest S values because of the strong intersolute affinity. Thickness profiles of C16BzPVP showed the central round hill type patterns. The significant influence of hydrophobicity of the polyelectrolytes on the macroscopic patterns is clear.

5.8 Biological Polymers

As is well known, natural polymers are grouped into three major categories: polypeptides, polysaccharides, and polynucleotides. Each polymer has amide linkages of amino acids, ether linkages of hexose units, and ester linkages of phosphate and glucose (attached with nucleic acid base), respectively. It is interesting to clarify these structural effects of the biopolymers on the drying patterns, because information on the conformation and chemical structures of the polymers in the initial solution state is transferred more or less into the drying patterns. Thus there is the possibility to know the chemical structures and the conformations of biopolymers from the systematic study on the drying patterns.

5.8.1 Polypeptides

Drying patterns of several biological polyelectrolytes, grouped into polypeptides have been studied. Drying patterns of sodium salts of poly(alpha-glutamic acid) NaPGA (NaP*d*GA, NaP*l*GA and NaP*d*GA + NaP*l*GA) were studied.[93] They were hygroscopic and did not dry up in the room-temperature atmosphere. Then the observation was made in a dry box containing silica gel desiccant. After 2 months, the clear broad ring pattern appeared. After 1.5 years, the fine structure of the broad rings and spoke lines appeared clearly for these samples. Concerning microscopic patterns, single crystals of NaGA (monomer analogs of NaPGA) are known to be needle-like.[94,95] However, bent, rodlike microstructures accumulate densely at the outside area near the broad ring and less densely at the central region, respectively. The bent rods did not orient on a substrate. These features are similar to the different optical isomer. The d_f/d_i values were quite small at 0.2 to 0.6, and increased as solute concentration increased. The values were also insensitive to the kind of optical isomer.

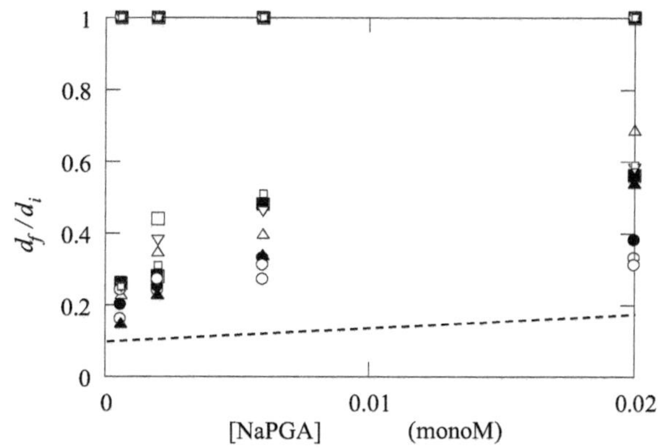

Figure 5.87
Value of d_f/d_i of NaPdGA (*open symbols*), NaPlGA2 (*solid symbols*) and NaPdGA + NaPlGA2 (*circled dot, inverted triangles and dotted square*) solutions on a cover glass (*circles*), a watch glass (*triangles*), and a glass dish (*squares*) as a function of polymer concentration in monoM. *Broken line:* Na*l*GA. *From Okubo T, Takahashi S, Tsuchida A:* Colloid Polym Sci *289:1729-1737, 2011.*

Macroscopic drying patterns of NaP*d*GA, NaP*l*GA, and their mixture NaP*d*GA + NaP*l*GA were observed on a cover glass.[93] The broad rings were observed clearly at the outside edges irrespective of polymer concentration. A secondary faint broad ring also appeared in the inner area. The d_f/d_i values are shown in Fig. 5.87. The symbols located at unity are the values corresponding to the primary broad rings at the outside edge, and the symbols at the values smaller than unity are the secondary broad rings. Values of d_f/d_i for the secondary rings were small and slightly increased with polymer concentration. These features of the broad rings indicate that strong solute-substrate affinity exists and intersolute interaction is also strong. NaPGA is group B in Table 5.6.

The sharpness parameters of NaPGA are shown in Fig. 5.88 with the data of other polypeptides (P*l*LHBr, P*d*LHBr, and COL31) and also polysaccharides (hydroxypropyl cellulose [HPC] and αCD). Concentrations of these biopolymers are given in *monoM* (molar concentration of monomer units). The diffusivity of NaPGA during drying processes is considered to be similar to that of PLHBr from the *S* values.

Fig. 5.89 shows the typical examples of the microscopic drying patterns of NaP*l*GA on a cover glass (*A, B*). Fig. 5.89, *C* through *F*, are the patterns of NaP*d*GA + NaP*l*GA on a watch glass. Spherulite and hedrite crystal structures are observed clearly in Fig. 5.89, *A* through *F*, respectively. The Maltese crosses, although incomplete, appeared in *E* and *F* from polarizing microscopy. The spherulite and hedrite crystals have been observed and described

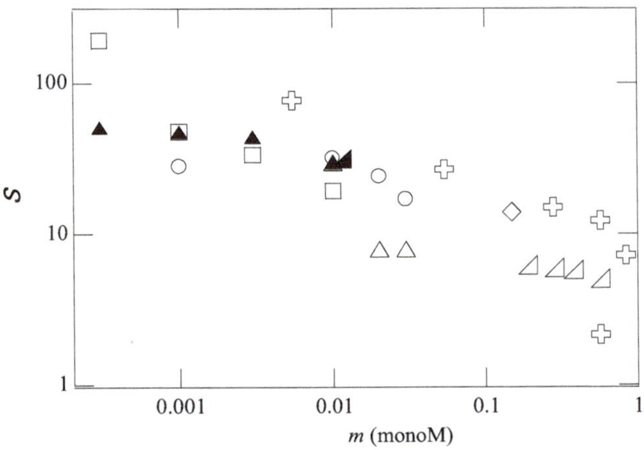

Figure 5.88
Sharpness of the broad rings of biopolymers. 0.1 ml; 25° C; *circles:* NaPGA; *triangles, solid triangles:,* P*l*LHBr; *squares:* P*d*LHBr; *plus:* COL31; *flag:* HPC1; *solid flag,* HPC3; *diamond* αCD.

in Section 5.7. It should be noted here that the rodlike crystal structures formed mainly in the broad ring area (Fig. 5.90, *A*, for example). Iizuka et al. reported the rodlike liquid crystals for the concentrated aqueous solution of NaP*ld*GA.[96] The rodlike crystals observed in NaP*l*GA and others are similar to the liquid crystal patterns of HPC, sodium salt of deoxyribonucleic acid (NaDNA), and polynucleotide complexes described in Sections 5.8.2 and 5.8.3.

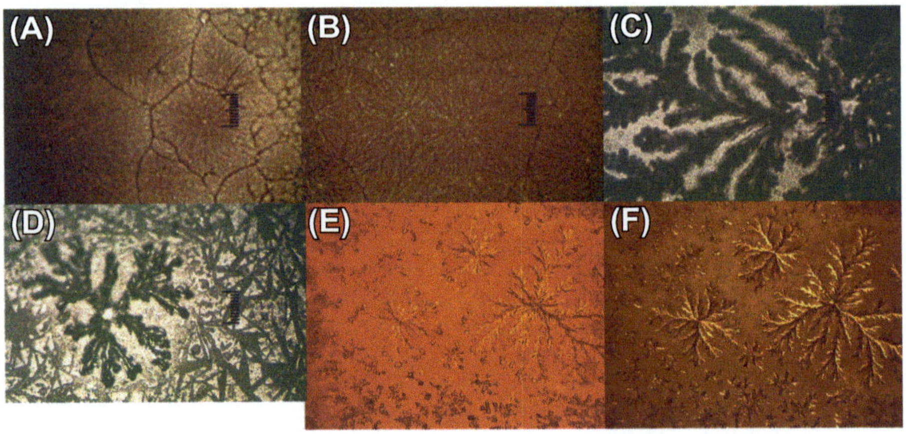

Figure 5.89
Optical (*A-D*) and polarizing microscopic drying patterns (*E, F*) of aqueous NaP*l*GA (*A, B*) and NaP*d*GA + NaPlGA1 (1:1) (*C-F*) solutions on a cover glass (*A, B*) and a watch glass (*C-F*) at 25° C. 0.1 ml (*B, C*), 0.7 ml (*A, D-F*), [NaP*l*GA1] = 0.02 monoM, [NaP*d*GA + NaP*l*GA1] = 0.02 monoM, full scale is 100 μm (*A-D*), length of the picture frame (*E, F*) is 2.6 × 3.4 mm. *From Okubo T, Takahashi S, Tsuchida A:* Colloid Polym Sci *289:1729-1737, 2011.*

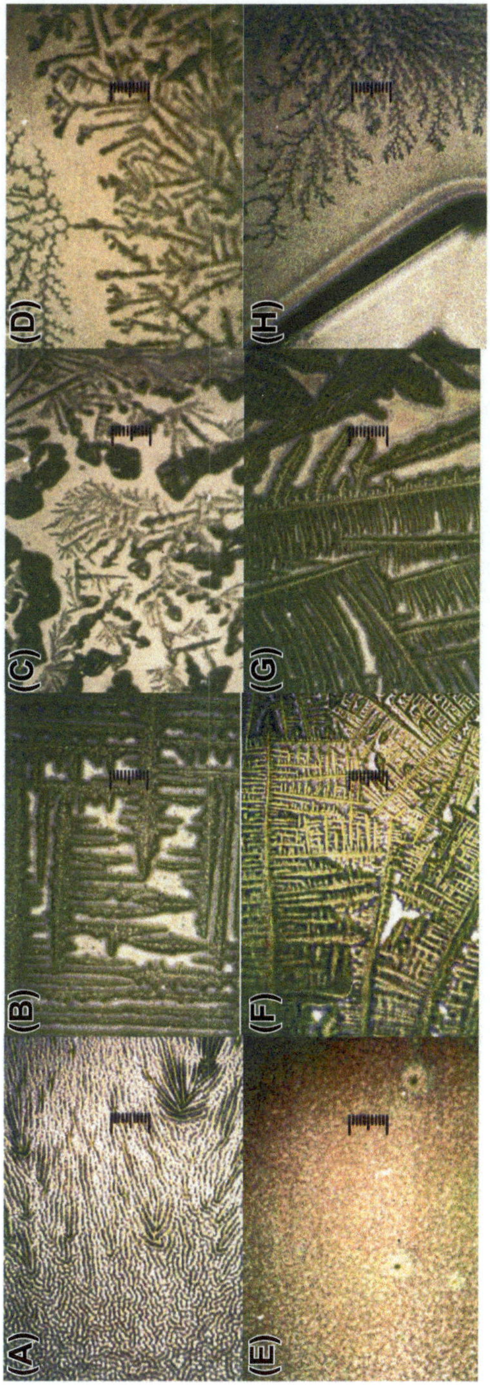

Figure 5.90 Microscopic drying patterns of aqueous NaP*d*GA + NaCl (*A-D*) and NaP*l*GA2 + NaCl (*E-H*) solutions on a watch glass at 25° C. 0.7 ml, [NaP*d*GA] = [NaP*l*GA] = 0.008 monoM, **A, E** [NaCl] = 0 M; **B, F,** 0.004 M; **C, G,** 0.04 M; **D, H,** 0.2 M; full scale is 100 μm.

From Okubo T, Takahashi S, Tsuchida A: Colloid Polym Sci 289:1729-1737, 2011.

Salt effects on the macroscopic and microscopic drying patterns of NaPGA were studied.[93] Fig. 5.90 shows the microscopic patterns of Na*PdG*A + Na*PlG*A on a watch glass. Interestingly, the microscopic patterns shifted from the rodlike assemblies toward the hedrite structures by the addition of NaCl. This pattern shift is due to (1) the conformational change from the stretched helical rod to the coil-like main chains, (2) cooperative interaction between NaPGA and the salt, and (3) the segregation effect between the polymers and the salt. It should be mentioned here that the drying patterns of NaPGA were also reported from my laboratory.[97]

Macroscopic and microscopic drying patterns of poly(*D*-lysine hydrobromide) (P*d*LHBr), poly(*L*-lysine hydrobromide) (P*l*LHBr), and their low-molecular-weight analogs were observed.[41,97] Fig. 5.91 shows macroscopic (*A-E*) and microscopic drying patterns (*F-J*) of P*d*LHBr + P*l*LHBr with various mixing ratios in a watch glass at 24° C.[41] The concentration ratios of P*d*LHBr against P*l*LHBr changed from 1:0 (*A, F*) to 0:1 (*E, J*). The broad rings formed clearly at the outside edge ($d_f/d_i = 1$) and the inner area ($d_f/d_i < 1$). Microscopic patterns were often "city road–like" and dendritic at the outside edge and inner region, respectively. Any essential differences did not appear by mixing the stereo-isomers in the macroscopic and microscopic drying patterns.

The d_f/d_i values of P*l*HBr and their low-molecular-weight analogs LHCl are shown in Fig. 5.92, *A*. The ratios of LHCl were similar to each other among the various stereospecific solutes, although those seem to increase in the order of cover glass < watch glass < glass dish, depending on the substrates used. Furthermore, the small size of the broad ring was clear compared with the corresponding polymers. This is mainly due to the very small excluded volumes of the small solutes compared with the polymeric ones. Fig. 5.92, *B*, shows the plots of d_f/d_i of P*d*LHBr and P*l*LHBr. Several important findings are clear. First, the d_f/d_i values are fairly large compared with those of low-molecular-weight analogs. This is due to the very large excluded volumes of the extended polyelectrolytes compared with those of the monomer units. Second, the ratios increased as polymer concentration increased. It is recalled that below the critical polymer concentration, m^* ($\approx$0.01 monoM), the broad ring size of the drying pattern (d_f) of P*l*LHBr was small compared with that of the initial solution (d_i) and increased with polymer concentration. Furthermore, the m^* value agreed excellently with the critical polymer concentration where the surface tensions decrease sharply as polymer concentration increased. Third, the ratios were insensitive to the isomers, when the large experimental errors are taken into account. The sharpness parameters *S* of P*d*LHBr and P*l*LHBr are shown in Fig. 5.88. These values are not so different from those of NaPGA, when the experimental errors included in the data are taken into account. Recently,

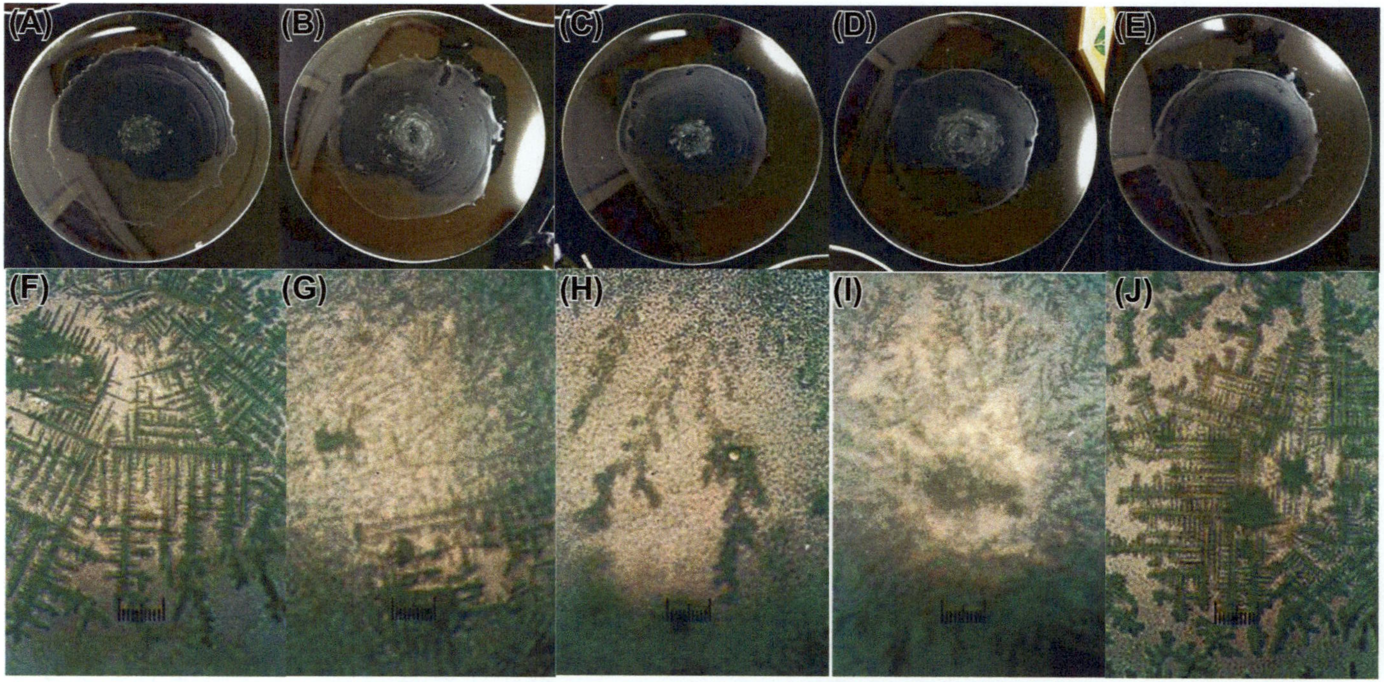

Figure 5.91 Macroscopic (*A-E*) and microscopic drying patterns (*F-J*) of PdLHBr + PlLHBr in a watch glass at 24° C. [P*d*LHBr] + [P*l*LHBr] = 0.001 monoM, pH = 6.6, 0.1 ml, **A, F,** [P*d*LHBr]:[P*l*LHBr] = 1:0; **B, G,** 3:1; **C, H,** 1:1; **D, I,** 1:3; **E, J,** 0:1; **F-J,** central area; full scale is 100 μm. *Modified from Okubo T, Okamoto J, Tsuchida A:* Colloid Polym Sci *288:981-989, 2010.*

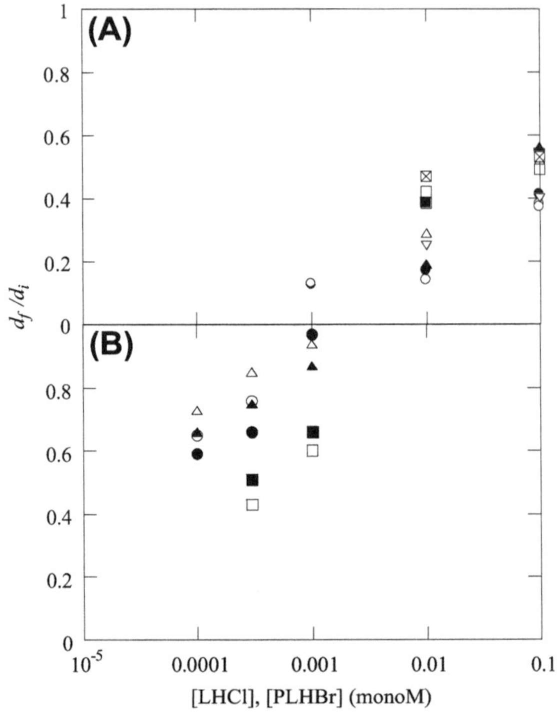

Figure 5.92
A, d_f/d_i of *d*LHCl (*open symbols*), *l*LHCl (*solid symbols*), and d*l*LHCl (*dotted, inverted, and crossed symbols*) of *l*LHCl concentration on a cover glass (*circles*) a watch glass (*triangles*) and a glass dish (*squares*) at 24° C. **B,** d_f/d_i of P*d*LHBr (*open symbols*) and P*l*LHBr (*solid symbols*) on a cover glass (*circles*), a watch glass (*triangles*), and a glass dish (*squares*) at 24° C. *From Okubo T, Okamoto J, Tsuchida A: Colloid Polym Sci 288:981-989, 2010.*

drying cracks and ring patterns in spherulite of poly(*L*-lactic acid) were studied in detail.[98]

5.8.2 Polysaccharides

Sodium salts of carboxymethyl cellulose (NaCMC),[99] dextran sulfate (NaDSS)[100], NaHLA[101]; HPC[42], α-cyclodextrin (CD), β-CD, and γ-CD[102] have been studied. NaCMC1 (≈50-200 cP at 4 wt% at 25° C), NaCMC2 (≈70-130 cP at 1 wt%) and NaCMC3 (≈1300-2200 cP at 1 wt%) were used.[99] These three samples were slightly hygroscopic and did not dry up completely on the substrate in a room-temperature atmosphere. Fig. 5.93 shows the macroscopic drying patterns of NaCMC coexisted with NaCl. The broad rings appeared in the presence of the salt and their sizes increased with increasing salt concentration, as is clear in the figure. At least two broad rings formed: a large primary ring at the outside edge of the dried film and a smaller secondary one at the adjacent inner area from the outside edge. The tertiary smallest broad rings also appeared at the central area, as shown in Fig. 5.93, *L*, on a cover glass. Interestingly, main broad rings

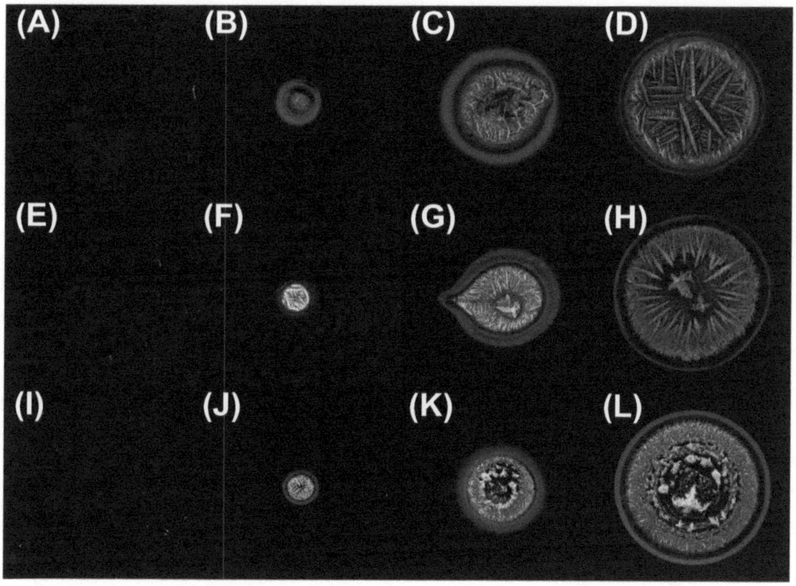

Figure 5.93 Drying patterns of NaCMC1 (*A-D*), NaCMC2 (*E-H*), and NaCMC3 (*I-L*) on a cover glass at 25° C. 0.1 ml, [NaCMC1] = 0.032 monoM, [NaCMC2] = 0.008 monoM, [NaCMC3] = 0.0048 monoM; **A, E, I,** [NaCl] = 0 M; **B, F, J,** 0.004 M; **C, G, K,** 0.04 M; **D, H, L,** 0.2 M. *From Okubo T:* Colloid Polym Sci *289:1205-1213, 2011.*

shifted from the primary to tertiary as salt concentration increased. The main cause for the shift is the depression of the polymer-substrate affinity by the salt addition.

The d_f/d_i values in the absence of salt increased as NaCMC concentration increased, and saturated at unity above the critical concentration, m^* as is shown in Fig. 5.94, *A*. The m^* value on a watch glass ($\approx$0.02 monoM) were highest compared with those on a cover glass and a glass dish. Fig. 5.94, *B*, also shows that d_f/d_i values decreased in the order on the substrates.

$$\text{Cover glass} > \text{glass dish} > \text{watch glass} \qquad 9$$

Furthermore, d_f/d_i values seem to decrease in the following order:

$$\text{NaCMC3} \approx \text{NaCMC2} > \text{NaCMC1} \qquad 10$$

The influence of the molecular weight of polymers on the d_f/d_i values shown in Equation 10 is consistent with the fact that the excluded volume increases as the molecular weight increases. On an inclined watch glass, the outward flow of the polymers will be retarded strongly compared with the flat substrates. Furthermore, substantial increase in d_f/d_i values with increasing salt concentration is clear (see Fig. 5.94, *B*). NaCMC is categorized in group B, as shown in Table 5.6.

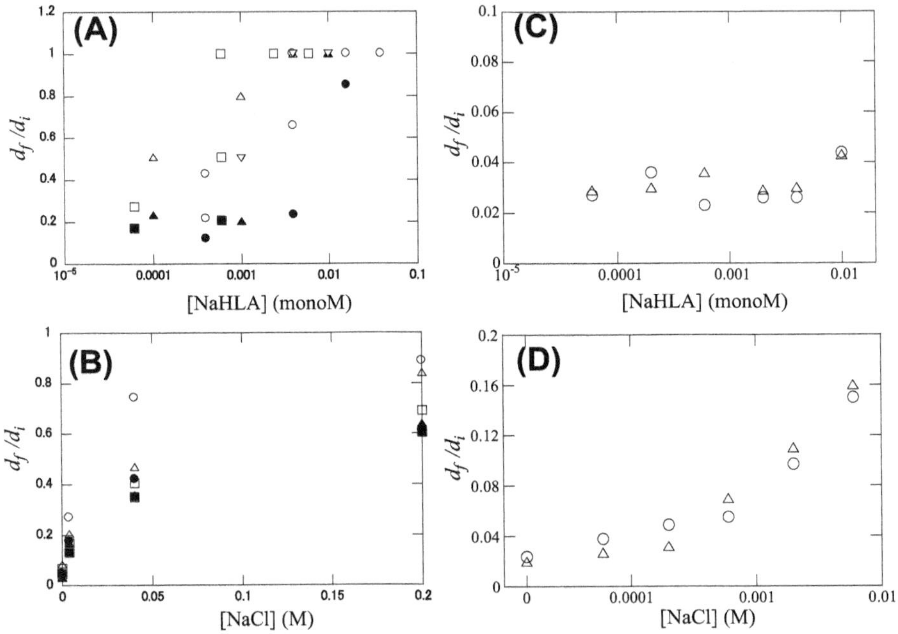

Figure 5.94
A, B, d_f/d_i of NaCMC1 (*circles*), NaCMC2 (*triangles*), and NaCMC3 (*squares*) solutions on a cover glass (*open symbols*), a watch glass (*solid symbols*), and a glass dish (*dotted or inverted symbols*) as a function of NaCMC concentration (*A*) and coexisted NaCl concentration (*B*) at 25° C. **C,** d_f/d_i of NaHLA on a cover glass (*circles*) and a watch glass (*triangles*) at 25° C. **D,** d_f/d_i of NaHLA + NaCl on a cover glass (*circles*) and watch glass (*triangles*) at 25° C. [NaHLA] = 0.004 monoM. *Modified from Okubo T:* Colloid Polym Sci *289:1205-1213, 2011; Okubo T, Tsuchida A:* Colloid Polym Sci *292:381-389, 2014.*

Microscopic drying patterns of NaCMC2 without coexistent salt were not observed so clearly. However, small rodlike and bent, rodlike patterns were recognized even without salt, which are similar to the patterns of other saccharides, HPC and NaDSS. Conformation of NaCMC chains are deduced to be extended like typical polyelectrolytes. The polymer chains of the hexose units of NaCMC contain the helical rod segments in part, and are stiff from their bulky hexose moieties compared with typical polyelectrolytes. It is plausible that NaCMC chains are assemblies into the rodlike single crystals. Then the short, rodlike crystal shape of NaCMC without salt is due to the hexose groups; their drying patterns are similar to those of HPC and NaDSS.

Fig. 5.95 shows typical examples of microscopic drying patterns of NaCMC2 in the presence of NaCl. The long dendritic rod crystals grew by the addition of NaCl in the inner area of the secondary broad ring. The salt enhanced clearly the stable rodlike assemblies of NaCMC polymer chains. The rather

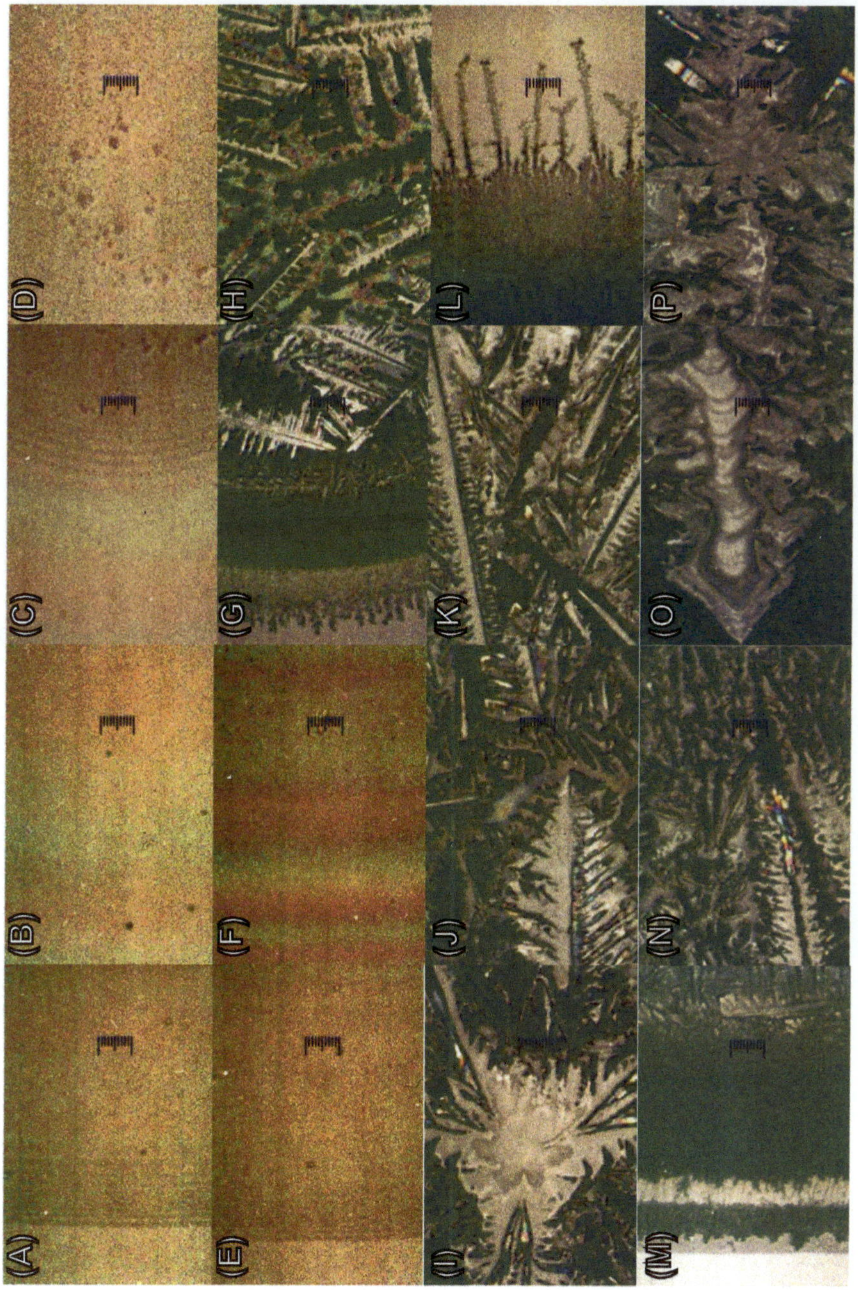

Figure 5.95 Microscopic drying patterns of NaCMC2 on a cover glass at 25° C. 0.1 ml, [NaCMC2] = 0.008 monoM; **A-D,** [NaCl] = 0 M; **E-H,** 0.004 M; **I-L,** 0.04 M; **M-P,** 0.2 M, pictures from the left edge to the center (*A-D, E-H, M-P*) and from the center to the right edge (*I-L*); full scale is 100 μm. *From Okubo T:* Colloid Polym Sci *289:1205-1213, 2011.*

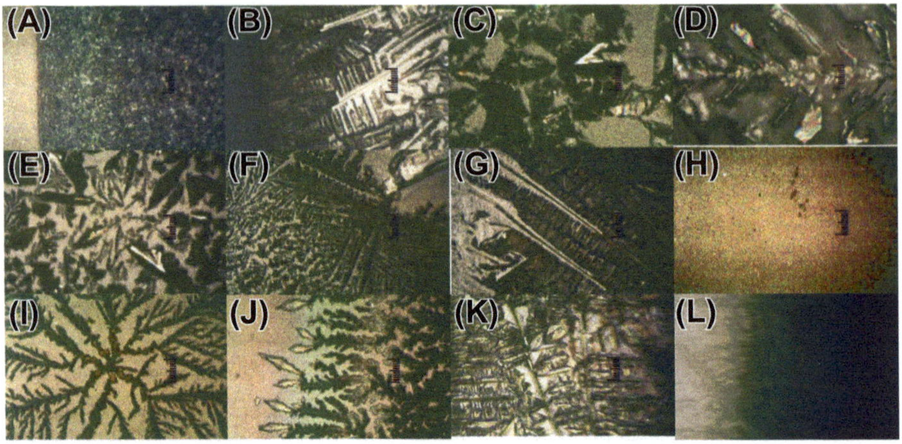

Figure 5.96
Microscopic drying patterns of NaCMC3 with NaCl (0.2 M) on a cover glass (*A-D*), a watch glass (*E-H*), and a glass dish (*I-L*) at 25° C. 0.1 ml, [NaCMC] = 0.0048 monoM, pictures from the left edge to the center (*A-D*) and from the center to the right edge (*E-H, I-L*); full scale is 100 μm. *Modified from Okubo T:* Colloid Polym Sci *289:1205-1213, 2011.*

stretched conformation of NaCMC molecules even in the presence of NaCl is supported strongly.

Fig. 5.96 shows typical microscopic drying patterns of NaCMC3 with 0.2 M NaCl on a cover glass (*A-D*), a watch glass (*E-H*), and a glass dish (*I-L*). Most patterns were long dendritic rods on a cover glass and a watch glass. However, hedrite type (i.e., primitive forms of spherulites) formed especially at the central area of a glass dish. The width of the dried film in the inner area is quite thin compared with a cover glass and a watch glass. Thus the hedrite crystals form with the help of the strong polymer-substrate affinity and the convectional flow forces on the polymer molecules.

The macroscopic and microscopic drying patterns of NaDSS1 (molecular weight is 5000), NaDSS2 (≈36,000-51000) and NaDSS3 (5×10^5) were observed on a cover glass, a watch glass and a petri glass dish.[100] Clearly, d_f/d_i increased as polymer concentration increased, and they also increased as molecular weight increased (Fig. 5.97). And the critical concentrations m^* decreased substantially as molecular weight of NaDSS increased. The d_f/d_i values were large for NaDSS3 and very low for NaDSS1 and in the following order:

$$\text{NaDSS3} > \text{NaDSS2} > \text{NaDSS1} \qquad 11$$

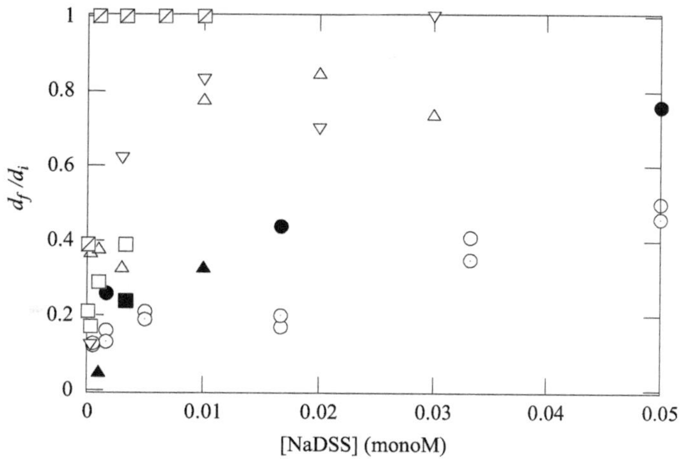

Figure 5.97 d_f/d_i of NaDSS1 (circles), NaDSS2 (triangles) and NaDSS3 (squares) on a cover glass (open symbols), a watch glass (dotted, inverted and dashed symbols) and a glass dish (solid symbols) at 25° C. *From Okubo T:* Colloid Polym Sci *289:159-167, 2011.*

NaDSS1 did not dry up and showed a liquid state only. On the other hand, NaDSS2 and NaDSS3 dried in part, but liquid parts still remained. These three polyelectrolytes are hygroscopic and in their magnitudes of strength are in the following order:

$$\text{NaDSS1} > \text{NaDSS2} > \text{NaDSS3} \qquad 12$$

Lower-molecular-weight polymer was a higher hygroscopic. It is highly plausible that the polymer-substrate and interpolymer affinities are in the order shown in Equation 12. Equation 11 also supports the order in the hygroscopic strength. The order of Equation 12 in hygroscopic strength is an example of the transference of information on molecular weight to the broad ring size.

Microscopic drying patterns of NaDSS2 on a cover glass and a watch glass are shown in Fig. 5.98, *A* and *B*, respectively. The rodlike patterns support the formation of some ordering of NaDSS in an aqueous phase. However, it is hard to believe at present that the ordering is similar to the cholesteric liquid-crystalline structure resembling that of HPC, because NaDSS is a branched polymer, which is difficult to crystallize.

Now the drying patterns of NaHLA are discussed.[101] The monomer unit of NaHLA is the disaccharide of sodium salt of *D*-glucuronic acid and β-*N*-acetyl-*D*-glucosamine. Salts of HLA exist in several soft connective tissues, including skin, the umbilical cord, synovial fluid, and the vitreous humor. HLA is found in lung, kidney, brain, and muscle tissue. The stock solution of NaHLA was viscous and hygroscopic. Low values of d_f/d_i, approximately 0.03, were observed as shown in Fig. 5.94, *C*. Small broad rings were

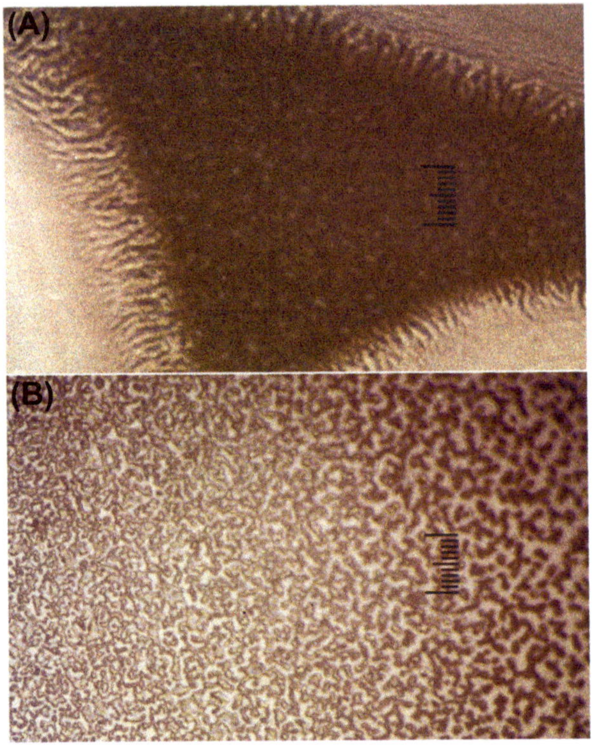

Figure 5.98 Microscopic drying patterns of NaDSS2 on a cover glass (*A*) and a watch glass (*B*) at 25° C. 0.1 ml (*A*) and 0.7 ml (*B*), [NaDSS2] = 0.01 monoM (*A*) and 0.001 monoM (*B*); full scale is 100 μm. *From Okubo T: Colloid Polym Sci 289:159-167, 2011.*

observed for hydrogels, as described in Chapter 4. However, the ratios of NaHLA without salt are smallest among the solutes examined hitherto. These small d_f/d_i values support the finding that there is very strong intersolute and weak solute-substrate affinity. NaHLA is categorized in group D (see Table 5.6). The d_f/d_i of NaHLA increased also by the addition of NaCl, but not so significantly; d_f/d_i increased from 0.03 to approximately 0.18 through 0.2 as salt concentration increased from 0 M to 0.01 M (see Fig. 5.94, *D*).

Both the macroscopic and microscopic drying patterns of NaHLA coexisted with NaCl on a cover glass are displayed in Fig. 5.99. The drying pattern areas on a cover glass and a watch glass were extended sharply by the addition of NaCl. In a glass dish, the drying patterns were observed mainly at the outside edge of the cell. By the addition of NaCl, the long and large dendritic rodlike single crystals grew on a cover glass (see Fig. 5.99) and also on a watch glass. The spokelike orientation of the dendritic rods was observed clearly but in part (e.g., see Fig. 5.99, *H*). It is highly plausible that the salt enhances the bundlelike twisting ribbon assemblies of the extended chains, resulting in the

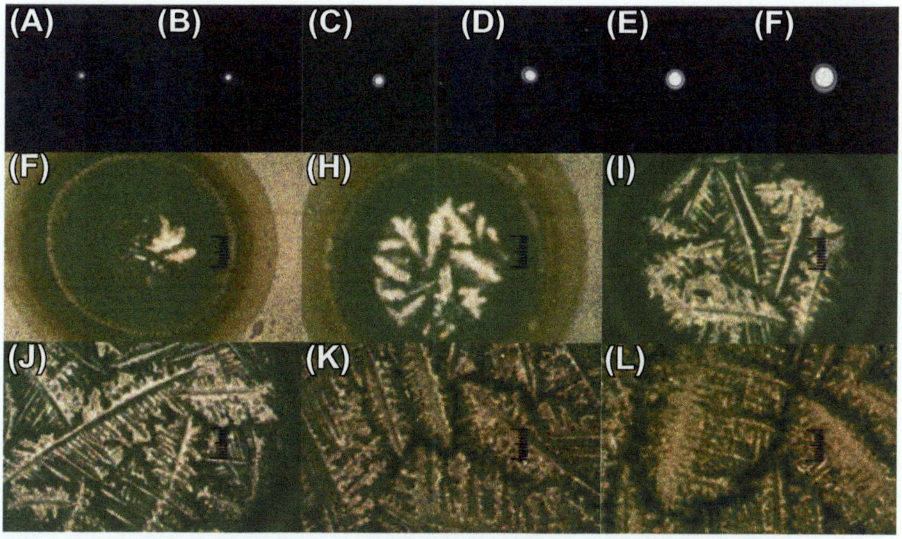

Figure 5.99
Macroscopic (*A-F*) and microscopic drying patterns (*G-L*) of NaHLA + NaCl on a cover glass at 25° C. 0.1 ml, humidity 33%, [NaHLA] = 0.004 monoM, **A, G,** [NaCl] = 0 M; **B,** 1×10^{-5} M; **H,** 6×10^{-5} M; **C, I,** 2×10^{-4} M; **D, J,** 0.0006 M; **E, K,** 0.002 M; **F, L,** 0.006 M; full scale of **G** to **L** is 100 μm. *Modified from Okubo T, Tsuchida A:* Colloid Polym Sci *292:381-389, 2014.*

long rods.[102-104] It should be recalled that salt effects on the morphologic characteristics of NaHLA single crystals resemble those of NaCMC.[99] However, the morphologic characteristics of NaHLA are different from that of other polysaccharides, NaCMC,[99] NaDSS,[100] and HPC.[42]

HPC is well known to form a cholesteric liquid crystalline structure in water at polymer weight fraction higher than 0.55.[105-108] Bluish ring structures appeared and their size changed during dryness. Colors disappeared in the final stage of drying, because the peak wavelengths of light reflected by the liquid crystalline structures still in the film state are shorter than the range of light wavelengths detected by the naked eye. Fig. 5.100 compares the macroscopic drying patterns of HPC1, HPC2, HPC3, and HPC4 on a cover glass, a watch glass, and a petri glass dish.[42] The broad rings on a cover glass and a watch glass were observed clearly at the outside edges of the dried film. Here, small spots of HPC2, HPC3, and HPC4 on a cover glass and a watch glass correspond to the areas of the low, central round hill. The small spots were faintly bluish. This supports the finding that the spot areas are not dried up completely in the experimental conditions of room temperature and humidity.

Fig. 5.101 shows d_f/d_i values of HPC1, HPC2, HPC3, and HPC4 on a cover glass and a watch glass as a function of polymer concentration. Sizes of the broad rings (d_f) were small compared with that of the initial liquid (d_i) and increased

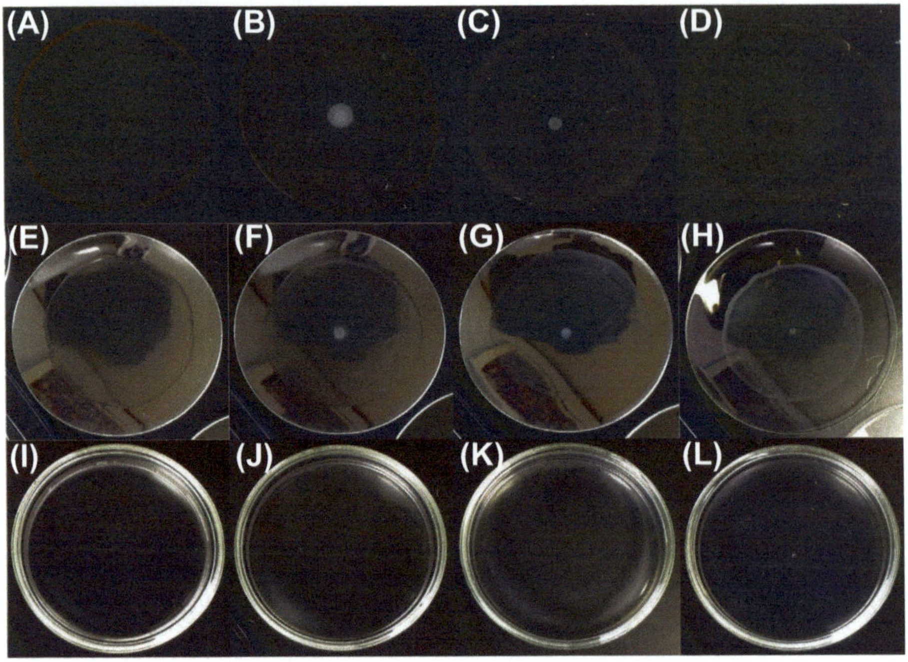

Figure 5.100 Macroscopic drying patterns of HPC1 (*A, E, I*), HPC2 (*B, F, J*), HPC3 (*C, G, K*) and HPC4 (*D, H, L*) on a cover glass (*A-D*), a watch glass (*E-H*), and a glass dish (*I-L*) at 22° C. [HPC1] = 0.05 g/ml, [HPC2] = 0.02 g/ml, [HPC3] = 0.005 g/ml, [HPC4] = 0.003 g/ml. *From Okubo T, Mizutani M, Takahashi S, Tsuchida A:* Colloid Polym Sci *288:1551-1559, 2010.*

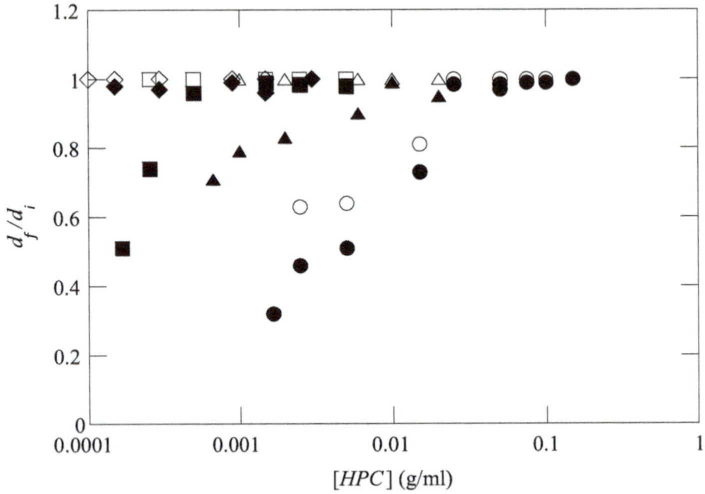

Figure 5.101 Value of d_f/d_i of HPC1 (*circles*), HPC2 (*triangles*), HPC3 (*squares*), and HPC4 (*diamonds*) on a cover glass (*open symbols*) and a watch glass (*solid symbols*) at 22° C. *From Okubo T, Mizutani M, Takahashi S, Tsuchida A:* Colloid Polym Sci *288:1551-1559, 2010.*

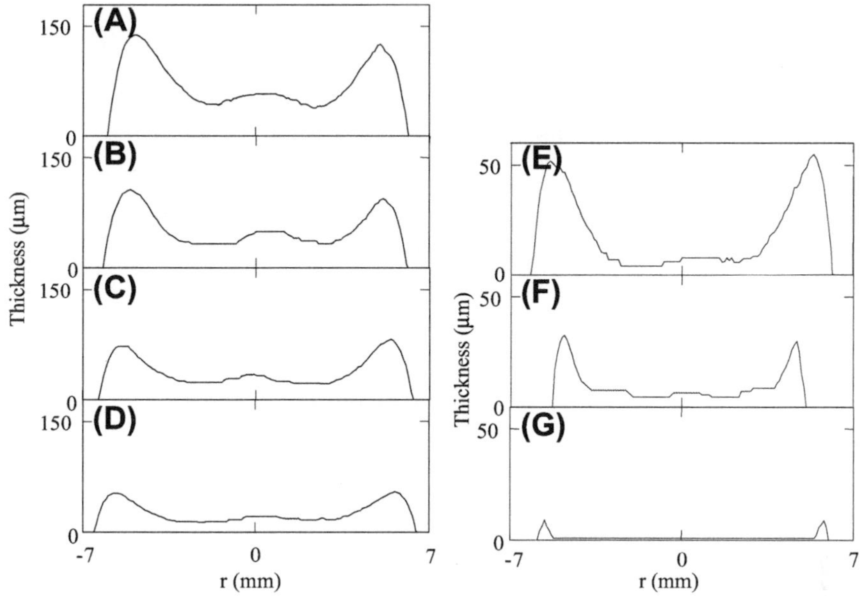

Figure 5.102 Thickness profiles of HPC1 (*A-D*), HPC2 (*E*), HPC3 (*F*), and HPC4 (*G*) as a function of the distance from the center on a cover glass at 22° C. 0.1 ml, **A,** [HPC] = 0.15 g/ml; **B,** 0.1 g/ml; **C,** 0.075 g/ml; **D,** 0.05 g/ml; **E,** 0.02 g/ml; **F,** 0.005 g/ml; **G,** 0.003 g/ml. *From Okubo T, Mizutani M, Takahashi S, Tsuchida A:* Colloid Polym Sci *288:1551-1559, 2010.*

as solute concentration or molecular weight of polymer increased. These observations have often been observed for NaCMC, NaDSS, NaPGA, and PLHBr, for example, and are categorized as group B in Table 5.6. However, for HPC1 the polymer-substrate is strong compared with the typical solutes of group B.

Fig. 5.102 shows the thickness profiles of HPC1, HPC2, HPC3, and HPC4. The round hill at the central area disappeared gradually as the concentrations decreased. The appearance of the central round hill is consistent with the formation of the local liquid crystal structures even with low concentrations of HPC.

Fig. 5.103 shows the typical examples of the microscopic patterns of HPC1 on a cover glass, a watch glass, and a glass dish. Rodlike microstructures are observed. These patterns are similar to the cholesteric liquid crystal structures observed by polarized microscopy.[109,110] These patterns also support the finding that cholesteric structures still remain in the dried film.

αCD, βCD, and γCD are composed of six, seven, and eight hexose units making ring structure by 1,4-α-glucosidic bonds. The structure is a shallow, truncated cone shape with six, seven, and eight primary hydroxyl groups lined at the other rim. A cavity forms containing rows of CH groups and a row of glucosidic oxygens. The cavities are nonpolar relative to their outer surface. Then the cavities act as hosts for a great variety of guest molecules.[111]

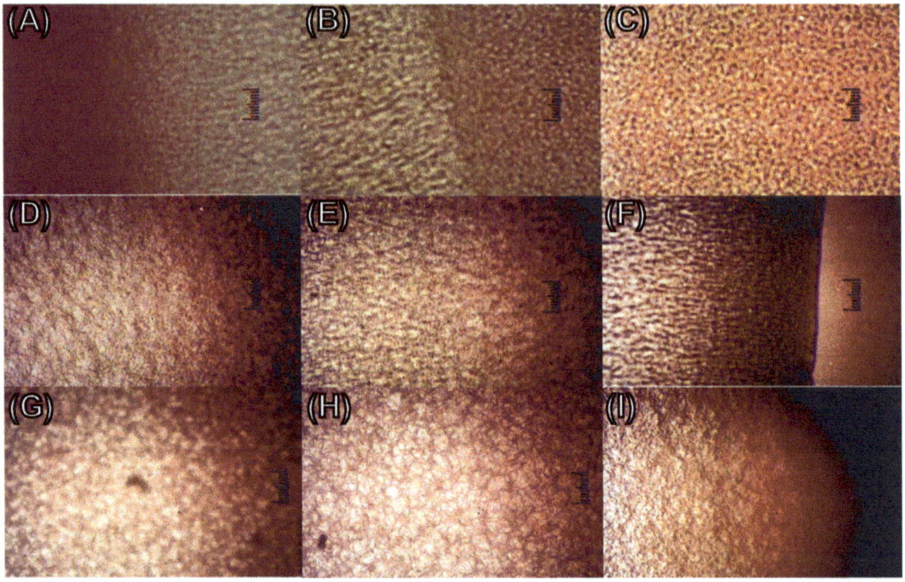

Figure 5.103 Microscopic drying patterns of HPC1 on a cover glass (*A-C*), a watch glass (*D-F*), and a glass dish (*G-I*) at 22° C. 0.1 ml (*A-C*), 4 ml (*D-F*), and 5 ml (*G-I*), [HPC1] = 0.15 g/ml, **A** to **C** are from the left edge to the center, **D** to **F** and **G** to **I** are from the center to the right edge; full scale is 100 μm.

As is well known, keen attention has been paid for the structure and function including the inclusional properties of CD molecules.[112-117] Inclusional association of CD with ionic detergent molecules has been found in my laboratory for the first time.[118]

Fig. 5.104 shows the drying patterns of βCD molecules on a cover glass (*A-D*), a watch glass (*E-H*), and a glass dish (*I-L*).[102] The outer single and inner multiple broad rings formed in the macroscopic patterns irrespective of the substrates. Interring distance in the multiple broad ring regions decreased as CD concentration decreased. The strong solute-substrate affinity is supported for *β*-CD molecules.

The macroscopic drying patterns of αCD and γCD were not observed as clearly, because their drying processes did not occur at room temperature and humidity because of the slightly hygroscopic nature of the dried crystals. The broad ring area of αCD showed impressive patterns like a laurel crown. The d_i values were insensitive to the concentration of CD molecules, which supports αCD, βCD, and γCD being surface inactive. Fig. 5.105 shows one of the typical examples of the macroscopic drying patterns of αCD. Further extended pictures are also given in Fig. 5.105, *A* through *D*.

The values of d_f/d_i are shown in Fig. 5.106 as a function of CD concentration. Here, the size of the multiple inner rings was represented by the main ring among the multiple rings. The outer rings kept unity irrespective of CD

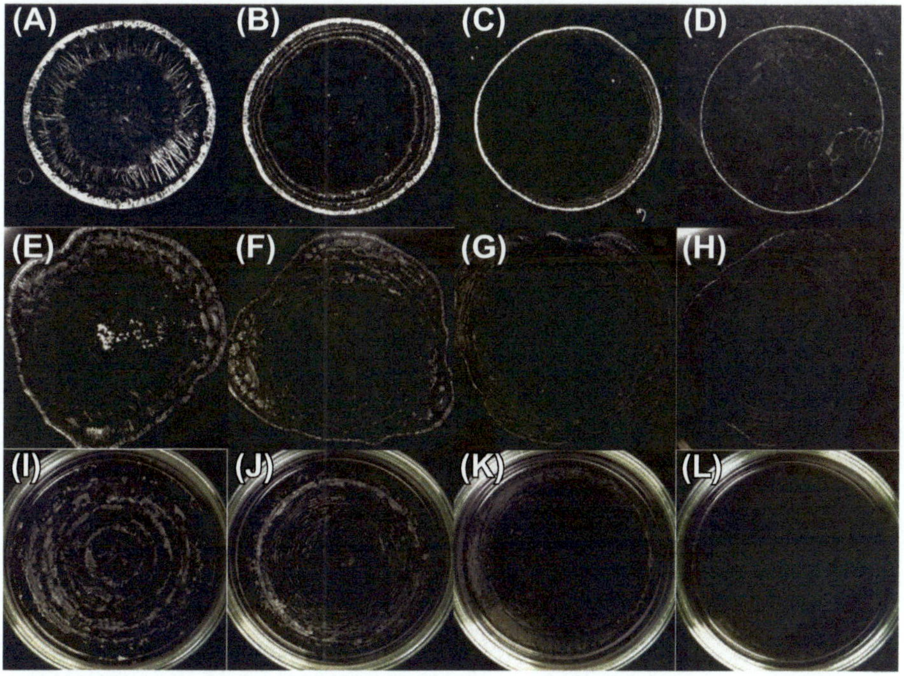

Figure 5.104 Drying patterns of βCD on a cover glass (0.1 ml, *A-D*), a watch glass (0.7 ml, *E-H*), and a glass dish (0.7 ml, *I-L*) at 20° C. **A, E, I,** 0.01 M; **B, F, J,** 0.004 M; **C, G, K,** 0.001 M; **D, H, L,** 0.0001 M. *Taken from Ref. 102. Copyright 2013, Springer.*

concentration. This means that most CD molecules remained at the outside edge of the initial liquid by the strong solute-substrate affinity forces. A part of CD, however, formed the multiple inner rings. The size of the main inner ring is given by the sum of the excluded volumes of each CD molecules during the course of drying. CD molecules are classified to group B (i.e., strong hydrophilic affinity between solutes and substrate and weak intersolutes affinity).

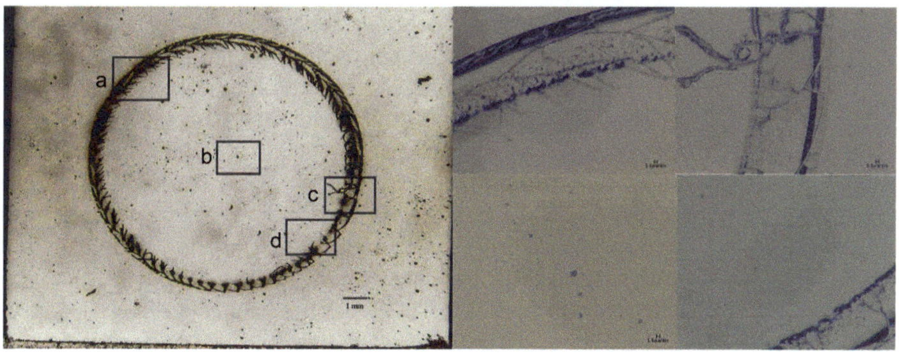

Figure 5.105 Drying patterns of αCD on a cover glass at 25° C. 0.1 mL, 0.01 M. *Courtesy Professor A. Tsuchida.*

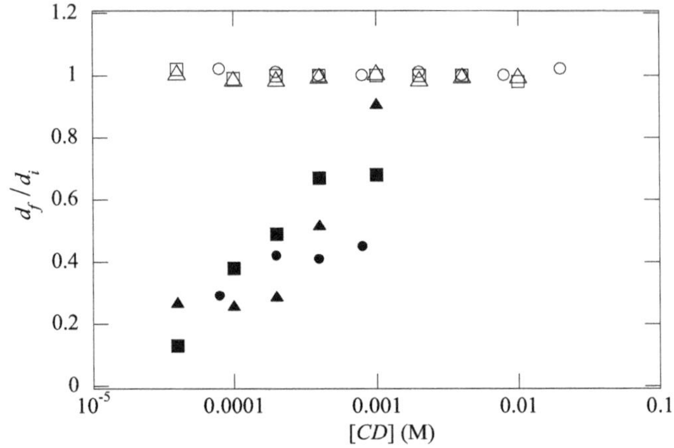

Figure 5.106 Values of d_f/d_i of αCD (*open and closed circles*), βCD (*open and closed triangles*) and γCD (*open and closed squares*) on a cover glass at 20° C. *From Okubo T:* Colloid Polym Sci *291:2447-2454, 2013.*

Fig. 5.107 shows typical examples of the microscopic drying patterns of βCD on a cover glass (*A-D*), a watch glass (*E-H*), and a glass dish (*I-L*). The patterns are similar to each other irrespective of the substrates used. The rodlike or swordlike patterns are oriented symmetrically around the center and along the spoke lines, especially in the central area. The broad rings at the outside edges of Fig. 5.107, *A, H,* and *L*, are composed of the densely accumulated single crystals of βCD molecules. The dendritic crystals are also observed, especially on a cover glass at the central area. Existence of colorful pink and green patterns are ascribed to the fact that their dried films contain a small amount of water molecules, because the drying processes proceed at room temperature and the film is hydrophilic in nature. Rodlike single crystals originate from the line-up of CD molecules with cone head-to-head followed by cone tail-to-tail or cone head-to-tail combinations in the crystal state. These orientation effects are driven with the increase of the entropy caused by the release of the water molecules from the CD hydration shell to the bulk of the solution.[119] The localized symmetrical accumulation of the single crystals of CD molecules is important to the dissipative effects in the drying processes. The main cause is the distortional forces on the CD solutes that results in the single crystal formation raised by the convectional flow of solvent and solutes during the cooperative drying processes. CDs are safely grouped also into typical polysaccharides from the drying patterns. Most polysaccharides are categorized in group B. Recently, drying patterns of inclusion complexes of βCD with sodium *n*-alkylsulfates were studied.[120] Association constants of the inclusion associates were estimated from the macroscopic drying patterns.

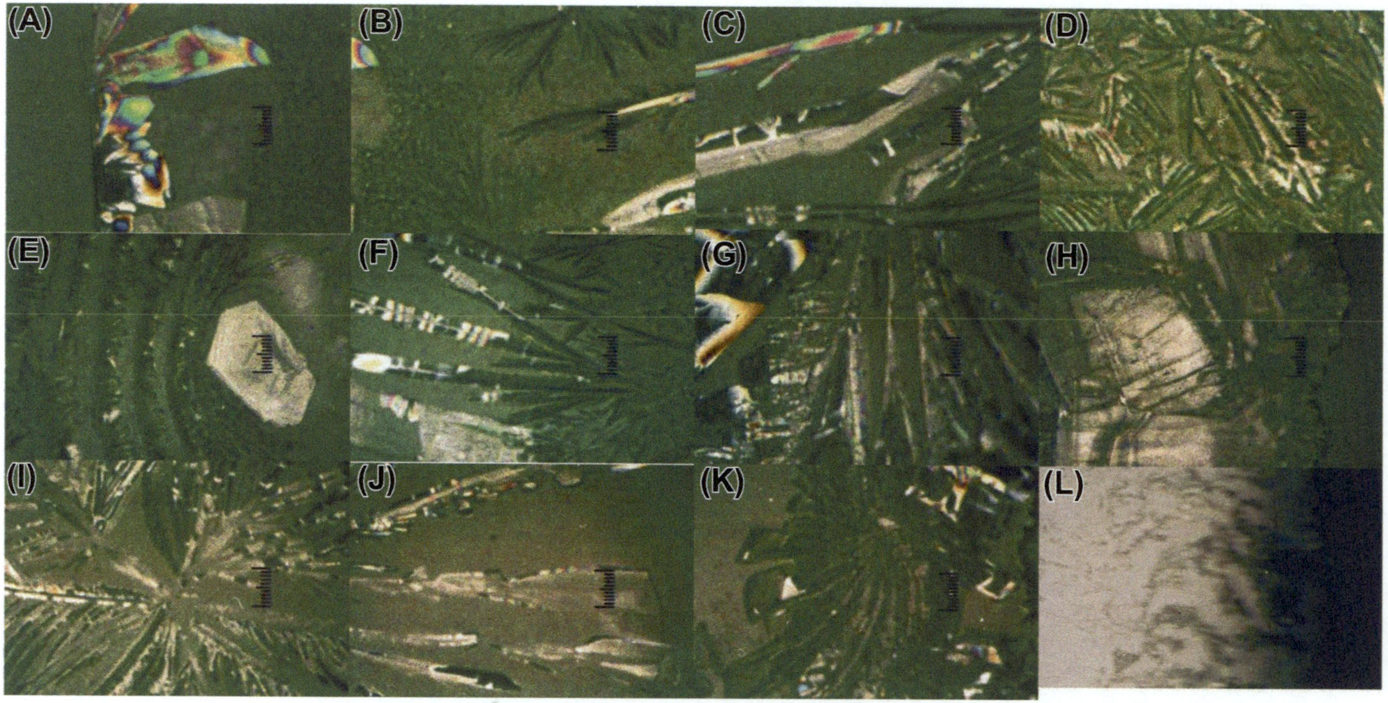

Figure 5.107 Microscopic drying patterns of βCD on a cover glass (0.1 ml, *A-D*), a watch glass (0.7 ml, *E-H*), and a glass dish (0.7 ml, *I-L*) at 20° C. 0.01 M. **A-D** are from the left edge to the center; **E-H** and **I-L** are from the center to the right edge. Full scale is 100 μm. *From Okubo T:* Colloid Polym Sci *291:2447-2454, 2013.*

5.8.3 Polynucleotides

Polynucleotides such as DNA and PolyA are the convenient polymers for studying the dissipative effects (orientation and accumulation of single crystals, for example) in the drying processes, because monodispersed single crystals form in the dried film. Aqueous solutions of NaDNA,[43] KPolyA,[121,122] KPolyG,[123] KPolyC,[123] and KPolyU[122] were used for observation of the macroscopic and microscopic drying patterns. Furthermore, influences of the double-stranded or triple-stranded complex formation of KPolyA + KPolyU[122] and KPolyG + KPolyC[123] on the drying patterns have been studied. Fig. 5.108 shows typical examples of the macroscopic drying patterns of NaDNA without salt (*A-C*) and with coexistent NaCl (*D-F*). Long, rodlike single crystals accompanied many, but a short rod grown from the main rod with a right angle were packed densely within the broad ring area at the outside edge. Interestingly, the long rods were arrayed in the radial direction at low polymer concentrations. The large rods were oriented in the monolayer state on the substrate and not piled up in most cases. Table 5.9 shows the mean absolute values of the sloping angle (θ) of the main rods from the radial direction and the orientation index, which shows the mean values of the slope angles, including their directions (right- and left-side deviations against the radial direction given by the angles with plus or minus signs). Clearly, θ values decreased sharply as the polymer concentration decreased. The growing of the rodlike single crystals takes place under the convectional flow of water in the radial direction. The main cause for the concentration dependency of the orientation is the difficulty of the orientation of the large, long rods along the radial direction in the restricted space. At low concentrations, the rodlike single crystals become short and the orientation will occur much more easily.

Table. 5.9 Sloping Angle (θ) and Orientation Index of the Crystal Rods of NaDNA from the Radial Direction

[NaDNA] (monoM)	[NaCl] (M)	Cover Glass θ (degree)	*o.i.*	Watch Glass θ (degree)	*o.i.*
0.01	0	71 ± 11	−23	34 ± 10	+21
0.0067	0	50 ± 11	−16	40 ± 12	+1.8
0.005	0	50 ± 14	−4	75 ± 7	−15
0.0033	0	15 ± 5	+5	42 ± 11	−20
0.002	0	25 ± 5	−11	49 ± 16	+4
0.001	0	22 ± 5	−3	45 ± 12	+2
0.005	0	70 ± 11	−28	47 ± 14	−7
0.005	0.002	53 ± 10	−28	21 ± 7	+4
0.005	0.01	38 ± 12	−13	31 ± 10	+11
0.005	0.0333	—	—	20 ± 7	+2
0.005	0.1	—	—	12 ± 4	+2
0.005	0.2	14 ± 9	+1.4	10 ± 4	−0.5

NaCl, Sodium chloride; *NaDNA*, sodium salt of deoxyribonucleic acid, *o.i.*, orientation index.

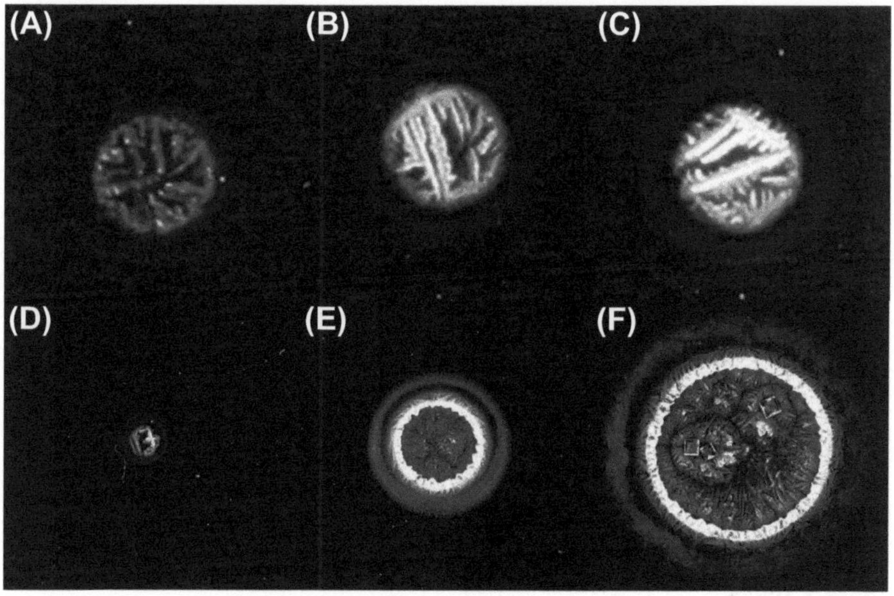

Figure 5.108 Macroscopic drying patterns of NaDNA without (*A-C*) and with NaCl (*D-F*) on a cover glass at 25° C. *A*, [NaDNA] = 0.002 monoM; *B*, 0.0067 monoM; *C*, 0.01 monoM; *D*, [NaDNA] = 0.005 monoM, [NaCl] = 0 M; *E*, 0.1 M; *F*, 0.2 M. *From Okubo T, Mizutani M, Takahashi S, Tsuchida A:* Colloid Polym Sci *288:1435-1444, 2010.*

In the presence of NaCl, bluish and whitish broad rings formed and their sizes increased sharply as the salt concentration increased, as is clear in Fig. 5.108, *D-F*. The main composition of the whitish broad ring is NaCl. Bluish ring structures were faint but recognized even in the absence of the salt, as is seen in Fig. 5.108, *A* through *C*. It is highly plausible that the main compositions of the bluish rings are low-molecular-weight analogs of DNA molecules included in the sample. The analogs are considered to be segregated to the most outside area.

Fig. 5.109, *A*, shows the d_f/d_i values of the broad ring on a cover glass (*circles*) and a watch glass (*crosses*). The ratios were very small, even smaller than 0.1, which indicates strong intersolute affinity in the absence of foreign salt. Single- or double-stranded helix structure formation in solution state is most plausible. The d_f/d_i values increased sharply as salt concentration increased and were insensitive to the substrates used, as is shown in Fig. 5.109, *B*. In Fig. 5.108, *E* and *F*, the cooperative complexation of DNA molecules with NaCl ions were observed at the broad ring areas and the inner regions. In both areas, rodlike DNA crystals were oriented around the single crystals of NaCl. It should be noted here that the cooperative drying patterns were observed also for colloidal silica sphere +NaCl, where the crosslike structure of NaCl was surrounded cooperatively by the dendritic-arrayed silica spheres (see Fig. 5.44).

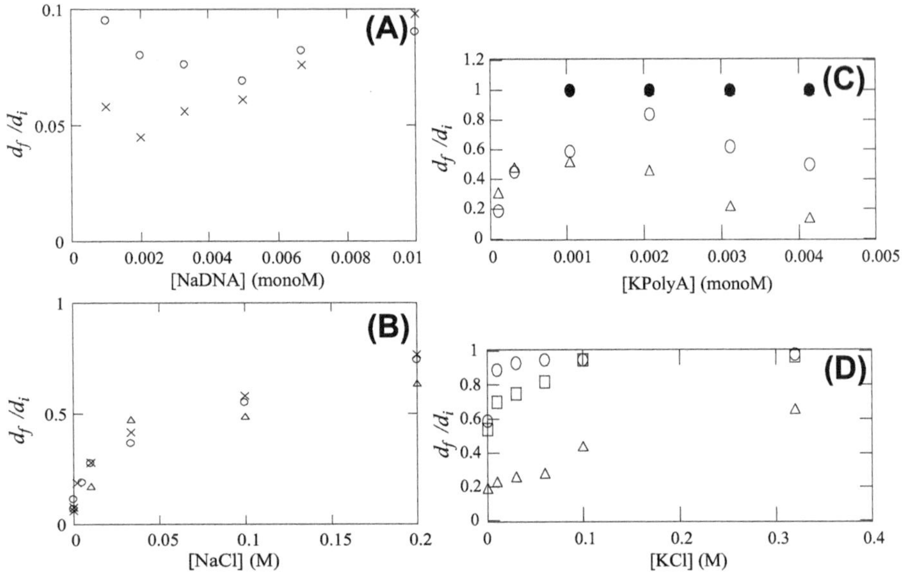

Figure 5.109
A, d_f/d_i values of NaDNA on a cover glass (*circles*) and a watch glass (*crosses*) as a function of polymer concentration at 25° C. **B,** d_f/d_i values of NaDNA on a cover glass (*circles*), a watch glass (*crosses*), and a glass dish (*triangles*) as a function of NaCl concentration at 25° C. [NaDNA] = 0.005 M. **C,** d_f/d_i values of KPolyA on a cover glass (*circles:* inner broad ring; *solid circles:* outer) and a watch glass (*triangles*) as a function of polymer concentration at 20° C. **D,** d_f/d_i values of KPolyA on a cover glass (*circles*), a watch glass (*triangles*), and a glass dish (*squares*) as a function of KCl concentration at 20° C. [KCl] = 0.00389 monoM. *Modified from Okubo T, Mizutani M, Takahashi S, Tsuchida A:* Colloid Polym Sci *288:1435-1444, 2010; Okubo T, Takahashi S, Tsuchida A:* Colloids Surf B Biointerf *87:11-17, 2011.*

The macroscopic patterns of NaDNA showed that most single crystals were long rods accompanied with many short rods. On a cover glass, small, blocklike single crystals were also distributed in the vacant spaces between the rods. On the outside edges of the broad ring areas, small dendritic single crystals were observed, especially at high polymer concentrations. On a watch glass, dendritic crystals grew up at the wide areas of the outside regions of the dried film, whereas the thick rods were observed in the central area. Fig. 5.110 shows the microscopic patterns with coexistent NaCl. Cooperative complexation between NaDNA and NaCl are observed (see Fig. 5.110, *E* and *I*). In most cases, blocklike or rodlike crystals of NaCl were surrounded by the arrayed DNA rods or dendrites cooperatively.

In the macroscopic patterns of KPolyA, two neighbored broad rings were observed at the outside edge of the dried film on a cover glass and a watch glass. Decreasing d_f/d_i values with increasing polymer concentration was observed in the absence of salt. Existence of the high affinity between solute-substrate and solute-solute is clear (see Fig. 5.109, *C* and *D*). By the

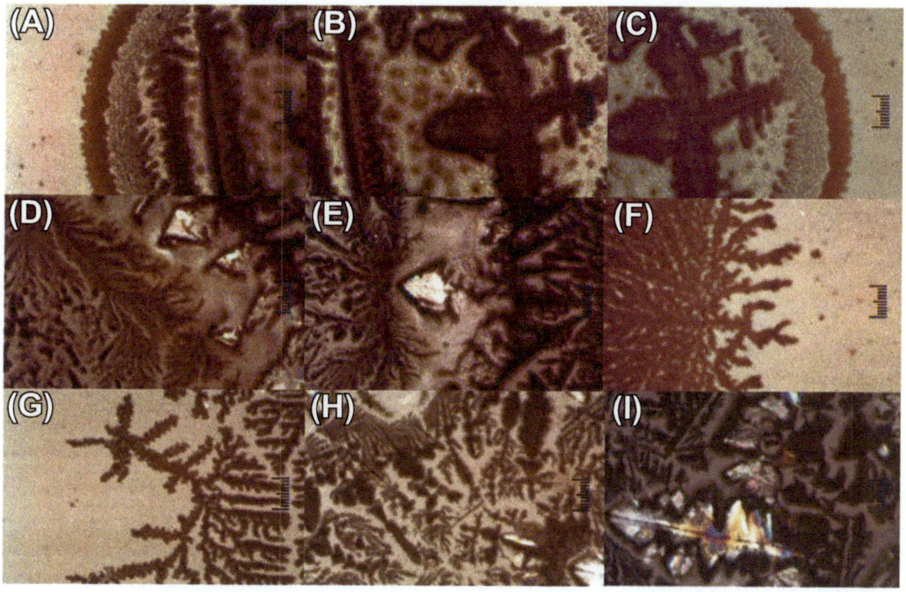

Figure 5.110 Microscopic drying patterns of NaDNA on a cover glass at 25° C. [NaDNA] = 0.005 monoM, **A-C**, [NaCl] = 0 M; **D-F**, 0.0333 M; **G-I**, 0.2 M; **A**, **G** to **C**, **I**, from the left edge to the right, **D** to **F** from the center to the right edge; full scale is 100 μm. *Modified from Okubo T, Mizutani M, Takahashi S, Tsuchida A:* Colloid Polym Sci *288:1435-1444, 2010.*

addition of potassium chloride (KCl), the d_f/d_i values increased sharply from 0.2 toward unity, which is similar qualitatively to the behavior of the ratios of NaDNA + NaCl systems. Fig. 5.111 shows the typical polarizing microscopic patterns of KPolyA at pH = 2.8 (*A*, *B*), 5.7 (*C*, *D*), 12.2 (*E*), and 13 (*F*). Large dendritic single crystals and the distorted spherulites are seen in the neutral and alkaline pH regions. Faint Maltese crosses were recognized in Fig. 5.111, *C* through *F*. It should be mentioned that typical spherulite patterns similar to *E* and *F* have been observed for PEG,[38] NaPMA,[40] and NaPGA.[93]

Several kinds of single crystals formed as a function of the distance from the center in the dried film, which is induced by the segregation, orientation, and accumulation effects during the course of wetting, evaporation, convection, sedimentation, and solidification. The dissipative effects appeared for many kinds of polymers, including polynucleotides. It should be noted that the dissipative crystallization means the change in morphologic characteristics of the single crystals during drying. The crystal lattice structures, however, never change during drying.

Now, let's discuss the drying patterns of PolyA + PolyU[122] and PolyG + PolyC mixtures.[123] Two kinds of broad rings appeared at the outer and inner regions. Fig. 5.112 shows the d_f/d_i values of the inner broad rings. The decreasing tendency of the ratios by mixing KPolyA and KPolyU means a decrease in the total excluded volume of the solute mixtures. The main cause is the formation of the double-stranded or triple-stranded complexation between

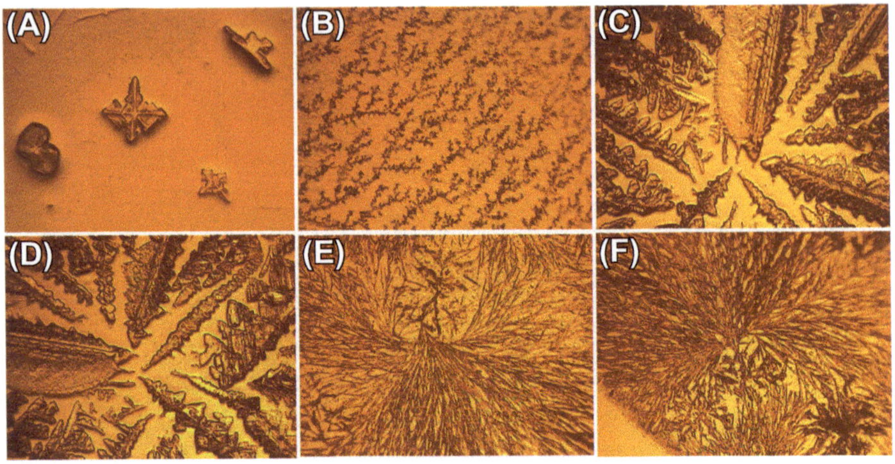

Figure 5.111
Polarizing microscopic drying patterns of KPolyA at different pH values on a watch glass at 20° C. [PolyA] = 0.00389 monoM, [KCl]= 0.01 M; **A,** pH= 2.8 ([HCl]= 0.01 M), central area; **B,** 2.8, broad ring area; **C,** 5.7 ([HCl]= 0.001 M), bright; **D,** 5.7 ([HCl]= 0.001 M), dark; **E,** 13.0 ([NaOH]= 0.01 M), broad ring area; **F,** 13.0 ([NaOH]= 0.01 M), broad ring area. *From Okubo T, Okamoto J, Takahashi S, Tsuchida A:* Colloid Polym Sci *287:933-942, 2009.*

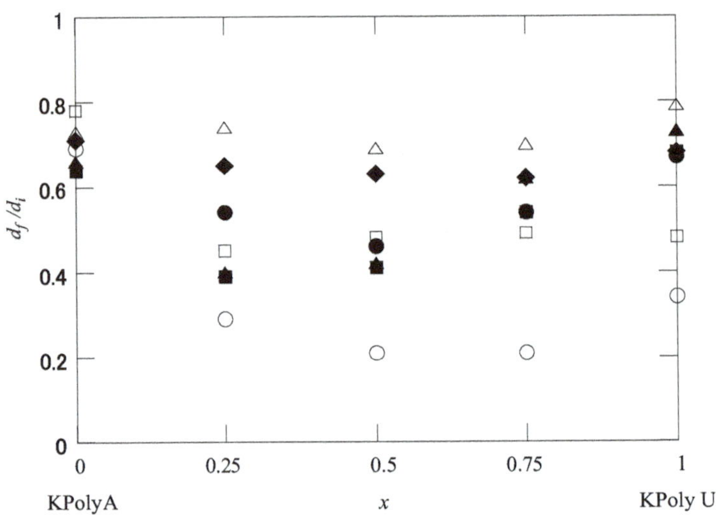

Figure 5.112
Value of d_f/d_i as a function of mixing ratio (x) in KPolyA + KPolyU mixtures at 25° C. *Open symbols:* on a cover glass; *solid symbols:* on a watch glass. *Open and solid circles:* [KCl] = 0 M; *open and solid triangles:* 0.01 M; *open and solid squares:* 0.03 M; *open and solid diamonds:* 0.1 M. *From Okubo T:* Colloids Surf B Biointerf *87:439-446, 2011.*

KPolyA and KPolyU. However, the experimental errors contained in the d_f/d_i values are too large to estimate the accurate mixing ratio from the figure. It is well known that the double-stranded A:U complex formation takes place at A:U mole ratio of unity or greater, while at A:U ratios less than 1 a transition occurs to the triple-stranded A:2U species, formation of which is complete at an A:U ratio of 1:2.[124-127]

Fig. 5.113 shows typical examples of the microscopic patterns of the mixtures KPolyA + KPolyU at the ratios of A:0U (*A-E*) from the left edge to the right of the dried film, 2A:U (*F-J*), A:U (*K-O*), A:2U (*P-T*), and 0A:U (*U-Y*).[122] Fig. 5.113, *C*, supports the formation of dendritic long rods originating from the single-stranded or double-stranded helices, although it is hard to distinguish these two helix structures from the drying patterns. It has been reported that the single-stranded helices form by van der Waal interactions (not hydrogen-bonding forces) at alkaline pH region.[128] On the other hand, at acidic pH PolyA polymers form double helices with a different conformation from that of DNA.[129-131] The pH limits of stability depend on the ionic strength and temperature. In 0.1 M KCl at 25° C, the transition from the alkaline to acid forms is reported to occur at approximately pH 6.[129,130] The stable helix conformation induced by KCl will reduce the affinitive interaction of KPolyA with the glass substrate, and it is plausible that the KPolyA polymers move easily toward the central area by the convectional forces during convectional processes. The extent of the movement depends on the change in the total excluded volume of the polymer during drying. It should be noted here that the dendritic long rod patterns originated from the single-stranded or double-stranded helices are observed only in the central area of the dried film, as is shown in Fig. 5.113, *A* through *E*. The segregation effect of the helices from the other conformations is significant. In other words, drying patterns changed drastically depending on the location in the dried film. Fig. 5.113, *U* through *Y*, shows the drying patterns of KPolyU on a cover glass. Rodlike patterns appeared at the several places, although they appear slightly different from the dendritic rod and are not organized so completely compared with those of KPolyA. PolyU is known to exist as a structureless extended coil at ordinary temperatures. However, PolyU undergoes a transition to an organized structure at low temperature and at neutral pH.[38] It is highly plausible that the helical structure forms on the glass substrate during dryness in our experimental condition, at pH = 6.6 and at 25° C. Thus, the rodlike patterns observed in Fig. 5.113, *V*, *W*, and *X*, are from the extended structureless coil, single-stranded helix chain, or double-stranded helix chain.

Fig. 5.113, *F* through *T*, show the microscopic drying patterns on a cover glass from the left edge to the right of the dried film at the mixing ratios 2:1, 1:1, and 1:2 between KPolyA and KPolyU. Highly ordered dendritic long rods were observed at the inner area from the outside edge. Fig. 5.113, *G* through *I* (2A:U), *L* through *N* (A:U), and *Q* (A:2U) are the patterns of single crystals

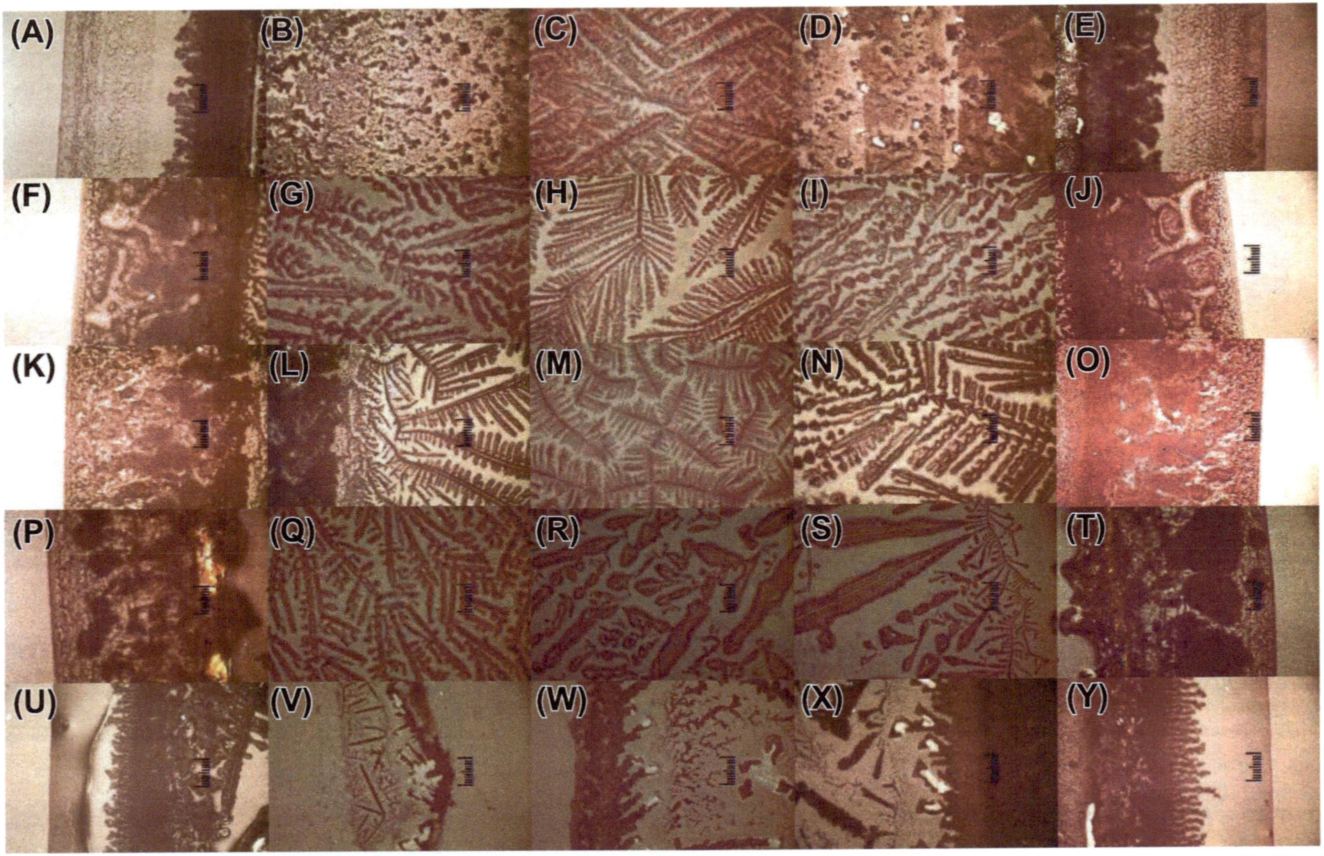

Figure 5.113 Microscopic drying patterns of aqueous KPolyA + KPolyU solution on a cover glass at 25° C. 0.1 ml, pH= 6.6, [KCl]= 0.01 M; **A-E,** [KPolyA] = 0.004 monoM, [KPolyU]= 0 monoM; **F-J,** 0.00267 monoM, 0.00133 monoM; **K-O,** 0.002 monoM, 0.002 monoM; **P-T,** 0.00133 monoM, 0.00267 monoM; **U-Y,** 0 monoM, 0.004 monoM; **A-E, F-J, K-O, P-T** and **U-Y** are from the left edge to the right; full scale is 50 μm. *From Okubo T:* Colloids Surf B Biointerf *87:439-446, 2011.*

originated from the single-stranded or double-stranded helix chains. Interestingly, very large sword (halberd)-like rods were grown at the mixing ratio of A:2U (see Fig. 5.113, *R* and *S*). It is highly plausible that these characteristic shapes originated from the A:2U type triple-stranded complexation between PolyA and PolyU. It should be noted here that the drying patterns, especially on a watch glass, change depending on the location in a dried film. It is interesting to note that the coexistence of the patterns originated from single-stranded or double-stranded helix chains with the triple-stranded chains observed, but they exist separately and are segregated from each other on a dried film.

For KPolyG + KPolyC mixtures, two kinds of broad rings appeared at the outside edge and the inner area.[123] The fine multiple ring structures formed, which supports that the affinity between the polymer and the substrate is strong. The microstructures of KPolyG, KPolyC, and KPolyG + KPolyC were spherulites, dendritic long rods, and many sword (halberd)-like rods, respectively. The patterns changed depending on the location in the dried film. The dendritic long rods and swordlike rods were assigned to the crystals of double-stranded or triple-stranded helices of the G:C and 2G:C complexes. Crosslike drying patterns originated from the salt-polymer interaction are also observed. The relationship between the polymer complexation of KPolyG + KPolyC systems and the drying patterns are similar to that of KPolyA + KPolyU systems. Typical drying patterns of polypeptides, polysaccharides, and polynucleotides are summarized in Table 5.10.

5.8.4 Biological Gel Solution

The relationship of the drying patterns of gel solutions, such as gelatin of collagen and ARS with their physicochemical properties is studied. It is well known that gelatin is composed of collagen protein and its solution is well known to show the thermally reversible helix-coil transition, and makes up the network structures (i.e., gelation) mainly by the double-helix (or triple-helix) complex formation at low temperatures.[132-137] The melting point is approximately 30° C, but changes slightly depending on the kind of animal extracted. The melting temperature decreased sharply in the presence of the various foreign salts.[133] Maeda studied the crystal structure of collagen.[138,139]

Fig. 5.114 compares the drying patterns of COL31 as a function of gelatin concentration and temperature.[139] Several important findings are clear in the pictures. First, the clear-cut broad rings formed at the outside edges of the dried films in all the experimental conditions. Broad rings appeared in the inner area, too, and their size increased as gelatin concentration decreased in most cases. Second, the broad and flamelike spoke lines formed especially at high polymer concentration or temperature, which supports the finding that the weak interpolymer and further polymer-substrate interaction exists even at

Table 5.10 Typical Drying Patterns of Polypeptides, Polysaccharides, and Polynucleotides

Polymers	Macroscopic Patterns		Microscopic Patterns Crystal Shape
	Group	d_f/d_i	
Polypeptides			
NaPGA	B	≈0.2-0.6	Spherulite, hedrite, dendritic, rodlike
PLHBr	A	≈0.4-1.0	Dendritic, rodlike, hedrite
Polysaccharides			
NaHLA	D	≈0.02-0.04	Dendritic, rodlike
NaCMC	B	≈0.2-0.8	Dendritic, rodlike, swordlike
NaDSS	B	≈0.1-0.8	Rodlike
HPC	B	≈0.3-0.9	Short rodlike
α-, β-, γ-CD	B	≈0.2-0.8	Dendritic, hedrite, rodlike, swordlike
Polynucleotides			
NaDNA	D	≈0.05-0.1	Dendritic, rodlike, swordlike
KPolyA	C	≈0.2-0.8	Spherulite, dendritic, rodlike, swordlike
KPolyA +KPolyU	C	≈0.2-0.8	Dendritic, rodlike, swordlike
KPolyG	C	≈0.6-0.8	Spherulite
KPolyC	C	≈0.3-0.8	Dendritic long rodlike
KPolyG +KPolyC	C	≈0.3-0.7	Swordlike

CD, Cyclodextrin; *HPC,* hydroxypropyl cellulose; *KPolyA,* potassium salt of poly(riboadenylic acid); *KPolyC,* potassium salt of poly(ribocytidilic acid); *KPolyG, potassium salt of poly(riboguanylic acid); KPolyU,* potassium salt of poly(ribouridylic acid); *NaCMC,* sodium salts of carboxymethyl cellulose; *NaDNA,* sodium salt of deoxyribonucleic acid; *NaDSS,* sodium salt of dextran sulfate; *NaHLA,* sodium salt of hyaluronic acid; *NaPGA,* sodium poly(alpha-glutamate); *PlLHBr,* poly(*L*-lysine hydrobromide).

high temperatures. Fine, flickering spokelines formed, as shown in Fig. 5.114, *3D, 4C,* and *4D.*

Fig. 5.115 shows the size ratio, d_f/d_i, as a function of COL31 concentration. The broad rings appeared at the outside edges ($d_f/d_i = 1$) in addition of the broad rings at the inner region, which supports the strong affinity of COL31 polymers with the surface of the glass substrate irrespective of temperature. The size of the inner rings decreased with increasing polymer concentration, which is unusual compared with the typical polymers and colloidal particles but observed for small cationic gel spheres,[64,65] NaDNA[43] and KPolyA.[121] These results support the finding that both the interpolymer and polymer-substrate attractive interactions are strong for COL31. It should be recalled here that the broad ring at the outside edge disappeared and the very small inner broad rings formed instead for NaHLA.[101] The d_f/d_i values of NaHLA were small, approximately 0.03, irrespective of polymer concentration. The macroscopic broad ring profiles teach us clearly that for NaHLA, very strong

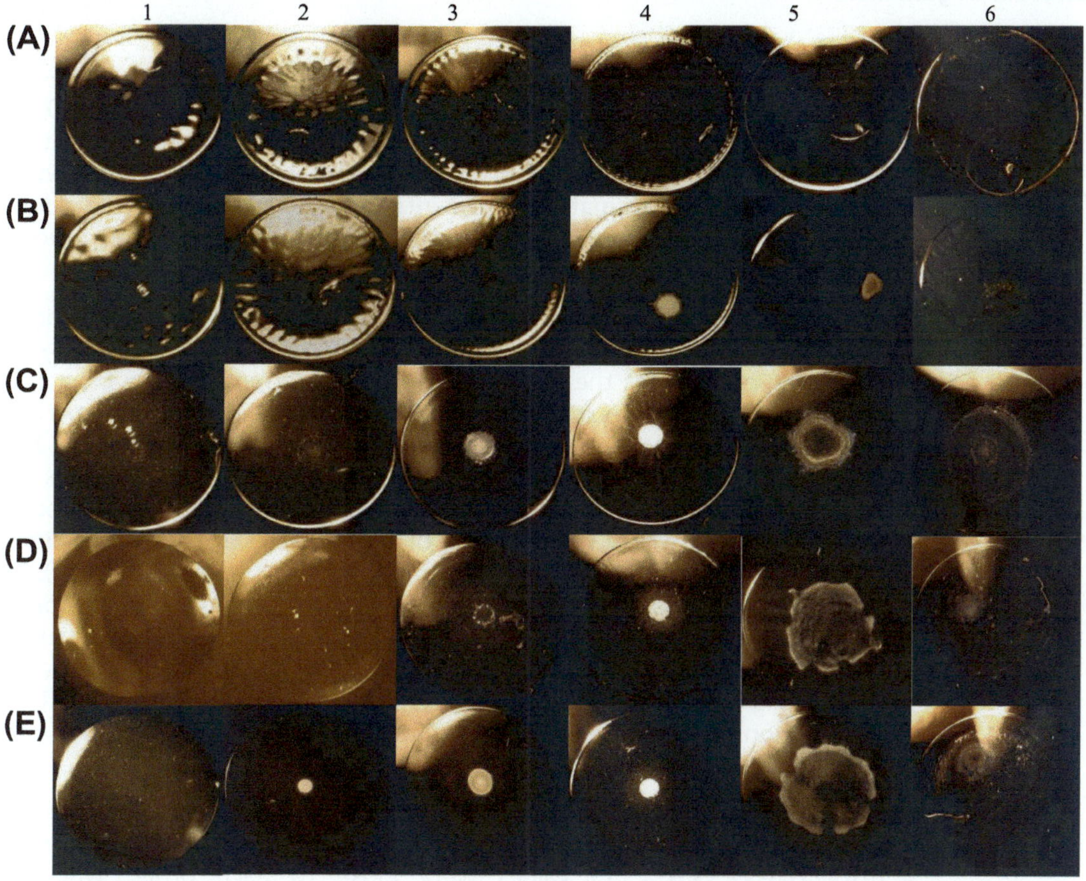

Figure 5.114 Drying patterns of COL31on a cover glass. **A,** 50° C; **B,** 30° C; **C,** 20° C; **D,** 12° C; **E,** 5° C; **1,** [COL31] = 7.73 wt%, **2,** 5.27 wt%; **3,** 2.60 wt%; **4,** 0.52 wt%; **5,** 0.052 wt%; **6,** 0.0052 wt%. *Okubo T, Tsuchida A, Kokufuta E:* Colloid Polym Sci *2015, in press. http://dx.doi.org/10.1007/s00396-015-3558-z.*

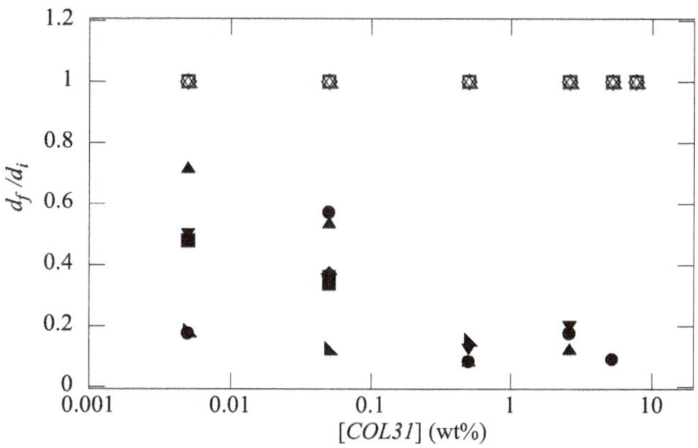

Figure 5.115

Value of d_f/d_i of COL31. *Outer rings:* circles 5° C; *open triangles:* 12° C; *open inverted triangles:* 20° C; *open flag:* 30° C; *open squares:* 40° C; *open diamond:* 50° C. *Inner rings:* solid circles 5° C; *solid triangles:* 12° C; *solid inverted triangles:* 20° C; *solid flag:* 30° C; *solid squares:* 40° C; *solid diamond:* 50° C. *Okubo T, Tsuchida A, Kokufuta E:* Colloid Polym Sci *2015, in press. http://dx.doi.org/10.1007/s00396-015-3558-z.*

interpolymer affinity exists, but the polymers are not as interactive with substrate. At the concentrations of COL31 higher than approximately 0.2 wt %, d_f/d_i values were quite low (between 0.1 and 0.2), especially at low temperatures, as is shown in Fig. 5.115. These observations are quite consistent with the easy formation of the interpolymer network by the double-helix formation reaction, resulting in the lowered convectional diffusion of COL31 at low temperatures.

Fig. 5.116 shows the thickness profiles of the dried film of COL31 as a function of the distance from the center of the dried film at 30° C. Interestingly, the sharp ring patterns at the low polymer concentrations became broad, and the broad ring and a round, hill-shaped patterns coexisted as polymer concentration increased. A round, hill-shaped profile is unique and has been observed often for the aggregates of the anisotropic-shaped particles such as plate-shaped bentonites[28] and the colloidal crystal systems of spherical rubber particles, where the diffusion of the particles are restricted significantly during the convectional diffusion processes (see Section 5.6.1). The thickness profiles of the dried films of COL31 as a function of the drying temperature keep concentration of COL31 at 2.60 wt% were also studied. Interestingly, sharp broad rings appeared at high temperature. However, broadening of the peak profiles took place as temperature decreased. These observations are consistent with the fact that the network formation takes place more easily at lower temperatures.

Fig. 5.117 shows the S (or S') values of COL31 as a function of gelatin concentration (open and solid circles) and temperature (open and solid triangles, respectively). S (or S') values were close to each other, when the

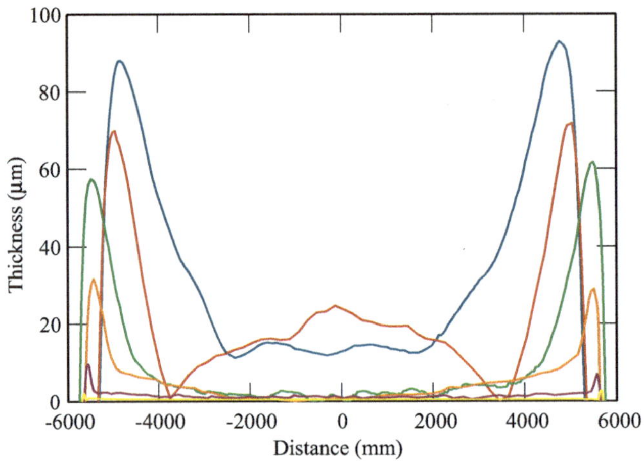

Figure 5.116
Thickness profiles of the dried film of collagen as a function of the distance from the center. Dried at 30° C, *blue curve:* [COL31] = 7.73 wt%; *red:* 5.27 wt%; *green:* 2.60 wt%; *orange:* 0.50 wt%; *purple:* 0.050 wt%; *yellow:* 0.0050 wt%. *Okubo T, Tsuchida A, Kokufuta E:* Colloid Polym Sci *2015, in press. http://dx.doi.org/10.1007/s00396-015-3558-z.*

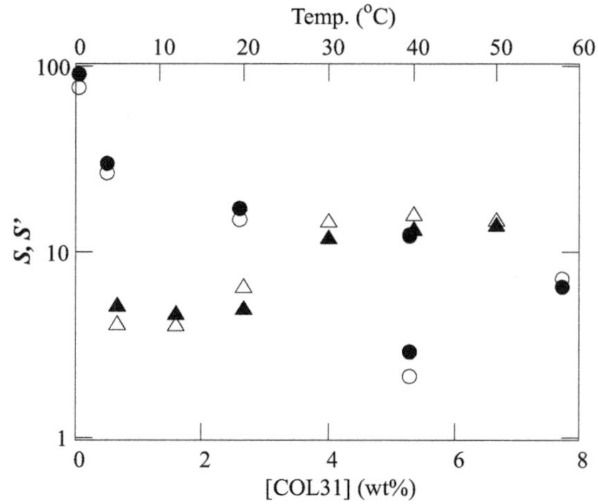

Figure 5.117 Sharpness parameters S and S' in the thickness profiles of the dried films of COL31 as a function of concentration (*open circles: S; solid circles: S'*) and temperature (*triangles*). *Okubo T, Tsuchida A, Kokufuta E:* Colloid Polym Sci *2015, in press. http://dx.doi.org/10.1007/s00396-015-3558-z.*

experimental errors were taken into account. S (or S') values decreased with COL31 concentration, and they increased as temperature increased. These observations of S are understandable, when the network structures form as gelatin concentration increases or temperature is raised.

Fig. 5.88 in Section 5.8.1 shows the sharpness parameter, S values of the broad ring or round hill as a function of monomer molal concentration of COL31 (broad plus) with a part of the biological polymers observed previously: sodium salt of poly(α,*L*-glutamic acid) (NaPlGA)[93] (*open circles*), P*l*LHBr[41,97] (*triangles*), P*d*LHBr[41] (*squares*), HPC1 and HPC3[42] (*flag* and *solid flag*), and αCD (Fukami H, Kimura H, Tsuchida A, Okubo T, unpublished data) (*diamond*). The S values do not scatter so much within the experimental error among the biopolymers. They decreased from approximately 50 to approximately 5 as the polymer concentration increased from 0.001 to 1 monoM. COL31 is considered to form gel structure at the high concentrations. Thus the low values of COL31 support clearly the formation of some associative assemblies, which disturb the free diffusion during the course of drying. It should be mentioned here that nonspherical colloidal particles such as bentonite,[28] TA particles,[77] and PGK,[77] gave the round-hill type thickness profiles in the macroscopic drying patterns, and their S (or S') values were small between unity and three. S (or S') are clearly one of the convenient parameters indicating the diffusivity of solutes during dryness.

Fig. 5.118 shows the typical examples of the microscopic drying patterns of COL31 on a cover glass at 50° C (*A, B*), 30° C (*C, D*), 20° C (*E-F*), and 5° C (*G-H*). Clearly, dendritic spherulite single crystals of COL31 formed especially at low temperatures (20° C and 5° C in the figure) and at the central area of a cover

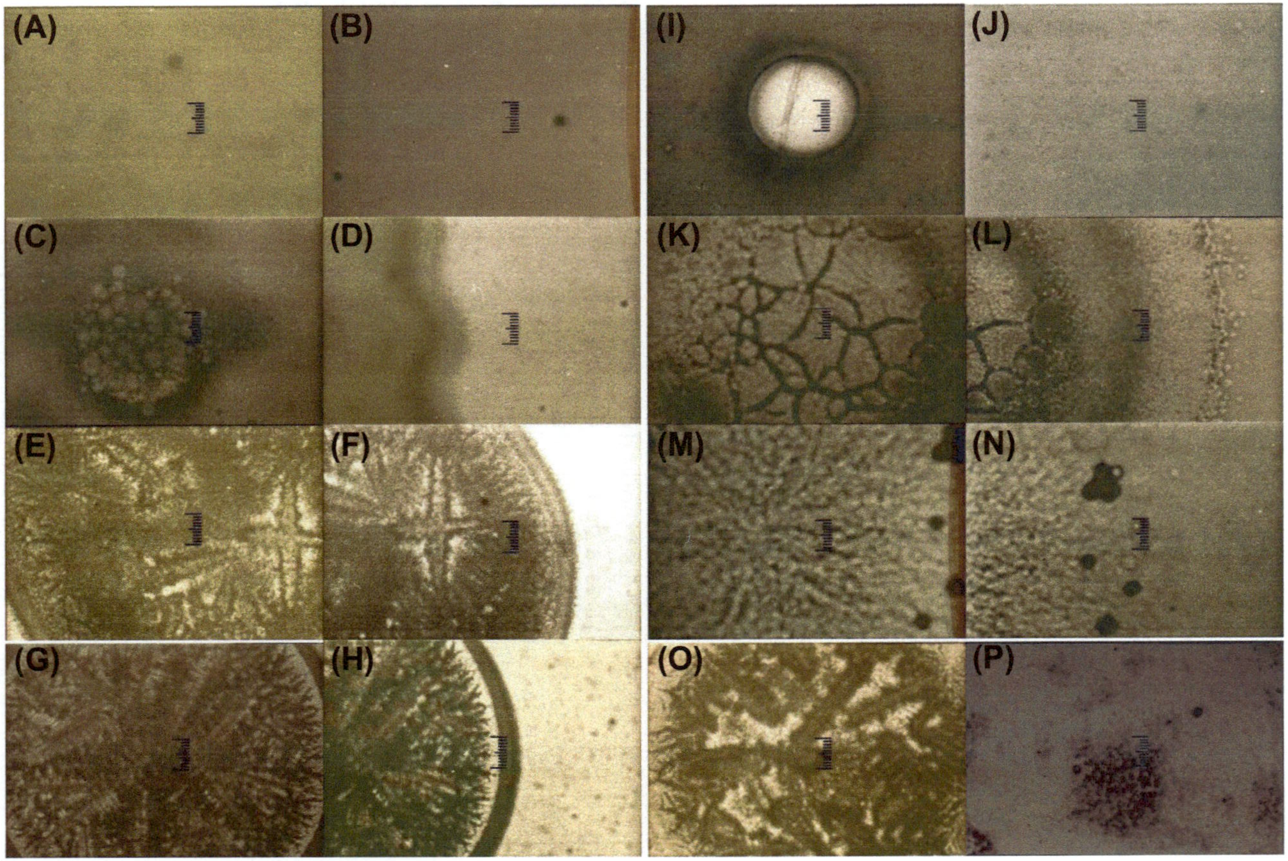

Figure 5.118 Microscopic drying patterns of COL31 on a cover glass. 0.1 ml, [COL31] = 0.50 wt%; **A, B,** at 50° C; **C, D,** at 30° C; **E, F,** at 20° C; **G, H,** at 5° C. At 20° C, **I, J,** [COL31] = 7.73 wt%; **K, L,** 2.60 wt%. At 12° C, **M, N,** 2.60 wt%; **O, P,** 0.50 wt%. **A-P** are the around the central area; full scale is 100 μm. *Okubo T, Tsuchida A, Kokufuta E:* Colloid Polym Sci *2015, in press. http://dx.doi.org/10.1007/s00396-015-3558-z.*

glass substrate. At 30° C, the hedrite-like crystals still formed at the central area of the dried film (see Fig. 5.118, *C*), but no single crystals were observed microscopically at 50° C. These results support that gelation takes place for gelatin certainly. However, it is highly plausible that the network structures do not disturb the spherulite crystallization so much, especially in the central area of the dried film where the convectional flow must be symmetrical. Furthermore, the spherulite crystallization will take place favorably at the central area compared with the outward areas, where the convectional outward flow of solvents and solutes will disturb formation of the symmetrical crystallization during the course of dryness.

Fig. 5.118, *I* through *P*, are the typical examples of the change in the microscopic drying patterns depending on the polymer concentration at 20° C and at 12° C, respectively. The patterns at the central area at 7.73 wt% (e.g., see Fig. 5.118, *I*), show the rodlike crystals. Picture Fig. 5.118, *K*, at 2.60 wt% shows the assembly of the small spherulites. At 0.50 wt%, morphologic characteristics of the single crystals appears to be dendritic and spherulite crystals. When the initial polymer concentration is lowered at 0.05 wt%, any clear-cut single crystals were not recognized microscopically. Interestingly, morphologic characteristics of the single crystals of COL31 changed depending on the initial polymer concentration. Fig. 5.118, *M* and *O*, demonstrate that the single crystals of COL31 dried at low temperatures are spherulites or dendritic rods. These observations support the finding that the interaction between COL31 molecules and the substrates plays an important role in the drying patterns during the course of drying. It should be recalled here that the initial conformational information of cationic polyelectrolyte solution was transferred beautifully to the drying patterns with the help of the macroion-substrate interaction during the dissipative drying processes (see Fig. 5.85).[37]

Macroscopic and microscopic drying patterns of ARS were observed on a cover glass at various concentrations and temperatures, and the patterns observed were compared with those of gelatin and colloidal crystal suspensions of rubber-network forming PBA spheres in the dried state.[67] ARS (*kudzu* in Japanese) is a food used in traditional Japanese dishes. ARS is composed of amylose molecules, and their solutions show the thermally reversible helix-coil transition and make up the network structures; gelation takes place by the formation of double helix complexes, especially at low temperatures like gelatin.[140-142]

Fig. 5.119 shows the concentration dependencies of the thickness profiles of the dried film of ARS as a function of the distance from the center of the dried film. Interestingly, most ARS molecules accumulated around the central area very broadly at concentrations lower than 0.227 wt%. It is recalled that most solutions and dispersions including gelatin solution showed the sharp ring patterns only at the outside edge. This observation suggests the strong

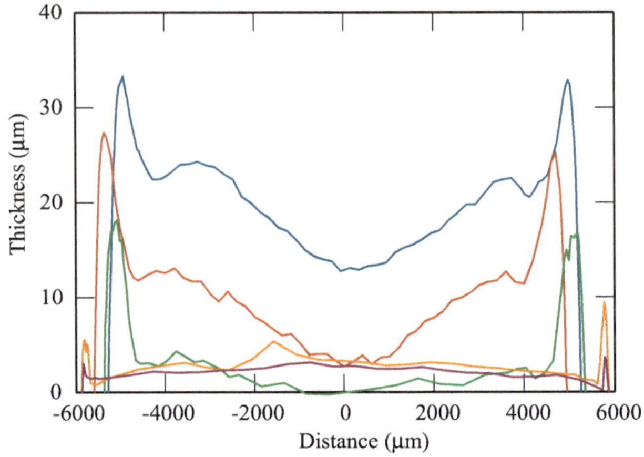

Figure 5.119
Thickness profiles of ARS dried films as a function of the distance from the center. Concentration dependence at 20° C. *Blue:* [ARS] = 2.88 wt%; *red:* 2.30 wt%; *green:* 1.13 wt%; *yellow:* 0.227 wt%; *purple:* 0.0378 wt%. *From Fujii S, Nakamura Y, Tsuchida A, Okubo T:* Colloid Polym Sci *292:2303-2310, 2014.*

affinitive interaction of ARS with the wall of the glass substrate. The gelation of ARS at these low concentrations will not occur, because the accumulation of ARS molecules at the central area is not so significant even at the higher ARS concentrations (see Fig. 5.119). A round, hill-shaped profile is unique and has been observed often for the aggregates of the anisotropic-shaped particles such as plate-shaped bentonites[28] and the colloidal crystal systems of spherical rubber particles, where the diffusion of the particles are restricted significantly during the convectional diffusion processes. In this study, however, strong solute–substrate interaction was found to be another important factor for the broad, round hill or rather flat thickening in the inner area. In a range of ARS concentrations between 1.13 and 2.88 wt%, double broad rings grew at the outside edge and the inner region as ARS concentration increased. However, the thickness at the inner region also increased with increasing concentration, as is shown in the figure. These observations support the finding that the gelation of ARS molecules takes place as ARS concentration increases, although the ARS gels formed are diffusible during the course of drying compared with gelatin gels and rubber-forming PBA, where the broad rings almost disappeared and only the round hill grew significantly instead. It is plausible that the networks of ARS are loosely bound to each other compared with those of gelatin and PBA. The distance between the cross-linking points of ARS will be longer than those of the gelatin and PBA.

Thickness profiles of the dried films of ARS were also studied as a function of the drying temperature, keeping the concentration of ARS at 2.30 wt%. Interestingly, sharp double broad rings formed at high temperatures.

However, thickness at the central area also increased as temperature increased. These observations are consistent with the fact that the network structures by the gelation are insensitive to temperature. The thickness profiles show the comparative strengths of the diffusive and associative (network-forming) factors during the course of drying.

Fig. 5.120 shows the *S* values of ARS as a function of ARS concentration (*open circles*) and temperature (*solid circles*), respectively. *S* values decreased sharply with ARS concentration, although the small *S* values were observed in addition to the large *S* values at low concentrations. At low concentrations, the affinitive interactions between ARS and the substrate enhance the adsorption of part of ARS on the substrate surface. The *S* values increased first transitionally and then kept constant at temperatures higher than 20° C. The transitional changes with temperature were for gelatin, as is shown in the figure. Fig. 5.120 shows the concentration and temperature dependencies of gelatin and PBA in addition to those of ARS. All the data connected with curved lines originate from the broad rings at the outside edges. The other three data are from the round hills. The *S* values of ARS were high compared with those of gelatin when comparison was made at the same wt% of concentrations lower than approximately 2 wt%. Especially at ARS concentrations lower than 0.2 wt%, high (≈100) and low (≈1-2) *S* values were observed simultaneously. These dual data were not obtained for gelatin and PBA. These observations support the finding that ARS molecules interact

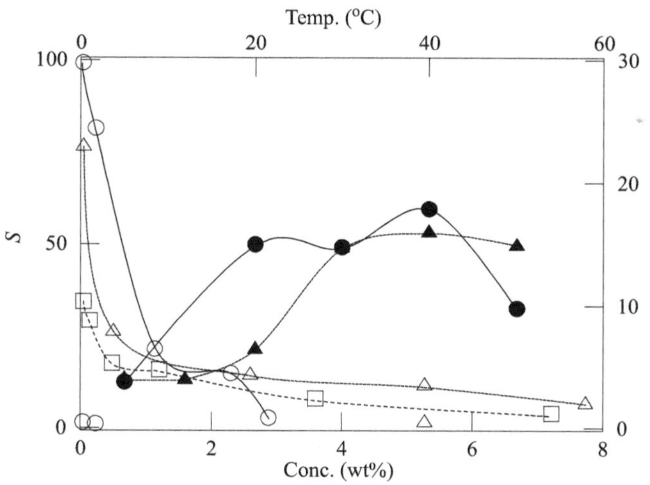

Figure 5.120
S values of ARS as a function of polymer concentration at 20° C (*circles*) and temperature at 2.30 wt% (*solid circles*). Concentration and temperature dependencies of gelatin (*triangles* at 30° C and *solid triangles* at 2.60 wt%, respectively) and concentration dependency of PBA (squares at 22° C) are also shown. *From Fujii S, Nakamura Y, Tsuchida A, Okubo T:* Colloid Polym Sci *292:2303-2310, 2014.*

affinitively with the glass substrate, but the network gel structures are loose and move freely. Gelation of ARS took place rapidly above approximately 2 wt%. When temperature was raised from 5° C to 20° C, *S* increased sharply from 3 to 15, and kept a constant 15 of *S* at higher temperatures. The convectional diffusion strength was in the order ARS > gelatin > PBA, when comparison was made at the same concentration in weight percent.

Most extended (×1000) microscopic drying patterns of ARS from the central area to the right edge were observed on a cover glass at 50° C, 40° C, 20° C, and 5° C. Regular network structures of ARS formed, especially at high temperatures. Furthermore, the size of the net structures seems to increase with rising temperature. It should be noted that no single crystals of ARS molecules were observed, which is consistent with the fact that amylose structure of ARS does not crystallize in general. It is recalled that the dendritic-, hedrite-, and spherulite-microscopic patterns were observed for gelatin. It is highly plausible that the helical regions of ARS are short and incomplete.

5.8.5 Biological Colloidal Particles

The drying patterns of biological colloidal particles, lycopodium spore particles (31 μm in average diameter) were studied.[143] SEM pictures of lycopodium spore particles are shown in Fig. 5.121, *A*. A deep hexagonal wall appears to be broken by the mechanical stress and a small part of surface

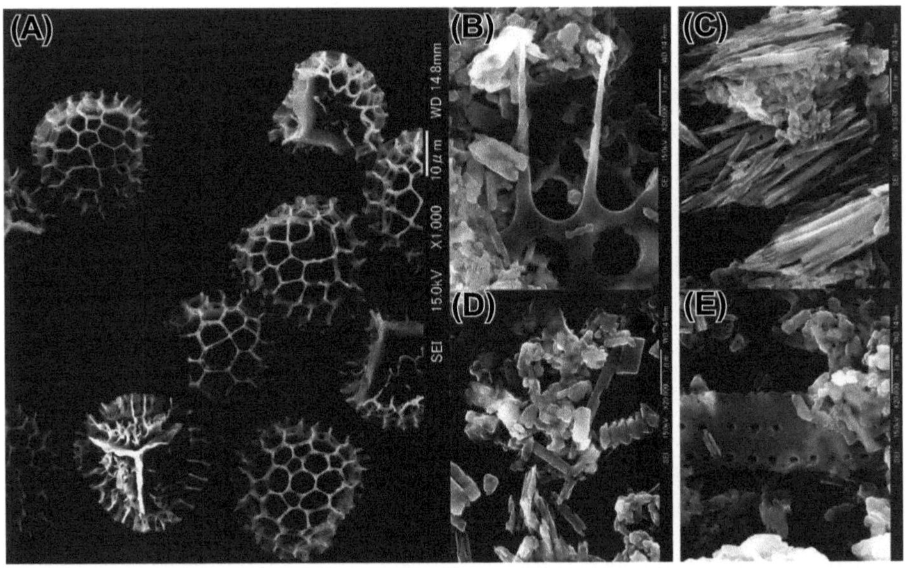

Figure 5.121
A, SEM picture of lycopodium spore particles; **B-E,** SEM pictures of OC(1) particles.

structures of the main bodies are also observed in the drying patterns. Lycopodium particles are composed of the hard organic resins and are not swollen in an aqueous media. The vague broad ring structure was observed, and their size on a cover glass decreased as particle concentration decreased. Microscopic drying patterns almost disappeared except for the short, chainlike patterns. These observations support the finding that the convectional contribution on the drying patterns is still effective in the lycoposium suspension, although the convectional forces are weak compared with the sedimentary forces. The vague broad ring structure was observed, and their size on a cover glass decreased as particle concentration decreased. Microscopic drying patterns almost disappeared except for the short, chainlike patterns. The d_f/d_i values decreased as the particle concentration decreased, which is the same as the observations for other typical small colloidal particles. These observations support the finding that the convectional contribution on the drying patterns is still effective in the lycopodium suspension, although the convectional forces are weak compared with the sedimentary forces.

Fig. 5.121, *B-E*, shows SEM pictures of colloidal particles of marine clay (OC1) (Okubo T, et al., unpublished data). Highly polydispersed particles (≈0.1-10 μm) are recognized in the pictures. There were two main broad rings; small particles were in the outside edge and large ones were in the inner area (Fig. 5.122, *A*). The d_f/d_i values for the inner broad rings are shown in Fig. 5.122. They decreased sharply as particle concentration decreased. The importance of the excluded volume effects on the drying patterns are shown clearly. Surprisingly, the dissipative effects are clearly important even for these large and anisotropic particles.

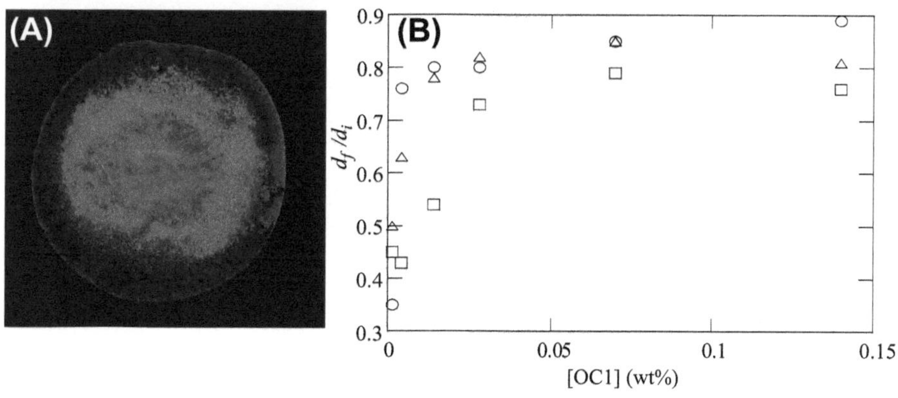

Figure 5.122
A, Macroscopic drying patterns of OC(1) on a cover glass at 24° C. 0.1 ml, 5.0 wt%. **B,** d_f/d_i as a function of concentration of OC(3) on a cover glass (*circles*), a watch glass (*triangles*), and a petri glass dish (*squares*) at 25° C.

5.9 Surface-Active Agents

Microscopic patterns of cationic surfactant, *n*-dodecyltrimethylammonium chloride (12TMAC) were observed.[144] Typical patterns are shown in Fig. 5.123. Basic structures were crosslike. However, crystal size and shape changed as a function of the distance from the film center. Dissipative effects are tremendous.

Drying patterns of a series of *n*-alkyltrimetylammonium bromides (alkyl: *n*-decyl (10TMAB), *n*-dodecyl (12TMAB), *n*-tetradecyl (14TMAB) and *n*-hexadecyl (16TMAB)) were studied.[145] Fig. 5.124 shows typical macroscopic patterns of 10TMAB (*A-C, G-I*) and 12TMAB (*D-F, J-L*) on a cover glass and a watch glass. The broad rings were observed. Interestingly, the final size of the drying patterns were sensitive to the carbon number of the alkyl group n^* and the initial concentration of the surface-active agents. These results clearly show the importance of the wetting process for the drying pattern formation.

Fig. 5.125 shows changes in the d_f/d_i values as a function of the initial surfactant concentration for a series of the cationic surfactants, n^*TMAB (n^*-values are 10, 12, 14 and 16). The curves showed the clear-cut bent points. Interestingly, the initial surfactant concentrations corresponding to the bent points were very close to *cmc* values of each surfactant. The similar observations were obtained also for d_i and d_f values. Qualitatively, this correspondence is ascribed to the fact that the areas of the initial liquid increases as the surface tension of the liquid decreases and the bent points appear in the surface tension versus surfactant concentration at the *cmc*

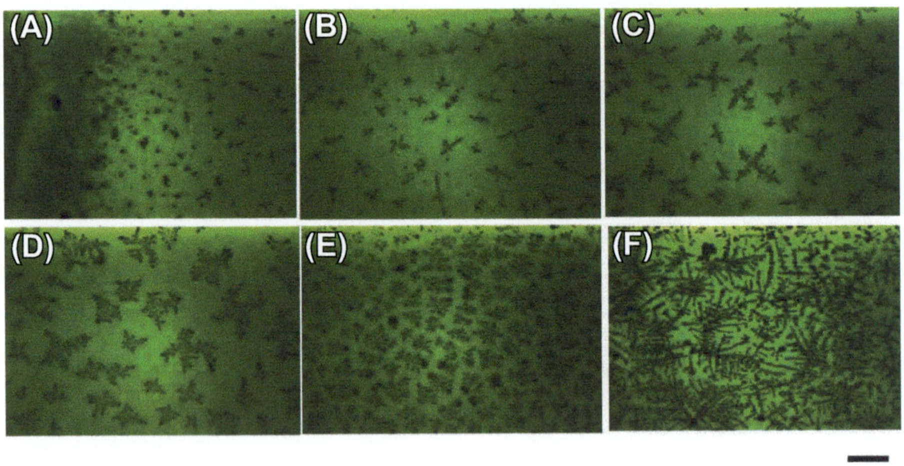

Figure 5.123 Typical patterns formed for 12TMAC at 25° C. 0.1 mL, 2×10^{-5} M; length of the bar is 20 μm. *From Okubo T, Kanayama S, Kimura K:* Colloid Polym Sci *282:486-494, 2004.*

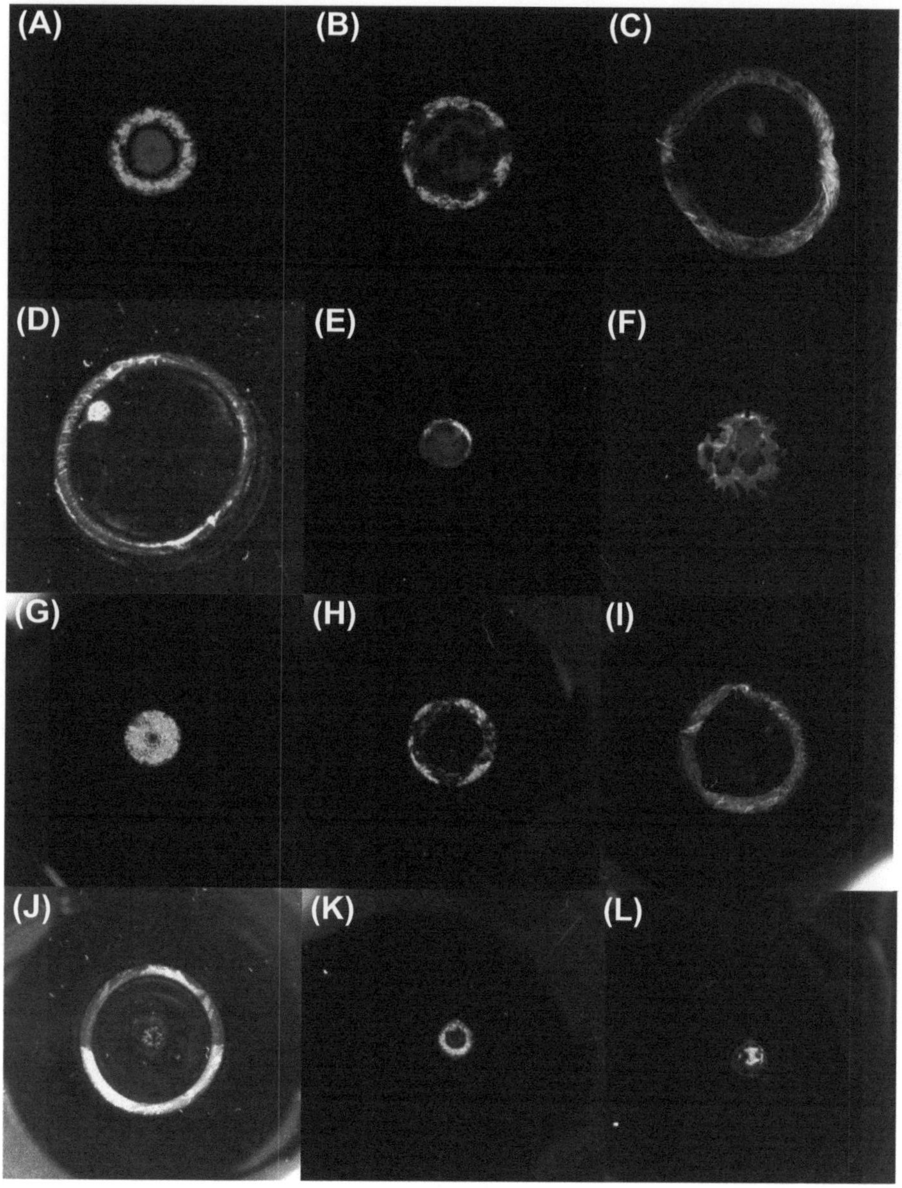

Figure 5.124 Dissipative drying patterns of 10TMAB (*A-C, G-I*) and 12TMAB (*D-F, J-L*) on a cover glass (0.1 ml, *A-F*) and a watch glass (0.7 ml, *G-L*) at 20° C. **A, G,** [10TMAB] = 0.01 M; **B, H,** 0.001 M; **C, I,** 0.0001 M; **D, J,** [12TMAB] = 0.005 M; **E, K,** 0.0002 M; **F, L,** 0.00005 M. *From Okubo T:* Colloid Polym Sci *291:2455-2462, 2013.*

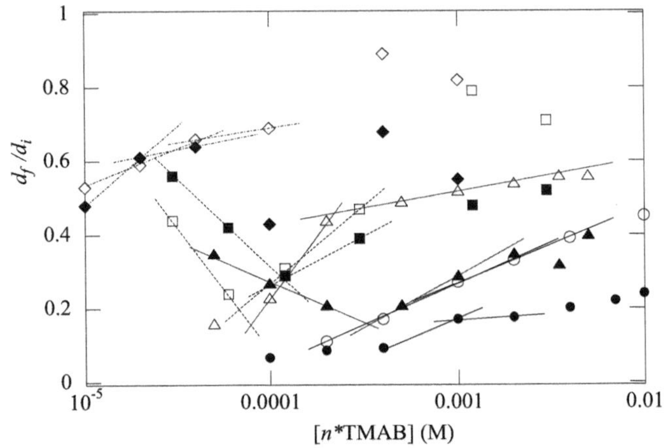

Figure 5.125
Values of d_f/d_i of 10TMAB (*circles*), 12TMAB (*triangles*), 14TMAB (*squares*), and 16TMAB (*diamonds*) on a cover glass (*open symbols*) and a watch glass (*solid symbols*) at 20° C. *From Okubo T:* Colloid Polym Sci *291:2455-2462, 2013.*

values. This is because the liquid at the *cmc* value shows the lowest surface tension in general.

Quantitatively, the d_i values should increase as the contact angles, θ between the liquid and the substrate decreases. The contact angle is given by Young's equation, Equation 13, as a function of the surface tension of liquid, σ_l, surface tension of the substrate, σ_s, and the interface tension between the substrate and liquid, σ_{ls}.

$$\text{Cos}\,\theta = (\sigma_s - \sigma_{ls})/\sigma_l \qquad 13$$

Thus d_i value should increase when θ decreases and then cos θ increases. In the range of surfactant concentrations below the *cmc*, both σ_{ls} and s_l decreases as the concentration increases. On the other hand, at the concentrations above the *cmc* both σ_{ls} and σ_l keep constant or increase as the concentration increases. Thus cos θ should increase as surfactant concentration increases above the *cmc* and the bent point should appear, because the numerator of the right side of Equation 13 increases and the denominator decreases in the condition of the term $(\sigma_s - \sigma_{ls})$ being positive. The deviation from the linearity of the d_i values especially at the surfactant concentrations far from *cmc* is due to the fact that the increase of the nominator and the decrease of denominator become independent of each other. It should be mentioned here that the d_f values contain the information of the drying processes and substantially smaller than the d_i, as is clear from the fact that d_f/d_i is very small in general compared with unity (see Fig. 5.125). It is highly plausible that d_f and d_f/d_i values also have bent point as observed because the d_i curves have bents. In general, d_i values are greatly influenced by the interfacial activities of the initial surfactant solution with air and the substrate. On the other hand, d_f values contain information of the surfactant solution (e.g., shape, size, and conformation of the surfactant molecules during the course of drying, interaction of the surfactant molecules with the substrate, and further atmospheric temperature and humidity). The most important factor determining the size of the drying broad ring is the sum of the excluded volume of each surfactant molecules in the whole drying area during drying. The excluded volume of each surfactant molecules is likely to be influenced by the interaction with the other surfactants and the substrate surfaces. It should be noted here that solubility of each surfactant molecules is likely to influence the microscopic drying pattern. However, the solubility is not so important a factor for the broad ring size compared with the excluded volume effect mentioned previously.

The *cmc* values from the broad ring size contain large experimental errors. However, the cmc_{obs} values are close to the reference values of cmc, cmc_{ref}.[146-149] It is well known that the *cmc* values of *n*-alkyl-trimethylammonium bromides change following Equation 14.

$$\text{Log}\ cmc = 0.1128 - 0.3074\ n^* \qquad 14$$

Here, n^* is the number of carbon atoms of the alkyl group.

It should be mentioned here that d_f/d_i values were smaller than unity, as is shown in Fig. 5.125. This means that the size of the drying pattern decreases during the course of drying. This fact supports the finding that the interaction between the cationic surfactants and the substrate glass is not so strong that the broad rings shift inward during drying.

Microscopic drying patterns such as rodlike, dendritic, and granulelike patterns were observed, and they changed as a function of the carbon number of the alkyl group, surfactant concentration, and the distance from the center of the pattern. Drying patterns of aqueous solution of sodium *n*-alkylsulfates on a cover glass were also observed.[150] Branched, stringlike microscopic patterns appeared, irrespective of the number of alkyl groups.

Film surfaces of POEmCn changed from the rough to smooth as the hydrophile-lyophile balance (HLB) values of the surfactants increased (Fig. 5.126).[151] Furthermore, film thickness profiles as a function of the distance from the center of the film shifted from the center-accumulated hill shape to the broad ring–like with increasing HLB values, as is shown in Fig. 5.127. The sharpness parameters were low, between 2 and 13, and increased as surfactant concentration increased. The main cause for these tendencies is, of course, micellization (i.e., large ordered structural

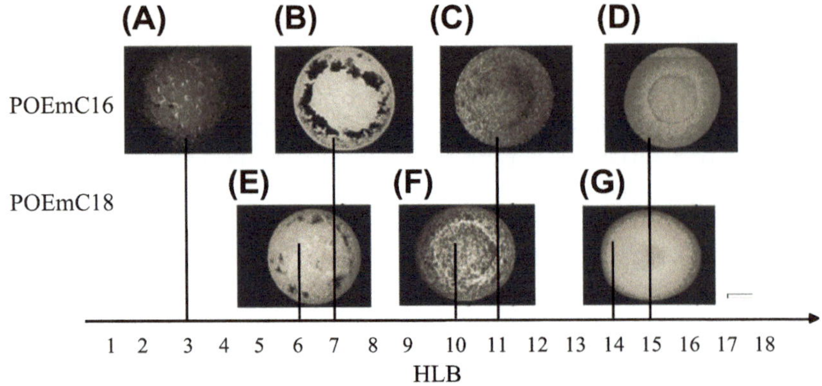

Figure 5.126
Drying dissipative structures of aqueous solutions of polyoxyethylene-*n*-cetylethers (POEmCn), $CH_3(CH_2)_n$-$O(CH_2CH_2O)_mH$, $n = 16$, m = 2, 5, 10, 20, and polyoxyethylene-*n*-stearylethers (POEmC18), n = 18, m= 5, 10, 20. *From Okubo T, Shinoda C, Kimura K, Tsuchida A:* Langmuir *21:9889-9895, 2005.*

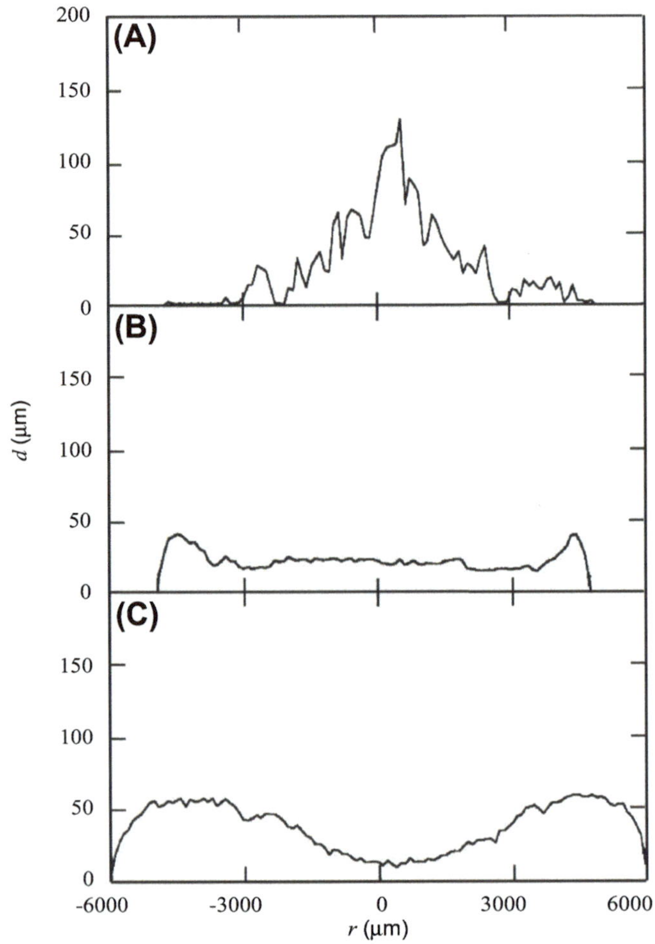

Figure 5.127
Thickness profiles of the dried film as a function of distance from the center at 25° C. 0.06 M, 0.05 mL; **A,** POE5C18; **B,** POE10C18; **C,** POE20C18. *From Okubo T, Shinoda C, Kimura K, Tsuchida A:* Langmuir *21:9889-9895, 2005.*

formation). Macroscopic and microscopic drying patterns of surfactant molecules were found to give important information on the association and dissociation phenomena. Adsorbed and dried patterns of ionic surfactants were studied by several researchers[152-155] hitherto.

5.10 Dyes, Low-Molecular-Weight Organic Substances and Simple Electrolytes

Drying patterns of fluorescence dyes, rhodamine 6G, uranine, 7-hydroxy coumarin, and 7-amino-4-(trifluoromethyl) coumarin were studied.[156] The macroscopic broad ring patterns formed for all the ethanol solutions were examined. Fig. 5.128 shows typical macroscopic drying patterns of rhodamine 6G in ethanol on a cover glass at 25° C. The critical concentration m^*, where the d_f/d_i starts to decrease with decreasing concentration, is near 0.001 M. Broad ring size is determined by the excluded-volume effects of crystallized solutes clearly. Typical patterns for group A are observed for rhodamine 6G. The S values at 0.1 M of rhodamine 6G and uranine are approximately 60 and 34, respectively.

Microscopic fine patterns including the city road–like, needlelike, and flowerlike crystals formed for the ethanol solution of the fluorescence dyes. The typical microscopic patterns are shown in Fig. 5.129 for 7-amino-4-trifluoromethyl cumarin in ethanol solution. The beautiful city road–like crystal patterns formed (see Fig. 5.129, C).

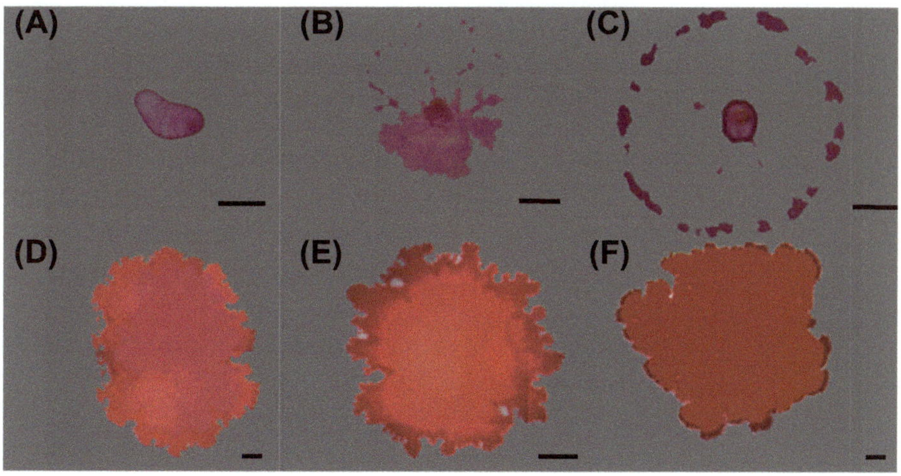

Figure 5.128 Patterns formed in the drying process of rhodamine 6G on a cover glass at 25° C. **A,** 0.0001M; **B,** 0.0005M; **C,** 0.001M; **D,** 0.005M; **E,** 0.01M; **F,** 0.1M. In ethanol, 10 ml; length of the bar is 2 mm. *From Okubo T, Yokota N, Tsuchida A:* Colloid Polym Sci *285:1257-1265, 2007.*

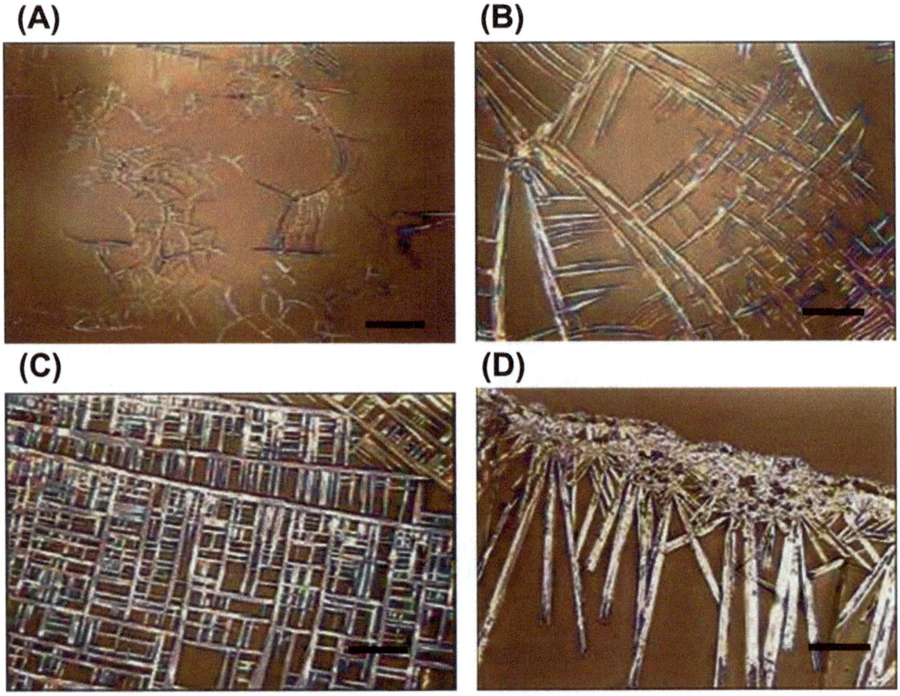

Figure 5.129 Drying patterns of 7-amino-4-trifluoromethyl coumarin on a cover glass at 25° C. **A,** 0.001 M; **B,** 0.004 M; **C, D,** 0.04 M, in ethanol, 10 ml; length of the bar is 20 μm (*A, B, C*), 100 μm (*D*). *From Okubo T, Yokota N, Tsuchida A:* Colloid Polym Sci *285:1257-1265, 2007.*

Droplets of the aqueous caffeine solutions were dried on various substrates (glass, polystyrene, and polyester).[157] Coexistence of α- and β-caffeine was observed and locally separated. Drying patterns of porphyrin derivatives were observed on silica.[158]

Drying patterns of simple electrolytes such as KCl, NaCl, $CaCl_2$, and lanthanum chloride ($LaCl_3$) were studied.[159] The macroscopic broad rings formed and the size of the broad ring (d_f) was very small compared with that of the initial solution (d_i). The d_f/d_i values of KCl to $LaCl_3$ are shown in Fig. 5.130. The values are smaller than unity and increased with concentration. The d_f/d_i seems to increase in the order shown in Equation 15 at 25° C and 40° C, although the experimental errors are large.

$$NaCl < KCl < CaCl_2 < LaCl_3 \quad (15)$$

The order simply means that the inter-solute affinity increases in the order of Equation 16:

$$NaCl < KCl < CaCl_2 < LaCl_3 \quad (16)$$

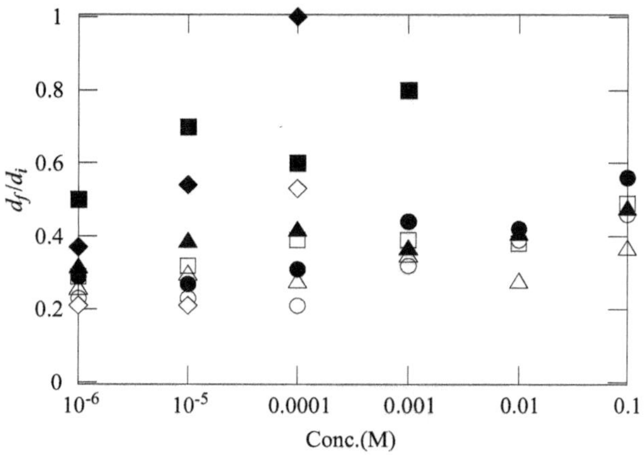

Figure 5.130 Value of d_f/d_i of NaCl (*circles*), KCl (*triangles*), $CaCl_2$ (*squares*), and $LaCl_3$ (*diamonds*) on a cover glass at 25° C (*open symbols*) and 40° C (*solid symbols*).

The cause for the observed order is not clear at present. The d_f/d_i values of NaCl, KCl, $CaCl_2$, and $LaCl_3$ were also observed at 40° C. The values increased clearly at high temperature. The microscopic blocklike and dendritic crosslike crystal patterns were observed for all the salts examined. The size of single crystals dried on a cover glass increased as salt concentration increased. The drying patterns of the binary mixtures of the salts were also observed. The size of the broad ring increased sharply by mixing, whereas the microscopic patterns were insensitive to mixing.

Many kinds of macroscopic and microscopic drying patterns observed hitherto are compiled in Tables 5.11 and 5.12.

5.11 Unique Drying Patterns

Professor D. Jishiashvil of Georgian Technical University made interesting drying experiments (D. Jishiashvil, J. Okamoto, A. Tsuchida, and T. Okubo, unpublished results). A drop of water was put on the Ge:GeO_2 film on an Si wafer, where Ge and GeO_2 are in the crystal state and amorphous matrix, respectively. Furthermore, Ge and GeO_2 are water insoluble and soluble, respectively. A water drop was added to the Ge:GeO_2 film carefully and then dried (Fig. 5.131). The beautiful drying patterns appeared at the dried spot (see Fig. 5.131). Two examples of the microscopic pictures are shown from the central to the outside edge in Figs. 5.132 and 5.133. Colorful microscopic patterns appeared. Interestingly, aqueous GeO_2 solution did not show similar microscopic patterns on any substrates including, the Si

Table 5.11 Macroscopic Drying Patterns of Suspensions and Solutions

Patterns			Solutes
BR	Normal BR	Single BR	Polystyrene, silica, PMMA spheres, PLHBr, PAL, PEG, NaPMA, dyes
		Multiple BR	pNIPAm, PEG, green tea
	Notched BR		Silica-spheres + NaCl, dyes in EtOH
	Inner and outer Multiple BR		
Wave			Silica spheres, PEG
SL	SL crack		Polystyrene, silica, PMMA spheres, NaPGA, PLHBr
	SL hill		pNIPAm (high temperature)
CRH	CRH with BR		Bentonite with NaCl, POEmCn, PEG
	Inner hill, outer BR		Coffee
Hat-shape	Water-soluble polymer (high concentration), NaCl		
Column		Silica-spheres (DC electric field)	
Laurel crown		αCD	

BR, Broad ring; *CD,* cyclodextrin; *CRH,* central round hill; *DC,* direct current; *EtOH,* ethanol; *NaCl,* sodium chloride; *NaPGA,* sodium poly(alpha-glutamate); *NaPMA,* sodium polymethacrylate; *PAL,* poly(allylamine hydrochloride); *PEG,* polyethylene glycol; *PLHBr,* poly(L-lysine hydrobromide); *PMMA,* poly(methyl methacrylate); *pNIPAm,* poly(*N*-isopropylacrylamide); *POEmCn,* polyoxyethylene alkyl ether; *SL,* spoke line.

Table 5.12 Microscopic Drying Patterns of Suspensions and Solutions

Patterns	Solutes
Fine ordered rings	Silica, gel spheres,
Fine ordered spokes	Silica, Gel spheres, PMMA
Branchlike	POEmCn, silica spheres
Earth worm–like	Polystyrene-spheres
Stringlike	PAL, *n*-Alkyl sulfate, POEmCn,
Dendritic	Silica, polystyrene spheres, polynucleotides, polysaccharides
Starlike	PAL, bentonite,
City road–like	PAL, PLHBr, dyes,
Flowerlike	PAL, DTAC, gel-spheres, NaPGA, Ge:GeO_2,
Blocklike	Normal bentonite, NaPGA, n-Alkyl sulfate, POEmCn
Crosslike	Silica sphere + NaCl, NaPGA,
Rodlike	NaPGA
Arclike	DTAC
Needlelike	Dyes
Wrinkled pattern	Bentonite, PEG
Chapped pattern	Bentonite PEG
Netting	Silica, gel spheres
Latticelike	Gel spheres

DTAC, Decyltrimetylammonium chloride; *GeO_2,* germanium oxide; *NaCl,* sodium chloride; *NaPGA,* sodium poly(alpha-glutamate); *PAL,* poly(allylamine hydrochloride); *PLHBr, poly(L-lysine hydrobromide)*; *PMMA,* poly(methyl methacrylate); *POEmCn,* polyoxyethylene alkyl ether.

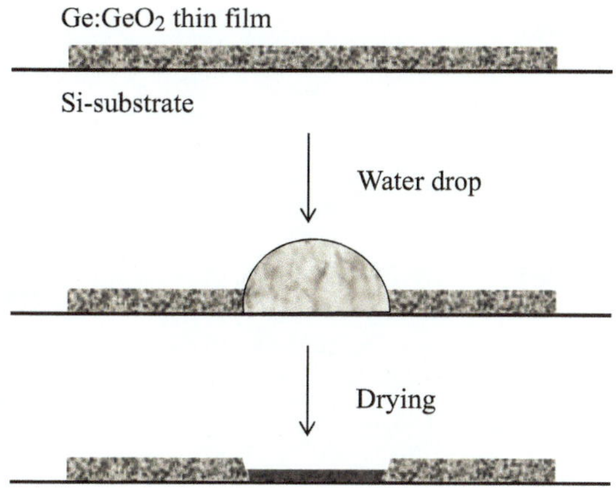

Figure 5.131 Schematic representation of drying pattern formation of $Ge:GeO_2$ film.

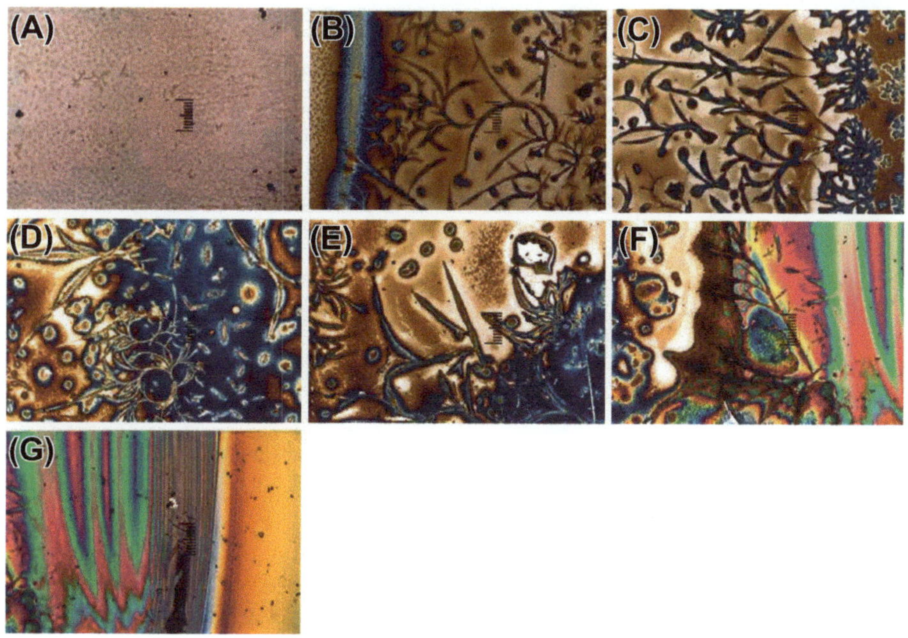

Figure 5.132 Macroscopic drying patterns of $Ge:GeO_2$ film. Code 23b, on Si, from the center to the right edge, ×5, full scale: 200 μm.

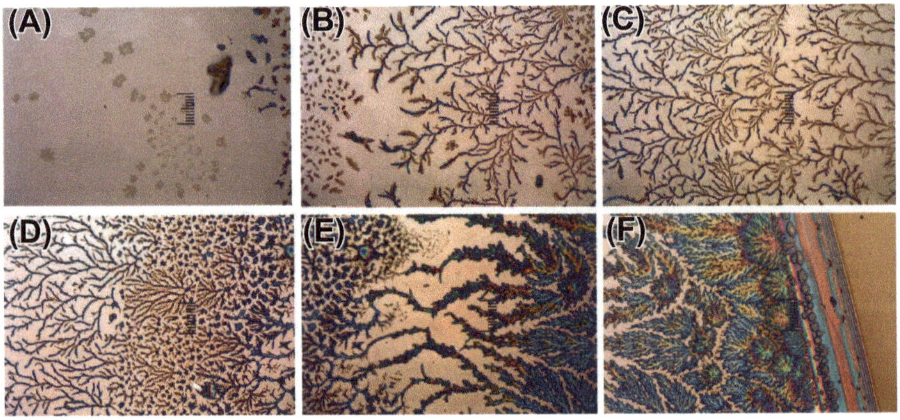

Figure 5.133 Macroscopic drying patterns of $Ge:GeO_2$ film. Code 24b, on Si, from the center to the right edge, ×40, full scale: 25 μm.

wafer. Interestingly, a macroscopic broad ring also formed. The mechanisms of the drying processes are not yet clarified.

5.12 Microgravity Experiments on the Drying Patterns

Microgravity experiments on the drying patterns were made using the free-fall techniques in my group. Fig. 5.134 shows the macroscopic drying patterns of colloidal silica spheres, CS91 in ethanol-aqueous mixtures at 0 *g* (*A-D*) and at 1 *g* (*E-H*) (Tsuchida A, et al., unpublished results). Surprisingly, the broad rings at the outside edges appeared clearly even in microgravity condition, where gravitational convection must be safely neglected.

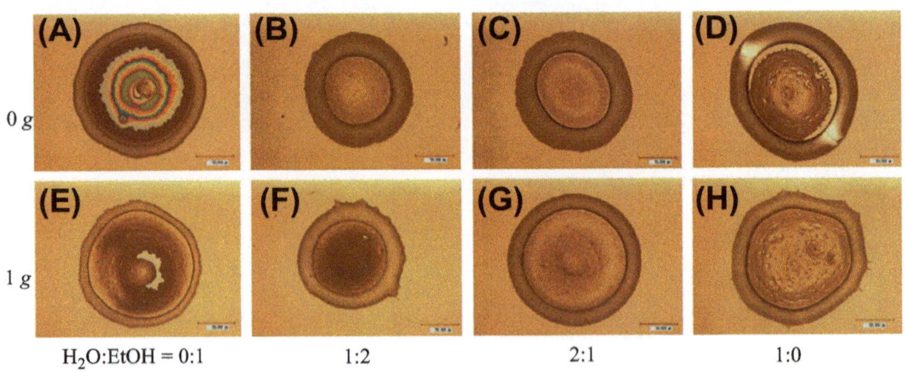

Figure 5.134 Microgravity experiments of drying patterns of CS91 spheres on a cover glass at 25° C. **A-D,** 0 *g*; **E-H,** 1 *g*; **A, E,** H_2O:EtOH = 0:1; **B, F,** 1:2; **C, G,** 2:1; **D, H,** 1:0, ϕ = 0.015, length of the bar = 50 μm.

However, we should recall that the contribution of Marangoni convection still remain even in microgravity. Drying patterns of the corresponding suspensions at normal gravity were similar to the patterns at microgravity. The widths of the broad rings at 0 *g* were clearly larger than those at normal gravity. S' values were evaluated from Fig. 5.134 as 21, 11, 11, and 13 at 0 *g* for H_2O:EtOH = 0:1, 1:2, 2:1, and 1:0, respectively. The corresponding values at 1 *g* were 29, 13, 18, and 18, respectively. S' values decreased sharply at microgravity for any mixtures. Larger widths should result in smaller S values. Thus the results support the finding that the Marangoni convections of solutes and solvents still remain in microgravity. However, the magnitude of the strength of convection in microgravity is small compared with that in normal gravity, where both the gravitational and Marangoni convections are effective.

5.13 Transferring Information during the Course of Drying Suspensions and Solutions on the Substrates

One of the main purposes of the drying experiments is to know the information of solutes, solvent, and atmosphere from the drying patterns. It is now clarified that information of suspensions and solutions (e.g., shape and size of solutes and temperature and humidity) is transferred, more or less, to the drying patterns by the cooperative and successive processes of wetting, evaporation, convection, sedimentation, and solidification. Therefore it is highly plausible that a part of the information is available from the macroscopic and microscopic drying patterns. The relationship between the information and drying patterns is summarized in Table 5.13 mainly using the data of the drying patterns found hitherto.

For example, an increase in the solute concentration of the initial suspensions and solutions results in an increase of the size and height of the broad ring and decrease of number of spoke lines. Interestingly, single crystal size in the broad ring area increases with increasing solute concentration in the dried film, which is opposite the tendency to of the solution and suspension to crystalize in the bulk phase. We should note here that the macroscopic and microscopic drying patterns are originated from the integrated information of solutes and atmosphere. The assignment of the solutes and atmosphere from the drying patterns will be made correctly, when the specific drying patterns assigned to the special substances appeared. For example, when the ordering of the agglomerates was observed, the solutes were likely to be gel spheres.

Table 5.13 Transferring Information of Suspensions and Solutions to the Drying Patterns

Information		Drying Patterns	
Concentration	***up***	Broad-ring size	***up***
		Broad-ring shape, *S* (*S′*)	***broaden*** (***down***)
		Broad-ring height	***up***
		Spoke-line number	***down***
		Fractal dimension	***up*** or ***insensitive***
		Single crystal size	***up***
Salt	***up***	Broad ring	***up*** or ***down***
		Bumpy-shaped broad-ring	
		Central round hill, from *broad ring*	
		Spoke line number	***down***
Temperature	***up***	Broad-ring shape, *S* (*S′*)	***sharp*** , ***up***
		Broad-ring size	***up***
		Spoke-line number	***down***
Humidity	***up***	Broad-ring size	***down***
		Broad-ring shape, *S* (*S′*)	***broaden***, ***down***
		Spoke-line number	***down***
Surface tension	***down***	Pattern area, Broad-ring size	***up***
Polarity of solvent	***up***	Pattern area	***down***
		Drying time	***up***
Particle size	***up***	Broad-ring size	***down***
		Spoke-line number	***down***
Nonspheres, from spheres		***Central round-hill,*** from ***broad-ring***	
		Spoke line disappear	
Gel spheres		***Flamelike*** (***Flickering***) ***spoke line***	
		Ordering (net, lattice, etc.) of ***agglomerates***	
Mixture of different size spheres		***Multiple broad rings***	
		Segregation, small outer, large inner	
Hydrophobicity	***up***	Broad-ring size	***down***
Molecular weight, polymer	***up***	Broad-ring size (d_f/d_i)	***up***
Conformation ***extended,*** *from coil*		***Cross-***, ***star-***, from *arc-*, *string-*	
Ordered structure (helix or β-sheet)		Broad-ring size	***up***
HLB, surfactant	***up***	***Broad ring,*** from *round hill*	
		Smooth surface, from *rough*	

HLB, Hydrophile-lipophile balance.

5.14 Film Formation and Drying Patterns

Film formation mechanisms from polymer colloidal dispersions have been studied intensively for more than 50 years, and are correlated with the drying patterns.[160-176] Three processes from the dispersions into the film have been proposed: (1) close-packing processes of polymer colloids by evaporation of solvent, (2) deformation of the particles followed by disappearing voids, although the original particles are still distinguishable, and (3) continuous film formation by the diffusion of polymer across particle boundaries. Cracking patterns of the films induced by the shrinkage during dryness were also investigated for the paste layers of clay, sand, and water deposited on a rigid substrates.[177-179] Characteristic pattern formation in the dried thin film on a substrate has been reported. Wormlike and hexagonally packed domains were obtained from asymmetric organic-organometallic diblock copolymers.[180] Honeycomb patterns of polymers were studied intensively by Shimomura et al.[181] The mechanism of the multibranched and honeycomb network drying pattern formation of gold suspension on substrates was discussed by Haidara et al.[182] and Korgel et al.[183] Dissipative patterns during drying of colloidal suspension ejected by ink-jet printing were studied by Park and Moon.[184] Self-organization of a chromophore-labeled double-chain amphiphile during drying was reported.[185] Superhelix-to-tubule transformation of J-aggregates was studied by TEM.[186]

5.15 Conclusions and Future Perspectives

Most of the self-organization phenomena in nature are known to take place accompanied by dissipation of free energy and in the nonequilibrium state. Among the huge number of dissipative self-organization events, study of the drying processes is one of the most convenient and simple. I described the drying processes of convection, sedimentation, and solidification as systematically as possible in this chapter, although the wetting and evaporation processes were not described in detail. Several important factors to determine the drying patterns are pointed out. First, the importance of the cooperativity among the five processes during the course of dryness are discussed. We should note that the primitive patterns have information of suspensions and solutions form already from the wetting to convection processes. Furthermore, the primitive patterns are further transferred to the sedimentation and then solidification structures with selection and amplification of the initial information of suspension and solution. Second, the important role of the affinity between solutes and substrates must be pointed out. For example, extended conformation of macroions is maintained during the drying through the interaction of the solutes with the substrate. The stretched conformation of the macroions toward coil changes easily in the bulk solution state during the course of concentration. However, when the stretched polymer chains interact with the substrate, change of their conformation os retarded during the whole course of drying. Thus the initial conformation is transferred to the drying pattern. Third, the important role of the concentration of solutes by the evaporation of a solvent during drying must be pointed out for the organization of solutes. Solute concentration corresponds to enhancement of the intersolute repulsion forces, which are inevitable for the organization of colloidal systems as understood from colloidal crystallization. Intersolute repulsion is inevitable for the ordering phenomena. Fourth, the important role of the

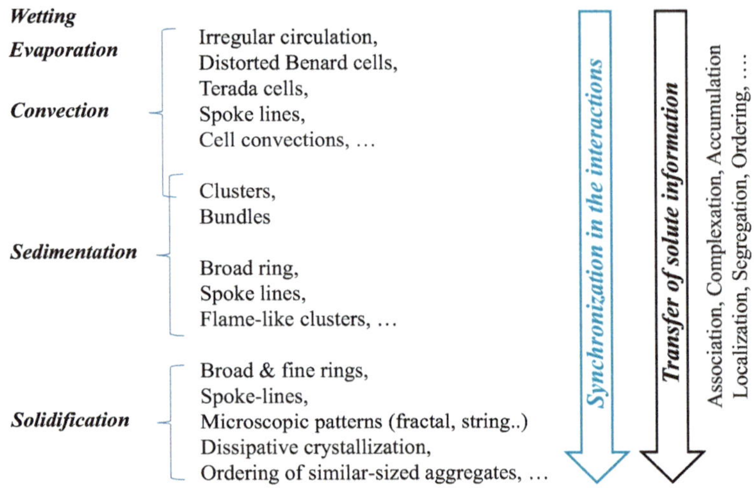

Figure 5.135 Processes of synchronized pattern formation during the course of dryness of suspensions and solutions.

gravitational effect for convectional flow of solvents and solutes was discussed. Marangoni convection is also important. Fifth, the most important factor for self-organization during drying is the translational thermal movements of all the relating elementary substances (e.g., solutes, solvents, components of the substrate surfaces, etc.). Each of the components have their own thermal energy k_BT. Analyses of information included in the initial suspensions and solutions from the drying dissipative patterns are still unsatisfactory at present. Further experimental and theoretical detailed studies are desired for understanding of the dissipative structural formation mechanisms in nature. The total drying processes of suspensions and solutions are compiled in Fig. 5.135. It should be noted that synchronization of each process is one of the most important factors determining the actual drying patterns.

Keen attention has been paid especially by physics researchers to the coffee ring in drying patterns. A broad ring appears at the outside edge of the initial liquid drop for coffee. Capillary flow induced by the differential evaporation rates from the central and outside edge areas and the pinning effects explained by Deegan et al. is famous.[47,48] Several researchers proposed further additional factors based on Deegan's theory, such as Marangoni flow,[187,188] interionic interactions,[189] fast evaporation effects,[190] particle-shaped effects,[191] and others. The experimental results described in this chapter on the drying patterns, however, teach us the discussion of the drying patterns using Deegan's theory is successful for the solidification processes, but is not satisfactory for explaining the total drying processes, because the wetting, evaporating, convectional, and sedimentary processes are almost neglected.

Furthermore, coffee is one of the specific solutes in which strong affinitive interaction takes place in the solidification process (i.e., type E solutes in the grouping [see Table 5.6]). Further theoretical discussion of these observations are eagerly expected in the future.

References

1. Ball P. *The self-made tapestry. Pattern formation in nature.* Oxford, UK: Oxford University Press; 1999.
2. Okubo T. In: Stoylov SP, Stoimenova MV, editors. *Molecular and colloidal electro-optics.* New York: Taylor & Francis; 2006. p. 573–89.
3. Okubo T. In: Nagarajan R, Hatton TA, editors. *Nanoparticles: syntheses, stabilization, passivation and functionalization.* Washington, DC: ACS Book; 2008. p. 256–70.
4. Okubo T. *Macromol Symp* 2010;**288**:67–77.
5. Hisatake K, Tanaka S, Aizawa Y. *J Appl Phys* 1993;**73**:7395–401.
6. Shahidzadeh-Bonn N, Rafai S, Azouni A, Bonn D. *J Fluid Mech* 2006;**549**:307–13.
7. Gribbin G. *Almost everyone's guide to science. The universe, life and everything.* New Haven, CN: Yale University Press; 1999.
8. Ball P. *The self-made tapestry. Pattern formation in nature.* Oxford, UK: Oxford University Press; 1999.
9. Terada T, Yamamoto R, Watanabe T. *Sci Paper Inst Phys Chem Res Jpn* 1934;**23**:173–84.
10. Terada T, Yamamoto R, Watanabe T. *Sci Paper Inst Phys Chem Res Jpn* 1935;**27**:75–92.
11. Nakaya U. *Memoirs of Torahiko Terada (Japanese).* Tokyo: Kobunsya; 1947.
12. Okubo T, Kimura H, Kimura T, Hayakawa F, Shibata T, Kimura K. *Colloid Polym Sci* 2005;**283**:1–9.
13. Okubo T. *Colloid Polym Sci* 2006;**285**:225–31.
14. Okubo T. *Colloid Polym Sci* 2009;**287**:167–78.
15. Okubo T, Okamoto J, Tsuchida A. *Colloid Polym Sci* 2009;**287**:351–65.
16. Okubo T, Okamoto J, Tsuchida A. *Colloid Polym Sci* 2009;**287**:645–57.
17. Palmer HJ. *J Fluid Mech* 1976;**75**:487–511.
18. Anderson DM, Davis SH. *Phys Fluids* 1995;**7**:248–65.
19. Pouth AF, Russel WB. *AIChEJ* 1998;**44**:2088–98.
20 Burelbach JP, Bankoff SG. *J Fluid Mech* 1998;**195**:463–94.
21. Fischer BJ. *Langmuir* 2002;**18**:60–7.
22. Okubo T. *Colloid Polym Sci* 2006;**284**:1191–6.
23. Okubo T. *Colloid Polym Sci* 2006;**284**:1395–401.
24. Okubo T, Okamoto J, Tsuchida A. *Colloid Polym Sci* 2007;**285**:967–75.
25. Okubo T. *Colloid Polym Sci* 2007;**285**:1495–503.
26. Okubo T, Okamoto J, Tsuchida A. *Colloid Polym Sci* 2008;**286**:385–94.
27. Okubo T, Okamoto J, Tsuchida A. *Colloid Polym Sci* 2008;**286**:941–9.
28. Yamaguchi T, Kimura K, Tsuchida A, Okubo T, Matsumoto M. *Colloid Polym Sci* 2005;**283**:1123–30.
29. Okubo T. *Colloid Polym Sci* 2006;**285**:331–7.
30. Vanderhoff JW. *J Polym Sci Symp* 1973;**41**:155–74.
31. Nicolis G, Prigogine I. *Self-organization in non-equilibrium systems.* New York: Wiley; 1977.
32. Ohara PC, Heath JR, Gelbart WM. *Angew Chem* 1997;**109**:1120–2.
33. Maenosono S, Dushkin CD, Saita S, Yamaguchi Y. *Langmuir* 1999;**15**:957–65.
34. Nikoobakht B, Wang ZL, El-Sayed MA. *J Phys Chem* 2000;**104**:8635–40.
35. Ung T, Litz-Marzan LM, Mulvaney P. *J Phys Chem B* 2001;**105**:3441–52.
36. Okubo T, Okamoto J, Tsuchida A. *Colloid Polym Sci* 2010;**288**:189–97.

37. Okubo T, Kanayama S, Ogawa H, Hibino M, Kimura K. *Colloid Polym Sci* 2004;**282**:230–5.
38. Okubo T, Okamoto J, Takahashi S, Tsuchida A. *Colloid Polym Sci* 2009;**287**:933–42.
39. Choi BK. *Solid State Ionics* 2004;**168**:123–9.
40. Okubo T, Hagiwara A, Kitano H, Okamoto J, Takahashi S, Tsuchida A. *Colloid Polym Sci* 2009;**287**:1155–65.
41. Okubo T, Okamoto J, Tsuchida A. *Colloid Polym Sci* 2010;**288**:981–9.
42. Okubo T, Mizutani M, Takahashi S, Tsuchida A. *Colloid Polym Sci* 2010;**288**:1551–9.
43. Okubo T, Mizutani M, Takahashi S, Tsuchida A. *Colloid Polym Sci* 2010;**288**:1435–44.
44. Terada T, Yamamoto R. *Proc Imper Acad Tokyo* 1935;**11**:214–5.
45. Okubo T, Yamada T, Kimura K, Tsuchida A. *Colloid Polym Sci* 2005;**283**:1007–15.
46. Okubo T, Nozawa M, Tsuchida A. *Colloid Polym Sci* 2007;**285**:827–32.
47. Deegan RD, Bakajin O, Dupont TF, Huber G, Nagel SR, Witten TA. *Nature* 1997;**389**:827–9.
48. Deegan RD, Bakajin O, Dupont TF, Huber G, Nagel SR, Witten TA. *Phys Rev E* 2000;**62**:756–65.
49. Okubo T, Okamoto J, Tsuchida A. *Colloid Polym Sci* 2008;**286**:1123–33.
50. Okubo T. *Colloid Polym Sci* 2008;**286**:1307–15.
51. Cachile M, Benichou O, Cazabat AM. *Langmuir* 2002;**18**:7985.
52. Cachile M, Benichou O, Poulard C, Cazabat AM. *Langmuir* 2002;**18**:8070.
53. Maillard M, Motte L, Ugo AT, Pileni MP. *J Phys Chem B* 2000;**104**:11875–7.
54. Matar OK, Craster RV. *Phys Fluids* 2001;**13**:1869–83.
55. Schatz MF, Neitzel GP. *Ann Rev Fluid Mech* 2001;**33**:93–129.
56. Martincigh BS, Simoyi RH. *J Phys Chem* 2002;**106**:482–9.
57. Hu H, Larson RG. *J Phys Chem B* 2002;**106**:1334–44.
58. Truskett JN. *Langmuir* 2003;**19**:8271–9.
59. Savino R, Fico S. *Phys Fluids* 2004;**16**:3738–54.
60. Hu H, Larson RG. *Langmuir* 2005;**21**:3963–71.
61. Girard F, Antoni M, Faure S, Steinchen A. *Langmuir* 2006;**22**:11085–91.
62. Stone HA. *Phys Rev Lett* 2007;**99**(234502):1–4.
63. Girard F, Antoni M, Faure S, Steinchen A. *Colloids Surf A* 2008;**323**:36–49.
64. Fujii S, Yamashita Y, Nakamura Y, Okubo T. *Colloid Polym Sci* 2014;**292**:2621–31.
65. Okubo T, Fujii S, Nakamura Y. *Colloid Polym Sci* 2013;**291**:2805–13.
66. T.Okubo, D.Suzuki, A.Tsuchida. 2012;**290**:1901–1911.
67. Fujii S, Nakamura Y, Tsuchida A, Okubo T. *Colloid Polym Sci* 2014;**292**:2303–10.
68. Okubo T, Kimura K, Kimura H. *Colloid Polym Sci* 2002;**280**:1001–8.
69. Kralchevsky PA, Nagayama K. *Adv Colloid Interf Sci* 2000;**85**:145–92.
70. Maenosono S, Duskin CD, Salta S, Yamaguchi Y. *Langmuir* 1999;**15**:957–65.
71. Ida T. *J Appl Crystallography* 2008;**41**:393–401.
72. Okubo T. *J Chem Soc Faraday Trans* 1 1986;**82**:3175–83.
73. Okubo T. *J Chem Soc Faraday Trans* 1 1986;**82**:3185–96.
74. Okubo T, Nakagawa N, Tsuchida A. *Colloid Polym Sci* 2007;**285**:1247–55.
75. Okubo T. *Colloid Polym Sci* 2008;**286**:1527–35.
76. Okubo T, Okuda S, Kimura H. *Colloid Polym Sci* 2002;**280**:454–60.
77. Okubo T, Otake A, Tsuchida A. *Colloid Polym Sci* 2009;**287**:1435–44.
78. Sasano T, Huang Y, Ohno K, Fukuda T, Tsujii Y. *Polym Prep Jpn* 2010;**59**:2821–2.
79. Ohno K, Huang Y, Tsuchida A, Okubo T. *Colloid Polym Sci* 2014;**292**:723–31.
80. Ohno K, Huang Y, Okubo T. *Colloid Polym Sci* 2014;**292**:1143–51.
81. Okubo T. *Colloid Polym Sci* 2013;**291**:1623–9.
82. Okubo T. *Colloid Polym Sci* 2013;**291**:1887–93.
83. Okubo T, Yamada T, Kimura K, Tsuchida A. *Colloid Polym Sci* 2006;**284**:396–404.
84. Choi BK. *Solid State Ionics* 2004;**168**:123–9.

85. Doulia D, Tzia K, Gekas V. *Intern J Food Prop* 2000;**3**:1–14.
86. Mehrafarin M, Faghihi M. *Physica A* 2001;**301**:163–8.
87. Chai LH. *Intern J Thermal Sci* 2007;**46**:802–11.
88. Xu P, Mujumdar AS, Yu B. *Drying Tech* 2008;**26**:640–50.
89. Okubo T, Ogawa H, Tsuchida A. *Colloid Polym Sci* 2010;**288**:245–56.
90. Holtz JH, Asher SA. *Nature* 1997;**389**:829–32.
91. Pelton R. *Adv Colloid Interf Sci* 2000;**85**:1–33.
92. Xia Y, Gates B, Yin Y, Lu Y. *Adv Mater* 2000;**12**:693–713.
93. Okubo T, Takahashi S, Tsuchida A. *Colloid Polym Sci* 2011;**289**:1729–37.
94. Groen H, Borissova A, Roberts KJ. *Ind Eng Chem Res* 2003;**42**:198–206.
95. Daub CD, Leung K, Luzar A. *J Phys Chem B* 2009;**113**:7687–700.
96. Iizuka E, Kondo Y, Ukai Y. *Polym J* 1977;**9**:135–44.
97. Okubo T, Onoshima D, Tsuchida A. *Colloid Polym Sci* 2007;**285**:999–1007.
98. Nurkhamidah S, Woo EM. *Colloid Polym Sci* 2012;**290**:275–88.
99. Okubo T. *Colloid Polym Sci* 2011;**289**:1205–13.
100. Okubo T. *Colloid Polym Sci* 2011;**289**:159–67.
101. Okubo T, Tsuchida A. *Colloid Polym Sci* 2014;**292**:381–9.
102. Okubo T. *Colloid Polym Sci* 2013;**291**:2447–54.
103. Schneider N, Doty P. *J Phys Chem* 1954;**58**:762–9.
104. Brown W, Henley D. *Makromol Chem* 1964;**79**:68–88.
105. Werbowyj RS, Gray DG. *Mol Cryst Liq Cryst* 1976;**34**:97–103.
106. Bheda J, Feller JF, White JL. *Colloid Polym Sci* 1980;**258**:1335–42.
107. Fortin S, Charlet G. *Macromolecules* 1989;**22**:2286–92.
108. Okubo T. *J Chem Soc Faraday Trans* 1991;**87**:1361–3.
109. Onogi Y, White JL, Feller JF. *J Polym Sci Polym Phys Ed* 1980;**18**:663–82.
110. Suito S, Hasegawa S. *J Mater Sci* 2002;**37**:4857–79.
111. Connors KA. *Chem Rev* 1997;**97**:1325–58.
112. Bender ML, Komiyama K. *Cyclodextrin chemistry*. New York: Springer; 1978.
113. Saenger W. *Angew Chem Intern Ed Engl* 1980;**19**:344–62.
114. Breslow R. *Acc Chem Res* 1980;**13**:170–7.
115. Szejtli J. *Chem Rev* 1998;**98**:1743–54.
116. Harada A. *Acc Chem Res* 2001;**34**:456–64.
117. Dodziuk H. *Cyclodextrins and their complexes: chemistry, analytical methods, applications*. Weinheim: Wiley; 2006.
118. Okubo T, Kitano H, Ise N. *J Phys Chem* 1976;**80**:2661–4.
119. Sabadini E, Cosgrove T, Egidio FC. *Carbohydrate Res* 2006;**341**:270–4.
120. Okubo T. *Colloid Polym Sci* 2013;**291**:2599–605.
121. Okubo T, Takahashi S, Tsuchida A. *Colloids Surf B Biointerf* 2011;**87**:11–7.
122. Okubo T. *Colloids Surf B Biointerf* 2011;**87**:439–46.
123. Okubo T. *Colloid Polym Sci* 2014;**292**:1419–27.
124. Steiner RF, Beer RF. *Biochim Biophys Acta* 1959;**33**:470–81.
125. Stevens CL, Felsenfeld G. *Biopolymers* 1964;**2**:293–314.
126. Massoulie J. *Eur J Biochem* 1968;**2**:428–39.
127. Ross PD, Sturtvant JM. *Proc Natl Acad Sci* 1960;**46**:1360–5.
128. Steiner RF, Beers RF. *Polynucleotides*. Amsterdam: Elsevier; 1961.
129. Steiner RF. *Beers: Biochim Biophys Acta* 1959;**32**:166–76.
130. Holcomb DN, Timasheff SN. *Biopolymers* 1968;**6**:513–29.
131. Janik B, Sommer RG, Bobst AM. *Biochim Biophys Acta Nucleic Acids Protein Synth* 1972;**281**:152.
132. Morawetz H. *Macromolecules in solution*. New York: John Wiley & Sons; 1965.
133. von Hippel P, Schleich T. In: Timasheff SN, Fasman GD, editors. *Structure and stability of biological macromolecules*. New York: Marcel Dekker; 1969. p. 417–574. ch 6.

134. Ward AG, Courts A. *The science and technology of gelatin.* New York: Academic Press; 1977.
135. Ramachandran GN, Kartha G. *Nature* 1954;**174**:269–70.
136. Hall CE, Doty P. *J Am Chem Soc* 1958;**80**:1269–74.
137. Hulmes DJS, Miller A. *Nature* 1979;**282**:878–80.
138. Maeda H. *Langmuir* 1999;**15**:8505–13.
139. Okubo T, Tsuchida A, Kokufuta E. *Colloid Polym Sci.* 2015, in press. http://dx.doi.org/10.1007/s00396-015-3558-z.
140. Hoover R. *Carbohydrate Polym* 2001;**45**:253–67.
141. Srichwong S, Sunarti TC, Mishima T, Isono N, Hisamatsu M. *Carbohydrate Polym* 2005;**60**:529–38.
142. Peroni TS. *Food Sci Tech Intern* 2006;**12**:505–13.
143. Okubo T, Kokufuta E, Nakamuro M, Yoshinaga K, Mizutani M, Tsuchida A. *Colloid Surf Biointerf B* 2010;**80**:193–9.
144. Okubo T, Kanayama S, Kimura K. *Colloid Polym Sci* 2004;**282**:486–94.
145. Okubo T. *Colloid Polym Sci* 2013;**291**:2455–62.
146. Bemardes JS, Rezende CA, Galembeck F. *Langmuir* 2010;**26**:7824–32.
147. Scott AB, Tartar HV. *J Am Chem Soc* 1943;**65**:692–8.
148. Debye P. *Ann NY Acad Sci* 1949;**51**:575–92.
149. Jungermann E, editor. *Cationic surfactants.* New York: Marcel Dekker; 1970.
150. Kimura K, Kanayama S, Tsuchida A, Okubo T. *Colloid Polym Sci* 2005;**283**:898–904.
151. Okubo T, Shinoda C, Kimura K, Tsuchida A. *Langmuir* 2005;**21**:9889–95.
152. Patrick HN, Warr GG, Manne S, Aksay IA. *Langmuir* 1999;**15**:1685–92.
153. Wall JF, Zukoski CF. *Langmuir* 1999;**15**:7432–7.
154. Velegol SB, Fleming BD, Biggs S, Wanless EJ, Tilton RD: *Langmuir* 2000; **16**:2548–56.
155. Li Bai C. *Langmuir* 2002;**18**:657–60.
156. Okubo T, Yokota N, Tsuchida A. *Colloid Polym Sci* 2007;**285**:1257–65.
157. Leiterer J, Emmerling F, Panne U, Christen W, Rademann K. *Langmuir* 2008;**24**:7970–8.
158. Latterini L, Blossey R, Hofkens J, Vanoppen P, deSchryver FC, Rowan AE, Nolte RJM. *Langmuir* 1999;**15**:3582–8.
159. Okubo T, Tsuchida A, Togawa H. *Colloid Polym Sci* 2009;**287**:443–54.
160. Dillon RE, Matheson LA, Bradford EB. *J Colloid Sci* 1951;**6**:108–17.
161. Henson WA, Taber DA, Bradford EB. *Ind Eng Chem* 1953;**45**:735–9.
162. Brown GL. *J Polym Sci* 1956;**22**:423–34.
163. Sheetz DP. *J Appl Polym Sci* 1965;**9**:3759–73.
164. Vanderhoff JW, Tarkowski HL, Jenkins MC, Bradford E. *J Macromol Chem* 1966;**1**:361–97.
165. Vanderhoff JW, Bradford EB, Carrington WK. *J Polym Sci Symp* 1973;**41**:155–74.
166. Croll SG. *J Coatings Tech* 1986;**58**:41–9.
167. Eckersley ST, Rudin AJ. *Coatings Tech* 1990;**62**:89–100.
168. Dobler F, Holl Y. *TRIP* 1996;**4**:145–51.
169. Winnik MA, Feng J. *J Coatings Tech* 1996;**68**:39–50.
170. Winnik MA. *Curr Opin Colloid Interf Sci* 1997;**2**:192–9.
171. Winnik MA. *Emulsion polymerization and emulsion polymers.* New York: John Wiley; 1997. pp. 467–517. [Chapter 14].
172. Routh AF, Russel WB. *AIChEJ* 1998;**44**:2088–98.
173. Routh AF, Russel WB. *Langmuir* 1998;**15**:7762–73.
174. Petersen C, Heldmann C, Johannsmann D. *Langmuir* 1998;**15**:7745–51.
175. Routh AF, Russel WB, Tang J, El-Aasser MS. *J Coatings Tech* 2001;**73**:41–8.
176. Shmuylovichi L, Shen AQ, Stone HA. *Langmuir* 2002;**18**:3441–5.
177. Hutchinson JW, Suo Z. *Adv Appl Mechanics* 1992;**29**:63–191.

178. Colina H, Roux S. *Eur Phys J E* 2000:1189–94.
179. Tirumkudulu MS, Russel WB. *Langmuir* 2005;**21**:4938–48.
180. Lammertink RGH, Hempenius MA, Vancso GJ. *Langmuir* 2000;**16**:6245–52.
181. Karthaus O, Maruyama N, Cieren X, Shimomura M, Hasegawa H, Hashimoto T. *Langmuir* 2000;**16**:6071–6.
182. Mougin K, Haidara H. *Langmuir* 2002;**18**:9566–9.
183. Stowell C, Korgel BA. *Nano Lett* 2001;**1**:595–600.
184. Park J, Moon J. *Langmuir* 2006;**22**:3506–13.
185. Luo G, Liu T, Ying L, Zhao XS, Huang Y, Wu D, Huang C. *Langmuir* 2000;**16**:3651–9.
186. von Berlepsch H, Boettcher C, Ouart A, Regenbrecht M, Akari S, Keiderling H, Schnablegger H, Dahne S, Kirstein S. *Langmuir* 2000;**16**:5908–16.
187. Hu H, Larson R. *J Phys Chem B* 2006;**110**:7090–4.
188. Eval HB, Mampallil-Agustine D, Duits MHG, Mugele F. *Soft Matter* 2011;**7**:7090–4.
189. Bhardwaj R, Fang X, Somasundaran P, Attinger D. *Langmuir* 2010;**26**:7833–42.
190. Shen X, Ho C-M, Wong T-S. *J Phys Chem B* 2010;**114**:5269–74.
191. Yunker PJ, Still T, Lohr MA, Yodh AG. *Nature* 2011;**476**:308–11.

Index

Note: Page numbers followed by "f," and "t" indicate figures, and tables, respectively.

D

W

X

Y

Z

FSC
www.fsc.org
MIX
Papier | Fördert gute Waldnutzung
FSC® C083411

Kolibri 360 GmbH
Ferdinand-Jühlke-Straße 7
99095 Erfurt, Deutschland
produktsicherheit@kolibri360.de